BIOACTIVES AND PHARMACOLOGY OF PTERIDOPHYTES

AAP Focus on Medicinal Plants

BIOACTIVES AND PHARMACOLOGY OF PTERIDOPHYTES

T. Pullaiah, PhD

Editor

APPLE ACADEMIC PRESS

First edition published 2027

Apple Academic Press Inc.
1265 Goldenrod Circle, NE,
Palm Bay, FL 32905 USA

760 Laurentian Drive, Unit 19,
Burlington, ON L7N 0A4, Canada

CRC Press
2385 NW Executive Center Drive,
Suite 320, Boca Raton FL 33431

4 Park Square, Milton Park,
Abingdon, Oxon, OX14 4RN, UK

Library and Archives Canada Cataloguing in Publication

CIP data on file with Canada Library and Archives

Library of Congress Cataloging-in-Publication Data

CIP data on file with US Library of Congress

ISBN: 978-1-77964-272-1 (hbk)
ISBN: 978-1-77964-273-8 (ebk)
DOI: 10.1201/9781779642738

AAP Focus on Medicinal Plants Book Series

Book Series Editor:

Editor: T. Pullaiah, PhD
Former Professor, Department of Botany
Sri Krishnadevaraya University
Andhra Pradesh, India
Email: pullaiah.thammineni@gmail.com

For over a millennia, medicinal plants have been a valuable source of therapeutic agents, and still many of today's drugs are plant-derived natural products or their derivatives. Bioactive compounds typically occur in small amounts, and they have more subtle effects than nutrients. The bioactive compounds influence cellular activities that modify the risk of disease and cure and alleviate disease symptoms. They have potentially immense health benefits, and can act as antioxidants, enzyme inhibitors and inducers, inhibitors of receptor activities, and inducers and inhibitors of gene expression among other actions. A wide array of biological activities and potential health benefits of medicinal plants have been reported, which include antiviral, antimicrobial, antioxidant, anti-cancer, anti-inflammatory, antidiabetic properties as well as protective effects on the liver, kidney, heart, and nervous system.

The volumes in this book series discuss each species' bioactive compounds along with their chemical structures and pharmacological activities, which include antiviral, antibacterial, antifungal, antioxidant, anticancer, anti-inflammatory, anti-diabetic, hepatoprotective, cardioprotective, nephroprotective, etc.

These volumes aim to be comprehensive desk references on bioactives and pharmacology of medicinal plants as well as sourcebooks for the development of new drugs.

BOOKS IN THE SERIES:

- **Bioactives and Pharmacology of Medicinal Plants** (2 volumes)
  Editor: T. Pullaiah, PhD
- **Biomolecules and Pharmacology of Medicinal Plants** (2 volumes)
  Editor: T. Pullaiah, PhD
- **Phytochemistry and Pharmacology of Medicinal Plants** (2 volumes)
  Editor: T. Pullaiah, PhD
- **Bioactives and Pharmacology of Legumes**
  Editor: T. Pullaiah, PhD
- **Phytochemical Composition and Pharmacy of Medicinal Plants** (2 volumes)
  Editor: T. Pullaiah, PhD
- **Bioactives and Pharmacology of Lamiaceae**
  Editor: T. Pullaiah, PhD
- **Frankincense—Gum Olibanum: Botany, Oleoresin, Chemistry, Extraction, Utilization, Propagation, Biotechnology, and Conservation**
  Editors: T. Pullaiah, PhD, K. Venkata Ratnam, PhD, Mallappa Kumara Swamy, PhD, and Lepakshi Md. Bhakshu, PhD
- **Bioactives and Pharmacology of Pteridophytes**
  Editor: T. Pullaiah, PhD
- **Bioactives and Pharmacology of Gymnosperms**
  Editor: T. Pullaiah, PhD

Book Series Editor:
T. Pullaiah, PhD
Former Professor, Department of Botany
Sri Krishnadevaraya University
Andhra Pradesh, India

About The Editor

T. Pullaiah, PhD
Department of Botany, Sri Krishnadevaraya University,
Anantapur, Andhra Pradesh, India

Dr. T. Pullaiah is a former Professor at the Department of Botany, Sri Krishnadevaraya University, Andhra Pradesh, India, where he taught for more than 35 years. He has held several positions at the university, including Dean, Faculty of Biosciences; Head, Department of Botany; Head, Department of Biotechnology; and Member, Academic Senate. He served as President of the Indian Botanical Society (2014), President of the Indian Association for Angiosperm Taxonomy (2013), and was elected Fellow of the Andhra Pradesh Akademi of Sciences. He was awarded the Panchanan Maheshwari Gold Medal, the Prof. P. C. Trivedi Medal, the Dr. G. Panigrahi Memorial Lecture Award of the Indian Botanical Society, the Prof. Y. D. Tyagi Gold Medal of the Indian Association for Angiosperm Taxonomy, and a Best Teacher Award from the Government of Andhra Pradesh. Under his guidance, 54 students obtained their doctoral degrees. He has authored 80 books, edited 50 books, and published over 340 research papers. His books include *Mangroves: Biodiversity, Livelihoods and Conservation* (Springer); *Ethnobotany of India* (five volumes, Apple Academic Press); *Global Biodiversity* (four volumes, Apple Academic Press); *Monograph on Brachystelma and Ceropegia in India* (CRC Press); *Apomixis in Angiosperms* (CRC Press); *Invasive Alien Species* (four volumes, Wiley International); *Advances in Cell and Molecular Diagnostics* (Elsevier); *Camptothecin and Camptothecin Producing Plants* (Elsevier); *Paclitaxel* (Elsevier); *Red Sanders: Silviculture and Conservation* (Springer); *Genetically Modified Crops* (two volumes, Springer); *Flora of Andhra Pradesh* (5 volumes); *Flora of Eastern Ghats* (four volumes); *Flora of Telangana* (three volumes); *Encyclopedia of World Medicinal Plants* (seven volumes, 2nd edition); and *Encyclopedia of Herbal Antioxidants* (three volumes). Dr. Pullaiah is also the editor of the book series Biodiversity Hotspots of the World. He was a member of the Species Survival Commission of the International Union for Conservation of Nature (IUCN). Professor Pullaiah received his PhD from Andhra University, India, attended Moscow State University, Russia, and worked as a Postdoctoral Fellow during 1976–1978.

Contents

Contributors

Reenu Agarwal
Department of Botany, IIS (Deemed to be University), Jaipur, Rajasthan, India

Veeramaneni Alekhya
Department of Pharmacology, GIET School of Pharmacy, Chaitanya Knowledge City, Rajahmundry, Andhra Pradesh, India

Alex Philip Alen
Department of Botany, Government Victoria College, Palakkad, University of Calicut, Kerala, India

Shailesh G. Atram
Department of Chemistry, Arts, Commerce and Science College, Maregaon, Yavatmal, Maharashtra, India

Richa Bhardwaj
Department of Botany, IIS (Deemed to be University), Jaipur, Rajasthan, India

Pradeep Bhat
ICMR–National Institute of Traditional Medicine, Belagavi, Karnataka, India

Thiyagarajan Deepan
Department of Pharmacy, GIET School of Pharmacy, Chaitanya Knowledge City, Rajahmundry, Andhra Pradesh, India

Sagar A. Deshmukh
Department of Botany, Anandibai Raorane Arts, Commerce and Science College, Vaibhavwadi, Sindhudurg (affiliated to the University of Mumbai); The New College, Kolhapur (affiliated to Shivaji University, Kolhapur), Maharashtra, India

Vishal P. Deshmukh
Department of Botany, Jagadamba Mahavidyalaya, Achalpur City, Amravati, Maharashtra, India

N. Sarojini Devi
Centre of Excellence for Biodiversity and Climate Resilience, Dharmavana Nature Ark, Hyderabad, Telangana, India

Suraj P. Devkar
Department of Botany, The New College, Kolhapur (affiliated to Shivaji University, Kolhapur), Maharashtra, India

Suvarna V. Gaikwad
Post Graduate Research Centre in Botany, Rayat Shikshan Sanstha's Dada Patil Mahavidyalaya, Karjat, Ahilyanagar, Maharashtra, Indi

Savaliram G. Ghane
Department of Botany, Plant Physiology Laboratory, Shivaji University, Kolhapur, Maharashtra, India

Lalremkimi Hauhulh
Department of Zoology, Mizoram University (A Central University), Aizawl, Mizoram, India

M. Indira
Department of Biotechnology, Vignan's Foundation for Science, Technology and Research (Deemed to be University), Vadlamudi, Andhra Pradesh, India

Bharatkumar Jadhav
Department of Botany, Balasaheb Desai College, Patan, Maharashtra, India

Santoshkumar Jayagoudar
Department of Botany, G. S. S. College and Rani Channamma University P. G. Centre, Belagavi, Karnataka, India

Siddappa B. Kakkalameli
Department of Studies in Botany, Davangere University, Shivagangotri, Davangere, Karnataka, India

Prashant Karadakatti
Department of Studies in Botany, Davangere University, Shivagangotri, Davangere, Karnataka, India

H. N. Karthik
Department of Forest Products and Utilization, College of Forestry, University of Agricultural Sciences, Dharwad, Sirsi, Karnataka, India

Manpreet Kaur
Department of Pharmacognosy, G.H.G. Khalsa College of Pharmacy, Gurusar Sadhar, Ludhiana, Punjab, India

Bhupedra S. Kholia
Northern Regional Center, Botanical Survey of India, Dehra Doon, Uttarakhand, India

Shivani V. Khorate
Department of Botany, Shivaji University, Kolhapur, Maharashtra, India

R. Kripasree
Department of Botany, Carmel College (Autonomous), Mala, Thrissur, Kerala, India

K. Raja Kullayiswamy
Centre of Excellence for Biodiversity and Climate Resilience, Dharmavana Nature Ark, Hyderabad, Telangana, India

Sunil Kumar
Department of Chemistry, Sant Kavi Baba Baijnath Government P.G. College, Harakh, Barabanki, Uttar Pradesh, India

Lalremkimi
Department of Zoology, Mizoram University (A Central University), Aizawl, Mizoram, India

Lalremruati
Department of Zoology, Mizoram University (A Central University), Aizawl, Mizoram, India

F. Lalsangpuii
Department of Botany, Mizoram University (A Central University), Aizawl, Mizoram, India

Lakshmanan Leeja
Department of Botany, Government Victoria College, Palakkad, University of Calicut, Kerala, India

K. V. Madhusudhan
Department of Biological Sciences, Government IASE, Kurnool, Andhra Pradesh, India

Bince Mani
Department of Botany, St. Thomas College Palai (Autonomous), Kottayam, Kerala, India

Surya Prabha Matangi
Department of Pharmaceutical Sciences, School of Biotechnology and Pharmaceutical Sciences, Vignan's Foundation for Science, Technology and Research, Guntur, Andhra Pradesh, India

K. Sajitha Menon
Department of Botany, Government Victoria College, Palakkad, University of Calicut, Kerala, India

Divakar K. Mesta
Department of Botany, Carmel College of Arts, Science and Commerce for Women, Nuvem, Goa, India

V. N. Koteswararao Nerella
Department of Pharmacy, GIET School of Pharmacy, Chaitanya Knowledge City, Rajahmundry, Andhra Pradesh, India

Mansinghraj S. Nimbalkar
Department of Botany, Shivaji University, Kolhapur, Maharashtra, India

Shreedhar Otari
Department of Botany, Plant Physiology Laboratory, Shivaji University, Kolhapur, Maharashtra, India

Sandeep R. Pai
Post Graduate Research Centre in Botany, Rayat Shikshan Sanstha's Dada Patil Mahavidyalaya, Karjat, Ahilyanagar, Maharashtra, India

Suraj B. Patel
Department of Botany, Plant Physiology Laboratory, Shivaji University, Kolhapur, Maharashtra, India

Sachin B. Patil
Department of Botany, Anandibai Raorane Arts, Commerce and Science College, Vaibhavwadi, Sindhudurg (affiliated to the University of Mumbai), Maharashtra, India

Sachin M. Patil
Department of Botany, Dr. Ghali College, Gadhinglaj (affiliated to Shivaji University, Kolhapur), Maharashtra, India

Yogeshree Patil
Department of Botany, Plant Physiology Laboratory, Shivaji University, Kolhapur, Maharashtra, India

Nilesh V. Pawar
Department of Botany, The New College, Kolhapur, Maharashtra, India

M. Surya Prabha
Department of Pharmaceutical Sciences, Vignan's Foundation for Science, Technology and Research, Vadlamudi, Andhra Pradesh, India

M. Swetha Priya
Sri Padmavati Mahila Visvavidyalayam (Women's University), Tirupati, Andhra Pradesh, India

Gautam Raj Puri
School of Pharmaceutical Sciences, RIMT University, Mandi Gobindgarh, Punjab, India

R. Radha
Department of Pharmaceutical Chemistry, MB School of Pharmaceutical Sciences, Mohan Babu University, A. Rangampet, Tirupati

Mitta Raghavendra
Department of Pharmaceutical Sciences, School of Biotechnology and Pharmaceutical Sciences, Vignan's Foundation for Science, Technology and Research, Guntur, Andhra Pradesh, India

V. V. Rajesham
Department of Pharmacology, CMR College of Pharmacy, Hyderabad, Telangana, India

Kishore S. Rajput
Department of Botany, Faculty of Science, The Maharaja Sayajirao University of Baroda, Vadodara, Gujarat, India

Ajay V. Rajurkar
Department of Botany, Lokmanya Tilak Mahavidyalaya, Wani, Yavatmal, Maharashtra, India

Somasundaram Ramachandran
Department of Pharmacology, GIET School of Pharmacy, Chaitanya Knowledge City, Rajahmundry, Andhra Pradesh, India

A. Shobha Rani
Sri Padmavati Mahila Visvavidyalayam (Women's University), Tirupati, Andhra Pradesh, India

Sharabu Ravichandra
Department of Pharmaceutical Chemistry, Chebrolu Hanumaiah Institute of Pharmaceutical Sciences, Guntur, Andhra Pradesh, India

S. Siva Sankar
Department of Pharmaceutical Chemistry, Sri Padmavati School of Pharmacy, Tiruchanoor, Tirupati, Andhra Pradesh, India

Manoj L. Sankpal
Department of Botany, Shivaji University, Kolhapur, Maharashtra, India

J. M. Sasikumar
Department of Microbiology, Karpagam Academy of Higher Education, Coimbatore, Tamil Nadu, India

Nataru Savithramma
Department of Botany, Sri Venkateswara University, Tirupati, Andhra Pradesh, India

G. Savithri
Sri Padmavati Mahila Visvavidyalayam (Women's University), Tirupati, Andhra Pradesh, India

Bhasha Shanmugam
Department of Education in Science and Mathematics (DESM), North East Regional Institute of Education, NCERT, Shillong, Meghalaya, India

Pasupuleti Sivaramakrishna
Department of Botany, Sri Venkateswara University, Tirupati, Andhra Pradesh, India; Department of Botany, Government College for Men, Kadapa, Andhra Pradesh, India

K. S. Shanthi Sree
Sri Padmavati Mahila Visvavidyalayam (Women's University), Tirupati, Andhra Pradesh, India

Sibbala Subramanyam
Department of Pharmaceutical Sciences, School of Biotechnology and Pharmaceutical Sciences, Vignan's Foundation for Science, Technology and Research, Guntur, Andhra Pradesh, India

B. Sudarsini
Department of Biotechnology, Vignan's Foundation for Science, Technology and Research, Vadlamudi, Andhra Pradesh, India

P. Sujathamma
Sri Padmavati Mahila Visvavidyalayam (Women's University), Tirupati, Andhra Pradesh, India

Veerankutty Suresh
Department of Botany, Government Victoria College, Palakkad, University of Calicut, Kerala, India

Sinjumol Thomas
Department of Botany, Carmel College (Autonomous), Mala, Thrissur, Kerala, India

Lalchhandami Tochhawng
Mizoram Science, Technology and Innovation Council, Aizawl, Mizoram, India

Vinayak Upadhya
Department of Forest Products and Utilization, College of Forestry, University of Agricultural Sciences, Dharwad, Sirsi, Karnataka, India

Alekhya Veeramaneni
Department of Pharmacy, GIET School of Pharmacy, Chaitanya Knowledge City, Rajahmundry, Andhra Pradesh, India

Pulala Raghuveer Yadav
Department of Biotechnology, Indian Institute of Technology Hyderabad, Sangareddy, Telangana, India

Pulicherla Yugandhar
Survey of Medicinal Plants Unit, Regional Ayurveda Research Institute, Itanagar, Arunachal Pradesh, India

Stella C. Zothankhumi
Department of Zoology, Mizoram University (A Central University), Aizawl, Mizoram, India

Zothansiama
Department of Zoology, Mizoram University (A Central University), Aizawl, Mizoram, India

Abbreviations

AChE	acetylcholinesterase
ASA	acetylsalicylic acid
CCl_4	carbon tetrachloride
CEDL	chloroform leaf extract
CNS	central nervous system
EC	ethyl cellulose
EMT	epithelial-mesenchymal transition
FIC	ferrous ion chelating
HCC	hepatocellular carcinoma
LDL-C	low-density lipoprotein cholesterol
LPS	lipopolysaccharide
MBC	minimum bactericidal concentration
MEAR	methanolic extract of *Actiniopteris radiata*
MEDL	methanolic extract of *Dicranopteris linearis*
MIC	minimum inhibitory concentration
MRSA	methicillin-resistant *S. aureus*
NBT	nitroblue tetrazolium
NMR	nuclear magnetic resonance
NSAID	non-steroidal anti-inflammatory drug
PBS	phosphate buffer solution
PCM	paracetamol
PPI	protein-protein interaction
PUFA	polyunsaturated fatty acid
ROS	reactive oxygen species
SOD	superoxide dismutase
TNBC	triple-negative breast cancer
ZOI	zone of inhibition

Preface

Pteridophytes are non-flowering plants comprising over 12,000 species spread across 337 genera. They are distributed in both tropical and temperate forests. Pteridophytes include lycophytes and ferns. These lycophytes and ferns have been sources of remedies for the treatment of human diseases since time immemorial. They are rich in bioactive phytochemicals such as phenolic acids, flavonoids, phenylpropanoids, lignans, terpenoids, glycosides, alkaloids, coumarins, chromones, quinones, and xanthones.

Pteridophytes exhibit antioxidant, anticancer, antimicrobial, anti-inflammatory, antimalarial, antidiabetic, neuroprotective, cardioprotective, hepatoprotective, and nephroprotective properties. In this book, the bioactives and pharmacology of selected medicinal plants of pteridophytes are presented. A brief introduction is provided for each species. Under each species, bioactive compounds are listed along with their chemical structures, followed by their pharmacological activities. All the published literature on the phytochemistry and pharmacological activities of each species is reviewed.

I hope this will serve as a sourcebook for the development of new drugs for the benefit of mankind. I am thankful to all the contributors for their cooperation and erudition.

CHAPTER 1

Overview of the Medicinal Fern Genus *Aleuritopteris* Fée (Pteridaceae) from India

SACHIN M. PATIL,[1,*] BHUPEDRA S. KHOLIA,[2] and KISHORE S. RAJPUT[3,*]

[1]*Department of Botany, Dr. Ghali College, Gadhinglaj (affiliated to Shivaji University, Kolhapur), Maharashtra, India*

[2]*Northern Regional Center, Botanical Survey of India, Dehra Doon, Uttarakhand, India*

[3]*Department of Botany, Faculty of Science, The Maharaja Sayajirao University of Baroda, Vadodara, Gujarat, India*

*Corresponding authors

ABSTRACT

The Cheilanthoid fern genus *Aleuritopteris* is an important medicinal fern genus distributed mainly in Southeast Asia; however, only three species have reached Africa, and two extend to the neotropics. It has been reported from lowland to highland areas, tropical to temperate regions, and dry, semiarid regions to high rainfall regions. The present chapter focuses on the diversity, distribution, ethnomedical properties, phytochemistry, and pharmacology of the genus *Aleuritopteris* from India, along with a focus on the future scope of the genus in pharmacology and medical biology.

1.1 INTRODUCTION

The Cheilanthoid ferns are generally placed into the subfamily Cheilanthoideae of the family Pteridaceae (PPG – I, 2016). Many of them are

Bioactives and Pharmacology of Pteridophytes. T. Pullaiah (Ed.)

DOI: 10.1201/9781779642738-1

characterized by small habits with an indument of farina, scales, or hairs on the fronds and are adapted to xeric or semixeric environmental conditions. The different species of *Aleuritopteris* (under the genus *Cheilanthes* s.l.) have widely been used as medicines in traditional therapy, and their medicinal properties are listed in various dictionaries or encyclopedias of medicinal plants (Ambast, 2000; Jain, 1991; Patil et al., 2020). The present chapter emphasizes the distribution, ethnomedicinal value, phytochemistry, and pharmacognosy of the genus *Aleuritopteris* (silver fern) from India. It also focuses on the future scope of the genus in medicinal biology.

1.2 PHYTOGEOGRAPHICAL DISTRIBUTION OF *ALEURITOPTERIS* SPECIES IN INDIA

The genus *Aleuritopteris* generally grows as a lithophyte and rupestral plant and is found along roadsides, hilly slopes, moist rocks, and walls. It has some drought-resistant capacity, and fronds curl up in the summer season or whenever atmospheric moisture goes below 50%. Most of the species of *Aleuritopteris* are distributed in the subtropical and temperate regions; however, a few species also extend to both higher and lower altitudes in tropical as well as subalpine and alpine zones. Hence, they occur between 100 m to 4500 m altitudes in almost all climates and ecological zones, from drier regions to wet regions, low rainfall areas to high rainfall areas (300 to 5000 mm), and low-temperature areas to high-temperature areas (up to 35°C). As far as the fern biogeography of India is concerned, the maximum taxa are nestled in the Himalayan region or northern India; the same rule is also followed by the genus *Aleuritopteris*. The phytogeography and distribution of *Aleuritopteris* in India are more interesting. Nearly 50% of global species of *Aleuritopteris* are present in India and are distributed in almost all the phytogeographical regions except in the Andaman and Nicobar Islands. However, there is no endemism because the Indian landmass is well connected with neighboring countries like Pakistan, Nepal, Bhutan, Myanmar, and China; hence, all the north Indian species are common with these countries. The diversity of the genus *Aleuritopteris* in north India (Western Himalaya, Eastern Himalaya, and Northeast India) is the highest; out of 19 species and seven hybrids, 16 species (88.46%) and all (100%) hybrids occur in this region. The remaining three species are confined to southern and western India with restricted distribution, and there are no hybrids in these regions. Among these, *A. scioana* occurs only in Rajasthan; however, *A. bullosa* and *A. wollenweberi* are restricted to very few areas in Kerala and Tamil Nadu and also in Sri Lanka; hence, these two are typical examples of Hindu Lankan endemics (Fraser-Jenkins, 2012). In northern India, the maximum diversity is in the Himalayas (16 spp. plus 5 hybrids), followed by Northeast India (9 spp. and 2 hybrids). Within the Himalayas, the eastern Himalayas (with 16 spp. and one hybrid) are richer and more diverse than the western Himalayas (with 14 spp. and 4 hybrids). In Himalayan species, *A. argentea* and *A. duclouxii* do not occur in the western Himalayas; they have restricted distribution in Sikkim and Arunachal Pradesh, respectively, and these are considered the rarest species in India. Beyond Northern India, Peninsular India is the richest in *Aleuritopteris* diversity, having a maximum of seven species, followed by Western India with four species. Central India has minimum diversity with only three species. Hybrids are confined to the Himalayas and Northeastern India. Out of seven Indian hybrids of *Aleuritopteris*, four (*A. x confundans, A. x godavariensis, A. x pangteyi,* and *A. x vermae*) occur in the western Himalayas, two (*A. x khasiana* and *A. x nepalensis*) in Northeast India, and one (*A. x hamiltonii*) in the Sikkim Himalayas. Among these hybrids, five also occur in Nepal; however, *A. x khasiana* is restricted to Meghalaya, and *A. x vermae* is confined to western Himalayan states. The distribution of *Aleuritopteris* species in India in different climatic and phytogeographical regions is shown in Table 1.1.

A. anceps, A. bicolor and *A. formosana* are distributed throughout mainland India. Though they are very common in northern India, their distribution in southern India or peninsular India is very limited. On the other hand, *A. argentea, A. duclouxii,* and *A. scioana* have restricted distributions in Sikkim, Arunachal Pradesh, and Rajasthan respectively, hence they are considered the rarest *Aleuritopteris* species in India. The other rare Indian *Aleuritopteris* species are *A. duthiei* and *A. tamburii*. Though *A. duthiei* (Uttarakhand, Nepal, Sikkim, Bhutan, China) and *A. tamburii* (Arunachal Pradesh, Meghalaya, Sikkim, Uttarakhand, Nepal, China, Myanmar) have a wide range of distribution, their populations are very low and rare to collect in the field.

1.3 ETHNOMEDICINAL USES OF *ALEURITOPTERIS* SPECIES

Plants are used by various ethnic communities as folklore medicine. The ancient knowledge of plants used by ethnic communities and local

TABLE 1.1 Distribution of *Aleuritopteris* Species in Different Phytogeographical Zones of India

Name of the Species	W.H	E.H	NE India	Central India	Western India	Peninsular India
A. albomarginata (C.B.Clarke) Ching	+	+	+	–	–	+
A. anceps (Blanf.) Panigrahi	+	+	+	+	+	+
A. argentea (S.G.Gmel.) Fee	–	+	–	–	–	–
A. bicolor (Roxb.) Fraser-Jenk.	+	+	+	+	+	+
A. bullosa (Kunze) Ching	–	–	–	–	–	+
A. chrysophylla (Hook.) Ching	+	+	+	–	–	–
A. dealbata (C.Presl) Fee	+	+	+	–	–	–
A. dubia (C.Hope) Ching	+	+	–	–	–	+
A. duclouxii (Christ) Ching	–	+	–	–	–	–
A. duthiei (Baker) Ching	+	+	–	–	–	–
A. formosana (Hayata) Tagawa	+	+	+	+	+	+
A. grisea (Blanf.) Panigrahi	+	+	–	–	–	–
A. leptolepis (Fraser-Jenk.) Fraser-Jenk.	+	+	–	–	–	–
A. rufa (D.Don) Ching	+	+	+	–	–	–
A. scioana (Chiov.) Fraser-Jenk. & Dulawat	–	–	–	–	+	–
A. stenochlamys Ching	+	+	+	–	–	–
A. subdimorpha (C.B. Clarke & Baker) Fraser-Jenk.	+	+	+	–	–	–
A. tamburii (Hook.) Ching	+	+	–	–	–	–
A. wollenweberi Fraser-Jenk.	–	–	–	–	–	+
Hybrids						
A. x *confundans* Fraser-Jenk.	+	–	–	–	–	–
A. x *godavariensis* Fraser-Jenk. & Khullar	+	–	–	–	–	–
A. x hamiltonii Fraser-Jenk. & Khullar	–	+	–	–	–	–
A. x *khasiana* Fraser-Jenk.	–	–	+	–	–	–
A. x *nepalensis* Fraser-Jenk.	–	–	+	–	–	–
A x *pangteyi* Fraser-Jenk. & E.Wollenw.	+	–	–	–	–	–
A. x *vermae* (Fraser-Jenk. & Viane) Fraser-Jenk. & Khullar	+	–	–	–	–	–
Total number of species (including hybrids)	18	17	11	3	3	7

practitioners is referred to as ethnic therapeutics or ethnomedicinal plants. Generally, plants produce a variety of chemical compounds and are sorted by their chemical class, biosynthetic origin, and functional groups.

The medicinal value of plants depends on the chemical composition or groups of compounds present in plants (secondary metabolites), and they produce a definite physiological action in the human body (Johnson et al., 2021; Patil et al., 2022). Plants synthesize a wide variety of secondary metabolites with complex chemical compositions, which are produced in response to different forms of biotic stresses, as well as to fulfill important physiological processes, like attracting pollinators, establishing symbiosis, and providing structural components to the lignified cell walls of vascular tissue. Medicinal plants contain active ingredients (viz., alkaloids, phenolics, terpenoids, and tannins) that can be used for therapeutic purposes and are precursors of chemotherapeutical synthesis. They also provide temporary relief to symptomatic problems and health-promoting characteristics and have curative properties. The discovery of clinically useful compounds from plants has been an effective approach in recent times. They are under chemical scrutiny due to the advent of modern scientific methods, which lead to the isolation of active compounds. After their isolation and characterization, these active compounds in pure form or well-characterized extracts became useful in several countries. Therefore, more recently, plant-based medicines are replacing modern medicines around the world. Pteridophytes, a neglected group, are found to provide food, medicine, fiber, crafts, and building materials or are used in decorations (Kholia, 2010; Kholia and Balakrishna, 2022; May, 1978; Nwosu, 2002). They synthesize a wide range of primary and secondary metabolites enriched with various bioactivities to defend and acclimatize to various ecological conditions that could potentially be used as alternative medicine for the treatment of many diseases. However, the available literature on the phytochemicals and their pharmacological applications in pteridophytes is limited. Pteridophytes have been ethnomedicinally used in different communities around the world. Until now, 60% of pteridophytes have been used by inhabitants around the world, depending upon the ethnomedicinal uses of plants. Therefore, pteridophyte-based medicines can replace traditional and other systems of modern medicine in this changing world.

There are several reports in encyclopedia compilations (Ambast, 2000; Jain, 1991; Kirtikar and Basu, 1935) and research articles published on ethnomedicinal pteridophytes in various journals, books, and conference proceedings, mentioning that various types of pteridophytes were widely used to treat fever, cold, cough, epilepsy, leprosy, stomach pain, gastrointestinal disorders, eradication of worms in children, and many more ailments in different parts of India (Gaur and Bhatt, 1994; Kholia, 2014; Kirn and Kapahi, 2001; Kumar and Ramesh, 2003; Nayar, 1959). A few of them are also sold in local markets by different tribal communities (Giri et al., 2021; Kholia and Punetha, 2005; Ojha and Devkota, 2021; Patil et al., 2020; Rana et al., 2015).

The genus *Aleuritopteris* (previously placed in *Cheilanthes*) is traditionally used in various parts of India. The stipe of *Aleuritopteris* was used as antiseptic studs in freshly pierced noses or ears, whereas fresh lamina juice is also applied if these pricked ears and nose get infected. Additionally, fresh lamina juice is also applied to cuts and wounds (Kholia and Punetha, 2005). However, its medicinal properties were not mentioned in classical British Indian literature on medicinal plants or plant products (Dymock, 1893; Watt, 1883). Furthermore, in a monumental work, only one Chelianthoid fern (*Cheilanthes tenuifolia*) was reported with medicinal properties (Caius, 1935). However, various species of *Aleuritopteris* (under the genus *Cheilanthes* s.l.) have widely been used as medicines in traditional therapy, and their medicinal properties are listed in various dictionaries or encyclopedias of medicinal plants (Ambast, 2000; Chopra et al., 1956; Jain, 1991; Kirtikar and Basu, 1935). Besides their traditional use against wounds and cuts, they are also mentioned to be used to cure various diseases, viz., peptic ulcer, gastritis, infertility in women, dysentery, cold, cough, fever, rheumatism, kidney stones, liver disorders, and fungal infections (Table 1.2). Therefore, an attempt has been made to explore indigenous and ethnomedicinally important species of *Aleuritopteris* and properly document their useful aspects. It has been screened for various biological activities by Wollenweber (1978), Wollenweber and Schneider (2000), Jiang et al. (2022), and Rimal et al. (2023). Phytochemical screening of frond, rhizome, and root extracts of *Aleuritopteris albormarginata, A. anceps, A. bullosa, A. grisea,* and *A. rufa* contains a significant amount of total phenolic content, total flavonoids content, total sugar content, and total tannin, indicating that the genus is rich in secondary metabolites. The radical scavenging activity of the extracts showed effective inhibition activity against the DPPH radical. Also, the white or yellow coatings are known to occur on the lower leaf surfaces of *Aleuritopteris*, known as wax, cera, powder, farina, resins, and 'pseudostearoptenes' (Wollenweber, 1976). It was found early that these coatings were excreted by glandular trichomes (De Bary, 1877) (Figure 1.1).

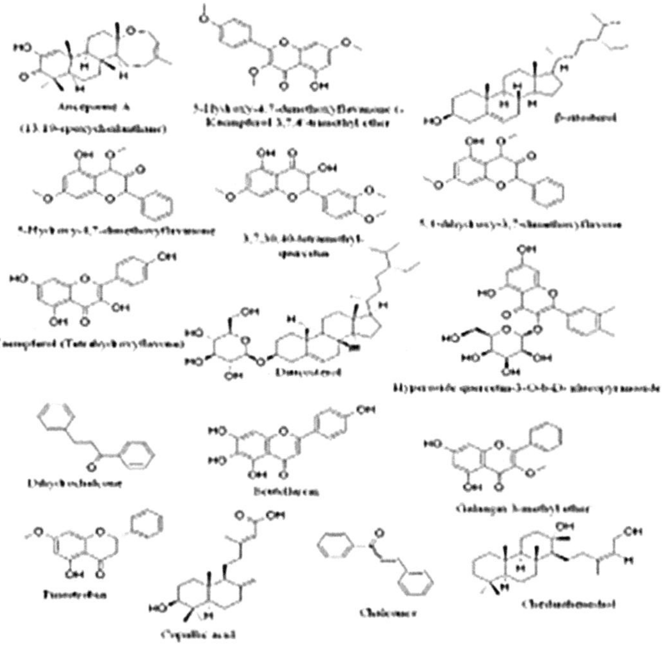

FIGURE 1.1 Chemical compounds from different species of *Aleuritopteris*.

TABLE 1.2 Ethnomedicinal Uses of *Aleuritopteris* Species

Name of the species	Part used	Ethnic value	Source
Aleuritopteris albomarginata	Whole plant	Frond and rhizome decoction of the fronds is used to cure peptic ulcers. Powdered dried leaves are mixed with crushed ginger rhizomes and used to treat infertility in women. A mixture of it is used orally against dysentery. The juice of the frond and rhizome is used to treat stomachaches and gastritis.	Budha-Magar et al. (2020); Giri et al. (2021); Nwosu (2002); Ojha and Devkota (2021); Patil et al. (2020); Rana et al. (2015); Upreti et al. (2010)
A. anceps	Whole plant	The juice of the plant part is mixed with *Drymaria diandra,* which is used to cure peptic ulcers. The stipe part is used for abdominal spasms, cuts, and wounds. The fronds and rhizomes of the plant are boiled and consumed three times a day to treat stomach problems. It is also used to cure colds, coughs, fever, diabetes, and kidney problems.	Giri et al. (2021); Joshi et al. (2020); Manandhar (2002); Ojha and Devkota (2021); Patil et al. (2020)

TABLE 1.2 *(Continued)*

Name of the species	Part used	Ethnic value	Source
A. argentea	Rhizome	The decoction of whole plants used by Tibetans and Mongolians to promote blood circulation, to cure menstruation problems, control the menstruation cycle, prevent cancer and remedy for tuberculosis or hemoptysis.	Wu et al. (2023); Zhang et al. (2020)
A. bicolor	Whole plant	The whole plant powder is taken orally for sinusitis or fever and is applied to cuts. The juice is taken during diarrhea, dysentery, and gastritis. Plant powder with animal ghee is used as incense to keep away fear in children.	Adhikari et al. (2019); Luitel et al. (2014); Patil et al. (2020); Tamang et al. (2017)
A. bullosa	Whole plant	Plant juice filtered and mixed with rice starch given thrice daily against the discharge of yellowish urine.	Lal et al. (1996); Padhy et al. (2016)
A. formosana	Fronds	Young frond extract mixed with honey is used against body weakness and as a herbal tonic.	Irfan et al. (2021)
A. grisea	Fronds	The young frond extract mixed with cow butter and sugar is used to enhance bone strength and reduce body weakness.	Irfan et al. (2021)
A. leptolepis	Whole Plant	Fronds and rhizomes have been used to cure fever or bacterial infections.	Singh (2003)
A. rufa	Whole Plant	The juice of rhizome and frond is used in the treatment of abdominal disorders, cuts, and wounds.	Dhami (2008); Ojha and Devkota (2021)

1.4 PHYTOCHEMISTRY OF ALEURITOPTERIS SPECIES

The genus *Aleuritopteris* has different types of secondary metabolites. A total of 20 chemical compounds, viz., Ancepsone A, 5-hydroxy-3,7,4′-trimethoxyflavone, β-sitosterol, 5-hydroxy-4,7-dimethoxyflavanone, 5,4-dihydroxy-3,7-dimethoxyflavone, Kaempferol (=Tetrahydroxyflavone), Daucosterol (=Sitogluside), Hyperoside quercetin-3-O-β-D-glucopyranoside, 2,17-diol (2β, 4α)-15,16-ene-beyerane diterpenoid (=aleuritopsis A), 2,18-diol (2β, 4α)-16-ene-ent-kaurane diterpenoid (=aleuritopsis B), dihydrochalcone, chalcones, scutellarein, galangin 3-methyl ether, pinostrobin, copallic acid, alepterolic acid, isoscoparin-7-olate, 7,4'-O-dimethylkaempferol, and cheilanthenediol were isolated from the species, viz., *Aleuritopteris albomarginata, A. anceps, A. argentea, A. bullosa, A. grisea*, and *A. rufa*.

However, many species remain unexplored regarding phytochemical constituents. The common phytochemical constituent kaempferol was found in all the studied *Aleuritopteris* species. Ancepsone A, 3,7,30,40-tetramethyl-quercetin, 5-hydroxy-3,7,4′-trimethoxyflavone, daucosterol, and hyperoside quercetin-3-O-β-D-glucopyranoside were found only in *A. anceps*; aleuritopsis A, aleuritopsis B, dihydrochalcone, chalcones, scutellarein, galangin 3-methyl ether, and pinostrobin were found in *A. argentea*; isoscoparin-7-olate, 7,4'-O-dimethylkaempferol, and cheilanthenediol were found in *A. bicolor* (Jiang et al., 2020, 2022; Li et al., 2017; Wollenweber, 1978; Wollenweber and Schneider, 2000).

1.5 PHARMACOLOGY OF *ALEURITOPTERIS* SPECIES

1.5.1 ANTIBACTERIAL ACTIVITY

The fractions in different solvents of *Aleuritopteris bicolor* showed antibacterial activity against various bacterial strains (gram-positive and gram-negative) that are pathogenic to human beings, causing several diseases. The ethyl acetate fraction exhibited the maximum activity against the studied gram-negative and gram-positive bacteria, followed by acetone. In contrast, benzene, chloroform, alcohol, petroleum ether, and aqueous extract showed no activity or negligible activity against the bacterial strains (Singh and Chand, 2020). The whole plant extract of *Aleuritopteris bicolor* showed a zone of inhibition (ZOI) of 15 mm and 18 mm at 100 mg/ml and 200 mg/ml, respectively, against *Streptococcus pneumoniae*. The formulation of small-sized iron particles (nanoparticles) from *A. bicolor* extract showed antimicrobial efficacy. The nanoparticles were amorphous in shape with an average size of 40–60 nm. These particles were tested against different bacterial strains, demonstrating antimicrobial efficacy and displaying a MIC value of 6.25 μg/ml and 3.125 μg/ml against clinical pathogens *Staphylococcus aureus* and *Escherichia coli*, respectively (Seth et al., 2022). *Aleuritopteris anceps* showed noteworthy antibacterial activity against *Staphylococcus aureus* with a MIC value of 649 μg/mL and *Pseudomonas aeruginosa* with a MIC value of 38 μg/mL. The aqueous and ethanol fractions of *Aleuritopteris albomarginata* frond have antibacterial activity against *Agrobacterium tumefaciens, Escherichia coli, Salmonella arizonae*, and *Staphylococcus aureus*. However, the highest antibacterial activity of *A. albomarginata* was shown against *E. coli* (Joshi et al., 2020; Parihar et al., 2010). The above studies indicate that the antibacterial substances present in the rhizome and frond are in good amounts. The epidermal glands present on the rhizome and fronds contain substances like phenolic compounds, glycosides, flavonoids, and alkaloids. These substances are largely responsible for antimicrobial activity and are soluble in organic solvents, easily extracted in methanol, ethanol, and acetone but less soluble in water (Parihar et al., 2010; Tiwari et al., 2021).

1.5.2 ANTIFUNGAL ACTIVITY

The antifungal activities of *Aleuritopteris albomarginata* against two fungal strains were tested using the well diffusion technique. An average ZOI against *Aspergillus niger* and *Fusarium oryzae* was observed (Jarial, 2017). *A. albomarginata* at 18 mg/ml showed significant activity against *A. niger. A. anceps* was tested against *Candida albicans*, showing an average zone (MIC of 35 μg/mL) against it (Joshi et al., 2022).

1.5.3 ANTI-INFLAMMATORY ACTIVITY

The methanol extract of *Aleuritopteris anceps* inhibited fresh egg albumin-induced acute paw edema in rats. After treatment with 200 mg/kg and 400 mg/kg methanol extract of *A. anceps* and 5 mg/kg indomethacin, a dose-dependent anti-inflammatory activity was observed. Therefore, it could be a good source of natural anti-inflammatory substances (Parajuli and Khanal, 2017). The *A. albomarginata* fraction of the rhizome has been used to treat peptic ulcers. Therefore, Lamichhane et al. (2014) studied the biological activities (antioxidant, anti-inflammatory, anti-adipogenic, and anti-obesity). The study revealed that ethyl acetate and butanol fractions showed anti-inflammatory activity; hence, it is suggested that the ethyl acetate and butanol extracts of *A. albomarginata* (=*Cheilanthes albomarginata* s.l.) are potential sources for anti-inflammatory activity. The anti-inflammatory and anti-nociceptive activities of *A. farinosa (Cheilanthes farinosa* s.l.) are used in many parts of Ethiopia to treat inflammatory skin disorders; therefore, it was studied by Yonathan et al. (2018). The potentiation test was done to analyze if the methanol extract of *A. farinosa*, which showed significant anti-inflammatory activity when given alone, could also potentiate the anti-inflammatory activity of ASA.

1.5.4 ANTIDIABETIC PROPERTIES

Phenolic and flavonoid compounds are widely distributed in plants and are included in the human diet, having antioxidant, antidiabetic, and various other medicinal properties. To understand the antidiabetic properties of the medicinal fern, *Aleuritopteris bicolor,* α-amylase enzyme inhibition activity was performed to evaluate the antidiabetic activity of plant extracts by Pandey and Sharma (2022). *A. bicolor* showed α-amylase inhibitory activity with IC_{50} 651.58 ± 10.32 μg/mL. Therefore, it could be a good source of natural antidiabetic compounds.

1.5.5 ANTICANCER PROPERTIES

Aleuritopteris bicolor (=*Cheilanthes farinosa* s.l.) has antiproliferative and apoptotic activity in human liver cancer cells and is not poisonous towards non-cancerous macrophage cell lines. The aqueous fraction of *A. bicolor* markedly reduced the hepatic cancer cell population, Hep3B, whereas only marginal cytotoxicity was produced towards non-cancerous murine macrophage RAW264.7 cells (Radhika et al., 2010). Aleuritopsis A and Aleuritopsis B (new diterpenoid compounds) were obtained from *Aleuritopteris argentea.* The MTT method was used to study the cytotoxicity of newly discovered compounds against MCF-7 and SKOV3 cells, as the results exhibited inhibitory effects on the tumor cells. The inhibitory effects were observed against MCF-7 breast cancer cells and SKOV3 ovarian cancer cells. A natural diterpenoid (alepterolic acid) was isolated from *Aleuritopteris argentea*, which has potential anti-cancer activity. Alepterolic acid was modified to construct a series of arylformyl piperazinyl derivatives with inhibition rates higher than 30% at 10 μM. Further, it was measured to obtain the IC50 values against four cancer cell lines (including hepatoma cell lines HepG2, lung cancer cell lines A549, estrogen receptor-positive cell lines MCF7, and triple-negative breast cancer (TNBC) cell lines MDA-MB-231) by MTT assay. It was found that alepterolic acid was more effective against HepG2 and MDA-MB-231 cells, while less toxic to A549 and MCF7 cells, and the most toxic derivative against MDA-MB-231 cell lines, with an IC50 value of 5.55±0.56 μM. Conclusively, alepterolic acid was demonstrated to be a potential anti-cancer lead compound for further investigation (Liu et al., 2023).

1.6 CONCLUSION AND RECOMMENDATION

Based on the above discussion, pteridophytes are important groups of plants and are equally important in traditional medicinal systems. Due to their restricted distribution, their ethnomedicinal significance is commonly known in tribal or rural areas, while urban dwellers are not much aware. Since several species of *Aleuritopteris*, as well as other genera of the pteridophytes, remain unexplored, there is an urgent need to work on this group of plants. There is a possibility of discovering several novel compounds from this unique plant group that may help develop new drugs for newly emerging diseases like coronavirus or other diseases that are still awaiting drug development.

ACKNOWLEDGMENT

The authors are thankful to the Head, Department of Botany, Shivaji University, Kolhapur, for providing the laboratory facilities.

KEYWORDS

- **cheilantoids**
- **ethnomedicine**
- **non-cancerous macrophage cell lines**
- **phytochemistry**
- **phytogeography**
- **pteridophytes**
- **secondary metabolites**
- **silver fern**

REFERENCES

Adhikari, M., Thapa, R., Kunwar, R. M., Devkota, H. P., & Poudel, P. (2019). Ethnomedicinal uses of plant resources in the Machhapuchchhre Rural Municipality of Kaski District, Nepal. *Medicines, 6*(2), 69.

Ambast, S. P. (2000). In K. Kashyap & R. Chand (Eds.), *The Useful Plants of India.* NISCOM. CSIR, New Delhi.

Budha-Magar, S., Bhandari, P., & Ghimire, S. K. (2020). Ethno-medicinal survey of plants used by Magar (Kham) community, Rolpa district, Western Nepal. *Ethnobotany Research and Applications, 19*(18), 1–29.

Caius, J. F. (1935). The medicinal and poisonous ferns of India. *Journal of the Bombay Natural History Society, 38*(2), 341–361.

Chopra, R. N., Nayar, S. L., & Chopra, I. C. (1956). *Glossary of Indian Medicinal Plants*. Council of Scientific and Industrial Research, New Delhi.

De Bary, A. (1877). *Vergleichende Anatomie der Vegetationsorgane der Phanerogamen und Farne*. W. Engelmann, Leipzig.

Dhami, N. (2008). Ethnomedicinal uses of plants in Western Terai of Nepal: A case study of Dekhatbhuli VDC of Kanchanpur district. In P. K. Jha, S. B. Karmacharya, M. K. Chettri, C. B. Thapa, & B. B. Shrestha (Eds.), *Medicinal Plants in Nepal: An Anthology of Contemporary Research*. Ecological Society (ECOS), Kathmandu, Nepal.

Dymock, W. (1893). *Pharmacographia Indica: A history of the principal drugs of vegetable origin met within British India, Part VI*. Kegan Paul, London.

Fraser-Jenkins, C. R. (2012). Rare and threatened pteridophytes of Asia 2. Endangered species of India — the higher IUCN categories. *Bulletin of the National Museum of Nature and Science Series B, 38*(4), 153–181.

Gaur, R. D., & Bhatt, B. P. (1994). Folk utilization of some pteridophytes of Deoprayag area in Garhwal Himalaya: India. *Economic Botany, 48*(2), 146–151.

Giri, P., Kumari, P., Sharma, P., & Uniyal, P. L. (2022). Ethnomedicinal uses of pteridophytes for treating various human ailments in India. In L. J. Singh & V. Ranjan (Eds.), *New Vistas in Indian Flora* (pp. 199–212). M/s Bishen Singh Mahendra Pal Singh, Dehra Dun, India. https://doi.org/10.1007/978-981-16-6170-9.

Irfan, M., Jan, G., Jan, F. G., & Murad, W. (2021). Taxonomy and spore morphology of selected taxa of Cheilanthoideae and Pteridoideae (Pteridaceae) from Pakistan. *Microscopy and Research Technology, 84*, 2867–2882.

Jain, S. K. (1991). *Dictionary of Indian Folk Medicine and Ethnobotany*. Deep Publications, New Delhi.

Jarial, R., Singh, L., Thakur, S., Zularisam, A. W., Sakinah, M., & Kanwar, S. S. (2017). Evaluation of antilipolytic, antioxidant and antibacterial activities of selected ferns. *Journal of Applied Pharmaceutical Science, 7*(6), 150–156. https://doi.org/10.7324/JAPS.2017.70621.

Jiang, C., Chen, J., Dai, X., Wang, Q., Cao, J., & Huang, G. (2020). *ent*-Pimaradiene and cyathane diterpenes from *Aleuritopteris albofusca*. *Phytochemistry Letters, 40*, 10–14. https://doi.org/10.1016/j.phytol.2020.09.008.

Jiang, T., Dai, X., Gao, T., Wanga, L., Yang, F., Zhang, Y., Wang, N., Huang, G., & Cao, J. (2022). Ancepsone A, a new cheilanthane sesterterpene from *Aleuritopteris anceps*. *Tetrahedron Letters, 100*, 153869. https://doi.org/10.1016/j.tetlet.2022.153869.

Johnson, M., Anwardeen, I., Periasamy, A., Siva, H. R., Dhinesh, K. M., Almeida, R. S., & Coutinho, H. D. M. (2021). Evaluation of phenolic constituents and toxicity of lycophytes and ferns of Shervarayan hills aqueous extracts. *Chemistry Africa, 4*, 513–523.

Joshi, B., Panda, S. K., Jouneghani, R. S., Liu, M., Parajuli, N., Leyssen, P., Neyts, J., & Luyten, W. (2020). Antibacterial, antifungal, antiviral, and anthelmintic activities of medicinal plants of Nepal selected based on ethnobotanical evidence.

Evidence-Based Complementary and Alternative Medicine, 2020, 1043471. https://doi.org/10.1155/2020/1043471.

Kholia, B. S. (2010). *Ferns and Fern-Allies of Sikkim: A Pictorial Handbook Part-1*. Sikkim State Biodiversity Board and Botanical Survey of India, Gangtok.

Kholia, B. S. (2014). Traditional uses of pteridophytes in Sikkim Himalaya. In S. Kumar (Ed.), *Ethnobotanical studies in India*. Deep Publications, Delhi.

Kholia, B. S., & Balakrishna, A. (2022). Pteridophytes used by peoples of Indian Himalayan Region and Northern India: An overview. In M. Johsen, H. Fernandez, A. Kumar, & S. Thangaiah (Eds.), *Ferns: Biotechnology, Propagation, Medicinal Uses and Environmental Regulation*. Springer Nature Singapore.

Kholia, B. S., & Punetha, N. (2005). Useful pteridophytes of Kumaon (central Himalaya, India). *Indian Fern Journal, 22*(1, 2), 1–6.

Kirn, H. S., & Kapahi, B. K. (2001). Ethnobotanical notes on some ferns and fern-allies of Jammu and Kashmir State, India. *Indian Fern Journal, 18*, 35–38.

Kirtikar, K. R., & Basu, B. D. (1935). *Indian Medicinal Plants* (Vols. 1–4). Lalit Mohan Basu, Allahabad.

Kumar, M., & Ramesh, M. (2003). Medicinal pteridophytes of Kerala, South India. *Indian Fern Journal, 20*, 1–28.

Lal, B., Vats, S., Singh, R. D., & Gupta, A. K. (1996). Plants used as ethnomedicine and supplementary food by Gaddis of Himachal Pradesh, India. In S. K. Jain (Ed.), *Ethnobiology in Human Welfare* (pp. 384–387). Deep Publications, New Delhi.

Lamichhane, R., Kim, S., Poudel, A., Sharma, D., Lee, K., & Jung, H. (2014). Evaluation of *in vitro* and *in vivo* biological activities of *Cheilanthes albomarginata*. *BMC Complementary and Alternative Medicine, 14*, 342. https://doi.org/10.1186/1472-6882-14-342.

Li, J., Yuan, X., Liu, Y., Li, Y., Cui, N., Li, L., & Ha, J. (2017). Two new diterpenoids from *Aleuritopteris argentea*. *Phytochemistry Letters, 20*, 22–25.

Liu, Z., Xia, C., Wang, N., Cao, J., Huang, G., & Ma, L. (2023). Synthesis and evaluation of piperazine-tethered derivatives of alepterolic acid as anticancer agents. *Chemistry and Biodiversity, 20*(5), e202300208. https://doi.org/10.1002/cbdv.202300208.

Luitel, D. R., Rokaya, M. B., Timsina, B., & Münzbergová, Z. (2014). Medicinal plants used by the Tamang community in the Makawanpur district of central Nepal. *Journal of Ethnobiology and Ethnomedicine, 10*(1), 1–11. https://doi.org/10.1186/1746-4269-10-5.

Manandhar, N. P. (2002). *Plants and People of Nepal*. Timber Press, Portland, Oregon, USA.

May, L. W. (1978). The economic uses and associated folklore of ferns and fern allies. *Botanical Review, 44*, 491–528.

Nayar, B. K. (1959). Medicinal ferns of India. *Bulletin Natural Botanical Garden (Lucknow) India, 29*, 1–36.

Nilsson, M. (1961a). Dihydrochalcones from the fronds of *Pityrogramma chrysophylla* var. *marginata* Domin. *Acta Chemica Scandinavica, 15*, 154–158.

Nilsson, M. (1961b). Chalcones from the fronds of *Pityrogramma chrysophylla* var. *heyderi*. *Acta Chemica Scandinavica, 15*, 211, 212.

Nwosu, M. (2002). Ethnobotanical studies on some pteridophytes of Southern Nigeria. *Economic Botany, 56*(3), 255–259.

Ojha, R., & Devkota, H. P. (2015). Edible and medicinal pteridophytes of Nepal: A review. *Ethnobotany Research and Applications, 22*, 1–16.

Padhy, R., Dash, S. K., & Padhy, S. (2016). Pteridophyta diversity of South Odisha, India, with special reference to medico folklore claims: A brief review. *Journal of Biodiversity, 7*(1), 25–32.

Pandey, L. K., & Sharma, K. R. (2022). Analysis of phenolic and flavonoid content, α-amylase inhibitory and free radical scavenging activities of some medicinal plants. *The Scientific World Journal, 2022*, 1–8. https://doi.org/10.1155/2022/4000707.

Parajuli, C., & Khanal, D. P. (2017). Anti-inflammatory activity and brine shrimp bioassay of *Aleuritopteris anceps*. *World Journal of Pharmaceutical Research, 6*(16), 160–168. https://doi.org/10.20959/wjpr201716-9669.

Parihar, P., Parihar, L., & Bohra, A. (2010). *In vitro* antibacterial activity of fronds (leaves) of some important pteridophytes. *Journal of Microbiology and Antimicrobials, 2*(2), 9–22.

Patil, S. M., Patil, M. N., & Patil, S. S. (2020). Diversity and ethnomedicinal pteridophytes of Kolhapur district, Maharashtra. *Indian Fern Journal, 37*(1, 2), 226–238.

Radhika, N. K., Sreejith, P. S., & Asha, V. V. (2010). Cytotoxic and apoptotic activity of *Cheilanthes farinosa* (Forsk.) Kaulf. against human hepatoma, Hep3B cells. *Journal of Ethnopharmacology, 128*(1, 2), 166–171. https://doi.org/10.1016/j.jep.2010.01.002.

Rana, S. K., Oli, P. S., & Rana, H. K. (2015). Traditional botanical knowledge (TBK) on the use of medicinal plants in Sikles area, Nepal. *Asian Journal of Plant Science and Research, 5*(11), 8–15.

Rimal, P., Kunwor, K., Baskota, M., Pokhrel, A., Khatri, J., Devkota, A., & Rajbhandar, M. (2023). Nature's healing palette: Unlocking the mysteries of *Cheilanthes bicolor* (Roxb) Fraser-Jenkins: A journey into phytochemical diversity and potent biological actions. *International Journal of Herbal Medicine, 11*(5), 101–109.

Seth, A., Devi, E., Thakur, K., Attri, C., Singh, V., Bhandari, A., Singh, M., & Seth, M. K. (2022). Himalayan fern *Cheilanthes bicolor* mediated fabrication and characterization of iron nanoparticles with antimicrobial potential. *BioNanoScience, 12*, 486–495. https://doi.org/10.1007/s12668-022-00969-z.

Singh, H. B. (2003). Economically viable pteridophytes of India. In S. Chandra & M. Srivastava (Eds.), *Pteridology in the New Millennium* (pp. 421–446). Springer, Dordrecht.

Singh, M., & Chand, V. (2020). Antibacterial potential of different extract of *Aleuritopteris bicolor* in Doon valley, Uttarakhand. *International Journal of Creative Research Thoughts, 8*(12), 7–13.

Tamang, R., Thakur, C., Koirala, D., & Chapagain, N. (2017). Ethno-medicinal plants used by Chepang community in Nepal. *Journal of Plant Resources, 15*(1), 21–30.

Tiwari, R., Baral, R., Parajuli, N., Shrestha, R., Pun, S., Pahari, A., & Gurung, S. (2021). Phytochemical screening, free radical scavenging and *in-vitro* antibacterial activity of ethanolic extracts of selected medicinal plants of Nepal and effort towards formulation of antibacterial cream from the extracts. *International Journal of Herbal Medicine, 9*(3), 39–47.

Upreti, K., Jalal, J. S., Tewari, L. M., Joshi, G. C., Pangtey, Y. P. S., & Tewari, G. (2009). Ethnomedicinal uses of pteridophytes of Kumaun Himalaya, Uttarakhand, India. *Journal of American Science, 5*(4), 167–170.

Watt, G. (1883). *Catalogue of the exhibits at the Calcutta International Exhibition Part. V. Medicinal Products*. Superintendent of the Government Printing India Hastings, Calcutta.

Wollenweber, E. (1976). Flavonoid exudations in farinose ferns. *Phytochemistry, 15*, 2013.

Wollenweber, E. (1978). The distribution and chemical constituents of the farinose exudates in gymnogrammoid ferns. *American Fern Journal, 68*, 13–28.

Wollenweber, E., & Schneider, H. (2000). Lipophilic exudates of Pteridaceae chemistry and chemotaxonomy. *Biochemical Systematics and Ecology, 28*, 75–777.

Wu, B., Wei, X. Y., Wang, N., Xia, C., Bao, R., Cao, J., Zong, Z. M., Liu, Z., Ma, L., & Huang, G. (2023). Synthesis, structural characterization and antiproliferative evaluation of phenylalkylamino-containing alepterolic acid derivatives. *Journal of Molecular Structure, 1284*, 135358.

Yonathan, M., Asres, K., Assefa, A., & Bucar, F. (2006). *In vivo* anti-inflammatory and anti-nociceptive activities of *Cheilanthes farinosa*. *Journal of Ethnopharmacology, 108*, 462–470. https://doi.org/10.1016/j.jep.2006.06.006.

Zhang, S., Feng, N., Huang, J., Wang, M., Zhang, L., Yu, J., Dai, X., Cao, J., & Huang, G. (2020). Incorporation of amino moiety to alepterolic acid improve activity against cancer cell lines: Synthesis and biological evaluation. *Bioorganic Chemistry, 98*, 103756.

CHAPTER 2

Bioactive Constituents and Pharmacological Importance of *Asplenium adiantum-nigrum* L. (Aspleniaceae): A Review

AJAY V. RAJURKAR,[1] SHAILESH G. ATRAM,[2] and VISHAL P. DESHMUKH[3,*]

[1]*Department of Botany, Lokmanya Tilak Mahavidyalaya, Wani, Yavatmal, Maharashtra, India*

[2]*Department of Chemistry, Arts, Commerce and Science College, Maregaon, Yavatmal, Maharashtra, India*

[3]*Department of Botany, Jagadamba Mahavidyalaya, Achalpur City, Amravati, Maharashtra, India*

*Corresponding authors

ABSTRACT

Asplenium adiantum-nigrum is an important small size fern belonging to the family Aspleniaceae and commonly known as "rock ferns." It is the richest source of diverse phytoconstituents, with a predominance of polyphenolics and flavonoids. It is used to treat ophthalmia, spleen ailments, and jaundice. It is also applied to treat cough and gynecological diseases. Plants have demonstrated antimicrobial, antioxidant, anti-tyrosinase, cytotoxic activity, and phytotoxicity activities. The current review emphasizes the bioactive constituents and pharmacological potential of *Asplenium adiantum-nigrum*.

Bioactives and Pharmacology of Pteridophytes. T. Pullaiah (Ed.)

DOI: 10.1201/9781779642738-2

2.1 INTRODUCTION

The forest ecosystem has generated important material resources for humanity for a long time. Ferns, an important class of plants, are a part of many old traditional medicine systems. A huge amount of diverse phytochemicals is recorded in fronds and is used to treat many disorders (Farràs et al., 2022b). Until now, very few explorations regarding bioactive phytocompounds and potential uses of ferns have been undertaken, and it has still been restricted to a very small number of fern species (Baskaran et al., 2018; Britto et al., 2012; Ding et al., 2010; Hanus et al., 2003; Langhansova et al., 2021; Santos et al., 2010). The family Aspleniaceae is commonly known by the name "spleenwort." It has a Greek origin, stating "a," which stands for "without," and "splen," which stands for spleen, directing its probable use in curing spleen ailments (Fiorin et al., 2019). The Aspleniaceae family mainly comprises two important genera, *viz., Asplenium* L. and *Hymenasplenium* Hayata, having 700 and 60 species, respectively, under them (Xu et al., 2019). Most of the genera belonging to this family dwell in rocky locations; they are commonly called "rock ferns." *Asplenium adiantum-nigrum* is an important small-sized fern species commonly recognized as black spleenwort. The height of *A. adiantum-nigrum* ranges between 15–40 cm, with fronds that are 2 to 4 pinnate. It has a rising rhizome covered with triangular dark brown scales that are narrowly triangular. The fronds usually have a swollen base with blackish hairy scales scattered all over them (Hammami et al., 2016). It has common names like *Cama Negra, Culantrillo negro*, and *Foguera borda* in Spain (Bonet and Valles, 2007). Globally, *A. adiantum-nigrum* is distributed in Africa, Europe (including Britain), and Eurasia. It is native to a few locations in Mexico and the United States. It stretches to the Himalayas, North Africa, and South Africa (Kramer and Viane, 1990). Ethnobotanically, *A. adiantum-nigrum* is recognized all over the world. In Iran, *A. adiantum-nigrum* has been used for the treatment of spleen, kidney, and bowel diseases (Bahadori et al., 2015). It is also used in treating gynecological diseases (Ibadullayeva et al., 2022). A mixture of frond extract and honey is utilized for treating eye issues. The fronds themselves act as diuretics and laxatives. The leaves and rhizome are beneficial in treating jaundice and spleen-related problems. Additionally, a decoction made from the leaves is effective in alleviating cough (Singh et al., 2021). Besides this, the ethnoveterinary use of *Asplenium adiantum-nigrum* fronds has been done to treat jaundice in hens through oral administration (Bonet and Valles, 2007).

2.2 BIOACTIVES

Preliminary phytochemical analysis of *Asplenium adiantum-nigrum* was performed using aqueous, ethanol, methanol, and acetone extracts. All these extracts qualitatively identified the presence of glycosides, carbohydrates, free amino acids, proteins, saponins, phenolic compounds, tannins, volatile oils, terpenoids, and phlobatannins. These extracts failed to detect alkaloids, resins, and flavonoids (Mir et al., 2013). Ethanol, methanol, acetone, and water extracts of *A. adiantum-nigrum* tested positive for glycosides, carbohydrates, proteins, and free amino acids, phenolic compounds, tannins, and terpenoids; however, all extracts did not report the presence of alkaloids and flavonoids (Shah et al., 2022). Chemical screening of leaves and rhizomes revealed the presence of triterpenoids, coumarins, and polyphenols. Total flavonoid content extracted from leaves and rhizomes measured 6.33±0.55 and 19.33±1.45 mg CE/g, respectively (Bahadori et al., 2015). Light petroleum extracts of *A. adiantum-nigrum* showed 0.13% wax, 0.82% fatty acids, 0.07% hydrocarbons, and 0.44% alcohols and sterols. The alkane fraction of this extract revealed the highest hentriacontane (C_{31}), followed by nonacosane (C_{29}) and tritriacontane (C_{33}) in proportion. The triterpenoid fraction revealed the highest percentages of 22(29)-hopene, followed by 13(18)-hopene (Figure 2.1) and 8-fernene, respectively (Bottari et al., 1972). Leaves of *A. adiantum-nigrum* reported around 2000 µg/g of free and 730 µg/g of bound phenolic, and the rhizome showed about 900 and 1000 µg/g of free and bound phenolic, respectively (Đurđević et al., 2007). HPLC-MS/MS analysis of *A. adiantum-nigrum* fronds' methanolic extract reported 1405.68 mg/kg dry extract of phenolic species. It includes eight phenolic compounds, viz., caffeic acid, chlorogenic acid, ferulic acid, gallic acid, hydroxybenzoic acid, p-coumaric acid, p-vanillic acid, and 3,5-dicaffeoylquinic acid. It also has nine flavonols, viz., hyperoside, isorhamnetin, isoquercitrin, kaempferol, kaempferol-3-glucoside, myricetin, quercitrin, quercetin, and rutin. Along with this, three flavan-3-ols like epicatechin, procyanidin B2, procyanidin A2, and two dihydrochalcones such as phloridzin and phloretin were also recorded. Under the flavanones category, compounds like hesperidin and trans-cinnamic acid were also observed (Farràs et al., 2022a).

Methanol and hexane extracts of *A. adiantum-nigrum* fronds reported 113.5 ± 5.8 and 96.0 ± 3.8 μGAE/mg total phenolic content, respectively (Farràs et al., 2019). Chemical investigation of *A. adiantum-nigrum* fronds' essential oils derived from the hydro distillation method was done through GC-MS analysis. A total of 35 components were reported in *A. adiantum-nigrum* essential oil, measuring 77.5% of the whole components. Palmitic acid is the prime component of essential oil, measuring 34.5%, followed by phytol (7.5%) and hexahydrofarnesyl acetone (4.4%). The essential oil consists of 19.9% terpenoids, with 12.5% being oxygenated diterpenes. It also contains a significant portion of non-terpenic derivatives, including aldehydes, carboxylic acids, hydrocarbons, and ketones, which make up 57.6% of the identified volatile compounds. Additionally, the ethyl acetate extract of *A. adiantum-nigrum* frond reported the highest 49.3 ± 1.7 mg GAE/g DW total phenolic content (Hammami et al., 2016). C-glycosylxanthones were successfully isolated from the 80% ethanol extract of *A. adiantum-nigrum* (Richardson and Lorenz-Liburnau, 1982). The 70% extract of *A. adiantum-nigrum* identified eight flavonoids, *viz.*, procyanidin, prodelphinidin, different kaempferol 3,7-di-O-glycosides, O-glycosides of apigenin di-C-glycosides, mangiferin, mangiferin O-glucoside, isomangiferin, isomangiferin O-glucoside, and an uncharacterized xanthone (Umikalsom et al., 1994). The methanolic extract of *A. adiantum-nigrum* leaves reported the least amount of total phenolic contents (21.85 ± 3.12 mg GAE/g) compared to the rhizome (Valizadeh et al., 2016). The methanolic extract (80%) of *A. adiantum-nigrum* rhizome recorded the highest total phenolic content as 68.62 mg GAE/g FW in comparison to this, fronds and roots showed lower amounts of total phenolic content. Frond and rhizome extracts reported almost equal 58 μg RE/g FW total flavonoid content; on the contrary, the root extract revealed the least total flavonoid contents as 9.10 μg RE/g FW. UHPLC/DAD/(-)HESI-MS/MS investigation of *A. adiantum-nigrum* frond and rhizome recorded 10 phenolic acids and derivatives, viz., aesculin, caffeic acid, chlorogenic acid, gallic acid, gentisic acid, protocatechuic acid, p-hydroxybenzoic acid, p-coumaric acid, ferulic acid, and rosmarinic acid. Along with these compounds, some flavonoids and derivatives, as well as xanthones such as epigallocatechin, epigallocatechin gallate, rutin, mangiferin glucoside, and mangiferin were also identified (Živković et al., 2020). Based on chemical and spectroscopic techniques, a novel xanthone-O-glycoside was extracted from *A. adiantum-nigrum* frond, and it was determined as 3,7,8-trihydroxyxanthone-1-O-β-laminaribioside (Imperato, 1980). Further, two mangiferin glycosides, viz., 1,3,6,7-tetrahydroxyxanthone-2-C-glucosylglucoside and 1,3,6,7-tetrahydroxyxanthone-2-C-rhamnosylglucoside, were also determined from the ethanolic extract of *A. adiantum-nigrum* aerial parts (Imperato, 1990). In addition to this, three novel xanthones, viz., xanthone 2,4-di-C-glycosides, 1-hydroxy-3,6,7-trimethoxyxanthone 2,4-di-C-glucoside, and two of its O-rhamnosides, were reported in *A. adiantum-nigrum* frond (Imperato, 1991).

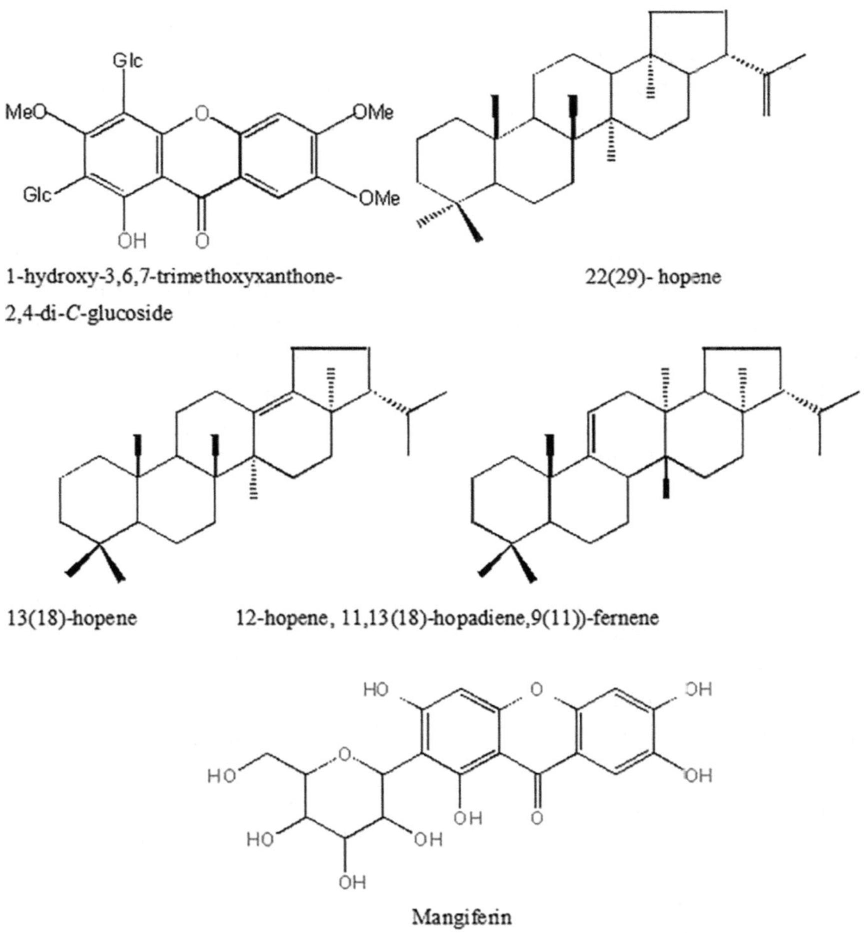

FIGURE 2.1 Structures of phytochemicals identified from *Asplenium adiantum-nigrum*.

2.3 PHARMACOLOGY

2.3.1 ANTIBACTERIAL ACTIVITY

Asplenium adiantum-nigrum rhizome revealed least 4 and 8 μg/ml MIC values respectively against *E. coli* and *S. aureus* compared to leaves exhibiting higher MIC >32 μg/ml against both microbes (Bahadori et al., 2015). Amongst fronds, rhizomes, and roots extract, rhizome extract of *A. adiantum-nigrum* showed strong bactericidal activity towards all evaluated bacteria with an MIC range of 375–750 μg/ml and an MBC range of 500–1000 μg/ml, followed by fronds with MIC and MBC ranges of 375–1000 and 500–2000 μg/ml, respectively. The least activity was reported by root extract with MIC and MBC ranges of 500–1000 and 1000–2000 μg/ml, respectively (Živković et al., 2020).

2.3.2 ANTIFUNGAL ACTIVITY

Amongst fronds, rhizomes, and roots extract, root extract of *A. adiantum-nigrum* showed the best antifungal activity against all evaluated fungi, with minimal inhibitory concentrations in the range of 125–750 μg/ml and minimal fungicidal concentrations at 250–1000 μg/ml, followed by fronds with MIC and MFC ranges of 125–750 μg/ml and 250–1000 μg/ml, respectively. Rhizome extract revealed the least antifungal activity with MICs of 250–750 μg/ml and MFC of 500–1000 μg/ml (Živković et al., 2020).

2.3.3 ANTIOXIDANT ACTIVITY

Methanolic extracts of *A. adiantum-nigrum* fronds revealed more potent DPPH reduction capacity compared to hexane extract, achieving 100% radical inhibition in a range of 0.01 to 0.1 mg/ml. In the ORAC assay, methanolic and hexane extracts of *A. adiantum-nigrum* fronds revealed substantial antioxidant activity with 2.25 ± 0.03 and 0.34 ± 0.11 μmol trolox equivalents/mg sample, respectively. Methanolic extracts of *A. adiantum-nigrum* fronds revealed superior activity compared to the reference compound trolox against superoxide radicals generated by X/XO. The IC50 values for methanolic and hexane extracts of *A. adiantum-nigrum* fronds for the above three assays are 0.008 and 0.044 mg/mL (DPPH assay); 0.011 and 0.128 mg/mL (O_2 − radical assay) (Farràs et al., 2019). DPPH free radical-scavenging activity assessment revealed that the ethyl acetate extract of *A. adiantum-nigrum* depicted strong antioxidant activity, DPPH IC50 = (0.88 ± 0.06) mg/mL (Hammami et al., 2016). Methanolic extract of leaves and rhizome depicted significant antioxidant capacity in the DPPH assay with IC50 = 71.33 ± 2.02 and 46.61 ± 1.21, respectively. However, rhizome extract revealed its superiority over leaf extract. In the ABTS radical scavenging assay, methanol extract of *A. adiantum-nigrum* leaf and rhizome extracts demonstrated the poorest free radical scavenging activity with 4.02 ± 0.29 and 3.24 ± 0.24 μmol Trolox/g, respectively (Valizadeh et al., 2016).

In DPPH and ABTS radical scavenging assays, rhizome extract of *A. adiantum-nigrum* revealed the strongest antioxidant capacity with IC50 = 1.53 ± 0.49 and 0.35 ± 0.02, respectively, followed by fronds with IC50 = 3.20 ± 0.43 and 0.64 ± 0.02, respectively. The lowest radical scavenging activity was depicted by root extract with IC50 = 3.31 ± 0.04 and 4.10 ± 0.25, respectively (Živković et al., 2020).

2.3.4 ANTI-TYROSINASE ACTIVITY

Methanolic extracts of *A. adiantum-nigrum* fronds exhibited promising tyrosinase inhibition activity with IC50 = 0.216 mg/ml. Surprisingly, the hexane extract of the same plant part failed to report any anti-tyrosinase activity (Farràs et al., 2019).

2.3.5 ANTIBIOFILM ACTIVITY

A. adiantum-nigrum fronds, rhizome, and root extracts showed substantial antibiofilm activity in the range of 27–46% at the sublethal concentration of 0.5 MIC, and 8–18% at the 0.25 MIC values, respectively. Interestingly, the rhizome extract of *A. adiantum-nigrum* failed to encourage *C. albicans* biofilm formation at the least evaluated concentration (0.25 MIC) (Živković et al., 2020).

2.3.6 CYTOTOXICITY ACTIVITY

A. adiantum-nigrum fronds methanolic extract at 0.01, 0.1, and 1 mg/ml concentrations was tested against non-tumoral cell lines like 3T3 and HaCaT, and tumoral cell lines such as HeLa, HepG2, and A549. The extract failed to show any significant cytotoxicity in non-tumoral cell lines. However, a little viability has been observed for fibroblasts compared to keratinocytes; the lowest cell viability is observed at 0.1 mg/mL methanolic extract in 3T3 (62.4% cell viability). Interestingly, *A. adiantum-nigrum* extract failed to show cytotoxicity activity against tumoral cell lines (Farràs et al., 2022a). *A. adiantum-nigrum* leaves and rhizome methanolic extract showed a moderate level of cytotoxic activity against brine shrimp larvae. Leaf and rhizome extract had LC_{50} values of 158.9 and 366.1 μg/ml, respectively, which was quite high compared to podophyllotoxin, showing 50% lethality at 30 μg/ml as a positive control (Valizadeh, 2016).

2.3.7 CYTOPROTECTIVE ACTIVITY

A. adiantum-nigrum fronds methanolic extract at 1 mg/mL showed statistically significant cytoprotective action in 3T3 and HaCaT (non-tumoral) cells against 2 mM H_2O_2 for 2.5 h and is measured at 56% (Farràs et al., 2022b).

2.3.8 CELLULAR REPAIR ACTIVITY

In the MTT assay, previously treated 3T3 cells with 2 mM H_2O_2, when exposed to 0.01, 0.1, and 1 mg/ml concentrations of *A. adiantum-nigrum* fronds methanolic extract, revealed a concentration-dependent increase in cell viability as 18.8%, 41.6%, and 77.5% for the above-mentioned concentrations, respectively. However, it is noteworthy to mention that the increase in cell repair activity is only statistically significant at 0.1 and 1 mg/ml (Farràs et al., 2022b).

2.3.9 PHOTOTOXICITY ACTIVITY

The sensitivity of 3T3 and HaCaT (non-tumoral) cells towards UVA light sensitivity was determined by treating cells with the methanolic extract of *A. adiantum-nigrum* fronds and irradiating them in the dark. 3T3 cells showed 63.3% and HaCaT cells exhibited 75% cell viability, reflecting fair sensitivity of the cells towards UVA light (Farràs et al., 2022b).

2.3.10 ANTI-ACETYLCHOLINESTERASE EFFECT

'Mangiferin' isolated from *A. adiantum-nigrum* recorded inhibitory action against the acetylcholinesterase enzyme with an IC50 value of 55.42 ±

1.52 μg/mL compared to the well-known standard drug galantamine with an IC50 value of 5.543 ± 0.213 μg/mL (Flores-González et al., 2022).

KEYWORDS

- **anti-tyrosinase**
- **Aspleniaceae**
- **bioactive molecules**
- **forest ecosystem**
- **non-tumoral cell lines**
- **pteridophytes**
- **rhizome methanolic extract**
- **spleenwort**
- **xanthones**

REFERENCES

Bahadori, M. B., Mahmoodi, K. F., Ahmadi, A. A., Bahadori, S., & Valizadeh, H. (2015). Antibacterial evaluation and preliminary phytochemical screening of selected ferns from Iran. *Research Journal of Pharmacognosy, 2*, 53–59.

Baskaran, X. R., Geo Vigila, A. V., Zhang, S. Z., Feng, S. X., & Liao, W. B. (2018). A review of the use of pteridophytes for treating human ailments. *Journal of Zhejiang University Science B, 19*, 1–35. https://doi.org/10.1631/jzus.B1600344.

Bonet, M. A., & Valles, J. (2007). Ethnobotany of Montseny biosphere reserve (Catalonia, Iberian Peninsula): Plants used in veterinary medicine. *Journal of Ethnopharmacology, 110*, 130–147.

Bottari, F., Marsili, A., Morelli, I., & Pacchiani, M. (1972). Aliphatic and triterpenoid hydrocarbons from ferns. *Phytochemistry, 11*, 2519–2523.

Britto, J. D., Gracelin, H. S., & Kumar, P. B. J. R. (2012). Phytochemical studies on five medicinal ferns collected from Southern Western Ghats, Tamilnadu. *Asian Pacific Journal of Tropical Biomedicine, 2*, 536–538.

Ding, Z. T., Fang, Y. S., Tai, Z. G., Yang, M. H., Xu, Y. Q., Li, F., & Cao, Q. E. (2008). Phenolic content and radical scavenging capacity of 31 species of ferns. *Fitoterapia, 79*, 581–583.

Đurđević, L., Mitrović, M., Pavlović, P., Bojović, S., Jarić, S., Oberan, L., Gajić, G., & Kostić, O. (2007). Total phenolics and phenolic acids content in leaves, rhizomes and rhizosphere soil under *Ceterach officinarum* DC, *Asplenium trichomanes* L. and *A. adiantum-nigrum* L. in the Gorge of Sićevo (Serbia). *Ekológia (Bratislava), 26*, 164–173.

Farràs, A., Cásedas, G., Les, F., Terrado, E. M., Mitjans, M., & López, V. (2019). Evaluation of anti-tyrosinase and antioxidant properties of four fern species for potential cosmetic applications. *Forests, 10*, 179.

Farràs, A., López, V., Maggi, F., Caprioli, G., Vinardell, M. P., & Mitjans, M. (2022a). Chemical composition and cytoprotective activities of methanolic extract of *Asplenium adiantum-nigrum* L. (Aspleniaceae). *Horticulturae, 8*, 815.

Farràs, A., Mitjans, M., Maggi, F., Caprioli, G., Vinardell, M. P., & López, V. (2022b). Exploring wild Aspleniaceae ferns as safety sources of polyphenols: The case of *Asplenium trichomanes* L. and *Ceterach officinarum* Willd. *Frontiers in Nutrition, 9*, 994215.

Fiorin, E., Saez, L., & Malgosa, A. (2019). Ferns as healing plants in medieval Mallorca, Spain? Evidence from human dental calculus. *International Journal of Osteoarchaeology, 29*, 82–90. https://doi.org/10.1002/oa.2718.

Flores-González, M., Simirgiotis, M., & Torres-Benítez, A. (2022). Systematic review: Antioxidant and neuroprotective capacity of species of the genus *Asplenium* (Monilophyta: Aspleniaceae). *Medical Sciences Forum, 8*, 14.

Hammami, S., Snène, A., El Mokni, R., Faidi, K., Falconieri, D., Dhaouadi, H., Piras, A., Mighri, Z., & Porcedda, S. (2016). Essential oil constituents and antioxidant activity of *Asplenium* ferns. *Journal of Chromatographic Science, 54*, 1341–1345.

Hanus, L. O., Rezanka, T., & Dembitsky, V. M. (2003). A trinorsesterterpene glycoside from the North American fern *Woodwardia virginica* (L.) Smith. *Phytochemistry, 63*, 869–875.

Ibadullayeva, S. J., Movsumova, N. V., Shiraliyeva, G. S., Askerova, N. A., & Mammadova, H. C. (2022). Pteridophytes: Ethnobotanical use and active chemical composition. *Indian Journal of Traditional Knowledge, 21*, 353–359.

Imperato, F. (1980). A xanthone-O-glycoside from *Asplenium adiantum-nigrum*. *Phytochemistry, 19*, 2030–2031.

Imperato, F. (1990). Two mangiferin glycosides from *Asplenium adiantum-nigrum* L. *American Fern Journal, 80*, 27–29.

Imperato, F. (1991). Xanthone 2,4-di-C-glycosides from *Asplenium adiantum-nigrum*. *Phytochemistry, 30*, 3839–3840.

Kramer, K. U., & Viane, R. L. (1990). Aspleniaceae. In K. U. Kramer & P. S. Green (Eds.), *Pteridophytes and Gymnosperms*. Springer-Verlag, Berlin.

Langhansová, L., Pumprová, K., Haisel, D., Ekrt, L., Pavičić, A., Zajíčková, M., et al. (2021). European ferns as rich sources of antioxidants in the human diet. *Food Chemistry, 356*, 129637. https://doi.org/10.1016/j.foodchem.2021.129637.

Mir, S. A., Mishra, A. K., Reshi, Z. A., & Sharma, M. P. (2013). Preliminary phytochemical screening of some pteridophytes from district Shopian (J & K). *International Journal of Pharmacy and Pharmaceutical Sciences, 5*, 632–637.

Richardson, P. M., & Lorenz-Liburnau, E. (1982). C-glycosylxanthones in the *Asplenium adiantum-nigrum* complex. *American Fern Journal, 72*, 103–106.

Santos, M. G., Kelecom, A., Paiva, S. R., Moraes, M. G., Rocha, L., & Garrett, R. (2010). Phytochemical studies in pteridophytes growing in Brazil: A review. *Americas Journal of Plant Science and Biotechnology, 4*, 113–125.

Shah, A. F., Rashid, S., & Kumarasamy, D. (2022). Preliminary phytochemical studies on some medicinally important fern species collected from Udhagamandalam, Tamil Nadu, India. *Indian Journal of Natural Sciences, 13*, 51361–51368.

Singh, C., Chauhan, N., Rani, A., Kishore, K., Singh, K., & Singh, R. (2021). Ethnobotanical and pharmacological studies of pteridophytes (ferns) from Mussoorie region, Dehradun district, Uttarakhand. *International Journal of Botany Studies, 6*, 694–702.

Umikalsom, Y., Grayer-Barkmeijer, R. J., & Harborne, J. B. (1994). A comparison of the flavonoids in Athyriaceae and Aspleniaceae. *Biochemical Systematics and Ecology, 22*, 587–594.

Valizadeh, H., Sonboli, A., Kordi, F. M., Dehghan, H., & Bahadori, M. B. (2016). Cytotoxicity, antioxidant activity and phenolic content of eight fern species, from north of Iran. *Pharmaceutical Sciences, 21*, 18–24.

Xu, K. W., Lorence, D., Wood, K. R., Liao, W. B., & Zhang, L. B. (2019). A revision of the *Hymenasplenium unilaterale* subclade (Aspleniaceae; Pteridophyta) with the description of nine new species. *Phytotaxa, 419*, 1–27.

Živković, S., Milutinović, M., Maksimović, V., Ćirić, A., Ivanov, M., Božunović, J., Banjanac, T., & Mišić, D. (2020). Antioxidant and antimicrobial activity of two *Asplenium* species. *South African Journal of Botany, 132*, 180–187.

CHAPTER 3

Blechnum orientale L.: Chemical Composition and Pharmacological Properties

LALREMRUATI,[1] F. LALSANGPUII,[2] LALREMKIMI,[1] LALCHHANDAMI TOCHHAWNG,[3] and ZOTHAN SIAMA[1,*]

[1]*Department of Zoology, Mizoram University (A Central University), Aizawl, Mizoram, India*

[2]*Department of Botany, Mizoram University (A Central University), Aizawl, Mizoram, India*

[3]*Mizoram Science, Technology and Innovation Council, Aizawl, Mizoram, India*

*Corresponding authors

ABSTRACT

Blechnum orientale is an evergreen fern species, belonging to the family Blechnaceae. This species is indigenous to Malaysia, India, China, and other Asian countries, as well as Australia and the Pacific Islands. Different parts of *B. orientale* have been utilized as a conventional remedy for treating several ailments, including blisters, boils, intestinal worm infections, ulcers, stomach pain, and urinary bladder issues. Secondary metabolites such as phenolic acids, lignans, flavonoids, anthocyanins, sterols, terpenoids, fatty acids, heterocyclic compounds, niacin, saponins, betacyanins, quinones, glycosides, tannins, and coumarins have been isolated from *B. orientale* extracts. *B. orientale* has also been reported to exhibit various biological activities, including antifungal, antibacterial, trematocidal activity, antioxidant, anti-inflammatory, anticancer,

Bioactives and Pharmacology of Pteridophytes. T. Pullaiah (Ed.)

DOI: 10.1201/9781779642738-3

antipyretic, antidiabetic, tyrosinase inhibitory activity, wound healing, and analgesic activities. The present review revealed that *B. orientale* is a rich source of naturally occurring phytochemicals, possessing multiple pharmaceutical applications.

3.1 INTRODUCTION

The diversity of plants on Earth is significantly enhanced by pteridophytes. They are the dominant component of plant ecosystems and belong to the second largest class of vascular plants (Benjamin and Manickam, 2007). Pteridophytes are a class of plants that are frequently utilized as food, medicine, ornaments, fiber supplies, bioremediation agents, and organic fertilizers. Numerous attempts have been made to investigate the potential medical benefits of pteridophytes since the emergence of ethnobotany (Singh et al., 2001). The fern *Blechnum orientale* L., commonly known as 'Centipede fern,' is native to Malaysia, India, China, and many other Asian nations, Australia, and the Pacific Islands (Lai et al., 2010). It is also found in Mizoram, a northeastern Indian state known locally as 'Vawm-ban' (Kar et al., 2013). It grows from a thick rhizome that rises to a short, erect trunk, producing a clump of arching fronds that range in height from 20 to 200 cm. The fronds are rarely simple and are pinnate. Pinnae can be dentate, whole, or linear. Sori are straight and oriented toward the midrib on the bottom of the pinnae. Narrow, dark brown scales are thickly covered on the rhizome. This species is terrestrial and mostly inhabits moderate to high altitudes. It is commonly found in arid and exposed areas and is frequently spotted in colonies. The plant often grows beside roadways in rainforests or arid regions and is occasionally cultivated for decorative purposes (Lai et al., 2016). *B. orientale* is locally known as 'Paku ikan' by the Malays and 'Kuan Chung' by the Chinese. The tribes of Malays and Chinese widely use *B. orientale* fronds as a traditional remedy for treating female sterilization, curing blisters and boils, stomach pain, and urinary bladder issues. Furthermore, the rhizomes of *B. orientale* are believed to have anthelmintic properties and are also used for the treatment of typhoid fever. Additionally, intestinal worm infections and diaphoretics are treated with fresh fronds of *B. orientale* (SathyaPriya et al., 2022). A paste made from young fronds and rhizomes along with powdered dried ginger has been used for the treatment of newly cut wounds by the Khamti tribe in the Lohit district of Arunachal Pradesh, India (Tag et al., 2007). The boiled rhizome of *B. orientale* has also been reported to be consumed by the Mizo tribe of northeast India (Kar et al., 2013).

The presence of several secondary metabolites, including alkaloids, cardioglycosides, phenolics, saponins, tannins, terpenoids, quinones, flavonoids, glycosides, anthocyanins, and betacyanins, was discovered through qualitative phytochemical analysis of *B. orientale* frond extracts (Thomas et al., 2023). The chemical composition of the ethanolic extract of the whole plant of *B. orientale* was analyzed by GC-MS, and it revealed the presence of bioactive compounds such as tetradecanoic acid, n-hexadecanoic acid, hexadecanoic acid ethyl ester, and octadecanoic acid (Kale and Dubal, 2015). The frond extracts were used as a source of bio-reducing agents for the reduction of silver ions to synthesize silver nanoparticles (SathyaPriya et al., 2022). Different types of extracts derived from this plant have been found to exhibit biological activities, including trematocidal activity, antioxidant, anti-inflammatory, antibacterial, anticancer, antipyretic, antidiabetic, tyrosinase inhibitory activity, and diabetic-induced ulcer wound healing activity.

3.2 BIOACTIVE COMPOUNDS OF *BLECHNUM ORIENTALE*

Bioactive compounds are naturally occurring essential and non-essential compounds that have the potential to positively impact human health (Biesalski et al., 2009). Qualitative phytochemical analysis of various extracts of *B. orientale* frond revealed the presence of several secondary metabolites including alkaloids, cardioglycosides, phenolics, saponins, tannins, terpenoids, quinones, flavonoids, glycosides, anthocyanins, and betacyanins (Thomas et al., 2023). Gas chromatography-mass spectroscopy (GC-MS) analysis of the ethanolic extract of the whole plant of *B. orientale* revealed the presence of four bioactive compounds such as n-hexadecanoic acid, octadecanoic acid, ethyl ester of hexadecanoic acid, and tetradecanoic acid. Among the fatty acids with documented potential for antibacterial and antifungal action are octadecanoic, hexadecanoic, and tetradecanoic acids (Kale and Dubal, 2015). The presence of eicos-5-enoic acid (eicosapentaenoic acid), an omega-3 fatty acid, was also reported from various solvent extracts (petroleum ether, chloroform, and methanol) in the fronds of *B. orientale* by spectral analysis, that is, IR, mass, C13 NMR, and 1H NMR (Deepa et al., 2017). Additionally, H-NMR and C-NMR spectroscopy analysis revealed the presence of lignan compounds, namely blechnic acid, 8-epiblechnic acid, and brainic acid from the methanolic extract of fronds of *B. orientale* (Wada et al., 1992). Thin-layer chromatographic analysis of the ethanolic extract of *B. orientale* fronds detected the presence of chlorogenic acid (Bohm, 1968).

Phytochemical constituents of petroleum ether, chloroform, ethyl acetate, butanol, and aqueous extracts partitioned from the crude methanolic extracts of *B. orientale* fronds were investigated for the presence of terpenoids, saponins, alkaloids, flavonoids, and tannins. The phytochemical analysis revealed the strong presence of flavonoids, terpenoids, and tannins from acetate, butanol, and aqueous extracts (Lai et al., 2010). Similarly, the bioactive constituents of the methanolic extract of *B. orientale* fronds were investigated using chromatographic analysis, and the results revealed the presence of derivatives of flavones such as luteolin, apigenin, acacetin, genkwanin, and flavanols such as quercetin, kaempferol, and isorhamnetin (Yusuf, 1994). The phytochemical evaluation of the rhizome of *B. orientale* showed that phenolic triterpenoids and flavonoids were present in the active acetone extract. The petroleum ether extract revealed the existence of flavonoids, whereas the methanol extract exhibited the occurrence of both flavonoids and phenolics (Thomas, 2010). The presence of proanthocyanidin, a potent antibacterial, antioxidant, and anticancer compound, was reported from the aqueous extract of *B. orientale* fronds by subjecting it to repeated MCI gel chromatography, Chromatorex C18, Sephadex-LH-20, and semi-preparative high-performance liquid chromatography, and the structure of proanthocyanidin was further elucidated using Nuclear Magnetic Resonance (NMR) and Electrospray Ionization Mass Spectrometry (ESI-MS) (Lai et al., 2017).

Qualitative phytochemical analysis of various extracts of the whole plant of *B. orientale* revealed that the ethanolic extract possessed a strong presence of tannin, phenol, terpenoid, flavonoid, steroids, quinones, coumarins, and saponins. Furthermore, GC-MS analysis of the methanolic extract revealed the presence of compounds such as quercetin 7'3'4' trimethoxy, lucenin 2, phytol isomer, dodecane, 1,2,3-propanetricarboxylic acid 2-hydroxy-triethyl ester, 2-hexadecen-1-ol, 3,7,11,15-tetramethyl, 3-benzoyl-4-methyl-6-ethyl-2(1H)-pyridone, and hexanedioic acid mono (2-ethylhexyl) ester (Devi et al., 2016) (Figure 3.1) (Table 3.1).

n- hexadecanoic acid

Tetradecanoic acid

Eicos-5-enoic acid

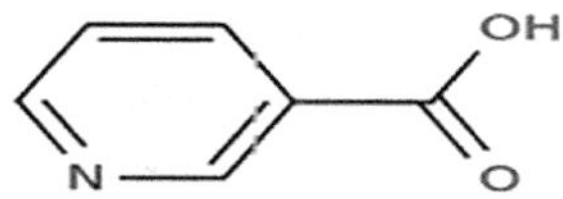

Nicotinic acid

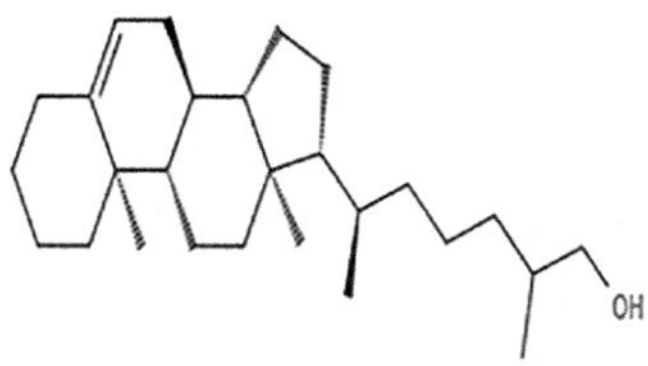

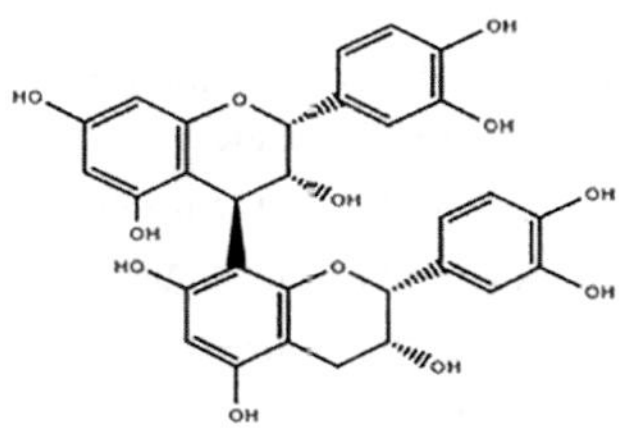

Proanthocyanidin

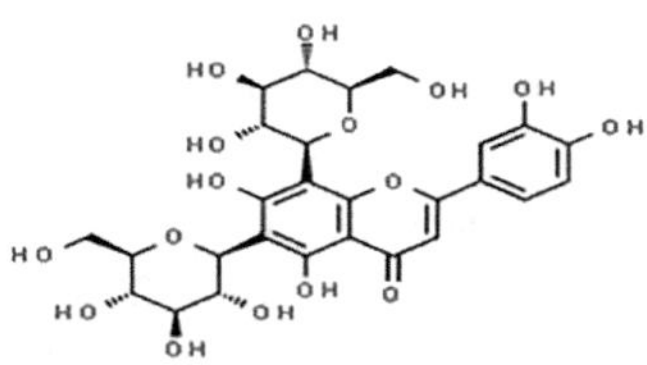

Lucenin-2

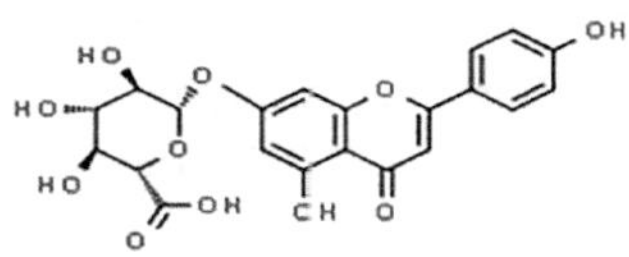

Apigenin-7-glucuronide

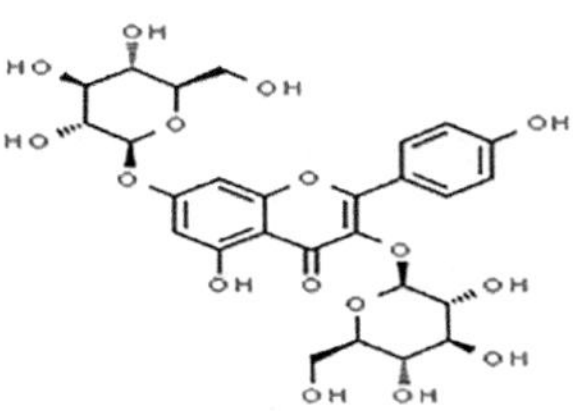

Kaempferol 3,7-diglucoside

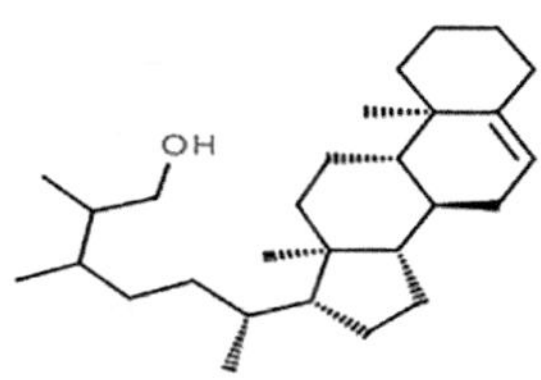

24-methylcholest-5-enol

FIGURE 3.1 Chemical structures of some major compounds isolated from *B. orientale*.

TABLE 3.1 The Details of Bioactive Compounds Identified from *B. orientale*

Sl. No	Bioactive Compounds	Solvent	Parts Screened	References
		Phenolic		
1	Proanthocyanidin	Aqueous	Fronds	Lai et al. (2017)
2	Chlorogenic acid	Ethanol	Fronds	Bohm (1968)
3	Blechnic acid	Methanol	Fronds	Wada et al. (1992)
4	8-Epiblechnic acid	Methanol	Fronds	Wada et al. (1992)
5	Brainic acid	Methanol	Fronds	Wada et al. (1992)
		Flavonoids		
6	Quercetin-3,7-digalactoside	Methanol	Fronds	Yusuf (1994)
7	Quercetin-7,4'-digalactoside	Methanol	Fronds	Yusuf (1994)
8	Quercetin-3,'4'dimethyl ether-3-glucoside	Methanol	Fronds	Yusuf (1994)
9	Quercetin-3, glucuronide	Methanol	Fronds	Yusuf (1994)
10	Kaempferol-3,7-digalactoside	Methanol	Fronds	Yusuf (1994)
11	Kaempferol-3,7-diglucuronide	Methanol	Fronds	Yusuf (1994)
12	Kaempferol-3,7-diglucoside	Methanol	Fronds	Yusuf (1994)
13	Luteolin-7-glucuronide	Methanol	Fronds	Yusuf (1994)
14	Apigenin-7-glucuronide	Methanol	Fronds	Yusuf (1994)
15	Apigenin-7,4'-diglucoside	Methanol	Fronds	Yusuf (1994)
16	Acacetin-7-galactoside	Methanol	Fronds	Yusuf (1994)
17	Genkwanin-4' -glucuronide	Methanol	Fronds	Yusuf (1994)
18	Isorhamnetin-3-glucoside	Methanol	Fronds	Yusuf (1994)
19	Quercetin 7'3'4'-Trimethoxy	Methanol	Whole Plant	Devi et al. (2016)
20	Lucenin-2	Methanol	Whole Plant	Devi et al. (2016)
21	Isoquercitrin	Ethyl acetate	Fronds	Rissyelly et al. (2022)
22	Trifolin	Ethyl acetate	Fronds	Rissyelly et al. (2022)
23	Anthocyanin	Ethanol, aqueous	Fronds	Thomas et al. (2023)
		Sterols		
24	Cholest-5-enol	Chloroform: methanol (2:1)	Whole Plant	Chiu et al. (1988)
25	24-methycholesta-5,22-dienol	Chloroform: methanol (2:1)	Whole Plant	Chiu et al. (1988)
26	24-methylcholest-5-enol	Chloroform: methanol (2:1)	Whole Plant	Chiu et al. (1988)
27	24-ethylcholesta-5,22-dienol	Chloroform: methanol (2:1)	Whole Plant	Chiu et al. (1988)

TABLE 3.1 *(Continued)*

Sl. No	Bioactive Compounds	Solvent	Parts Screened	References
28	24-ethylcholest-5-enol	Chloroform: methanol (2:1)	Whole Plant	Chiu et al. (1988)
		Terpenoids		
29	Phytol isomer	Methanol	Whole Plant	Devi et al. (2016)
30	2-Hexadecen-1-ol, 3,7,11, 15-tetramethyl	Methanol	Whole Plant	Devi et al. (2016)
31	Triterpenoid	Acetone	Rhizome	Thomas (2010)
		Fatty Acids		
32	Tetradecanoic acid	Ethanol	Whole Plant	Kale and Dubal (2015)
33	n-hexadecanoic acid	Ethanol	Whole Plant	Kale and Dubal (2015)
34	Hexadecanoic acid ethyl ester	Ethanol	Whole Plant	Kale and Dubal (2015)
35	Octadecanoic acid	Ethanol	Whole Plant	Kale and Dubal (2015)
36	Eicos-5-enoic acid	Methanol	Fronds	Deepa et al. (2017)
37	1,2,3-Propanetricarboxylic acid 2-hydroxy-, triethyl ester	Methanol	Whole Plant	Devi et al. (2016)
38	Hexanedioic acid, mono (2-ethylhexyl) ester	Methanol	Whole Plant	Devi et al. (2016)
		Polycyclic Aromatic		
39	3-benzoyl-4-methyl-6-ethyl-2(1H)- Pyridone	Methanol	Whole Plant	Devi et al. (2016)
		Niacin		
40	Nicotinic acid glucoside	Aqueous	Fronds	Ashihara et al. (2011)
		Alkane		
41	Dodecane	Methanol	Whole Plant	Devi et al. (2016)

3.3 PHARMACOLOGY OF *BLECHNUM ORIENTALE*

3.3.1 ANTICANCER ACTIVITY

The cytotoxic activities of aqueous, ethyl acetate, and n-butanol extracts of the fronds of *B. orientale* were investigated against human colonic adenocarcinoma HT-29, human colonic carcinoma HCT-116, human breast adenocarcinoma MCF-7, human leukemia K562, and liver Chang cells using the MTT assay. Among the various extracts, the butanol extract showed the highest toxicity against the HT-29 cell line with an IC50 of 27.5 μg/ml, followed by the aqueous extract and ethyl extract with IC50 values of 33.4 μg/ml and 42.8 μg/ml, respectively. However, the butanol extract displayed weak cytotoxicity towards the HCT-116 cell line with an IC50 of 72.7 μg/ml (Lai et al., 2010).

In another study, the cytotoxic potential of methylene chloride-methanolic extracts of the fronds, stems, and roots of *B. orientale* was tested against the MCF-7wt breast cancer cell line. The root extract induced the highest toxicity with 68.53% cell death, followed by the fronds and stems extracts that induced 18.56% and 10.95% cell death, respectively. Among the three extracts, the root extract demonstrated notable cytotoxicity with an IC50 value of 32.07 ± 7.85 μg/ml, suggesting the anticancer potential of *B. orientale* (Nor Aini et al., 2008).

The anticarcinogenic potential of *B. orientale* extract was also assessed by studying its inhibitory activity against oncogenic proteins isolated from two cancer cell lines (K562 and HeLa). A total of four differentially expressed proteins were identified and isolated using the NE-PER nuclear and cytoplasmic protein extraction kit. Based on matrix-assisted laser desorption ionization time-of-flight (MALDI-TOF-TOF) protein sequencing results, the protein bands were identified as an elongation factor 1-γ (EF1-γ), glyceraldehyde-3-phosphate dehydrogenase (GAPDH), HSP90-β, and heterogeneous nuclear ribonucleoprotein A2/B1 (hnRNP-A2/B1). Treatment with *B. orientale* extract significantly reduced the expression of the EF1 γ protein. It was hypothesized that the plant extract may exert its cytotoxic effect by interfering with the biosynthesis pathway of the protein. The level of GAPDH expression was also found to be downregulated, suggesting the inhibitory activity of the extract in the glycolytic pathway. Similarly, a decrease in the expression of HSP90 β was also observed, which was verified by western blot study. However, hnRNP was found to be upregulated after treatment with *B. orientale* extract, which may be due to the increased metabolic activities of the cancer cells in response to cytotoxic stress. Treatment with *B. orientale* extract led to the downregulation of almost all of the isolated proteins, suggesting its potential as an anticancer agent (Tan et al., 2018).

3.3.2 ANTIDIABETIC ACTIVITY

Diabetes is a serious illness that affects more than 500 million people globally. Estimates for the number of individuals with diabetes in urban and rural regions in 2021 were 360 million and 176.6 million, respectively, with estimates for prevalence at 12.1% and 8.3%. It is anticipated that by 2045, there will be 596.5 million diabetics residing in urban regions (Sun et al., 2022).

The antidiabetic activity of *B. orientale* was assessed by determining its inhibitory activity against α-glucosidase, which is a known target for the treatment of type 2 diabetes mellitus. The α-glucosidase inhibitory activity of the aqueous extract of *B. orientale* was evaluated by measuring the half-maximal effective concentration (EC_{50}) value in albino rats. The extract showed a concentration-dependent increase in α-glucosidase inhibition activity with an EC50 of 65.78 ± 1.45 μg/ml, which was found to be comparable with the standard myricetin having an EC50 of 53.21 ± 0.91 μg/ml (Chai et al., 2015).

3.3.3 ANTI-INFLAMMATORY ACTIVITY

Ethanolic extract of *B. orientale* fronds was tested for its potential to reduce inflammation by *in vivo* studies using egg albumin-induced edema in male albino rats. The ethanolic extract was intraperitoneally administered at doses of 400 mg/kg and 800 mg/kg. Within 6 hours of administration, the extract showed a significant dose-dependent reduction of the inflammation with 45.07% at 400 mg/kg and 65.49% at 800 mg/kg, which was found to be comparable to the standard drug, diclofenac sodium (50 mg/kg) (Amoroso et al., 2014).

3.3.4 ANTIBACTERIAL ACTIVITY

The antibacterial activity of various extracts (petroleum ether, acetone, methanol, and water) of *B. orientale* rhizome was assessed using MTCC human pathogenic bacteria strains and clinical isolates with the disc diffusion method. It was observed that the acetone extract showed potent antibacterial properties in *Pseudomonas aeruginosa*, with a minimum inhibitory concentration (MIC) and minimum bactericidal concentration (MBC) of 25 mg/ml and 50 mg/ml, respectively. This finding confirmed the ethnobotanical role of *B. orientale* in preventing a range of bacterial illnesses (Thomas, 2010).

Various extracts of *B. orientale* fronds were screened for their activity against gram-positive bacteria such as *Bacillus cereus, Micrococcus luteus*, methicillin-susceptible *Staphylococcus aureus* (MSSA), methicillin-resistant *S. aureus* (MRSA), *Staphylococcus epidermidis*, and gram-negative bacteria like *Escherichia coli*, *P. aeruginosa*, *Klebsiella pneumoniae*, *Salmonella choleraesuis*, and *Enterobacter aerogenes* by the disc diffusion method. Methanolic extract, ethyl acetate fraction, butanol fraction, and water fraction of *B. orientale* were found to possess antibacterial activities, and their activity was highest against *M. luteus* with a 12.8–14.3 mm zone of inhibition (ZOI). The antibacterial activity results validated the traditional application of *B. orientale* for the management of skin conditions, bladder problems, and gastrointestinal diseases of bacterial origin (Lai et al., 2010).

In a similar study, the antibacterial activity of ethanolic extract of *B. orientale* stem was assessed against *S. aureus* and MRSA. The antibacterial activity of *B. orientale* against both bacteria was found to be promising, with an average ZOI ranging from 13.00 mm–14.95 mm in *S. aureus* and 11.31 mm–14.48 mm in MRSA (Rahmadeni et al., 2019).

Similarly, the acetone extract of *B. orientale* frond was investigated for antibacterial activity using the plate cup method against bacteria such as *Bacillus subtilis*, *K. pneumoniae, Salmonella typhi, S. aureus*, *Streptococcus pyogenes*, *Proteus vulgaris*, *Pseudomonas* sp., and *Serratia* sp. The minimum concentration of *B. orientale* acetone extract (0.025 mg/ml) was found to possess strong antibacterial activity, showing maximum ZOI against *P. vulgaris* (22.0 ± 0.816 mm), followed by *K. pneumoniae* and *S. typhi* (16.33 ± 0.471 mm), *S. pyogenes* (15.0 ± 0.816 mm), *Serratia sp.* (13.0 ± 0.816 mm), *B. subtilis* (9.0 ± 0.816 mm), and *S. aureus* (8.33 ± 0.471 mm) (Maridass and Ghanthikumar, 2008).

Silver nanoparticles (AgNPs) were synthesized by the green synthesis method using aqueous extract of *B. orientale* fronds as an agent for reducing silver ions. The antibacterial activity of *B. orientale* AgNPs was tested against *E. coli*, *P. aeruginosa*, *S. typhi*, and *Shigella dysenteriae* using the agar well diffusion method, and the ZOI was measured. The most potent antibacterial activity was observed in *S. dysenteriae* with a ZOI value of 27 mm. AgNPs were also found to be significantly more effective than standard ampicillin for all strains of the tested bacteria (SathyaPriya et al., 2022).

3.3.5 ANTIFUNGAL ACTIVITY

The antifungal activity of petroleum ether, chloroform, methanolic, and aqueous extracts of *B. orientale* fronds was determined against *Aspergillus flavus* and *Candida albicans*. At 1000 μg/mL, the different extracts of *B. orientale* exhibited strong antifungal activity against *C. albicans*, with ZOI ranging from 11.0 ± 0.09 to 12.13 ± 0.09 mm, which were comparable to the standard fluconazole. However, no inhibition activity was observed against *A. flavus* (Deepa et al., 2013).

3.3.6 ANTIOXIDANT ACTIVITY

A study was conducted to evaluate the antioxidant activity of the methanolic extract of the fronds of *B. orientale* by four different methods: DPPH radical scavenging, ferric ion reducing power (FRP), beta-carotene bleaching (BCB), and ferrous ion chelating (FIC) assays. The DPPH radical scavenging assay showed dose-dependent inhibition with an IC50 of 0.14 ± 0.03 mg/mL and Ascorbic Acid Equivalent Antioxidant Capacity (AEAC) of 2821 ± 417 mg of ascorbic acid/100 g. The reducing power assay yielded similar results, showing high ferric ion reducing power of 1089 ± 157 mg of GAE/100 g. The antioxidative activity in terms of its ability to inhibit lipid peroxidation through the BCB assay showed a concentration-dependent increase in BCB activity, with significant inhibition of 73% at the concentration of 0.7 mg/mL. However, the extract showed low FIC activity of 22%, which may be due to the inexistence of correlation between total phenolic content and FIC, indicating that phenolic compounds are poor ion chelators, despite having strong antioxidant activity and reducing power (Lai et al., 2009). On the other hand, the BCB activity may be associated with the presence of lipophilic antioxidants in the extract (Lim et al., 2007).

In another study, the aqueous acetone extracts of young fronds, mature fronds, rhizome, and fiddlehead of *B. orientale*, collected from two different locations (UKM Fern Garden and Chini Forest), were tested for their DPPH radical scavenging activity and ferric reducing antioxidant power (FRAP). The DPPH radical scavenging assay showed high scavenging activity of *B. orientale* extracts for all samples at both locations, with percent scavenged ranging from 85.52% by the extract of fiddleheads to 90.24% by the extract of young fronds. The FRAP assay also showed that the ferric compound was significantly reduced by the extracts sampled from both locations. The highest ferric ion reducing ability was recorded in the young fronds (1,077 mg TE/100 g DW) sampled from Chini Forest, and the lowest recorded in the rhizome extract (537.71 mg TE/100 g DW) sampled from UKM Fern

Garden. The study showed that there was no significant variation in antioxidative activities of extracts collected from the two different locations, validating the potential of *B. orientale* as a promising antioxidant owing to its high phenolic content (Naik et al., 2013).

Similarly, the DPPH radical scavenging activity of various extracts of *B. orientale* fronds was evaluated using three solvent fractions obtained from the methanolic extract through successive partitioning with ethyl acetate, butanol, and water. The scavenging activity of the various fractions was then compared with the IC50 of four standards: ascorbic acid, Trolox-C, α-tocopherol, and butylated hydroxytoluene (BHT). Lower IC50 values correspond to stronger scavenging activity. The various extracts showed dose-dependent inhibition of DPPH radicals, with the ethyl acetate fraction showing the highest scavenging activity with an IC50 of 8.6 ± 0.5 μg/mL, which was comparable to the standards ascorbic acid and Trolox-C. Similarly, the methanolic extract, butanol fraction, and aqueous fraction of *B. orientale* also showed significant scavenging activity with an IC50 of 10.9 ± 1.6 μg/mL, 10.1 ± 1.1 μg/mL, and 13.0 ± 1.3 μg/mL respectively, which were comparable with that of α-tocopherol and BHT (Lai et al., 2010).

3.3.7 ANTIPYRETIC ACTIVITY

The antipyretic activity of ethanolic extract of *B. orientale* fronds was investigated in pyrexia-induced male albino rats (using 1 ml/kg body weight of 15% baker's yeast suspension). As pyrexia is marked by an increase in body temperature, rectal temperature of pyrexia-induced rats was recorded to assess the antipyretic activity of the extract. Although the onset of decrease in rectal temperatures was delayed, rats treated with the *B. orientale* frond extract at the dose of 800 mg/kg showed a significantly gradual recovery which was comparable to that of the standard paracetamol (Amoroso et al., 2014). This may be attributed to the presence of flavonoids in *B. orientale* frond extract, which have been reported to possess antipyretic activity (Mutalik et al., 2003).

3.3.8 TREMATOCIDAL ACTIVITY

An *in vitro* bioassay was conducted to evaluate the trematocidal potency of the various extracts of *B. orientale* against *Gastrothylax crumenifer*, a known pathogenic trematode. Various concentrations of the extracts, ranging from 1 mg to 5 mg, were prepared for each test group and incubated with 25 trematodes each, which received 25 ml of the extract fractions. The movement of the trematodes in each group was evaluated at regular intervals. The ethanol extracts demonstrated significant trematocidal activities against *G. crumenifer* in a concentration-dependent manner. The highest concentration (5 mg) was found to exhibit significant trematocidal activity, immobilizing all the trematodes within 15 minutes of incubation (Devi et al., 2016). The trematocidal activity of ethanolic extract of *B. orientale* is in agreement with previous research, showing that the presence of tannins is responsible for its antiparasitic activity (Mithraja et al., 2012).

3.3.9 INHIBITION OF TYROSINASE

B. orientale fronds have been found to have inhibitory activity against tyrosinase, which is an essential enzyme to modulate the biosynthesis of melanin, overproduction of which may lead to the development of malignant melanoma (Murray et al., 1999). Using the dopachrome test, the methanolic extract of *B. orientale* fronds was investigated for its tyrosinase inhibition activity. The study revealed that the extract possessed significant tyrosinase-inhibition activity with $50.7 \pm 6.0\%$ inhibition, which was found to be 51% equivalent to 143 mg of quercetin/g of extract and 15 mg of kojic acid/g of extract (Lai et al., 2009). The tyrosinase inhibition activity may be attributed to the presence of antioxidants (Masuda et al., 2005).

3.3.10 INHIBITION OF 15-LIPOXYGENASE

Various extracts (n-hexane, ethyl acetate, ethanol, and aqueous) of *B. orientale* fronds were evaluated for their enzyme inhibitory activity against 15-lipoxygenase (15-LOX), an enzyme required for the synthesis of leukotrienes that induce constriction of bronchioles in asthma. Among the various extracts tested, the ethyl acetate extract was found to exhibit the highest enzyme inhibition activity (IC50: 129.40 ppm). Subsequently, trifolin and isoquercitrin were isolated from the ethyl acetate extract to analyze their interaction with 15-LOX and subjected to *in silico* study. Molecular docking analysis revealed that trifolin and isoquercitrin form hydrogen bonds with the C-terminus of the Ile 663 residue and His 366 residue, surrounding iron (Fe) in 15-LOX. This suggests that isoquercitrin and trifolin isolated from the ethyl acetate extract of *B. orientale* may act as effective inhibitors of 15-LOX, due to their interactions with residues surrounding Fe in 15-LOX, which is similar to known lipoxygenase inhibitors such as quercetin (Rissyelly et al., 2022).

3.3.11 TREATMENT FOR DIABETES-INDUCED ULCER WOUNDS

The crude methanolic extract of *B. orientale* was fractionated into an aqueous soluble fraction using flash column chromatography and then subjected to a wound healing assay. An open ulcer was induced on the back of a streptozocin-induced diabetic rat and then treated with the aqueous extract at concentrations of 2% and 4%. It was observed that treatment with the aqueous extract accelerated wound healing at both concentrations when compared with the control group. At the higher concentration of 4%, complete wound healing was observed on the 12th day, and the same was observed on the 14th day in groups treated with 2% of the aqueous extract, while the control group showed incomplete wound closure. Furthermore, the rats treated with 4% of the aqueous extract showed increased reepithelization, increased fibroblast proliferation, collagen production, and angiogenesis (Lai et al., 2016).

KEYWORDS

- **anticancer**
- **antidiabetic**
- **antioxidant**
- **bioactive compounds**
- **ethnomedicine**
- **pharmacology**

REFERENCES

Amoroso, V. B., Antesa, D. A., Buenavista, D. P., & Coritico, F. P. (2014). Antimicrobial, antipyretic and anti-inflammatory activities of selected Philippine medicinal pteridophytes. *Asian Journal of Biodiversity, 5*(1), 18–40.

Ashihara, H., Yin, Y., & Watanabe, S. (2011). Nicotinamide metabolism in ferns: Formation of nicotinic acid glucoside. *Plant Physiology and Biochemistry, 49*(3), 275–279.

Benjamin, A., & Manickam, V. S. (2007). Medicinal pteridophytes from the Western Ghats. *Indian Journal of Traditional Knowledge, 6*(4), 611–618.

Biesalski, H. K., Dragsted, L. O., Elmadfa, I., Grossklaus, R., Müller, M., Schrenk, D., & Weber, P. (2009). Bioactive compounds: Definition and assessment of activity. *Nutrition, 25*, 1202–1205.

Bohm, B. A. (1968). Phenolic compounds in ferns—III: An examination of some ferns for caffeic acid derivatives. *Phytochemistry, 7*(10), 1825–1830.

Chai, T. T., Yeoh, L. Y., Ismail, N. M., Ong, H. C., Manan, F. A., & Wong, F. C. (2015). Evaluation of glucosidase inhibitory and cytotoxic potential of five selected edible and medicinal ferns. *Tropical Journal of Pharmaceutical Research, 14*(3), 449–454.

Chiu, P. L., Patterson, G. W., & Salt, T. A. (1988). Sterol composition of pteridophytes. *Phytochemistry, 27*(3), 819–822.

Deepa, J., Parashurama, T. R., Krishnappa, M., & Nataraja, S. (2013). Antimicrobial efficacy of *Blechnum orientale* L. *International Journal of Pharma and Bio Sciences, 4*(2), 475–479.

Deepa, J., Parashurama, T. R., Krishnappa, M., & Nataraja, S. (2017). New report on eicos-5-enoic acid in *Blechnum orientale* L. *European Journal of Pharmaceutical and Medical Research, 4*(1), 457–461.

Devi, R. K., Vasantha, S., Panneerselvam, A., Rajesh, N. V., & Jeyathilakan, N. (2016). Phytochemical constituents and *in vitro* trematocidal activity of *Blechnum orientale* Linn. against *Gastrothylax crumenifer*. *Annals of Phytomedicine, 5*, 127–134.

Kale, M., & Dubal, K. (2015). Determination of bioactive components of *Blechnum orientale* L. whole plant by using GC-MS analysis. *Indian Journal of Science and Technology, 3*(11), 119–121.

Kar, A., Bora, D., Borthakur, S. K., Goswami, N. K., & Saharia, D. (2013). Wild edible plant resources used by the Mizos of Mizoram, India. *KUSET, 9*(1), 106–126.

Lai, H. Y., Lim, Y. Y., & Kim, K. H. (2010). *Blechnum orientale* Linn – A fern with potential as antioxidant, anticancer and antibacterial agent. *BMC Complementary and Alternative Medicine, 10*, 1–8.

Lai, H. Y., Lim, Y. Y., & Kim, K. H. (2017). Isolation and characterization of a proanthocyanidin with antioxidative, antibacterial and anticancer properties from fern *Blechnum orientale*. *Pharmacognosy Magazine, 13*(49), 31–37.

Lai, H. Y., Lim, Y. Y., & Tan, S. P. (2009). Antioxidative, tyrosinase inhibiting and antibacterial activities of leaf extracts from medicinal ferns. *Bioscience, Biotechnology, and Biochemistry, 73*(6), 1362–1366.

Lai, J. C. Y., Lai, H. Y., Rao, N. K., & Ng, S. F. (2016). Treatment for diabetic ulcer wounds using a fern tannin optimized hydrogel formulation with antibacterial and antioxidative properties. *Journal of Ethnopharmacology, 189*, 277–289.

Lim, Y. Y., & Quah, E. P. (2007). Antioxidant properties of different cultivars of *Portulaca oleracea*. *Food Chemistry, 103*(3), 734–740.

Maridass, M., & Ghantikumar, S. (2008). Antibacterial activity of leaves of *Blechnum orientale* L. *Pharmacology Online, 3*, 58–60.

Masuda, T., Yamashita, D., Takeda, Y., & Yonemori, S. (2005). Screening for tyrosinase inhibitors among extracts of seashore plants and identification of potent inhibitors from *Garcinia subelliptica*. *Bioscience, Biotechnology, and Biochemistry, 69*(1), 197–201.

Mithraja, M. J., Marimuthu, J., Mahesh, M., Paul, Z. M., & Jeeva, S. (2012). Chemical diversity analysis on some selected medicinally important pteridophytes of Western Ghats, India. *Asian Pacific Journal of Tropical Biomedicine, 2*(1), 34–39.

Murray, C., D'Intino, Y., MacCormick, R., Nassar, B., & Walsh, N. (1999). Melanosis in association with metastatic malignant melanoma: Report of a case and a unifying concept of pathogenesis. *American Journal of Dermatopathology, 21*(1), 28–30.

Mutalik, S., Paridhavi, K., Rao, C. M., & Udupa, N. (2003). Antipyretic and analgesic effect of leaves of *Solanum melongena* Linn. in rodents. *Indian Journal of Pharmacology, 35*(5), 312–315.

Naik, D. J., Ramappa, P. T., Maddappa, K., & Somalapura, N. (2013). Antioxidant activities of *Blechnum orientale* L. *International Journal of Biological and Pharmaceutical Research, 4*(2), 105–108.

Nor Aini, A. S., Merrina, A., Stanslas, J., & Sreeramanan, S. (2008). Cytotoxic potential on breast cancer cells using selected forest species found in Malaysia. *International Journal of Cancer Research, 4*(3), 103–109.

Rahmadeni, Y., Febria, F. A., & Bakhtiar, A. (2019). Potensi pakih sipasan (*Blechnum orientale*) sebagai antibakteri terhadap *Staphylococcus aureus* dan methicillin resistant *Staphylococcus aureus*. *Jurnal Biologi Sumatera, 6*(2), 224–229.

Rissyelly, Aziz, S., Sangande, F., Yodha, A. W. M., Budipramana, K., Elfahmi, & Sukrasno. (2022). The inhibition of 15-lipoxygenase by *Blechnum orientale* leaves and its glycoside-flavonoid isolates: *In vitro* and *in silico* studies. *HAYATI Journal of Biosciences, 29*(3), 353–359.

SathyaPriya, A., MohanaPriya, B., Kanagaraj, C., & Aravindh, S. (2022). Green synthesis of silver nanoparticles using medicinal fern *Blechnum orientale* leaves and its application as an antimicrobial agent against food borne pathogens. *International Journal of Health Sciences, 6*(5), 4619–4631.

Singh, L., Singh, S., Singh, K., & Jadu, E. (2001). Ethnobotanical uses of some pteridophytic species in Manipur. *Indian Fern Journal, 18*(1, 2), 14–17.

Sun, H., Saeedi, P., Karuranga, S., Pinkepank, M., Ogurtsova, K., Duncan, B. B., & Magliano, D. J. (2022). IDF diabetes atlas: Global, regional and country-level diabetes prevalence estimates for 2021 and projections for 2045. *Diabetes Research and Clinical Practice, 183*, 109–119.

Tag, H., Das, A. K., & Loyi, H. (2007). Anti-inflammatory plants used by the Khamti tribe of Lohit district in eastern Arunachal Pradesh, India. *Natural Product Radiance, 6*(4), 334–340.

Tan, S. T., Ong, H. C., Chai, T. T., & Wong, F. C. (2018). Identification of potential anticancer protein targets in cytotoxicity mediated by tropical medicinal fern extracts. *Pharmacognosy Magazine, 14*(54), 227–230.

Thomas, A. P., Sindu, N., Matthew, A. N., & Dillesh, S. (2023). Phytochemical profiling and antibacterial potency of *Blechnum orientale*. *Indian Fern Journal, 40*(1), 17–28.

Thomas, T. (2010). Anti-bacterial action of rhizomes of *Blechnum orientale* L. *Indian Fern Journal, 27*, 73–77.

Wada, H., Kido, T., Tanaka, N., Murakami, T., Saiki, Y., Chen, C. M. (1992). Chemical and chemotaxonomical studies of ferns. LXXXI. Characteristic lignans of *Blechnaceous* ferns. *Chemical and Pharmaceutical Bulletin, 40*(8), 2099–2101.
Yusuf, U. K. (1994). Flavonoid glycosides in the fern *Blechnum orientale* Linn. *American Fern Journal, 84*(2), 69–70.

CHAPTER 4

A Comprehensive Review on the Phytochemistry and Pharmacology of *Diplazium esculentum* (Retz.) Sw.

RICHA BHARDWAJ* and REENU AGARWAL

Department of Botany, IIS (Deemed to be University), Jaipur, Rajasthan, India

*Corresponding authors

ABSTRACT

Diplazium esculentum, belonging to the family Athyriaceae, is widely regarded as one of the most prominent species of wild edible ferns, esteemed for its therapeutic utility in consuming its circinate portion and juvenile leaves. Classified as an ethnomedicinal plant, *D. esculentum* exhibits a broad spectrum of pharmacological activities, including antibacterial, antioxidant, immunomodulatory, neuromodulatory, and cytotoxic effects. Despite its status as a neglected seasonal vegetable, *D. esculentum* harbors substantial potential as a functional food due to its rich reservoir of bioactive constituents such as terpenoids, flavonoids, alkaloids, and saponins. Traditionally consumed for its nutritional and medicinal attributes across diverse geographical regions, this fern species also serves as a commendable source of dietary fiber, calcium, iron, phosphorus, vitamins C and A, alongside various essential minerals. Consequently, *D. esculentum* holds promise in alleviating malnutrition to a certain extent. This review offers a comprehensive resource for future researchers endeavoring to explore novel bioactive compounds and their pharmacological properties associated with *D. esculentum*.

Bioactives and Pharmacology of Pteridophytes. T. Pullaiah (Ed.)

DOI: 10.1201/9781779642738-4

4.1 INTRODUCTION

Ferns, commonly regarded for their ornamental appeal, harbor several species with notable nutritional value, including *Diplazium esculentum* (Retz.) Sw., renowned for its culinary significance in the Himalayan and sub-Himalayan regions. Despite its prevalence in hilly terrains, *D. esculentum* predominantly thrives at lower elevations. Indigenous communities in the Himalayas utilize *D. esculentum* extensively as a traditional vegetable, with its distribution spanning across various countries, including Bangladesh, Singapore, Taiwan, Thailand, Vietnam, Papua New Guinea, China, India, Indonesia, Japan, Malaysia, and the Philippines.

As a member of the Pteridophyta class of vascular plants, *D. esculentum* reproduces via spores, exhibiting terrestrial growth primarily in marshy areas, stream banks, and canals, with occasional occurrences on limestone rocks, across altitudes ranging from sea level up to 2300 meters. Known as "dhekia" in Assam, "pucuk paku" in Malaysia, and "linguda" in northern India, *D. esculentum* is characterized by a diverse array of phytochemicals, encompassing steroids, triterpenoids, phenols, flavones, and flavonoids. Its rhizomes, when dried, serve as pesticides, while the leaves are utilized as vegetables. Decoctions of *D. esculentum* are administered to alleviate hemoptysis and cough, with vegetable shoots exhibiting antioxidant properties.

Despite the limited anticancer activity observed in the plant's crude ethanol extract, *D. esculentum* has been entrenched in traditional medicine, with historical usage extending to the remedy for various diseases such as diabetes, asthma, rheumatism, dysentery, fever, wounds, and skin-related diseases. Consequently, this chapter endeavors to provide a comprehensive analysis of the phytochemistry and pharmacology of *D. esculentum*, drawing upon the latest scientific insights to further elucidate its therapeutic potential.

Diplazium esculentum has historically held significant cultural and culinary importance across various regions, including China, Thailand, Indonesia, the Philippines, India, and Nepal. Documentation from literature reveals its enduring utilization within diverse communities for the management of a wide array of ailments. Moreover, rural and tribal populations in India actively harvest this species in substantial quantities, utilizing it as a source of income to enhance their livelihoods (Sarkar et al., 2018).

Notably, the Mishing people in Assam have traditionally incorporated *D. esculentum* into religious rituals associated with deceased rites (Kutum et al., 2011).

4.2 PHYTOCHEMISTRY

The nutrient profile of any food item can be assessed through an analysis of its mineral and proximate composition. Extensive literature screenings have unveiled the proximate composition of *Diplazium esculentum*, revealing the presence of carbohydrates, lipids, proteins, vitamins, fiber, and various other constituents. Additionally, the mineral composition encompasses essential micro and macro compounds (Archana et al., 2012; Chettri, 2018; Irawan et al., 2006; Koniyo et al., 2021).

Botanicals, owing to their rich repository of bioactive components, have been traditionally employed for the treatment of diverse ailments. Through literature-based screenings, *D. esculentum* has been identified to contain alkaloids, flavonoids, glycosides, polyphenols, tannins, terpenoids, steroids, carbohydrates, lipids, and oils. Furthermore, the chemical constitution of the essential oil derived from *D. esculentum* leaves has been investigated by Essien et al. (2019), revealing predominant volatile components such as β-pinene, α-pinene, caryophyllene oxide, sabinene, and 1,8-cineole. The essential oil comprises monoterpene hydrocarbons, sesquiterpene hydrocarbons, oxygenated monoterpenoids, oxygenated sesquiterpenoids, and nonterpene derivatives. Additional compounds isolated from *D. esculentum* include ascorbic acid, eriodictyol 5-O-methyl ether 7-O-β-D-xylosyigalactoside, tannins, phytates, α-tocopherol, quercetin, pterosin, ptaquiloside, terpene, hopan-triterpene lactone, and lutein (Caldwell, 1972; Ching and Mohamed, 2001; Pathania et al., 2012; Somvanshi et al., 2006; Srivastava et al., 1981; Wali et al., 2016). Moreover, Watanabe et al. (2021) identified four phenolic compounds and three ecdysteroids in fresh fronds of *D. esculentum* obtained from Japan, while methanolic extracts of young fronds from the Indian Himalaya revealed 26 bioactive compounds, including pentadecanoic acid, β-sitosterol, neophytadiene, methyl palmitate, α-linolenic acid, phytol, diisobutyl phthalate, and 10,12 hexadecadien-1-ol (Naik et al., 2021).

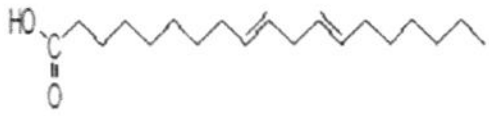

Linoleic acid

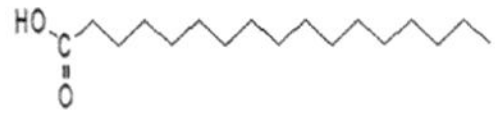

Palmitic acid

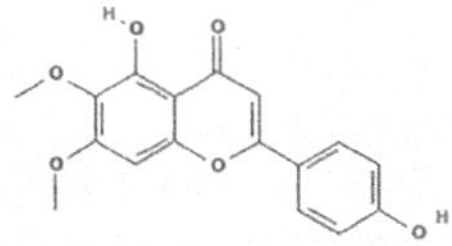

Quercetin

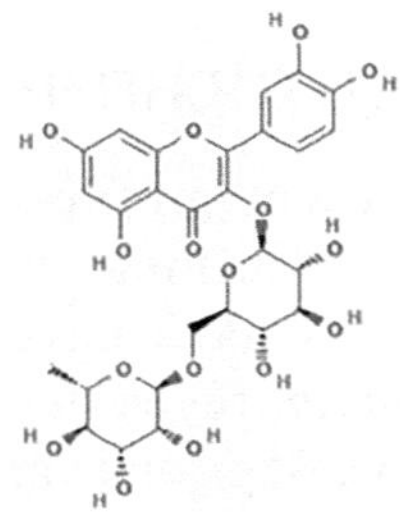

Rutin

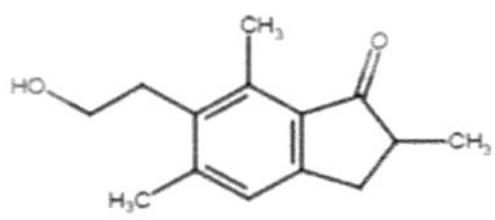

Pterosin B

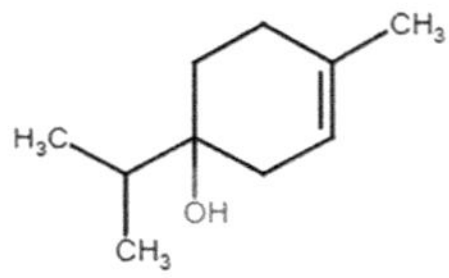

Terpene

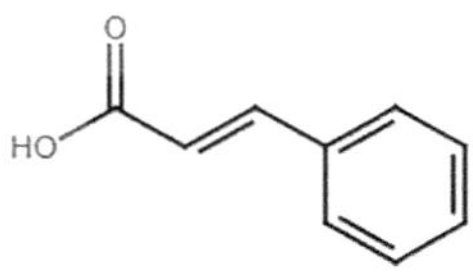

Trans-cinnamic acid

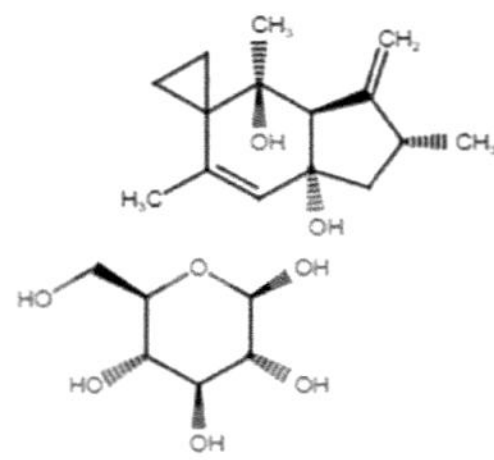

Ptaquiloside

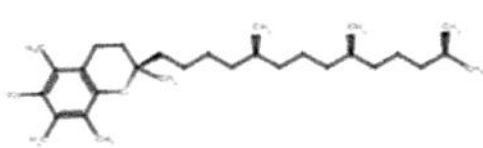

Alpha-tocopherol

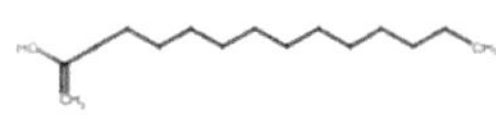

Pentadecanoic acid

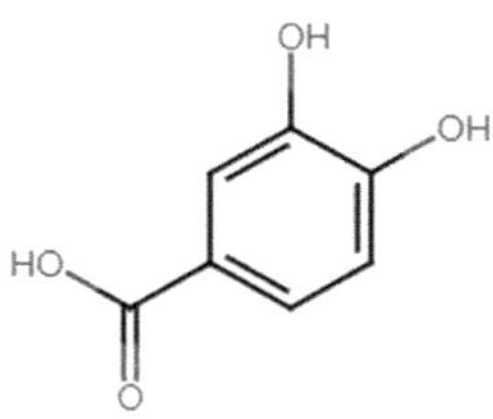

Protocatechuic acid

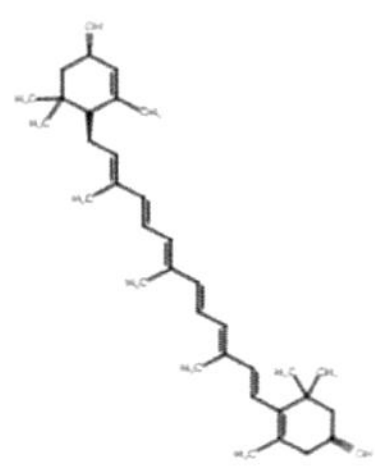

Lutein

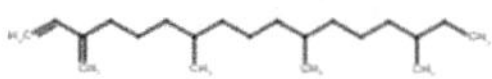

Neophytadiene

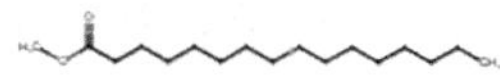

Methyl Palmitate

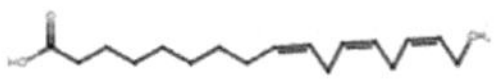

Alpha –Linolenic acid

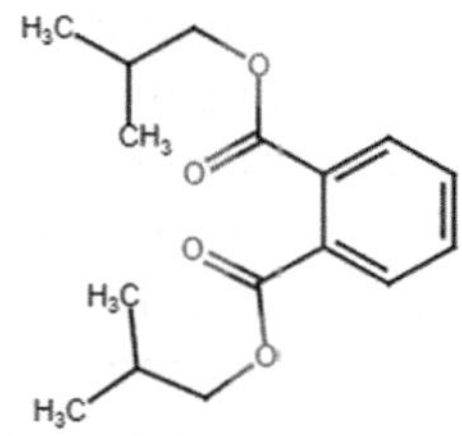

Diisobutyl phthalate

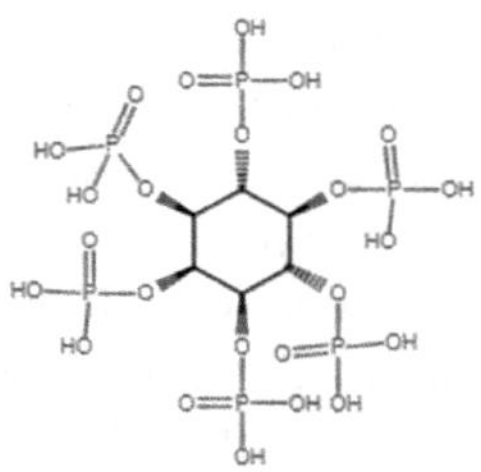

Phytate

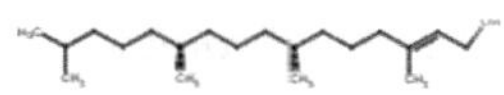

Phytol

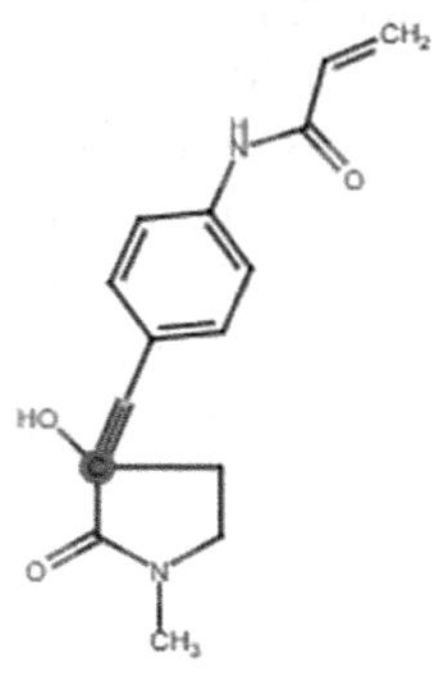

Amarasterone A

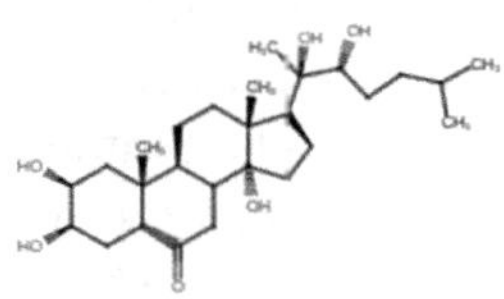

Ponasterone A

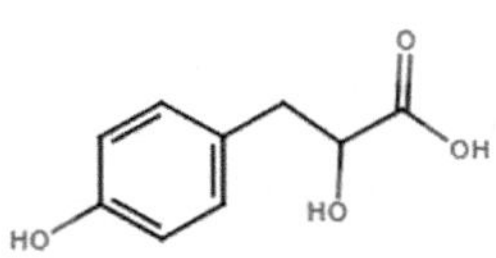

3-(4-hydroxyphenyl)lactic acid

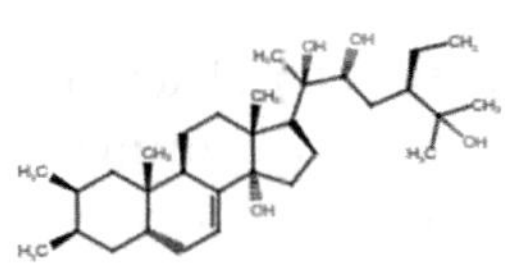

Makisterone C

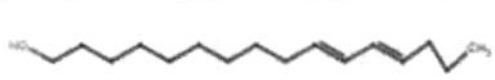

10,12-hexadecadien-1-ol

4.3 PHARMACOLOGICAL PROPERTIES OF *D. ESCULENTUM*

4.3.1 ANTI-OXIDANT ACTIVITY

The antioxidant efficacy of plant extracts was assessed utilizing various assays, including the 1,1-diphenyl-2-picrylhydrazyl (DPPH) radical scavenging assay, hydrogen peroxide (H_2O_2) radical scavenging assay, total antioxidant capacity assay, and ABTS (2,2′-azinobis (3-ethylbenzothiazoline-6-sulfonic acid)) assay (Akter et al., 2014; Das et al., 2013). Quantification of polyphenols and saponins, known contributors to the multifaceted medicinal attributes of *Diplazium esculentum*, was conducted via total polyphenol determination, with a particular focus on High-Performance Thin-Layer Chromatography (HPTLC) coupled with inferred antioxidant activity. Within the fiddlehead portion of this fern, three ecdysteroids—amarasterone A1, makisterone C, and ponasterone A—were identified. Noteworthy, Watanabe et al. (2021) provided the pioneering report on the isolation and characterization of four distinct phenolic compounds: (2R)-3-(4′-hydroxyphenyl) lactic acid, protocatechuic acid, trans-cinnamic acid, and rutin (Amna and Mardina, 2020). Kaushik et al. (2012) assessed the antioxidant potential of petroleum ether, chloroform, acetone, methanol, and aqueous extracts of *Diplazium esculentum*. Among the tested extracts, the aqueous extract exhibited the highest antioxidant activity (7.6 mM/dry weight). The central nervous system (CNS) stimulant properties of the aqueous extract were further evaluated in mice using an actophotometer. The extract demonstrated significantly enhanced CNS stimulation at all administered doses when compared to the control group and standard caffeine, a known CNS stimulant. The stimulatory effect was statistically significant across all tested doses and displayed a clear dose-dependent relationship.

4.3.2 ANTIBACTERIAL ACTIVITY

The antibacterial activity of *Diplazium esculentum* extracts was evaluated using the disc diffusion and broth dilution assays to determine the minimum inhibitory concentration (MIC) against five gram-negative and three gram-positive bacteria. Notably, *Salmonella arizonae*, *Escherichia coli*, *Salmonella typhi,* and *Staphylococcus aureus* were among the bacterial strains whose growth was significantly inhibited by nearly all alcoholic and aqueous extracts of *D. esculentum* (Amit et al., 2011).

Interestingly, when combined with tetracycline, the fern extract exhibited a more potent and synergistic effect on bacterial growth as compared to the antibiotic alone. Furthermore, the rhizome extract demonstrated higher inhibitory activity against *Salmonella arizonae* and *Staphylococcus aureus* compared to the leaves (Amit et al., 2011). Additionally, it was observed that the aqueous extracts of *D. esculentum* rhizomes inhibited the growth of certain bacteria, such as *Salmonella typhimurium* and *Klebsiella pneumoniae*, whereas the leaf extract did not exhibit similar inhibitory effects (Rios and Recio, 2005). Remarkably, the chloroform extract of *D. esculentum* displayed strong antibacterial activity, with maximum inhibition observed against *Sarcina lutea*, with *Salmonella typhimurium* showing the next highest level of inhibition. Moreover, the chloroform extract was effective against other bacterial strains, including *Shigella boydii, Vibrio cholera, Bacillus subtilis*, and *Klebsiella pneumoniae* (Rios and Recio, 2005).

4.3.3 LARVICIDAL ACTIVITY

The methanol extract of *Diplazium esculentum* exhibited potent larvicidal activity against *Culex* larvae, as initially observed. The larvicidal efficacy was evaluated on third or fourth instar *Culex* larvae. Halimatussakdiah et al. (2018) stated that the methanolic extract of *D. esculentum* possessed notable larvicidal properties, with an LC50 value of 149.279 ppm. In another study by Dash et al. (2017) and Umohata et al. (2020), the crude ethanolic extract derived from the aerial parts of *D. esculentum* was tested for larvicidal effects against third-instar larvae of *Anopheles gambiae* and *Culex quinquefasciatus*. Both mosquito species exhibited increased mortality with rising concentrations of the extract. Interestingly, the extract demonstrated higher efficacy against *A. gambiae* larvae compared to *Culex* larvae. This disparity in susceptibility may be attributed to variations in the physiological traits of the two mosquito species.

4.3.4 HEPATOPROTECTIVE ACTIVITY

Rats subjected to carbon tetrachloride (CCl_4) treatment exhibited a notable dose-dependent reduction in blood markers, including aspartate aminotransferase (AST), alanine aminotransferase (ALT), and alkaline phosphatase (ALP). This decline suggests a significant mitigation of hepatic injury by *Diplazium esculentum*. Notably, in groups pre-treated with 200 mg/kg and 400 mg/kg body weight of *D. esculentum* extract, the administration of CCl_4 did not induce a significant elevation in ALP and total bilirubin levels (Junejo et al., 2018).

4.3.5 IMMUNOMODULATORY ACTIVITIES

Boiling *Diplazium esculentum* administered to Swiss albino mice resulted in notable reductions in body weight and relative spleen weight. The immunosuppressive potential of *D. esculentum* was evidenced by a dose- and time-dependent decline in the number of peritoneal macrophages and plaque-forming cells, alongside a progressive reduction in the hemagglutination titer in treated mice. *In vitro* experiments revealed that *D. esculentum* induced both hemolysis and inhibited lymphocyte growth (Roy et al., 2013).

Moreover, chronic administration of *D. esculentum* to Swiss albino mice led to the induction of a Th1/Th2 imbalance, resulting in significant immunosuppression. This imbalance was dose-dependently mitigated by the reduction in Th1 (IL-2 and IFN-γ) and Th2 (IL-4 and IL-10) cytokine production by T cells in *D. esculentum* treated mice (Roy and Chaudhuri, 2017). Additionally, the chloroform and acetone extracts of *Diplazium esculentum* significantly attenuated carrageenan-induced paw edema in rats at a dose of 100 mg/kg orally, comparable to the conventional anti-inflammatory medication ibuprofen, suggesting anti-inflammatory properties of the plant (Kaushik et al., 2011).

Furthermore, the aqueous and ethanolic extracts of *D. esculentum* were investigated for their anti-anaphylactic and mast cell stabilizing properties. Both extracts exhibited anti-anaphylactic effects by reducing mast cell degranulation induced by horse serum and triple antigen (DPT) in Wistar rats. Rats pretreated with *D. esculentum* extracts showed decreased levels of nitric oxide in both serum and peritoneal fluid, attributed to mast cell membrane stabilization and inhibition of nitric oxide synthase synthesis (Das et al., 2012).

4.3.6 ANTIHELMINTHIC ACTIVITY

The anthelmintic properties of *Diplazium esculentum* aqueous, ethanolic, and petroleum ether extracts were evaluated against adult Indian earthworms, with piperazine citrate serving as the reference standard.

The ethanolic extract exhibited a dose-dependent paralytic effect, leading to eventual mortality. This effect manifested as diminished motility and decreased sensitivity to external stimuli. Conversely, the petroleum ether extract displayed the lowest activity among all the extracts, suggesting that the anthelmintic efficacy of *D. esculentum* increased progressively with polarity. Additionally, the aqueous extract demonstrated significant activity in this regard (Amit and Singh, 2012).

4.3.7 NEUROMODULATORY ACTIVITY

The impact of *Diplazium esculentum* on the CNS in mice was investigated, revealing that the aqueous leaf extract of the plant elicited a dose-dependent increase in locomotor activity. Moreover, the fern demonstrated potent analgesic properties, indicating its efficacy as a painkiller. Additionally, this extract exhibited stimulatory effects on the CNS compared to the control group (Kaushik et al., 2011). Further studies by Chawla et al. (2015) highlighted the significant effects of *D. esculentum* on peripheral and inflammatory pain models, in addition to its influence on the CNS. Additionally, Roy et al. (2015) suggested the potential of *D. esculentum* in treating neurodegenerative diseases attributed to oxidative stress.

4.3.8 ANTIDIABETIC ACTIVITIES

Irudayaraj et al. (2012) describe diabetes mellitus as a chronic metabolic disorder affecting the metabolism of fats, carbohydrates, and proteins, characterized by elevated blood glucose levels due to insufficient insulin secretion. Inhibiting carbohydrate-metabolizing enzymes such as α-glucosidase and α-amylase can effectively mitigate postprandial hyperglycemia, thus serving as a valuable strategy for managing blood sugar levels in individuals with type 2 diabetes and those at risk of developing the condition. It has been observed that *Diplazium esculentum* possesses antidiabetic properties attributed to its inhibition of α-glucosidase and α-amylase enzymes (Tabiano and Deliman, 2014). Experimental results indicated that the extract derived from *D. esculentum* exhibited notable inhibitory effects against α-amylase (92.09%) and α-glucosidase (70.01%). Additionally, a hydro-alcoholic extract of *D. esculentum* was evaluated for its potential to prevent streptozotocin (STZ)-induced diabetes (Junejo et al., 2018). 30 rats were administered the plant extract treatment for a duration of 21 days. Post-administration, it was seen that the plant extract (at a dose of 500 mg/kg) reduced blood glucose levels by 50.2% in STZ-induced diabetic rats. Moreover, significant reductions were noted in lipid profile ($p < 0.01$), serum marker enzyme activity ($p < 0.001$), beta cell necrosis, and regeneration in rats treated with the plant extract. The observed effects displayed dose-dependent characteristics across all experimental trials.

4.3.9 ANTI-INFLAMMATORY ACTIVITY

Zaini et al. (2016) evaluated the anti-inflammatory properties of *Diplazium esculentum's* ethanolic extract. The study involved 25 male mice divided into five groups. The treated group exhibited notable anti-inflammatory effects on hind paw edema, with inhibition percentages of 71.72%, 81.49%, and 92.60% at doses of 125 mg/kg b.w., 250 mg/kg b.w., and 250 mg/kg b.w., respectively. Furthermore, by the acetic acid-induced writhing method in mice, *D. esculentum* demonstrated significant analgesic activity (Chawla et al., 2015).

4.3.10 ANTIANAPHYLACTIC AND MAST CELL STABILIZING ACTIVITY

Aqueous and ethanolic solvents were utilized for the extraction of *Diplazium esculentum*, and their anti-anaphylactic and mast cell stabilizing properties were evaluated. Wistar rats weighing 150–170 g and Swiss albino mice weighing 18–20 g were utilized in this investigation. The number of degranulated mast cells in the models treated with plant extracts exhibited a significant reduction ($p < 0.001$). Following treatment with 250 and 500 mg/kg of both extracts, the percentage of intact mast cells was recorded as 72.83%, 76.67%, 69%, and 71.67%. The plant extract demonstrated anti-mast cell degranulation properties. The highest suppression of myeloperoxidase release from lung tissue was observed at 500 mg/kg for both extracts. Additionally, the plant extract reduced the levels of nitric oxide in serum and peritoneal fluid and stabilized the mast cell membrane (Das et al., 2012).

4.4 CONCLUSION

In contrast to other primitive vascular plants, *Diplazium esculentum* has endured since the Paleozoic era, adapting to a myriad of environmental fluctuations. Consequently, it is hypothesized that this species harbors a more diverse array of beneficial phytochemicals compared to its counterparts. Widely integrated into various traditional systems, further exploration and validation of its complete chemical profile and active constituents are imperative through meticulous bioassay-guided isolation. *D. esculentum* stands as the most frequently consumed fern globally, with its edible parts encompassing stems, leaves, rhizomes, young fronds, and, occasionally, the entire plant, as underscored by Liu et al. (2012). Despite an escalating documentation of its dietary and ethnomedical uses across diverse regions worldwide, the exploration of its pharmacological or toxicological effects on human health remains limited. Nonetheless, *D. esculentum* emerges as a prolific reservoir of diverse chemicals, each potentially endowed with varied pharmacological and therapeutic properties. This review endeavors to compile the extant information on the chemical profiles and pharmacological attributes of this edible fern, serving as a springboard for comprehensive research aimed at identifying potent compounds and undertaking in-depth investigations into their biological activities. Essential endeavors entail unraveling unknown chemicals, delineating their modes of action, and probing potential synergistic effects. Furthermore, sharing knowledge about the cultivation, conservation, and sustainable use of *D. esculentum* within local communities is essential for increasing its population and ensuring its long-term presence in the ecosystem.

KEYWORDS

- **anti-inflammatory medication ibuprofen**
- **antimicrobial**
- **antioxidant**
- **diabetes mellitus**
- **ethnomedicinal**
- **inflammatory pain models**
- **streptozotocin**
- **triterpenoids**

REFERENCES

Akter, S., Hossain, M. M., Ara, I., & Akhtar, P. (2014). Investigation of *in vitro* antioxidant, antimicrobial and cytotoxic activity of *Diplazium esculentum* (Retz.) Sw. *International Journal of Advanced Pharmaceutical Biological Chemistry, 3*(3), 723–733.

Amit, S., & Singh, F. M. (2012). *In vitro* anthelmintic activity of *Diplazium esculentum* (Retz.) Swiss rhizome extract. *Journal of Pharmacognosy and Phytochemistry, 1*(4), 84–87.

Amit, S., Sunil, K., Bhatt, S. P., & Arvind, N. (2011). Antibacterial activity of *Diplazium esculentum* (Retz.) Sw. *Pharmacognosy Journal, 3*(21), 77–79.

Amna, U., & Mardina, V. (2020). Antioxidant activity of methanol extract of *Diplazium esculentum* (Retz.) Sw. leaves collected from Aceh. In *IOP Conference Series: Materials Science and Engineering* (Vol. 725, No. 1, p. 012082). IOP Publishing.

Archana, G. N., Pradeesh, S., Chinmayee, M. D., Mini, I., & Swapna, T. (2012). *Prospects in Bioscience: Addressing the Issues*. Springer, New Delhi.

Caldwell, M. J. (1972). Ascorbic acid content of Malaysian leaf vegetables. *Ecology of Food and Nutrition, 1*(4), 313–317.

Chawla, S., Chawla, S., Ram, V., Semwal, A., & Singh, R. (2015). Analgesic activity of medicinally important leaf of *Diplazium esculentum*. *African Journal of Pharmacy and Pharmacology, 9*(25), 628–632.

Chettri, S. (2018). Nutrient and elemental composition of wild edible ferns of the Himalaya. *American Fern Journal, 108*(3), 95–106.

Ching, L. S., & Mohamed, S. (2001). Alpha-tocopherol content in 62 edible tropical plants. *Journal of Agricultural and Food Chemistry, 49*(6), 3101–3105.

Das, B., Paul, T., Apte, K. G., Chauhan, R., & Saxena, R. C. (2013). Evaluation of antioxidant potential & quantification of polyphenols of *Diplazium esculentum* Retz. with emphasis on its HPTLC chromatography. *Journal of Pharmaceutical Research, 6*(1), 93–100.

Das, B., Paul, T., Apte, K. G., Parab, P. B., Chauhan, R., & Saxena, R. C. (2012). Antianaphylactic and mast cell stabilizing activity of *Diplazium esculentum* Retz. on sensitized Wistar rats. *Inventi Impact Ethnopharmacology, 3*, 136–141.

Dash, K. G., Siti, K. J. K., & Ahmad, F. S. (2017). Physico-chemical analysis of the aerial parts of *Diplazium esculentum* (Retz.) Sw. (Family: Athyriaceae). *International Journal of Pharmacognosy and Phytochemical Research, 9*(6), 772–774.

Essien, E. E., Ascrizzi, R., & Flamini, G. (2019). Characterization of volatile compounds of *Diplazium esculentum*. *Chemistry of Natural Compounds, 55*, 958–959.

Halimatussakdiah, H., Amna, U., & Wahyuningsih, P. (2018). Preliminary phytochemical analysis and larvicidal activity of edible fern (*Diplazium esculentum* (Retz.) Sw.) extract against *Culex*. *Jurnal Natural, 18*(3), 141–147.

Irawan, D., Wijaya, C. H., Limin, S. H., Hashidoko, Y., Osaki, M., & Kulu, I. P. (2006). Ethnobotanical study and nutrient potency of local traditional vegetables in Central Kalimantan. *Tropics, 15*(4), 441–448.

Irudayaraj, S. S., Sunil, C., Duraipandiyan, V., & Ignacimuthu, S. (2012). Antidiabetic and antioxidant activities of *Toddalia asiatica* (L.) Lam. leaves in streptozotocin-induced diabetic rats. *Journal of Ethnopharmacology, 143*(2), 515–523.

Junejo, J. A., Gogoi, G., Islam, J., Rudrapal, M., Mondal, P., Hazarika, H., & Zaman, K. (2018). Exploration of antioxidant, antidiabetic and hepatoprotective activity of

Diplazium esculentum—A wild edible plant from North Eastern India. *Future Journal of Pharmaceutical Sciences, 4*(1), 93–101.

Kaushik, A., Jijta, C., Kaushik, J. J., Zeray, R., Ambesajir, A., & Beyene, L. (2012). FRAP (Ferric reducing ability of plasma) assay and effect of *Diplazium esculentum* (Retz.) Sw. (a green vegetable of North India) on central nervous system. *Indian Journal of Natural Products and Resources, 3*(2), 228–231.

Kaushik, A., Kaushik, J. J., Das, A., Gemal, S., & Gaim, D. (2011). Preliminary studies on anti-inflammatory activities of *Diplazium esculentum* in experimental animal models. *International Journal of Pharmaceutical Sciences and Research, 2*(5), 1251.

Koniyo, Y., Lumenta, C., Olii, A. H., Mantiri, R. O. S. E., & Pasisingi, N. (2021). Nutrition of local wild edible fern (*Diplazium esculentum*) leaves. In *IOP Conference Series: Earth and Environmental Science* (Vol. 637, No. 1, p. 012008). IOP Publishing.

Kutum, A., Sarmah, R., & Hazarika, D. (2011). An ethnobotanical study of Mishing tribe living in fringe villages of Kaziranga National Park of Assam, India. *Indian Journal of Fundamental and Applied Life Sciences, 1*(4), 45–61.

Liu, Y., Wujisguleng, W., & Long, C. (2012). Food uses of ferns in China: A review. *Acta Societatis Botanicorum Poloniae, 81*(4), 263–270.

Naik, B., Maurya, V. K., Kumar, V., Kumar, V., Upadhyay, S., & Gupta, S. (2021). Phytochemical analysis of *Diplazium esculentum* reveals the presence of medically important components. *Current Nutrition & Food Science, 17*(2), 210–215.

Pathania, S., Kumar, P., Singh, S., Khatoon, S., Rawat, A. K. S., Punetha, N., & Somvanshi, R. (2012). Detection of ptaquiloside and quercetin in certain Indian ferns. *Current Science, 102*(12), 1683–1691.

Rios, J. L., & Recio, M. C. (2005). Medicinal plants and antimicrobial activity. *Journal of Ethnopharmacology, 100*(1, 2), 80–84.

Roy, S., & Chaudhuri, T. K. (2017). Toxicological assessment of *Diplazium esculentum* on the reproductive functions of male Swiss albino mouse. *Drug and Chemical Toxicology, 40*(2), 171–182.

Roy, S., Dutta, S., & Chaudhuri, T. K. (2015). *In vitro* assessment of anticholinesterase and NADH oxidase inhibitory activities of an edible fern, *Diplazium esculentum*. *Journal of Basic and Clinical Physiology and Pharmacology, 26*(4), 395–401.

Roy, S., Tamang, S., Dey, P., & Chaudhuri, T. K. (2013). Assessment of the immunosuppressive and hemolytic activities of an edible fern, *Diplazium esculentum*. *Immunopharmacology and Immunotoxicology, 35*(3), 365–372.

Sarkar, B., Basak, M., Chowdhury, M., & Das, A. (2018). Importance of *Diplazium esculentum* (Retz.) Sw. (Athyriaceae) on the lives of local ethnic communities in Terai and Duars of West Bengal – A report. *Plant Archives, 18*(1), 439–442.

Somvanshi, R., Lauren, D. R., Smith, B. L., Dawra, R. K., Sharma, O. P., Sharma, V. K., & Gangwar, N. K. (2006). Estimation of the fern toxin, ptaquiloside, in certain Indian ferns other than bracken. *Current Science, 91*(11), 1547–1552.

Srivastava, S. K., Srivastava, S. D., Saksena, V. K., & Nigam, S. S. (1981). A flavanone glycoside from *Diplazium esculentum*. *Phytochemistry, 20*(4), 862.

Tabiano, J., & Deliman, Y. (2014). *In vitro* inhibitory activity of *Atuna racemosa, Euphorbia hirta* and *Diplazium esculentum* juices against α-amylase and α-glucosidase. In *International Seminar on Science and Technology* (p. 89).

Umohata, I. A., Ubulom, P. M. E., Udofa, E. J., Bala, D. N., & James, I. V. (2020). Potential of the ethanol extract of aerial parts of *Diplazium esculentum* (Retz.) Sw. as larvicide against *Anopheles gambiae* Giles and *Culex quinquefasciatus* Say. *International Journal of Tropical Disease & Health, 41*(3), 40–47.

Wali, A., Sharma, S., Walia, M., Kumar, P., Thakur, S., Kumari, A., & Agnihotri, V. K. (2016). Two edible ferns of Western Himalaya: A comparative *in vitro* nutritional assessment, antioxidant capacity and quantification of lutein by UPLC-DAD. *International Journal of Food and Nutritional Science, 5*, 9.

Watanabe, M., Miyashita, T., & Devkota, H. P. (2021). Phenolic compounds and ecdysteroids of *Diplazium esculentum* (Retz.) Sw. (Athyriaceae) from Japan and their chemotaxonomic significance. *Biochemical Systematics and Ecology, 94*, 104211.

Zaini, M., Biworo, A., & Anwar, K. (2016). Uji efek antiinflamasi ekstrak etanol herba lampasau (*Diplazium esculentum* Swartz) terhadap mencit jantan yang diinduksi karagenin-Λ. *Journal of Pharmascience, 3*(2), 119–130.

CHAPTER 5

Phytochemicals and Bioactivities in a Lesser-Known Pteridophyte: *Pteridium aquilinum*

SUVARNA V. GAIKWAD,[1] NILESH V. PAWAR,[2] and SANDEEP R. PAI[1,*]

[1]*Post Graduate Research Centre in Botany, Rayat Shikshan Sanstha's Dada Patil Mahavidyalaya, Karjat, Ahilyanagar, Maharashtra, Indi*

[2]*Department of Botany, The New College, Kolhapur, Maharashtra, India*

*Corresponding authors

ABSTRACT

Pteridium aquilinum (L.) Kuhn, one of the bracken ferns (Family: Dennstaedtiaceae), is found throughout Southern Asia and Northern Australia. The plant is regarded as significant because of its antidiabetic, antialzheimer, immunomodulatory, antitumor, anti-inflammatory, antioxidant, and antimicrobial efficacy. The genus is a source of a nor-sesquiterpene glucoside, called ptaquiloside, which is a known carcinogenic and mutagenic component of *Pteridium* taxa. On the contrary, it also includes other phytoconstituents such as pterosin A and ecdysone, which provide them anti-diabetic and insecticidal potential, respectively. This chapter provides insight into the distribution, phytoconstituents, and bioactivities in *Pteridium aquilinum*.

Bioactives and Pharmacology of Pteridophytes. T. Pullaiah (Ed.)

DOI: 10.1201/9781779642738-5

5.1 INTRODUCTION

Ferns are one of the underutilized plant resources by local populations and scientists, although they have significant ethnomedicinal benefits. Additionally, in several countries, ferns are employed as interesting and traditional delicacies (Bajracharya, 2022). Recently, studies on different varieties of ferns have become more prevalent. Bracken ferns, or *Pteridium* species, are among the most abundant plants on the globe. They are widely spread out in Asia, South America, Australia, and Europe. They can be found in and around grazing lands and forest areas in the Himalayan regions of India between 1800 and 2400 meters above sea level (Pathania et al., 2012). *Pteridium aquilinum* (L.) Kuhn [Syn. *Pteridium revolutum* (Blume) Nakai], one of the bracken ferns, is found throughout Southern Asia and Northern Australia (Wang et al., 2023). It belongs to the family Dennstaedtiaceae.

Because of the complexity of the genus *Pteridium*, its nomenclature is still disputed and unclear (Štefanić et al., 2022). Tryon (1941) rationalized 135 names that had been used to refer to bracken ferns worldwide, resulting in the single species *Pteridium aquilinum* with two subspecies, ss *P. aquilinum* and ssp. *caudatum*, which together contain 12 varieties. After DNA fingerprinting and morphometric studies, Thomson (2000) reported that Tryon's varieties (*revolutum, aquilinum, africanum, decompositum, esculentum, arachnoideum*, and *latiusculum*) may be treated as species. These findings are further supported by Bajracharya (2022) and Marimuthu et al. (2022), who mentioned that *Pteridium revolutum* (Blume) Nakai may be a taxonomic synonym of *Pteridium aquilinum* subsp. *wightianum* (J. Agardh) W.C. Shieh.

Pteridium aquilinum (L.) Kuhn is a polycarpic geophyte that reproduces by spores and underground stems. Roots are thin, black, and brittle, extending from the rhizome and penetrating deeper into the soil. The rhizome contains two types of shoots: long and short shoots. Long shoots form the main axis of the stem and will not produce leaves (fronds), whereas short shoots will produce annual fronds. Early in April, the coiled, silver-gray-haired young leaves emerge from the rhizomes. Large fronds measure between 0.3 and 1.3 meters in height and 15 and 45 centimeters in length. The blades are divided into pinnae or leaflets, and each pinna is further divided into oblong, smooth-edged pinnules. Sori are present beneath the outer margins of the leaflets in fertile fronds (Štefanić et al., 2022).

Pteridium revolutum is an ornamental fern that is cultivated for borders and interior decorating. Traditionally, it has been used to make animal bedding. Breads and brews are also made from the rhizomes.

Fronds of *P. revolutum* are used to treat chronic illnesses of the viscera and spleen. Delicate fronds are used to prepare soups as well as eaten as vegetables. The rhizome of *Pteridium revolutum* is helpful for diarrhea and inflammation of the mucous membranes lining the stomach and intestines (Bajracharya, 2022; Marimuthu et al., 2022). In addition, it removes dampness and serves as an insect-repellent, diuretic, and anti-pyretic (Wang et al., 2023). On the contrary, because it produces a variety of diseases in farm animals, such as thiamine deficiency, acute hemorrhagic syndrome, brilliant blindness, enzootic hematuria, and upper alimentary cancer, bracken fern poses a serious threat to livestock-based agriculture (Štefanić et al., 2022).

5.2 PHYTOCONSTITUENTS IN *PTERIDIUM AQUILINUM*

A list of phytoconstituents from *Pteridium aquilinum* is tabulated in Table 5.1 and structures in Figure 5.1 *Pteridium revolutum* is a source of a nor-sesquiterpene glucoside, called ptaquiloside. It belongs to an illudane family. It is a known carcinogenic and mutagenic component of *Pteridium* taxa. *Pteridium* has long been thought to be harmful to the health of animals. It produces a variety of disease syndromes in cattle, such as bright blindness in sheep, enzootic bovine hematuria due to urinary tract neoplasms, and aplastic anaemia in cattle and sheep. It makes bovine papillomavirus (BPV) infections in cattle more persistent. Significantly, exposure to bracken and its toxins has also been linked to a higher incidence of upper gastrointestinal cancers in human populations (Aranha et al. (2014); Bajracharya, 2022; Castillo et al., 1997; Santos et al., 2016) developed a new and selective method for the estimation of this ptaquiloside in cattle and reported that on hydrolysis, it produces a non-toxic and aromatized indanone known as pterosin B. Zhou et al. (2016) isolated about 14 compounds from methanolic fractions of *P. revolutum* as ptaquiloside and its derivatives (1), caudatoside (2), 25,26-didehydroponasterone A (3), 5-stigmastene-3β,7α-diol (4), 20-hydroxyecdysone (5), prunasin (6), ponasterone A (7), pterosin A (8), pterosin B (9), kaempferol-3-O-β-D-glucoside (10), β-sitosterol (11), and β-daucosterol (12). Interestingly, 5-stigmastene-3β, 7α-diol, prunasin, 25,26-didehydroponasterone A, and 20-hydroxyecdysone were isolated for the first time from the *Pteridium* genus. Another sesquiterpene with 1-indanone skeletons, known as pterosins, was first isolated from *Pteridium aquilinum* var. *latiusculum*. It was reported that pterosin A inhibited both type 1 and type 2 diabetes. Chen et al. (2015) provided isolation and structural elucidation of three novel compounds, namely 13-chloro-spelosin 3-O-β-D-glucopyranoside (13), (3R)-pterosin D 3-O-β-D-(3'-p-coumaroyl)-glucopyranoside (14), and (2R,3R)-pterosin L 3-O-β-D-(3'-p-coumaroyl)-glucopyranoside (15) from *Pteridium revolutum*. The rhizome of *P. aquilinum* was used in the isolation and characterization of pterosins and pterosides by Rizgar et al. (2016). The study revealed 13 novel pterosins and pterosides. It is interesting to note that rhedynosides C, rhedynosides D, and rhedynosin H, alongside the two known compounds, histiopterosin A and (2S)-pteroside A2, were isolated for the first time. Furthermore, rhedynosin A and rhedynoside A were the first examples of this class of pterosin-sesquiterpenoid. Additionally, the three previously described compounds, rhedynosin I, (2S)-2-hydroxymethylpterosin E, and (2S)-12-hydroxypterosin A, were also derived from plants for the first time. Aerial parts of *P. revolutum* are a source of novel polysaccharides such as PRP0, PRP1, and PRP2. These polysaccharides have significant anti-oxidant and anti-glycated potential. Wang et al. (2023) revealed the structural characterization of these polysaccharides from *P. revolutum* for the first time. Another novel polysaccharide, CPS, was isolated from the root, stem, and leaves of *Pteridium aquilinum*. Saimaiti and Genccelep (2023) studied the thermal and molecular characteristics of this polysaccharide and suggested that these polysaccharides can be used in high-temperature food preparation because of their exceptional thermal stability (up to 230°C). Song et al. (2017) studied the structural characterization of a novel heteropolysaccharide, PAP1-A, from the aerial parts of *Pteridium aquilinum*, which has been shown to possess immunostimulating potential. Ecdysteroids are a class of steroidal chemicals that share a close structural relationship with ecdysone (Sahayaraj, 2022). Another study isolated A ecdysones and B ecdysones from *Pteridium aquilinum* (L.) Kuhn and studied the insecticidal activity of different extracts against *Helicoverpa armigera* and *Spodoptera litura*. It has been shown that compared to *S. litura, H. armigera* had a greater sensitivity to the fern extracts, which suggested that the ethanolic extract of *P. aquilinum* should be further tested for field evaluation (Selvaraj et al., 2005). From the ethanolic extract of *Pteridium aquilinum*, Choi et al. (2018) isolated a new pterosin glycoside, (–)-pteroside N, and a new seco-illudoid sesquiterpene, pterosinone.

This group of compounds has been shown to possess marked anti-Alzheimer disease activity. In another study, Chen et al. (2013) isolated a novel bihomoflavanonol, known as pteridium III, from ethanolic fractions of *P. aquilinum* (L.) Kuhn.

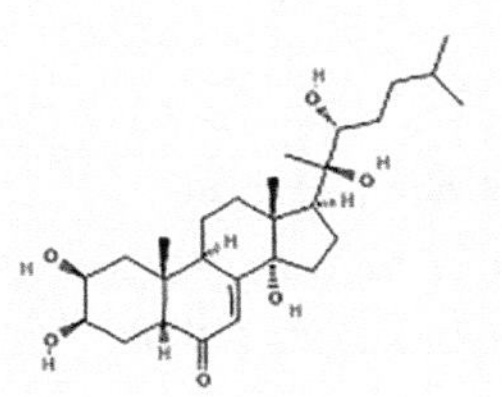

Ponasterone A

PubChem CID 115127

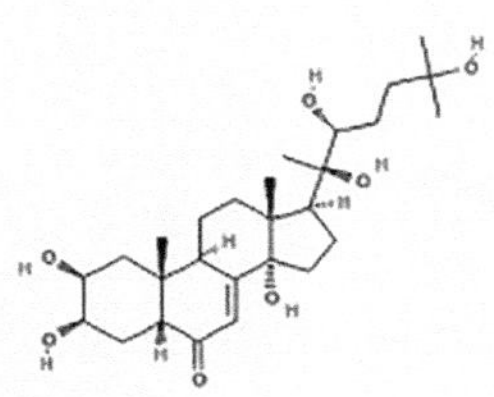

20-Hydroxyecdysone

PubChem CID 5459840

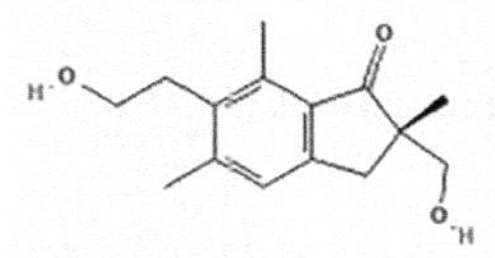

Pterosin A

PubChem CID 135017

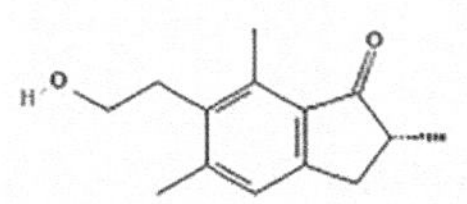

Pterosin B

PubChem CID 115049

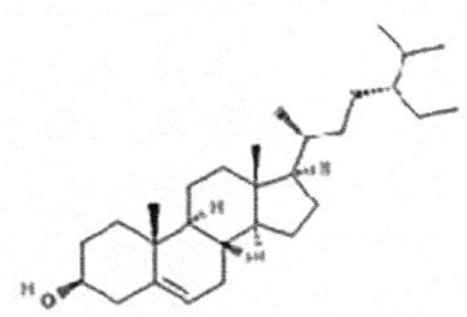

β -Sitosterol

PubChem CID 222284

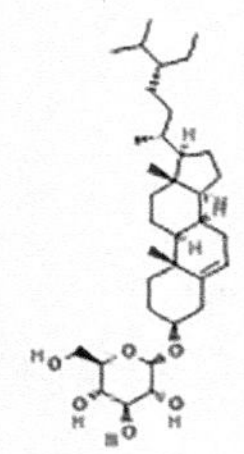

β -Daucosterol

PubChem CID 5742590

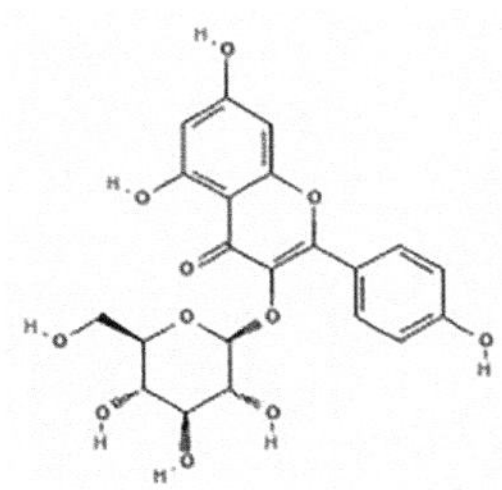

Kaempferol-3-*O*- β-D-glucoside

PubChem SID 481106910

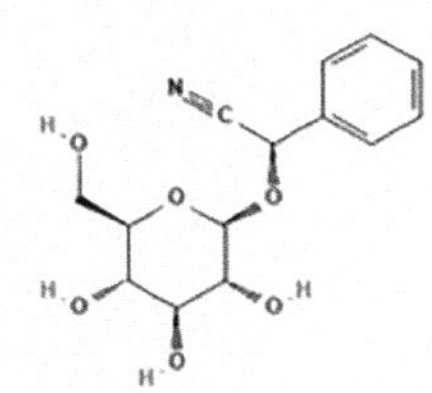

Prunasin

PubChem CID 119033

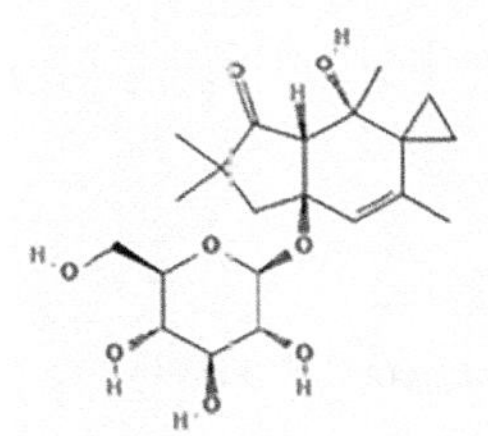

Ptaquiloside Z

PubChem CID 101242525

FIGURE 5.1 *(Continued)*

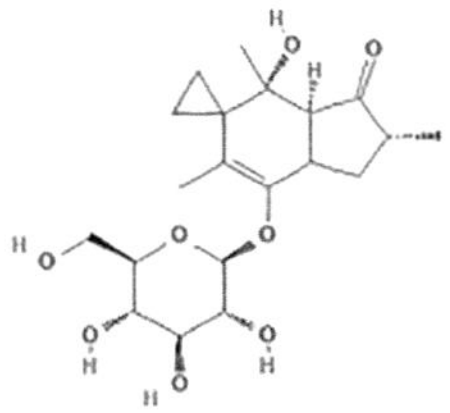

Isoptaquiloside

PubChem SID 275087478

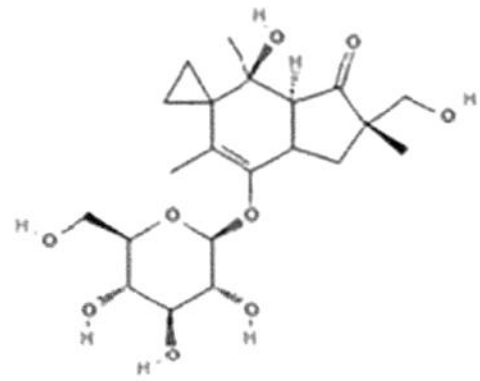

Caudatoside

PubChem CID 102065498

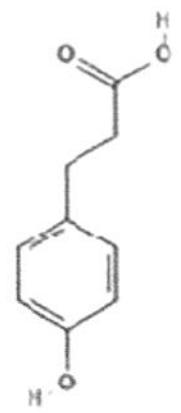

Dihydro-p-coumaric (phloretic) acid

PubChem CID 10394

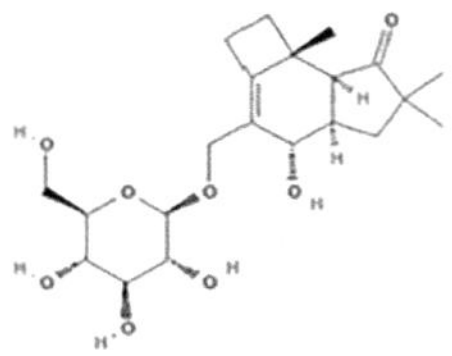

Pteridanoside

PubChem SID 272988570

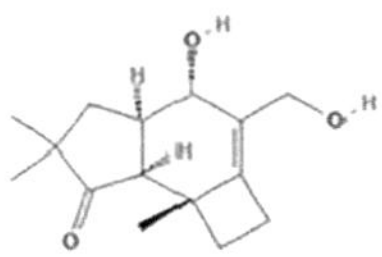

Pteridanone

PubChem SID 272988571

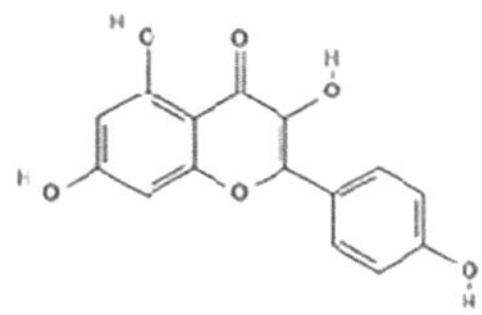

Kaempferol

PubChem CID 5280863

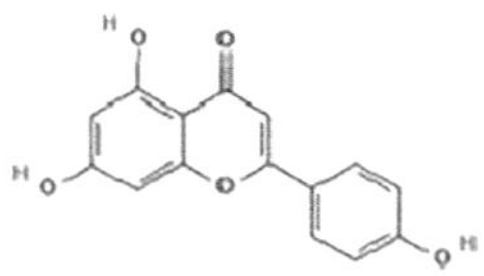

Apigenin

PubChem CID 5280443

FIGURE 5.1 Structure of various phytoconstituents identified from *P. aquilinum*.

Furthermore, previous studies have documented the isolation of several metabolites from this species, such as dihydro-p-coumaric (phloretic) acid, dihydroferulic acids (Méndez, 2005), pteridanoside and pteridanone (Castillo et al., 1999), kaempferol and apigenin (Imperato, 1997), illudane and illudalane sesquiterpenes (Castillo et al., 1998), pterosins K and pterosins Z (Castillo et al., 1997), steroids, terpenoids, and flavonoids.

TABLE 5.1 Phytoconstituents from Various Parts of *P. aquilinum* wrt Solvent Used for Extraction and Method Used for Its Identification and/or Quantification

Sr. No.	Metabolite	Plant Parts	Solvent	Method Used	References
1.	PRP0	Aerial parts	Ethanol	HPGPC, UV, IR and NMR	Wang et al. (2023)
2.	PRP1				
3.	PRP2				
4.	5-stigmastene-3β,7α-diol	Whole plant	Fractions	NMR, MS	Zhou et al. (2016)
5.	Ponasterone A				
6.	25,26-didehydroponasterone A				
7.	20-hydroxyecdysone				
8.	Pterosin A				
9.	Pterosin B				
10.	β -sitosterol				
11.	β -daucosterol				
12.	Kaempferol-3-*O*- β-D-glucoside				
13.	Prunasin				
14.	Ptaquiloside				
15.	Ptaquiloside Z				
16.	Isoptaquiloside				
17.	Caudatoside				
18.	13-chloro-spelosin 3-*O*-β-D-glucopyranoside	Whole plant	Organic	*LC-MS-MS*	Chen et al. (2015)
19.	(3*R*)-Pterosin D 3-*O*-β-D-(3'-*p*-coumaroyl)-glucopyranoside				
20.	(2*R*,3*R*)-Pterosin L 3-*O*-β-D-(3'-*p*-coumaroyl)-glucopyranoside				
21.	Polysaccharide CPS	Root, stem, leaves	NM	FTIR	Saimaiti and Genccelep (2023)
22.	(–)-Pteroside N	Whole plant	Ethanol	NMR	Choi et al. (2018)
23.	Pterosinone				
24.	PAP1-A	Aerial parts	Ethanol	NMR	Song et al. (2017)
25.	Ptaquiloside	NM	NM	NM	Santos et al. (2016)

TABLE 5.1 *(Continued)*

Sr. No.	Metabolite	Plant Parts	Solvent	Method Used	References
26.	Rhedynosides C	Rhizome	NM	NM	Rizgar et al. (2016)
27.	Rhedynosides D				
28.	Rhedynosin H				
29.	Histiopterosin A				
30.	(2S)-Pteroside A2				
31.	Rhedynosin A				
32.	Rhedynoside A				
33.	Rhedynosin I				
34.	(2S)-2-hydroxymethylpterosin E				
35.	(2S)-12-hydroxypterosin A)				
36.	Ptaquiloside(PTA)	Leaves		MS	Aranha et al. (2014)
37.	Pterosin B				
38.	Pteridium III	Aerial parts	Ethanol -fractions	NMR	Chen et al. (2013)
39.	Dihydro-p-coumaric (phloretic) acid	NM	NM	NM	Méndez (2005)
40.	Dihydroferulic acids				
41.	α-Ecdysones	NM	NM	NM	Selvaraj et al. (2005)
42.	β-Ecdysones				
43.	Pteridanoside	NM	NM	NM	Castillo et al. (1999)
44.	Pteridanone				
45.	Kaempferol 7-O-rhamnoside-4'-O-glucoside	Aerial parts	Ethanol	NMR	Imperato (1998)
46.	Rhamnetin 3-O-laminaribioside				
47.	Kaempferol				
48.	Apigenin				
49.	Kaempferol 3-0-(6"-feruloylglucoside				
50.	Kaempferol 3-O-(6"-p-coumaroylglucoxide)				
51.	Pterosins A	Leaves	NM	NMR	Castillo et al. (1997)
52.	Pterosins B				
53.	Pterosins K				
54.	Pterosins Z				
55.	Ptaquiloside				
56.	Isoptaquiloside				
57.	Caudatoside				

5.3 BIOACTIVITIES OF *PTERIDIUM AQUILINUM*

The plant is known for a wide array of biological activities. A list of distinct bioactivities is tabulated in Table 5.2.

TABLE 5.2 Biological Activities of *P. revolutum*

Sr. No.	Plant Part	Biological Activity	Extraction/ Solvent	References
1.	Aerial parts	Antioxidant, antiglycated	Ethanol	Wang et al. (2023)
2.	Whole plant	Antidiabetic	Organic	Chen et al. (2015)
		Biological activities of *P. aquilinum* (bracken fern)		
3.	Whole plant	Anti-Alzheimer disease (anti-AD) activity	Ethanol	Choi et al. (2018)
4.	Aerial parts	Immunomodulatory	Ethanol	Song et al. (2017)
5.	Leaves	Antibacterial	N-hexane and ethanolic	Awe and Amobi (2015)
6.	NM	Carcinogen	NM	Gomes et al. (2012)
7.	Purchased	Antioxidant	NM	Wang and Wu (2013)
8.	Leaves	Antioxidant and antibacterial	Ethanol, methanol, petroleum ether, chloroform and distilled water	Kardong et al. (2013)
9.	Aerial parts	Antitumor	Ethanol -fractions	Chen et al. (2013)
10.	Leaves	Antiinflammatory and antioxidant	Ethanol	Dion et al. (2015)

5.3.1 ANTIDIABETIC ACTIVITY

The antiglycated properties of polysaccharides viz. PRP0, PRP1, and PRP2 were assessed using the bovine serum albumin-glucose model. According to the findings, tested polysaccharides showed protective effects on protein thiol groups and demonstrated inhibitory efficacy on the production of Amadori products (intermediates during the early stage of glycation), dicarbonyl compounds, fluorescent Advanced Glycation End Products (AGE), pentosidine (a representative compound of fluorescent AGEs), and PCO (protein carbonyl content). These findings imply that the given polysaccharides have potential applications as AGE inhibitors (Wang et al., 2023).

A new perspective on the distribution of pterosins in Taiwanese ferns has been offered by Chen et al. (2015). They studied the distribution of pterosins (analogs), glucose-uptake efficiency, and protective effects of pterosin A from fern species viz. *Ceratopteris thalictroides, Pteridium revolutum,* and *Hypolepis punctata.* It was found that pterosin A protects cells against reactive oxygen species and palmitate-induced cell damage by influencing insulin secretion. However, additional functional investigations are necessary to study the possible anti-diabetic effect of these phytocompounds.

5.3.2 ANTI-ALZHEIMER DISEASE ACTIVITY

A study conducted by Choi et al. (2018) investigated the anti-Alzheimer effect of (–)-Pteroside N and pterosinone (glycoside and sesquiterpene) isolated from ethanolic fractions of *P. aquilinum*. They investigated anti-Alzheimer activity via enzyme inhibition of β-site amyloid precursor protein (APP) cleaving enzyme 1 (BACE1) and butyrylcholinesterase (BChE). According to the findings, (–)-Pteroside N had a strong inhibitory effect against AChE and BChE but very moderate BACE1 inhibitory activity, whereas pterosinone demonstrated a strong inhibitory effect against BACE1 but very modest inhibitory activity against AChE and BChE. These results provide support for the significant role of sesquiterpenoids from *P. aquilinum* in the treatment of Alzheimer disease.

5.3.3 IMMUNOMODULATORY ACTIVITY

Apart from the above, the plant is also valued for its immunomodulatory efficacy. Song et al. (2017) evaluated the immunomodulatory activity of PAP1-A (a novel polysaccharide isolated from *P. aquilinum*) by using RAW264.7 cells. It is well acknowledged that RAW264.7 cells can be used as a tool for examining the actions of macrophages that control immunity. By producing pro-inflammatory cytokines and inflammatory substances like NO, macrophages actively engage in immunological responses. The findings revealed that PAP1-A might considerably boost the proliferation of RAW264.7 cells and trigger those cells to produce NO, which justifies the immunomodulatory potential of *P. aquilinum.*

5.3.4 ANTI-TUMOR ACTIVITY

Ethanolic fractions of aerial parts of *P. aquilinum* exhibit anti-tumor properties. The anti-cancer property of Pteridium III (a bihomoflavanonol) was screened against six different cancer cell types using the MTT assay. Pteridium III has been reported to exhibit moderate anticancer activity against NCI-H46 lung cancer cells, A375 melanoma cells, and U-7MG glioma cells, with IC50 values of 22.6, 106.7, and 1540.5 µmol/L, respectively. Nevertheless, the study did not show any cytotoxic effects on the hepatoma carcinoma cell BEL-7402, gastric carcinoma SGC-7901, or prostatic carcinoma PC-3, which suggests more research and clarification (Chen et al., 2013).

5.3.5 ANTI-INFLAMMATORY ACTIVITY

In a separate study, Dion et al. (2015) examined the anti-inflammatory activity of *Matteuccia struthiopteris, Osmunda japonica, Matteuccia orientalis,* and *Pteridium aquilinum* from China. Young leaves of these fern species were tested for inhibiting the pro-inflammatory gene expression of interleukin-1 beta (IL1-) and interleukin-6 (IL6), as well as the gene expression of iNOS by LPS-elicited macrophages. Overall, the findings validated the anti-inflammatory potential of these Chinese fern species.

5.3.6 CARCINOGEN

The only higher plant known to cause cancer in animals is *Pteridium aquilinum*, or bracken fern. Its carcinogenic toxin, ptaquiloside, has been found in groundwater and cow's milk. In a mouse model and human gastric epithelial cells, the genotoxicity of *P. aquilinum* and ptaquiloside, including its effects on DNA damage and how cells respond to it, was studied by Gomes et al. (2012). Deregulation of the expression of a panel of genes linked to DNA damage signaling pathways and DNA repair in response to purified ptaquiloside revealed the role of ptaquiloside in *P. aquilinum's* DNA damage activities. These findings validate the carcinogenic characteristics of the fern and its toxin, ptaquiloside, and suggest their involvement in the promotion of stomach carcinogenesis.

In another study, Hirono et al. (1970) studied the carcinogenic ability of *P. aquilinum* in ACI rats. Three groups of rats were formed. Group I was given pellets containing raw bracken for four months; Group II was given pellets containing processed bracken that had been treated with boiling water for four months; and Group III was given bracken extract for 18 to 27 days. It was revealed that carcinogens in bracken can be partially eliminated by treating it with boiling water. In addition, exposure to a comparatively low amount of bracken carcinogen over an extended period increases the risk of intestinal and urinary bladder tumors in ACI rats.

5.3.7 ANTIOXIDANT ACTIVITY

Three polysaccharides from *P. revolutum* (Blume) Nakai named PRP0, PRP1, and PRP2 were tested for antioxidant efficacy by using 2,2-diphenyl-1-picryl-hydrazyl (DPPH) and hydroxyl radicals. According to the findings, PRP2 exhibited superior DPPH scavenging activity in comparison to PRP0 and PRP3, while PRP1 showed noteworthy hydroxyl radical scavenging capability in comparison to the other two (Wang et al., 2023). In another study, DPPH and ABTS (2,2'-azino-bis 3-ethylbenzothiazoline-6-sulfonic acid) radical scavenging activity of four fern species from China was assessed by Dion et al. (2015). In the results, all ferns tested (viz. *Matteuccia struthiopteris, Osmunda japonica, Matteuccia orientalis*, and *Pteridium aquilinum*) showed antioxidant properties, but *O. japonica* roots and *M. orientalis* fronds demonstrated the highest efficacy. Similar findings of DPPH and ABTS radical scavenging activities of leaves of *P. aquilinum* were reported by Kardong et al. (2013). In comparison to ascorbic acid, the results showed a significant amount of antioxidant efficacy of the plant.

5.3.8 ANTIMICROBIAL ACTIVITY

Studies on the plant's antimicrobial properties have also been conducted. *Pteridium aquilinum* leaf extracts in n-hexane and ethanolic solvents were tested for antibacterial activity against *Pseudomonas aeruginosa, Enterobacter aerogenes, Staphylococcus aureus, Escherichia coli*, and *Salmonella typhi.* It is evident from this research that *Pteridium aquilinum's* n-hexane extract was successful in inhibiting certain Gram-positive and Gram-negative bacteria at high concentrations (Awe and Amobi, 2015). Similar reports for the antibacterial potential of this plant were provided by Kardong et al. (2013). They have shown that ethanol and petroleum ether leaf extracts of *Pteridium aquilinum* possess antibacterial activity against *Bacillus subtilis, Streptococcus aureus, Proteus vulgaris*, and *Escherichia coli*, with zones of inhibition ranging from 16 to 20 mm.

5.4 CONCLUSION

In conclusion, it can be said that the plant has been studied for an array of pharmacological activities. There are a total of 57 phytoconstituents compiled in this chapter, reported by various authors, most of which belong to glycosides. Though less has been studied regarding its bioactivities, antitumor and antioxidant activities have been reported by a few. Reports on anti-inflammatory and anti-cancer activities suggest the potential role of *Pteridium aquilinum* in lifestyle disorders.

ACKNOWLEDGMENT

The authors are thankful to the heads of their respective departments. Thanks are also due to the principals of the respective colleges who provided necessary support.

KEYWORDS

- **bioactivities**
- **bovine serum albumin-glucose model**
- **ferns**
- **gram-negative bacteria**
- **phytoconstituents**
- **plant's antimicrobial properties**
- **pteridophyte**
- **taxonomy**

REFERENCES

Aranha, P. C. R., Hansen, H. C. B., Rasmussen, L. H., Strobel, B. W., & Friis, C. (2014). Determination of ptaquiloside and pterosin B derived from bracken (*Pteridium aquilinum*) in cattle plasma, urine and milk. *Journal of Chromatography B, 951*, 44–51. http://dx.doi.org/10.1016/j.jchromb.2014.01.022.

Awe, S., & Amobi, O. (2015). Anti-bacterial, phytochemical and proximate analysis of *Pteridium aquilinum. International Journal of Research, 2*(9), 1–7.

Bajracharya, G. B., & Bajracharya, B. (2022). A comprehensive review on Nepalese wild vegetable food ferns. *Heliyon, 8*, e11687.

Castillo, U. F., Ojika, M., Alonso-Amelot, M., & Sakagami, Y. (1998). Ptaquiloside Z, a new toxic unstable sesquiterpene glucoside from the neotropical bracken fern *Pteridium aquilinum* var. *caudatum*. *Bioorganic & Medicinal Chemistry, 6*, 2229–2233.

Castillo, U. F., Sakagami, Y., Alonso-Amelot, M., & Ojika, M. (1999). Pteridanoside, the first protoilludane sesquiterpene glucoside as a toxic component of the neotropical bracken fern *Pteridium aquilinum* var. *caudatum*. *Tetrahedron, 55*(42), 12295–12300. https://doi.org/10.1016/S0040-4020(99)00728-0.

Castillo, U. F., Wilkins, A. L., Lauren, D. R., Smith, B. L., Towers, N. R., Alonso-Amelot, M. E., & Jaimes-Espinoza, R. (1997). Isoptaquiloside and caudatoside, illudane-type sesquiterpene glucosides from *Pteridium aquilinum* var. *caudatum*. *Phytochemistry, 44*(5), 901–906.

Chen, C. Y., Chiu, F. Y., Lin, Y., Huang, W. J., Hsieh, P. S., & Hsu, F. L. (2015). Chemical constituents analysis and anti-diabetic activity validation of four fern species from Taiwan. *International Journal of Molecular Sciences, 16*(2), 2497–2516. https://doi.org/10.3390/ijms16022497.

Chen, N. D., Chen, N. F., Chen, C. W., & Zhang, L. (2013). A novel bihomoflavanonol with an unprecedented skeleton from *Pteridium aquilinum*. *Chinese Herbal Medicines, 5*(2), 96–100. https://doi.org/10.3969/j.issn.1674-6348.2013.02.002.

Choi, Y. H., Choi, C. W., Kim, J. K., Jeong, W., Park, G. H., & Hong, S. S. (2018). (−)-Pteroside N and pterosinone, new BACE1 and cholinesterase inhibitors from *Pteridium aquilinum*. *Phytochemistry Letters, 27*, 63–68. https://doi.org/10.1016/j.phytol.2018.06.021.

Dion, C., Haug, C., Guan, H., Ripoll, C., Spiteller, P., Coussaert, A., Boulet, E., Schmidt, D., Jianbing, W., Zhou, Y., & Lamottke, K. (2015). Evaluation of the anti-inflammatory and anti-oxidative potential of four fern species from China intended for use as food supplements. *Natural Product Communications, 10*(4), 1934578X1501000416.

Gomes, J., Magalhães, A., Michel, V., Amado, I. F., Aranha, P., Ovesen, R. G., Hansen, H. C., Gärtner, F., Reis, C. A., & Touati, E. (2012). *Pteridium aquilinum* and its ptaquiloside toxin induce DNA damage response in gastric epithelial cells, a link with gastric carcinogenesis. *Toxicological Sciences, 126*(1), 60–71. https://doi.org/10.1093/toxsci/kfr329.

Hirono, I., Shibuya, C., Fushimi, K., & Haga, M. (1970). Studies on carcinogenic properties of bracken, *Pteridium aquilinum*. *Journal of the National Cancer Institute, 45*(1), 179–188.

Imperato, F. (1997). Rhamnetin 3-O-laminaribioside from *Pteridium aquilinum*. *Phytochemistry, 45*(8), 1729–1730. https://doi.org/10.1016/S0031-9422(96)00801-1.

Imperato, F. (1998). Kaempferol-7-O-rhamnoside-4'-O-glucoside from *Pteridium aquilinum*. *Phytochemistry, 47*, 911–913.

Kardong, D., Upadhyaya, S., & Saikia, L. R. (2013). Screening of phytochemicals, antioxidant and anti-bacterial activity of crude extract of *Pteridium aquilinum* Kuhn. *Journal of Pharmacy Research, 6*(1), 179–182. https://doi.org/10.1016/j.jopr.2012.11.037.

Marimuthu, J., Fernández, H., Kumar, A., & Thangaiah, S. (2022). *Ferns: Biotechnology, propagation, medicinal uses and environmental regulation*. Springer, Singapore. https://doi.org/10.1007/978-981-16-6170-9.

Méndez, J. (2005). Dihydrocinnamic acids in *Pteridium aquilinum*. *Food Chemistry, 93*(2), 251–252. https://doi.org/10.1016/j.foodchem.2004.09.019.

Pathania, S., Kumar, P., Singh, S., Khatoon, S., Rawat, A. K. S., Punetha, N., Jensen, D. J., Lauren, D. R., & Somvanshi, R. (2012). Detection of ptaquiloside and quercetin in certain Indian ferns. *Current Science, 102*(12), 1683–1691.

Rizgar, H. M., Nur-E-Alam, M., Martina, L., Ifat, P., Tizzard, G. J., Coles, S. J., Fowler, M., Drake, A. F., Heyes, D., & Thoss, V. (2016). Isolation and characterization of 13 pterosins and pterosides from bracken (*Pteridium aquilinum* (L.) Kuhn) rhizome. *Phytochemistry, 128*, 82–94. https://doi.org/10.1016/j.phytochem.2016.05.001.

Sahayaraj, K. (2022). Ferns, a source of phytoecdysones, and their applications in pestiferous insect management. In J. Marimuthu, H. Fernández, A. Kumar, & S. Thangaiah (Eds.), *Ferns: Biotechnology, Propagation, Medicinal uses and Environmental Regulation* (pp. 181–198). Springer Nature, Singapore.

Saimaiti, M., & Genccelep, H. (2023). Thermal and molecular characteristics of novel water-soluble polysaccharide from the *Pteridium aquilinum* and *Dryopteris filix-mas*. *Food Chemistry Advances, 3*, 100401. https://doi.org/10.1016/j.focha.2023.100401.

Santos, C., Ferreirinha, P., Sousa, H., Ribeiro, J., Bastos, M. M., Neto, T., Oliveira, P. A., Medeiros, R., Vilanova, M., & Gil da Costa, R. M. (2016). Ptaquiloside from bracken (*Pteridium* spp.) inhibits tumor-infiltrating CD8+ T cells in HPV-16 transgenic mice. *Food and Chemical Toxicology, 97*, 277–285. https://doi.org/10.1016/j.fct.2016.09.019.

Selvaraj, P., De Britto, A. J., & Sahayaraj, K. (2005). Phytoecdysone of *Pteridium aquilinum* (L.) Kuhn (Dennstaedtiaceae) and its pesticidal property on two major pests. *Archives of Phytopathology and Plant Protection, 38*(2), 99–105. https://doi.org/10.1080/03235400400007517.

Song, G., Wang, K., Zhang, H., Sun, H., Wu, B., & Ju, X. (2017). Structural characterization and immunomodulatory activity of a novel polysaccharide from *Pteridium aquilinum*. *International Journal of Biological Macromolecules, 102*, 599–604. https://doi.org/10.1016/j.ijbiomac.2017.04.037.

Štefanić, E., Japundžić-Palenkić, B., Antunović, S., Gantner, V., & Zima, D. (2022). Botanical characteristics, toxicity and control of bracken fern (*Pteridium aquilinum* (L.) Kuhn). *Zbornik Veleučilišta u Rijeci, 10*, 467–478. https://doi.org/10.31784/zvr.10.1.26.

Thomson, J. A. (2000). Morphological and genomic diversity in the genus *Pteridium* (Dennstaedtiaceae). *Annals of Botany, 85*(Suppl. B), 77–99. https://doi.org/10.1006/anbo.1999.1101.

Tryon, R. M. (1941). A revision of the genus *Pteridium*. *Rhodora, 43*, 1–31, 37–67.

Wang, H., & Wu, S. (2013). Preparation and antioxidant activity of *Pteridium aquilinum*-derived oligosaccharide. *International Journal of Biological Macromolecules, 61*, 33–35. https://doi.org/10.1016/j.ijbiomac.2013.06.053.

Wang, K. W., Sheng, X. Y., Wu, B., Wang, H., Chen, J. B., & Wang, S. W. (2023). Structure characterization of novel heteropolysaccharides from *Pteridium revolutum* with anti-oxidant and anti-glycated activities. *Food Chemistry: X, 19*, 100826. https://doi.org/10.1016/j.fochx.2023.100826.

Zhou, M. Q., Wu, Q. Y., Han, Y. T., & Wang, K. W. (2016). Secondary metabolites of *Pteridium revolutum* and their immunosuppressive activity. *Chemistry of Natural Compounds, 52*, 1147–1150.

CHAPTER 6

Pharmacological Importance and Chemical Composition of *Tectaria coadunata* (J. Sm.) C. Chr.

LALREMKIMI HAUHULH,[1] STELLA C. ZOTHANKHUMI,[1] LALCHHANDAMI TOCHHAWNG,[2] F. LALSANGPUII,[3] and ZOTHAN SIAMA[1,*]

[1]*Department of Zoology, Mizoram University (A Central University), Aizawl, Mizoram, India*

[2]*Mizoram Science, Technology and Innovation Council, Aizawl, Mizoram, India*

[3]*Department of Botany, Mizoram University (A Central University), Aizawl, Mizoram, India*

*Corresponding author

ABSTRACT

Tectaria coadunata is a perennial evergreen edible fern species belonging to the family Tectariaceae. It is native to Eastern Tropical Africa, Tropical and Sub-tropical Asia. Different parts of *T. coadunata* are traditionally used to treat various ailments such as burns, diarrhea, malaria, stomach aches, gastrointestinal disorders, jaundice, urinary tract issues, and for the eradication of worms. The potential of *T. coadunata* as a medicinally relevant plant is due to its bioactive compounds, which include alkaloids, cardioglycosides, phenolics, saponins, tannins, terpenoids, steroids, quinones, flavonoids, and glycosides. Several studies have shown the medicinal characteristics of *T. coadunata*, revealing its antihyperlipidemic,

Bioactives and Pharmacology of Pteridophytes. T. Pullaiah (Ed.)

DOI: 10.1201/9781779642738-6

antimicrobial, antidiabetic, antidiarrheal, anthelmintic, analgesic, anti-inflammatory, and anticancer activities. Therefore, the main purpose of this review is to provide comprehensive information about the chemical composition and pharmacological potential of *T. coadunata*.

6.1 INTRODUCTION

Tectaria coadunata (J. Sm.) C. Chr. is a member of the fern family Tectariaceae. It grows in open areas, alongside roads, on wet, stony slopes, or in the cracks and fissures of rocks close to waterways (Bajracharya and Bajracharya, 2022). *T. coadunata* can be found in China, Madagascar, Bhutan, India, Laos, Malaysia, Myanmar, Nepal, Sri Lanka, Thailand, and Vietnam (Shrestha et al., 2019).

The rhizome of *T. coadunata* is erect, tapering to a point, lanceolate in form, and reaching up to 2 cm in diameter and 9 mm in length. The scales are dark brown with more white borders. The tuft-like, arching, sparsely herbaceous fronds can reach 1.8 meters in length, with proliferating buds on the rachis, costae, and costules on the top surface (Marahatta et al., 2019). The stipe is pale, glossy, and glabrescent, while the fronds are clustered. At the vein, a single row of sori is present on each side of the midrib. The indusia are globose, clypeate, brown, whole, membrane-bound, glabrous, or hairy (Wu et al., 2013). *T. coadunata* is commonly utilized as a wild vegetable owing to its excellent nutritional value and is known as an edible fern with high levels of dietary fiber. As a result, it is expected to be primarily comprised of water, proteins, carbohydrates, and an abundance of vitamins A and C. Young shoots bearing fiddleheads have low sodium (Na) content but high iron (Fe), magnesium (Mg), and potassium (K) content (Marahatta et al., 2019).

6.2 BIOACTIVE COMPOUNDS OF *TECTARIA COADUNATA*

Phytochemical fingerprinting and bioactivity analysis revealed the presence of flavonoids such as procyanidins, eriodoctyl glycosides, naringenin-7-O-glucuronide, eriodictyol-7-O-glucuronide, luteolin-7-O-glucuronide, and luteolin glycosides in the twigs and rhizomes of *T. coadunata* (Shrestha et al., 2019). Thomas et al. (2021) reported a wide range of secondary metabolites from various solvent extracts of the fronds and rhizome of *T. coadunata*, including alkaloids, cardioglycosides, phenolics, saponins, tannins, terpenoids, steroids, quinones, anthocyanins, flavonoids, and glycosides. It was also reported that, compared to petroleum ether extracts, ethyl acetate, methanolic, and aqueous extracts were found to have significantly higher contents of secondary metabolites. Additionally, Pawar et al. (2016) reported the presence of tannins, coumarins, flavonoids, steroids, phenols, and anthraquinones in the methanolic, petroleum ether, and aqueous extracts of the fronds of *T. coadunata.*

The constituents of the methanolic extract of rhizome of *T. coadunata* were determined using gas chromatography-mass spectroscopy (GC-MS) analysis (Dubal et al., 2013). The most prevalent compounds were found to be octadec-9-enoic acid (oleic acid), n-hexadecanoic acid (palmitic acid), octadecanoic acid (stearic acid), Di-n-octyl phthalate, hexadecanoic acid methyl ester, and hexadecanoic acid ethyl ester. Qualitative analysis of the rhizome extract also revealed the presence of tanin (0.67%), saponin (19.08%), proline (0.18g 100^{-1}g), phenol (21.0mg/100ml), alkaloids (17.77%), steroids (2.44%), and glycosides (0.60%) (Kale, 2014).

The following are some of the bioactive compounds reported from the rhizome of *T. coadunata* using GC-MS analysis (Dubal et al., 2013) (Figure 6.1).

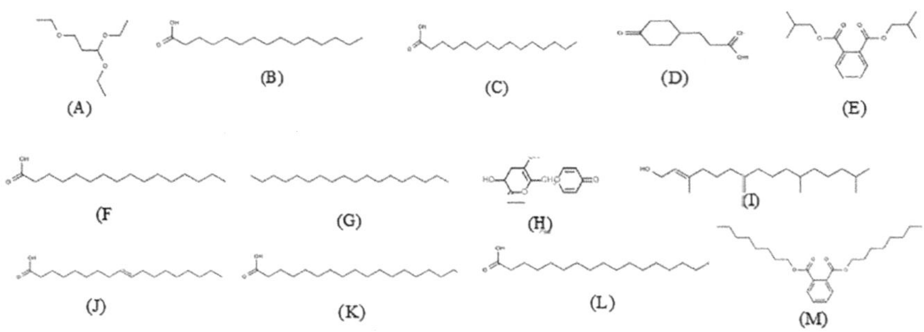

FIGURE 6.1 Chemical structures of some of the bioactive compounds isolated from the extracts of *T. coadunata* rhizome. (A) 1,1,3 triethoxy propane, (B) 4H-Pyran-4-one, 2,3-dihydro-3,5-dihydroxy-6-methyl, (C) Pentadecanoic acid, (D) 3-(4-Oxocyclohexyl)-propionic acid, (E) n-Hexadecanoic acid, (F) Pentadecanoic acid ethyl ester, (G) Octadecane, (H) 1,2-Benzenedicarboxylic acid, bis(2-methylpropyl) ester, (I) 3,7,11,15-Tetramethyl-2-hexadecen-1-ol, (J) Octadec-9-enoic acid, (K) Octadecanoic acid, (L) Heptadecanoic acid,15-methyl-ethyl ester, (M) Di-n-Octyl phthalate.

The total phenolic content (TPC) of methanolic extracts of the whole plant of *T. coadunata* was found to be 188.98 ± 3.31 mg per gram Gallic acid equivalent (mg/g GAE) (Nemkul, 2022). Similarly, the TPC of methanolic extracts of the fronds and rhizomes of *T. coadunata* was found to be 149.43 and 186.61 mg/g GAE, respectively, while that of the hexane extract of the fronds and rhizomes was calculated as 23.07 and 27.67 mg/g GAE, respectively. The total flavonoid contents (TFC) of methanolic and hexane extracts of the fronds and rhizomes were estimated as 119.62 and 143.72 mg per gram quercetin equivalent (mg/g QE) and 16.41 and 22.8 mg/g QE, respectively (Marahatta et al., 2019) (Table 6.1).

TABLE 6.1 List of Phytochemicals Isolated from Various Extracts of *T. coadunata*

Compound Nature	Compound	Solvent	Plant Part	References
Aromatic hydrocarbon	Benzene, (1-butylhexyl)-	Hexane	Rhizome	Nemkul (2022)
	Benzene, (1-propylheptyl)-	Hexane	Rhizome	Nemkul (2022)
	Benzene, (1-ethyloctyl)	Hexane	Rhizome	Nemkul (2022)
	Benzene, (1-methylnonyl)	Hexane	Rhizome	Nemkul (2022)
	Benzene, (1-pentylhexyl)-	Hexane	Rhizome	Nemkul (2022)
	Benzene, (1-butylheptyl)	Hexane	Rhizome	Nemkul (2022)
	Benzene, (1-propyloctyl)-	Hexane	Rhizome	Nemkul (2022)
	Benzene, (1-ethylnonyl)-	Hexane	Rhizome	Nemkul (2022)
	Benzene, (1-methyldecyl)	Hexane	Rhizome	Nemkul (2022)
	Benzene, (1-pentylheptyl)-	Hexane	Rhizome	Nemkul (2022)
	Benzene, (1-butyloctyl)	Hexane	Rhizome	Nemkul (2022)
	Benzene, (1-propylnonyl)-	Hexane	Rhizome	Nemkul (2022)
	Benzene, (1-ethyldecyl)-	Hexane	Rhizome	Nemkul (2022)
	Benzene, (1-methylundecyl)-	Hexane	Rhizome	Nemkul (2022)
	Benzene, (1-pentyloctyl)-	Hexane	Rhizome	Nemkul (2022)
	Benzene, (1-butylnonyl)-	Hexane	Rhizome	Nemkul (2022)
	Benzene, (1-propyldecyl)-	Hexane	Rhizome	Nemkul (2022)
	Benzene, (1-ethylundecyl)-	Hexane	Rhizome	Nemkul (2022)
	Benzene, (1-methyldodecyl)-	Hexane	Rhizome	Nemkul (2022)
Ester	1,1,3 triethoxy propane	Methanol	Rhizome	Dubal et al. (2013)
	Pentadecanoic acid, ethyl ester	Methanol	Rhizome	Dubal et al. (2013)
	3-(4-Oxocyclohexyl)-propionic acid, ethyl ester	Methanol	Rhizome	Dubal et al. (2013)
	N-Hexadecanoic acid, methyl ester	Methanol	Rhizome	Dubal et al. (2013)

TABLE 6.1 *(Continued)*

Compound Nature	Compound	Solvent	Plant Part	References
	1,2-Benzenedicarboxylic acid, disooctyl ester	Hexane	Rhizome	Nemkul (2022)
	N(alpha)-Benzyloxycarbonyl-L-leucine	Aqueous	Rhizome	Ram et al. (2023)
Procyanidin	A-type proanthocyanidin trimer with one unit of (epi)afzelechin Isomer 1	Ethyl acetate	Twigs & rhizome	Shrestha et al. (2019)
	A-type proanthocyanidin trimer with one unit of (epi)afzelechin Isomer 2	Ethyl acetate	Twigs & rhizome	Shrestha et al. (2019)
	B-type procyanidin dimer	Ethyl acetate	Twigs & rhizome	Shrestha et al. (2019)
	A-type procyanidin trimer	Ethyl acetate	Twigs & rhizome	Shrestha et al. (2019)
	A-type proanthocyanidin tetramer with one unit of (epi)afzelechin	Methanol	Twigs & rhizome	Shrestha et al. (2019)
	A-type procyanidin tetramer	Methanol	Twigs & rhizome	Shrestha et al. (2019)
	B-type proanthocyanidin decamer with two units of (epi)afzelechin	Ethyl acetate	Twigs & rhizome	Shrestha et al. (2019)
	A-type procyanidin octamer	Methanol	Twigs & rhizome	Shrestha et al. (2019)
	A-type procyanidin decamer with two A bonds	Methanol	Twigs & rhizome	Shrestha et al. (2019)
	B-type procyanidin pentamer	Ethyl acetate	Twigs & rhizome	Shrestha et al. (2019)
	A-type procyanidin pentamer	Methanol	Twigs & rhizome	Shrestha et al. (2019)
	B-type procyanidin decamer	Ethyl acetate	Twigs & rhizome	Shrestha et al. (2019)
	B-type procyanidin esamer	Methanol	Twigs & rhizome	Shrestha et al. (2019)
	B-type procyanidin heptamer	Methanol	Twigs & rhizome	Shrestha et al. (2019)
Fatty acid	Lauric anhydride	Methanol	Rhizome	Dubal et al. (2013)
	Pentadecanoic acid	Methanol	Rhizome	Dubal et al. (2013)
	Hexadecanoic acid, ethyl ester	Methanol	Rhizome	Dubal et al. (2013)
	Octadec-9-enoic acid	Methanol	Rhizome	Dubal et al. (2013)

TABLE 6.1 *(Continued)*

Compound Nature	Compound	Solvent	Plant Part	References
	Octadecanoic acid	Methanol	Rhizome	Dubal et al. (2013)
	Heptadecanoic acid,15-methyl-ethyl ester	Methanol	Rhizome	Dubal et al. (2013)
	n-Hexadecanoic acid	Methanol	Rhizome	Nemkul (2022)
	9, 12-Octadecadienoic acid, (Z,Z)-	Hexane	Rhizome	Nemkul (2022)
	Oleic acid	Hexane	Rhizome	Nemkul (2022)
Furan	2-furanmethanol	Methanol	Rhizome	Nemkul (2022)
	2-Furancarboxaldyhyde, 5- (hydroxymethyl)	Methanol	Rhizome	Nemkul (2022)
Phenol	Benzoic acid	Methanol	Rhizome	Ram et al. (2023)
	2,6-dihydroxybenzoicacid	Methanol	Rhizome	Ram et al. (2023)
	Resorcinol	Methanol	Rhizome	Ram et al. (2023)
	Gallic acid	Methanol	Rhizome	Ram et al. (2023)
	5,6-dihydroxy-3-(hydroxymethyl)-2-oxocyclohex-3-en-1-yl 2-hydroxy-6-methylbenzoate	Aqueous	Rhizome	Ram et al. (2023)
	Quinic acid	Aqueous	Rhizome	Ram et al. (2023)
	4-Hydroxyphenylpyruvic acid (HPPA)	Aqueous	Rhizome	Ram et al. (2023)
	2-caffeoylisocitrate	Methanol	Rhizome	Ram et al. (2023)
Terpenoid	Oryzalic acid B	Aqueous	Rhizome	Ram et al. (2023)
	Ponalactone A	Methanol	Rhizome	Ram et al. (2023)
	Nigakilactone B	Methanol	Rhizome	Ram et al. (2023)
	Lucidenic acid M	Methanol	Rhizome	Ram et al. (2023)
	Monomenthyl succinate	Aqueous	Rhizome	Ram et al. (2023)
	Cyclononasiloxane	Methanol	Rhizome	Dubal et al. (2013)
	3,7,11,15-Tetramethyl-2-hexadecen-1-ol	Methanol	Rhizome	Dubal et al. (2013)
	3,7,11,15-tetramethyl-2-hexadecen-1-ol	Hexane	Rhizome	Nemkul (2022)
Flavonoid	Naringenin-7-O-glucuronide	Ethyl acetate	Twigs & rhizome	Shrestha et al. (2019)
	Eriodictyol-7-O-glucuronide	Ethyl acetate	Twigs & rhizome	Shrestha et al. (2019)
	Luteolin-7-O-glucuronide	Ethyl acetate	Twigs & rhizome	Shrestha et al. (2019)

TABLE 6.1 *(Continued)*

Compound Nature	Compound	Solvent	Plant Part	References
	2,3-Dihydro-3,5-dihydroxy-6-methyl-4H-pyran-4-one	Methanol	Rhizome	Nemkul (2022)
Betaxanthin	Dopaxanthin	Methanol	Rhizome	Ram et al. (2023)
	Miraxanthin-I	Methanol	Rhizome	Ram et al. (2023)
Alkaloid	1,3-Dimethyluric acid	Aqueous	Rhizome	Ram et al. (2023)
Coumarin	Heliettin	Methanol	Rhizome	Ram et al. (2023)
Phenylpropanoid	Verimol A	Methanol	Rhizome	Ram et al. (2023)
Steroid	Triamcinolone diacetate	Methanol	Rhizome	Ram et al. (2023)
Methoxyflavanone	2-(4-hydroxyphenyl)-5,6,7,8-tetramethoxy-chroman-4-one	Methanol	Rhizome	Ram et al. (2023)
Pyrimidine nucleotide	Orotidne	Methanol	Rhizome	Ram et al. (2023)
Benzopyran	Citrinin	Methanol	Rhizome	Ram et al. (2023)
Lactone	2-Carboxy-5,7-dimethyl-4-octanolide	Aqueous	Rhizome	Ram et al. (2023)
Azaarene	Pteridine	Methanol	Rhizome	Ram et al. (2023)
Thiopurine	6-(Allylthio)purine	Methanol	Rhizome	Ram et al. (2023)
Hyadantoin derivative	Dantrolene	Methanol	Rhizome	Ram et al. (2023)
Fatty alcohol	4,6-heptadiyne-1,3-diol	Methanol	Rhizome	Ram et al. (2023)
Unclassified	Muramic acid	Methanol	Rhizome	Ram et al. (2023)
Unclassified	2,6-dihydroxyphenylacetate	Methanol	Rhizome	Ram et al. (2023)
Ketone	2-propanone, 1-3-dihydroxy	Methanol	Rhizome	Nemkul (2022)
Alkane	Octadecane	Methanol	Rhizome	Dubal et al. (2013)
Phthalate	1,2-Benzenedicarboxylic acid, bis(2-methylpropyl) ester	Methanol	Rhizome	Dubal et al. (2013)

6.3 PHARMACOLOGICAL IMPORTANCE OF *TECTARIA COADUNATA*

6.3.1 ANTI-HYPERLIPIDEMIC ACTIVITY

Rhizome extract of *T. coadunata*, at the dose of 540 mg/kg, has been shown to possess anti-hyperlipidemic activity in Wistar albino rats. The treatment with the rhizome extract resulted in a significant decrease in triglyceride and very low-density lipoprotein (VLDL) cholesterol levels when compared to untreated control groups. Furthermore, *T. coadunata*-treated rats were observed to show significantly lower total bilirubin, serum urea, serum glutamic pyruvic transaminase, and alkaline phosphatase when compared to untreated controls. However, no significant change was observed in cholesterol levels after treatment with the rhizome extract of *T. coadunata*. Based on the histopathological studies, the heart, liver, and kidneys exhibited recovery from the negative effects of the hyperlipidemic diet, demonstrating the antihyperlipidemic activity of *T. coadunata* rhizome in rats (Mori et al., 2016).

6.3.2 ANTI-DIABETIC ACTIVITY

The anti-diabetic activity of *T. coadunata* extracts was determined using the α-amylase inhibition assay (Marahatta et al., 2019). The computed IC50 values of the methanolic extract of fronds and rhizomes of *T. coadunata* against α-amylase activity were found to be 326.96 μg/ml and 164.76 μg/ml, respectively. These findings demonstrate a significant degree of enzyme inhibitory activity, or anti-diabetic activity, even though their inhibitory actions were found to be significantly lower than the standard acarbose, whose IC50 against α-amylase activity was 52.77 μg/ml.

6.3.3 ANTI-MICROBIAL ACTIVITY

In vitro anti-microbial tests using the agar well diffusion method revealed microbicidal activity of hexane and methanolic extracts of the whole plant of *T. coadunata* against gram-positive bacteria (*Staphylococcus aureus* and *Bacillus subtilis*) and gram-negative bacteria (*Escherichia coli, Salmonella typhi,* and *Pseudomonas aeruginosa*). The inhibition of bacterial growth by the extracts was evaluated by computing the zone of inhibition (ZOI) of the extracts with ampicillin and gentamicin as standards. The methanolic extract showed anti-microbial potential nearly equal to the standard drugs against *S. aureus*, *B. subtilis*, and *E. coli*. Furthermore, it was found that the methanolic extract elicited a higher ZOI of 16 ± 0 mm against *B. subtilis* than those of ampicillin and gentamicin, with ZOIs of 8.5 ± 0.5 mm and 15.5 ± 0.5 mm, respectively. On the other hand, the hexane extract was found to effectively inhibit the growth of *S. aureus* and *P. aeruginosa*, with ZOIs of 9 ± 0 mm and 15 ± 0 mm, respectively, with the latter showing microbicidal activity on par with gentamicin (ZOI of 14.66 ± 0.33 mm) (Nemkul, 2022).

A similar study was conducted using the fronds and rhizome extracts of *T. coadunata* by computing various metrics, including minimum bactericidal concentration (MBC), minimum inhibitory concentration (MIC), and ZOI. Both the fronds and rhizome extracts of *T. coadunata* showed potent antimicrobial activity against two gram-negative bacteria (*Klebsiella pneumoniae* and *E. coli*) and one gram-positive bacterium (*S. aureus*), as evidenced by the high values of ZOI. The inhibitory action of *T. coadunata* extracts against bacterial growth was found to be comparable with that of the standard drug ampicillin. Furthermore, no significant difference was observed between the fronds and rhizome extracts in their inhibitory activities against bacterial growth (Marahatta et al., 2019). Similar *in vitro* antimicrobial tests also revealed that methanolic and ethanolic extracts of *T. coadunata* fronds were found to be effective against *B. subtilis* and *E. coli*. The maximum and minimum ZOIs for both extracts were found to be 12 mm and 5 mm, respectively. The highest inhibition zone was found in the methanolic extract against *B. subtilis*, with a ZOI of 12 mm and 10 mm for *E. coli*. It was found that methanolic extracts were more effective in inhibiting the growth of the tested bacteria compared to the ethanolic extract (Saha et al., 2021).

Karki (2023) reported that various solvent extracts of *T. coadunata* possess antimicrobial activity against different bacterial species. Among the various solvent extracts, the methanolic extract of *T. coadunata* rhizome induced the highest inhibitory activity. Also, the anti-microbial activity of the methanolic extract of rhizome was found to be highest against *B. subtilis* when compared to *E. coli*, *S. typhi*, and *S. aureus*.

Antibacterial activity of methanolic and aqueous extracts of *T. coadunata* fronds against *Agrobacterium tumefaciens*, *E. coli*, *Salmonella arizonae*, *S. typhi*, and *S. aureus* was also reported using the disc diffusion method. The anti-bacterial activity of the extracts was determined by measuring the diameter of the ZOI expressed in millimeters (mm). The highest ZOI of 16 mm was recorded against *E. coli* and *S. aureus* in the aqueous extract. Both extracts were found to significantly inhibit the growth of the tested bacteria, which was comparable to the standard tetracycline (Parihar et al., 2010).

6.3.4 ANTI-CARCINOGENIC ACTIVITY

Shrestha et al. (2019) evaluated the cytotoxic activities of methanolic, ethyl acetate, and aqueous extracts of *T. coadunata* in BxPC3 (human pancreatic cancer) and 2008 (human ovarian cancer). Among the various solvent extracts, the ethyl acetate extract showed significantly higher toxicity against the two cancer cell lines with an IC50 of 12.5 μg/ml and 28.7 μg/ml, respectively. However, the methanolic and aqueous extracts showed moderate effects on these cell lines with an IC50 of >50 μg/ml. Furthermore, procyanidin isolated from *T. coadunata* has been reported to exhibit anti-cancer activity in BxPC-3 by arresting the cell cycle at the G1 phase through inhibition of MMP-2 expression, cyclin D1, E, A, and B1, and by increasing the levels of Cip1/p21 (Lee, 2017). Also, procyanidin isolated from *T. coadunata* has been demonstrated to exert cytotoxic activity through multiple mechanisms, including induction of apoptosis via DNA damage and caspase-3 mediation, and inhibition of pro-MMP-2 expression (Taparia and Khanna, 2016).

The anti-carcinogenic activity of the whole plant extract of *T. coadunata* was also studied using three distinct leukemic cell lines (KG1, MOLT-3, and K-562). The *in vitro* anti-proliferative activity of the plant extract occurred in a dose-dependent manner. The higher inhibitory activity was observed in KG1 with an IC50 of 42.85 ± 1.24 μg/ml (Reddy et al., 2020).

In silico studies using methanolic and aqueous extracts of *T. coadunata* rhizome revealed a few phytoconstituents that show potential inhibition of breast cancer cell growth. Upon sorting of phytoconstituents detected in the extracts by Lipinski's rule of five, 36 phytochemicals were obtained for molecular docking analysis. The phytochemicals were subjected to molecular docking to ascertain their binding affinities and mechanisms with selected target proteins (PGR, STK11, CHEK2, BRCA2, TPX2, 4ZTE, CDH1, and TP53) by using FlexX-Lead IT software. The target proteins were obtained through protein-protein interaction (PPI) network analysis and were identified as genes playing an essential role in the aetiology of breast cancer and other pathological processes, including endometrial cancer, bladder cancer, pancreatic cancer, thyroid cancer, and melanoma. Among all the docked compounds, it was found that dopaxanthin, dantrolene, and orotidine showed the highest binding affinities ranging between –18.45 kcal/mol to –34.21 kcal/mol with 3 target proteins (PGR, TPX2, and CDH1) that showed the maximum number of ligand interactions. The three ligands showed binding scores higher than those of their respective reference ligands with each target protein, that is, STR (Progesterone) for the PGR receptor, ACP (Phosphomethylphosphonic acid adenylate ester) for the TPX2 receptor, and PG4 (tetraethylene glycol) for the CDH1 receptor, with binding scores of –16.13 kcal/mol, –24.60 kcal/

mol, and –4.58 kcal/mol, respectively. Prediction of ADMET parameters also revealed that the top three docked compounds met the Lipinski rule of five and were found to be drug-like, with dopaxanthin emerging as the most potent isolated compound as it showed no toxicity with an LD50 of 750 mg/kg. According to molecular docking and pharmacokinetic analyses, all three of the top compounds have significant potential inhibitory activity against breast cancer proteins, especially the PGR receptor, due to their higher binding affinities. The docked data showed that the primary factors influencing the molecular interactions between ligands and optimized receptors were hydrophobic, electrostatic, and H-bond interactions (Ram et al., 2023).

6.3.5 ANTIOXIDANT ACTIVITY

The antioxidant activity of various extracts of the rhizome of *T. coadunata* was assessed using phosphomolybdenum or total antioxidant capacity, 2,2-diphenyl-1-picrylhydrazyl (DPPH) assay, 2,2'-azino-bis (3-ethylbenzothiazoline-6-sulphonic acid) (ABTS) assay, total reducing power, ferric-reducing antioxidant power (FRAP), and metal chelating assays (Shrestha et al., 2019). The DPPH radical scavenging activity of the methanolic extracts, aqueous extract, and ethyl acetate extract were 762.62 mg Trolox equivalent (TE)/g, 933.97 mg TE/g, and 948.59 mg TE/g, respectively. DPPH radical scavenging activity of various extracts of the fronds and rhizome were also measured by Marahatta et al. (2019). The methanolic and hexane extracts of fronds of *T. coadunata* showed a dose-dependent inhibition of DPPH radicals with an IC50 of 76.32 μg/mL and 143.49 μg/mL, respectively. Similarly, methanolic and hexane extracts of the rhizome of *T. coadunata* showed significant inhibition of DPPH radicals in a cell-free system with an IC50 of 50.81 μg/mL and 150.8 μg/mL, respectively. The inhibitory action of various solvent extracts of *T. coadunata* was comparable to that of the standard ascorbic acid.

T. coadunata varies in its ABTS radical scavenging activity from 1097.10 mg TE/g of the methanolic extract to 1661.21 mg TE/g of the ethyl acetate extract. In both the DPPH and ABTS assays, the ethyl acetate extract had the highest radical scavenging activity, which is consistent with its high polyphenol content (276.70 mg GAE/g), displaying in fact the highest quantity of procyanidin (PAC) dimers (11.13 mg/g), trimers (57.79 mg/g), tetramers, and polymers (29.08 mg/g) (Shrestha et al., 2019).

DPPH and H_2O_2 scavenging assays also revealed potent antioxidant activity of the methanolic extract of *T. coadunata*. The DPPH radical scavenging assay showed that the extract scavenged the DPPH by 28.94% at 100 μg/mL concentrations, and the IC50 was found to be 239.9 ± 6.9 μg/mL. On the other hand, the extract showed a higher percentage of H_2O_2 scavenging (43.42 ± 1.06%) than DPPH radical (Nemkul, 2022).

Electron-donating abilities of the different extracts were also evaluated with the help of reducing power assays, namely FRAP (from Fe^{3+} to Fe^{2+}), CUPRAC (from Cu^{2+} to Cu^{+}), and phosphomolybdenum (from Mo (VI) to Mo (V)) assays, which yielded results similar to the results of radical scavenging assays (ethyl acetate > aqueous > methanol). However, unlike the radical scavenging and reducing power assays, the ethyl acetate extract of *T. coadunata* was not active in the metal-chelating assay. The contrast in the results of the different assays may be attributed to the non-phenolic chelators such as peptides or polysaccharides (Shrestha et al., 2019). Moreover, it has been reported that metal-chelating ability is a minor antioxidant property of phenolics (Rice-Evans et al., 1996).

6.3.6 INHIBITORY EFFECT AGAINST DEGENERATIVE AND METABOLIC ENZYMATIC ACTIVITIES (CHOLINESTERASES, Α-AMYLASE, Α-GLUCOSIDASE, AND TYROSINASE)

Enzyme inhibitory property of various extracts of *T. coadunata* was tested against cholinesterase enzymes (acetylcholinesterase and butyrylcholinesterase), α-amylase, α-glucosidase, and tyrosinase (Shrestha et al., 2019). The inhibition of these enzymes is required for the management of chronic health conditions including Alzheimer's disease. Standard enzyme inhibitor compounds such as galantamine (GALAE, for cholinesterases), kojic acid (KAE, for tyrosinase), and acarbose (ACAE, for amylase and glucosidase) were used to express the inhibitory activity of the *T. coadunata* extracts against the concerned enzymes. It has been found that the ethanol extracts of *T. coadunata* had the highest inhibitory activity for acetylcholinesterase (AchE) (6.22 mg GALAE/g), butyrylcholinesterase (9.82 GALAE/g), and tyrosinase (153.89 mg GALAE/g), whereas the aqueous extracts showed the lowest activity against these enzymes. The methanolic and ethyl acetate extracts also showed significant inhibitory effects against AChE (6.22 mg GALAE/g for ethyl acetate extract, 5.58 mg GALAE/g for methanolic extract) (Shrestha et al., 2019), which is in line with the literature suggesting that compounds with higher phenolic contents have higher cholinesterase inhibitory properties (Orhan et al., 2007).

Amylase and glucosidase inhibitory activities were both very low for each of the studied extracts, with values of about 1 mmol acarbose equivalent (ACAE)/g for amylase inhibition and 5.5 mmol ACAE/g for glucosidase. On the other hand, except for the aqueous extract, every *T. coadunata* extract showed tyrosinase inhibitory action, ranging from 149.41 mg KAE/g for the methanolic extract to 153.89 mg KAE/g for the ethyl acetate extract (Shrestha et al., 2019).

KEYWORDS

- **anticarcinogenic**
- **antimicrobial**
- **antioxidant**
- **ethnomedicine**
- **ethyl acetate**
- **pharmacological activities**
- **tyrosinase inhibitory**

REFERENCES

Bajracharya, G. B., & Bajracharya, B. (2022). A comprehensive review on Nepalese wild vegetable food ferns. *Heliyon, 8*(11), 1–22.

Dubal, K. N., Ghorpade, P. N., & Kale, M. V. (2013). Studies on bioactive compounds of *Tectaria coadunata* (Wall. ex-Hook. & Grev.) C. Chr. *Asian Journal of Pharmaceutical and Clinical Research, 6*(2), 186–187.

Kale, M. V. (2014). *Ethnobotanical studies on some pteridophytes of South Western Ghats – A biodiversity hot spot* (Ph.D. thesis).

Karki, P. (2023). *Phytochemical analysis and antimicrobial analysis of Tectaria coadunata in Dharan, Eastern Nepal* (Doctoral dissertation, Department of Biology, Central Campus of Technology, Institute of Science and Technology, Tribhuvan University, Nepal).

Lee, Y. (2017). Cancer chemopreventive potential of procyanidin. *Toxicological Research, 33*, 273–282.

Marahatta, A. B., Poudel, B., & Basnyat, R. C. (2019). The phytochemical and nutritional analysis and biological activity of *Tectaria coadunata* Linn. *International Journal of Herbal Medicine, 7*, 42–50.

Mori, H., Nishteswar, K., Patel, B. R., & Nariya, M. (2016). Acute toxicity and anti-hyperlipidemic activity of rhizome of *Tectaria coadunata* (Kukkutnakhi): A folklore herb. *AYU, 37*(3, 4), 238–243.

Nemkul, C. M. (2022). *Ethnobotany and antimicrobial study of selected medicinal plants used by Magar community in Nawalpur District, Nepal* (Doctoral dissertation, Institute of Science and Technology).

Orhan, I., Kartal, M., Tosun, F., & Şener, B. (2007). Screening of various phenolic acids and flavonoid derivatives for their anticholinesterase potential. *Zeitschrift für Naturforschung C – Journal of Biosciences, 62*(11, 12), 829–832.

Parihar, P., Parihar, L., & Bohra, A. (2010). *In vitro* antibacterial activity of fronds (leaves) of some important pteridophytes. *Journal of Microbiology and Antimicrobials, 2*(2), 19–22.

Pawar, S. G., Kamble, S. Y., Patil, S. R., Sawant, P. S., & Singh, E. A. (2016). Preliminary phytochemical investigations of three species of traditional medicinal plants of tribal regions of Maharashtra (India). *International Journal of Pharmacognosy and Phytochemical Research, 8*(5), 742–745.

Ram, S., More-Adate, P., Tagalpallewar, A. A., Pawar, A. T., Nagar, S., & Baheti, A. M. (2023). An *in-silico* investigation and network pharmacology-based approach to explore the anti-breast-cancer potential of *Tectaria coadunata* (Wall.) C. Chr. *Journal of Biomolecular Structure and Dynamics, 42*(7), 1–12.

Reddy, M. N., Adnan, M., Alreshidi, M. M., Saeed, M., & Patel, M. (2020). Evaluation of anticancer, antibacterial and antioxidant properties of a medicinally treasured fern *Tectaria coadunata* with its phytoconstituents analysis by HR-LCMS. *Anticancer Agents in Medicinal Chemistry, 20*(15), 1845–1856.

Rice-Evans, C. A., Miller, N. J., & Paganga, G. (1996). Structure–antioxidant activity relationships of flavonoids and phenolic acids. *Free Radical Biology and Medicine, 20*(7), 933–956.

Saha, M., & Bhandari, J. B. (2021). Antibacterial activities of the plant extracts and biochemical characterization of associated microbes of *Angiopteris evecta* (G. Forst.) Hoffm. and *Tectaria coadunata* (J. Sm.) C. Chr. *Indian Fern Journal, 38*, 101–114.

Shrestha, S. S., Sut, S., Barbon Di Marco, S., Zengin, G., Gandin, V., De Franco, M., & Rajbhandary, S. (2019). Phytochemical fingerprinting and *in vitro* bioassays of the ethnomedicinal fern *Tectaria coadunata* (J. Smith) C. Christensen from Central Nepal. *Molecules, 24*(24), 1–20.

Taparia, S., & Khanna, A. (2016). Effect of procyanidin-rich extract from natural cocoa powder on cellular viability, cell cycle progression, and chemoresistance in human epithelial ovarian carcinoma cell lines. *Pharmacognosy Magazine, 12*(Suppl. 2), 109–115.

Thomas, A. P., Sindu, N., & Dileesh, S. (2021). Comparative phytochemical and antibacterial efficacy of *Tectaria coadunata* and *Tectaria wightii*. *International Journal of Research and Analytical Reviews, 8*(3), 152–158.

Wu, Z. Y., Raven, P., & Hong, D. Y. (2013). *Pteridophytes*. In *Flora of China* (1st ed., Vol. 2, 3, pp. 1–12). Science Press, Beijing, and Missouri Botanical Garden Press, St. Louis, MO, USA.

CHAPTER 7

Selaginella bryopteris (L.) Baker: A Review of Its Pharmacological, Phytochemical, and Ethnomedicinal Relevance

ALEX PHILIP ALEN, K. SAJITHA MENON, VEERANKUTTY SURESH, and LAKSHMANAN LEEJA*

Department of Botany, Government Victoria College, Palakkad, University of Calicut, Kerala, India

*Corresponding author

ABSTRACT

This review revolves around the phytochemical, pharmacological, and ethnomedicinal aspects of *Selaginella bryopteris*. The plant is globally known as the resurrection plant, attributing to its remarkable efficiency to sprout back after a long period of drought stress. It is utilized for ornamental, medicinal, and therapeutic applications throughout the world. Various combinations of the plant extract with various other plant extracts have also been used as folk medicine in many communities. Many traditional uses of the plant have also been scientifically verified through numerous pharmacological experiments. *S. bryopteris* has been the subject of numerous biochemical and pharmacological investigations, and a large number of its conventional uses have been verified by science. Antibacterial, growth-promoting, anti-protozoan, anti-stress, anticancer, anti-heat stroke, relieving burning when urinating, improving memory, relieving stomach ache, anti-hyperglycemic, and antidepressant are only a few of its biological qualities. *S. bryopteris* is one among the many medicinally

Bioactives and Pharmacology of Pteridophytes. T. Pullaiah (Ed.)

DOI: 10.1201/9781779642738-7

significant plants on the planet because of its enormous pharmacological and therapeutic potential.

7.1 INTRODUCTION

Selaginella P. Beauv. species differ greatly in terms of morphology, symmetry, and size. A few species are also annuals, but the majority are perennial herbaceous plants. However, among the various *Selaginella* species found worldwide, *Selaginella bryopteris* (L.) Baker occupies a noteworthy and distinct place. Often called the "resurrection plant," this species is remarkable not only for its ability to survive droughts but also for its rich phytochemical and pharmacological potential, as well as its many applications in ethnobotany. *Selaginella bryopteris* is globally distributed across Asia (Bhutan, India, Nepal, and Yemen) and Africa. The plant grows lushly and creates a lovely, velvety environment when it rains. Resurrection plants show a range of adaptations, including an increase in sugar content, such as sucrose, galactinol, and raffinose, which prevents the tissues from dying from osmotic stress and dehydration. These plants generally flourish in water-stressed conditions. They also include a range of alcohols and sugar acids that protect them from the harmful effects of cell death (Vashistha and Tejasvi, 2021). *Selaginella bryopteris* is a Pteridophyte often called the resurrection plant and one of the most well-known species in the global lycophyte family Selaginellaceae. The estimated number of species of plants under the family Selaginellaceae is 750 (Antony and Thomas, 2011). Spike moss, or *S. bryopteris*, is a type of plant that belongs to the Lycophytes or fern allies, often growing in rock crevices. Usually, they have simple, single-veined leaves, alternating, opposing, or whorled leaves, microphylls (small leaves), dichotomously branched stems, often dimorphic (two sizes), and scale-like ligules (early deciduous). The tiny, scale-like leaves of the creeping or ascending plants in the *Selaginella* genus are borne on branching stems that also generate roots. They are heterosporous (megaspores and microspores). Wiry rhizophores that emerge from stem forks carry the roots of the plant. In the axils of sporophylls, or fertile leaves, are sporangia. Spores in plants are heterosporous.

7.1.1 MYTHOLOGY BEHIND THE COMMON NAME 'SANJEEVANI'

It is believed that *Selaginella* is described in the Hindu mythological epic Ramayana by the poet Tulsidas as 'Sanjeevani Booti.' Because of its supposed ability to bring about the resurrection of life, *Sanjeevani* is sometimes referred to as a magical herb. *Sanjeevani* translates as something that provides life. Only three species were chosen from a preliminary list of 17 plant species that might be Sanjeevani, and *S. bryopteris* is currently regarded as Sanjeevani. According to Tejaswi and Ranjan (2018), *Sanjeevani* was a magical herb in Hindu mythology that could treat any kind of illness. The Ramayana has a description of the wonder herb *Sanjeevani* by the poet Tulsidas. Sushena, the Lankan Royal Physician, requested Hanuman to bring four plants from the Dronagiri Hills: *Mrithasanjeevani* (restorer of life), *Vishalyakarani* (remover of arrows), *Sandhanakarani* (restorer of the skin), and *Savarnyakarani* (restorer of skin color) after Lakshmana was struck by an arrow from Meghnath and became unconscious. When Hanuman could not make them out of the crowd, he repaired the whole hill, and Lakshmana, who was about to die, came back to life. The implication was that *Sanjeevani* could probably bring him back to life, or resurrect him, which would be the equivalent of raising him from the dead.

7.1.2 ETHNOMEDICINAL IMPORTANCE

It is a major ingredient in the regional medications prescribed in India to treat spermatorrhoea, venereal diseases, constipation, colitis, indigestion, and problems with the urine (diuretic). It is also used to treat comatose individuals and lower body temperatures in those who have fevers. The herb has long been used by Madhya Pradesh's Gond tribe members as a strength tonic. Dried powders of this herb are used by women of the tribes in the Bastar region of Chhattisgarh to treat gynecological problems such as irregular menstruation and leucorrhea, as well as to ease the agony of childbirth.

Indigenous people of Songhati, Sonbhadra, U.P., feed patients with cow's milk to revitalize them and treat beri-beri and dysentery using an oral herbal paste (Vashistha and Tejasvi, 2021). *Selaginella* species have been traditionally used to treat wounds and a range of illnesses, including tonsils, kidney stones, cancer, fever, pneumonia, bone fractures, skin conditions, paralysis, hepatitis, fever, blood coagulation, toothache, diarrhea, and gastrointestinal discomfort. They have also been used to neutralize the venom in snake and scorpion bites. Rupa and Bhawani (2014) reported that these uses have occurred historically.

Indian medicine has traditionally utilized *S. bryopteris* as a therapeutic herb. The plant's stem has demonstrated antibacterial qualities against *Neisseria gonorrhoeae*, and the leaf paste is used to treat spermatorrhea and leucorrhea, which both cause stomach aches and irritate children's urinary systems. Stomachaches can be eased by *S. bryopteris* ethanol extract. Inflammatory illnesses and venous issues are treated with plants. Amentoflavone and hinkoflavone from *S. bryopteris* show antiprotozoal activity against *Plasmodium falciparum, Leishmania donovani*, and *Trypanosoma* sp., claim Vasishtha and Tejasvi (2021).

The herb's widespread use by Indian tribal people as a food supplement for a number of medical disorders, according to Mishra et al. (2011), suggests that it has potential as a "panacea" for a wide range of illnesses.

7.2 PHYTOCHEMICAL CHARACTERISTICS OF *S. BRYOPTERIS*

Several studies have examined the bioactive components that confer therapeutic properties to various *Selaginella* species. It has been demonstrated that the high concentration of flavonoids in this herb serves a variety of biological purposes. The aqueous extract of *S. bryopteris* showed strong anti-cancer and chemopreventive effects at both the commencement and post-initiation stages of DMBA-induced mouse skin tumorigenesis, according to a medium-term anti-carcinogenesis bioassay. The combination of bioactive flavonoids in the benzene fraction was shown to be responsible for the effects that were seen. Flavonoids are well established to offer robust protection against a variety of potential diseases, including cancer. The fact that flavonoids provide a high level of protection against a number of possible diseases, including cancer, is fairly well documented. Because of their antioxidant qualities, flavonoids help the body build immunity and fight off dangerous free radical scavengers. They also possess anti-inflammatory properties (Vashistha and Tejasvi, 2021). Several significant secondary plant metabolites, including phenols, alkaloids, flavonoids, tannins, saponins, and terpenoids (triterpene, steroid), have been identified through chemical analysis of *S. bryopteris*, with bioflavonoids being the main secondary metabolite (Gautam et al., 2023b ; Pandey et al., 2010; Swamy et al., 2006).

A preliminary phytochemical examination of *S. bryopteris*'s aqueous and ethanolic extracts revealed the presence of reducing sugars, alkaloids, coumarin, anthocyanins, and flavonoids. The ethanolic extract lacked tannins and saponins, but the aqueous extract lacked glycosides (Prakash et al., 2023). Phytochemical investigation and pharmacological screening of *S. bryopteris* were carried out by Kumar et al. (2016), and various chemical constituents in the plant were identified through phytochemical analysis, which were evaluated for their pharmacological activities. The study found that *S. bryopteris* possesses significant antioxidant and antimicrobial properties, making it a potential source for the development of new drugs. The phytochemical analysis reported by Saxena and Verma (2022) from different extracts of *S. bryopteris* is shown in Figure 7.1.

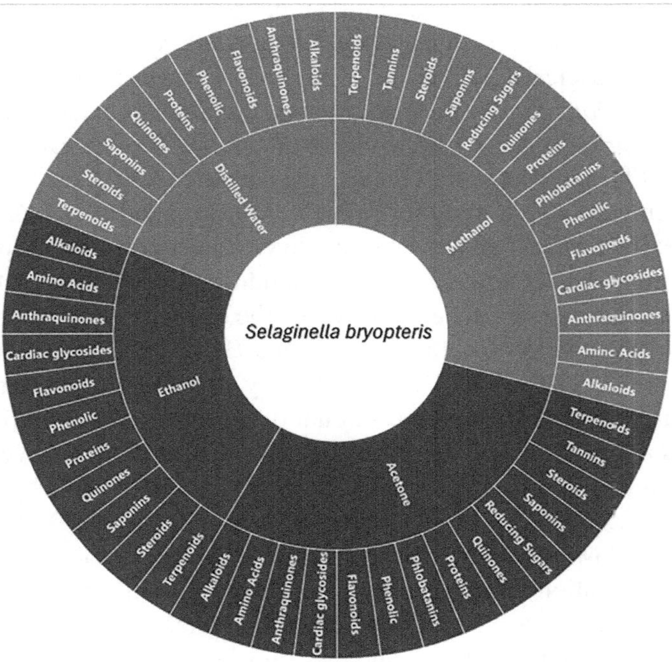

FIGURE 7.1 Graph showing phytochemicals of *Selaginella bryopteris* in Extracts on four different solvents.

Source: Drawn based on Data from Saxena and Verma (2022).

7.3 PHARMACOLOGICAL POTENTIAL OF *S. BRYOPTERIS*

In order to ascertain the phytochemical, analgesic, and anti-inflammatory properties of *S. bryopteris*, Paswan et al. (2021) conducted a study in which they found that the ethanol extract of the plant contains a variety of substances, including alkaloids, glycosides, tannins, saponins, flavonoids, carbohydrates, proteins and amino acids, steroids, vitamins, gums, and mucilages. The plant extract's active ingredients, gallic acid and rutin, were identified using HPTLC fingerprinting analysis. The hot plate method, the tail flick method, the carrageenan-induced paw edema model, and pain were used to assess the effectiveness of the plant extract against inflammation. Extract from *S. bryopteris* exhibited strong anti-inflammatory and analgesic effects. In light of this, the study suggested that *S. bryopteris* be used instead of non-steroidal anti-inflammatory medicines (NSAIDs) as a safer medication to treat pain and inflammation.

Positive results were found when Saxena and Verma (2022) investigated the antibacterial activity of *S. bryopteris* against various bacteria, including *E. coli*, *Staphylococcus aureus*, *Bacillus subtilis*, *Salmonella typhi*, *Aspergillus niger*, *Candida parapsilosis*, and *Trichophyton rubrum*. Using DPPH and H_2O_2 techniques, they also reported on the plant's antioxidant capacity. A study by Pal et al. (2022) found that *S. bryopteris* ethanol extract caused apoptosis in the liver cancer cell line (HepG2) via changes in nuclear condensation, mitochondrial membrane potential, and reactive oxygen species generation. When the extract was administered to rats with induced hepatocellular carcinoma (HCC), the increased levels of ALT, AST, GGT, and TBL considerably decreased. Additionally, in rats given NDEA, it lowered the amount of the stress marker (MDA) and increased antioxidant activity. By using qRT-PCR analysis on liver tissue, it was also possible to show that *S. bryopteris* inhibits HCC by upregulating the expression of pro-apoptotic genes (p53, caspase-3, caspase-9, Bax), while downregulating the expression of anti-apoptotic genes (Bcl-2). An examination of the extract's histology showed that it enhanced the liver's architecture. Imidazole, amentoflavone, gallic acid, catechin, palmitic acid, lupeol, L-fucitol, and myo-inositol were the main bioactive substances found by GC–MS analysis. In testing for acute toxicity, it was also shown to be safe. According to the research, *S. bryopteris* extract exhibits a strong antiproliferative effect and has the potential to be a safe, all-natural treatment option for hepatocellular cancer.

The possible wound healing activities of the flavonoid-enriched fraction of *S. bryopteris* extract were investigated by Gautam et al. (2023a) using streptozotocin-induced diabetic wounds in rats and human keratinocyte cell lines. The primary bioactive phytoconstituents identified by

GC–MS analysis of the ethyl acetate fraction's chemical profiling were amentoflavone, gallic acid, imidazole, palmitic acid, catechin, L-fucitol, lupeol, and myo-inositol. *S. bryopteris* has the ability to cause apoptosis in the HaCaT cell line, according to *in vitro* tests. *S. bryopteris* can improve epithelialization and decrease wound size, according to *in vivo* studies conducted on diabetic rats. Moreover, it raised the levels of antioxidant enzymes and lowered the expression of genes that promote apoptosis. According to the study, *S. bryopteris* may promote wound healing by activating the mitochondria-mediated apoptotic pathway, making it a viable treatment for diabetic wounds. Using the MTT test, Prakash et al. (2023) evaluated the hepatoprotective effect of *S. bryopteris* extracts in HepG2 cell line cultures treated with CCl_4 and APAP. Significant hepatoprotection was observed in the ethanol extract of *S. bryopteris* at 100 µg/mL, which was followed by the aqueous extract. *S. bryopteris*'s ability to scavenge free radicals has been linked to the presence of phenolic and flavonoid chemicals. According to the study's findings, *S. bryopteris* is a potential substitute hepatoprotective agent. The pharmacological potential of different compounds of *S. bryopteris* is shown in Figure 7.2.

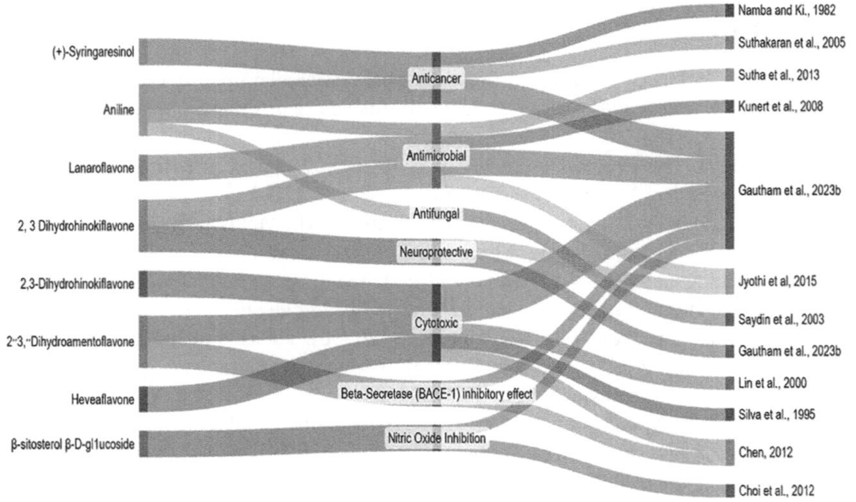

FIGURE 7.2 Flow Diagram showing pharmacological potential of compounds derived from *Selaginella bryopteris* with respective references.

7.4 CONCLUSION

Based on traditional knowledge and available scientific studies, it can be concluded that *Selaginella bryopteris* has a lot of latent potential that is yet to be discovered. Therefore, it is imperative to investigate the plant's numerous potential medicinal benefits and develop novel drugs that can be used to treat a variety of deadly illnesses.

KEYWORDS

- **hepatoprotective agent**
- **non-steroidal anti-inflammatory medicines**
- **pharmacology**
- **phytochemicals**
- **resurrection plant**
- **sanjeevani**
- **Selaginellaceae**

REFERENCES

Antony, R., & Thomas, R. (2011). A mini review on medicinal properties of the resurrection plant *Selaginella bryopteris* (L.) (Sanjeevani). *International Journal of Pharma and Life Sciences, 2*(7), 933–939.

Chen, L. (2012). The worldwide epidemiology of type 2 diabetes mellitus—Present and future perspectives. *Nature Reviews Endocrinology, 8*, 228–233.

Gautam, A., Kumar, V., Azmi, L., Rao, C. V., Khan, M. M., Mukhtar, B., Kamal, M., Arif, M., Mehdi, S., Alsanad, S. M., Al-Khamees, O. A., Jawaid, T., & Alam, A. (2023a). Wound healing activity of the flavonoid-enriched fraction of *Selaginella bryopteris* Linn. against streptozotocin-induced diabetes in rats. *Separations, 10*(3), 166–182.

Gautam, A., Pal, L. C., Rao, C. V., & Kumar, V. (2023b). The role of Indian magical herb *Selaginella bryopteris* L. (Selaginellaceae) in pharmacotherapeutic perspective: An overview. *Pharmacognosy Journal, 15*(1), 14–20.

Jyothi, C., Pratima, M., & Sundar, M. (2015). Phytochemical and antidepressant activities of *Selaginella bryopteris* (L.) Baker on albino mice. *International Journal of Applied Biology and Pharmaceutical Technology, 6*(4), 14–20.

Kumar, A., Kumar, A., & Singh, V. (2016). Phytochemical investigation and pharmacological screening of *Selaginella bryopteris* (L.) Baker. *Journal of Applied Pharmaceutical Science, 6*(11), 1–6.

Kunert, O., Swamy, R. C., & Kaiser, M. (2008). Anti-plasmodial and leishmanicidal activity of bioflavonoids from Indian *Selaginella bryopteris*. *Phytochemistry Letters, 1*(4), 171–174.

Mishra, P. K., Raghuram, G. V., Bhargava, A., Ahirwar, A., Samarth, R , Upadhyay, R., Jainand, S. K., & Pathak, N. (2011). *In vitro* and *in vivo* evaluation of the anti-carcinogenic and cancer chemopreventive potential of a flavonoid-rich fraction from a traditional Indian herb *Selaginella bryopteris* L. *British Journal of Nutrition, 106*, 1154–1168.

Pal, L. C., Gautam, A., Pande, V., & Rao, C. V. (2022). Anticancer property of *Selaginella bryopteris* (L.) Bak. against hepatocellular carcinoma *in vitro* and *in vivo*. *Phytomedicine Plus, 2*(1), 100201.

Pandey, S., Shukla, A., Pandey, S., & Pandey, A. (2017). An overview of resurrecting herb 'Sanjeevani' (*Selaginella bryopteris* L.) and its pharmacological and ethnomedicinal uses. *The Pharma Innovation Journal, 6*(2), 11–14.

Pandey, V., Ranjan, F., Deeba, F., Pandey, A. K., Singh, R., Shirke, P. A., & Pathre, U. V. (2010). Desiccation-induced physiological and biochemical changes in resurrection plant, *Selaginella bryopteris* (L.). *Journal of Plant Physiology, 167*(16), 1–9.

Paswan, S. K., Verma, P., Azmi, L., Srivastava, S., & Venkateswara, R. C. (2021). Phytochemical investigation and HPTLC fingerprinting analysis of *Selaginella bryopteris* ethanolic plant extract for analgesic and anti-inflammatory activities in animals. *Journal of Biologically Active Products from Nature, 11*(4), 395–405.

Prakash, D., Bisht, D., Yadav, D., Shakya, A. K., Shrivastava, S., Shukla, S., & Kumar, R. (2023). *In vitro* investigation of phytochemicals, antioxidant and hepatoprotective activity of *Selaginella bryopteris*. *Arabian Journal of Medicinal and Aromatic Plants, 9*(1), 23–45.

Rupa, P., & Bhavani, N. L. (2014). Preliminary phytochemical screening of desiccated fronds of *Selaginella bryopteris* (L.). *World Journal of Pharmaceutical Research, 3*(9), 1370–1378.

Saxena, P. A., & Verma, A. (2022). Phytochemical analysis and pharmacological actions of *Selaginella bryopteris* and *Opuntia dillenii*. *International Journal of Novel Research and Development, 7*(3), 899–909.

Saydin, A., Beis, K. D., & Can, D. (2003). Analgesic and antispasmodic activities of 2-(2-nitrophenyl)-1H-benzimidazole-5-carboxylic acid: Evidence for the importance of the 2-(o-substituted phenyl) group. *Pharmazie, 58*(6), 405–408.

Sutha, S., Shobana, S., Dharmaraja, J., & Selvaraj, S. (2013). Mixed ligand complexation of some transition metal ions in solution and solid state: Spectral characterization, antimicrobial, antioxidant, DNA cleavage activities and molecular modelling. *Spectrochimica Acta Part A: Molecular and Biomolecular Spectroscopy, 107*, 117–132.

Suthakaran, R., Nagarajan, G., Balasubramaniam, V., & Suganthi, K. (2005). Synthesis and anti-tumor activity of 4-aryl/alkyl-2-D aldosugar, disubstituted bis-benzimidazoles. *Indian Journal of Heterocyclic Chemistry, 15*, 173–176.

Swamy, R. C., Kunert, O., & Wolf, G. (2006). Structurally unique bioflavonoids from *Selaginella chrysocaulos* and *Selaginella bryopteris*. *Chemistry & Biodiversity, 3*(4), 404–413.

Tejaswi, J. K. D., & Rajan, R. G. (2018). A review on Indian magical herb Sanjeevani. *World Journal of Advanced Healthcare Research, 2*(5), 118–122.

Vashistha, N., & Tejasvi, A. (2021). *Selaginella bryopteris* (L.): A wonder herb. *Plant Archives, 21*(1), 701–706.

CHAPTER 8

A Review of the Biochemical Diversity and Pharmacology of the Genus *Selaginella* P. Beauv

SHIVANI V. KHORATE,[1] MANOJ L. SANKPAL,[1] and SACHIN M. PATIL[2,*]

[1]*Department of Botany, Shivaji University, Kolhapur, Maharashtra, India*

[2]*Department of Botany, Dr. Ghali College, Gadhinglaj (affiliated to Shivaji University, Kolhapur), Maharashtra, India*

*Corresponding authors

ABSTRACT

The genus *Selaginella* belongs to the family Selaginellaceae, and several species have been traditionally used for medicinal purposes worldwide. The phytochemistry of some has been investigated. More than 130 compounds have been isolated from different species of *Selaginella*, which are classified into appropriate chemical groups. The isolated compounds have been identified as belonging to the classes of alkaloids, benzenoids, carbohydrates, chromones, coumarins, flavonoids, lignans, oxygen heterocycles, phenylpropanoids, pigments, quinoids, and steroids. The pharmacological activities of the species of the genus *Selaginella* are reviewed.

8.1 INTRODUCTION

Selaginella P. Beauv. is commonly known as spike moss and is found in different habitats (i.e., temperate, tropical, arctic, and desert) around the world. More than 700 species have been described worldwide, and many of them are used as sources of wild food and medicines (PPG-I,

Bioactives and Pharmacology of Pteridophytes. T. Pullaiah (Ed.)

DOI: 10.1201/9781779642738-8

2016; Patil et al., 2020). Many species are adapted to xeric regions by their surface composition (the presence of waxes) and the ability of shoots to roll themselves during drought conditions. Members of the *Selaginellaceae* are mostly herbaceous, erect or creeping, either terricolous or saxicolous, annuals or perennials, reaching up to 30 cm in height. Roots are adventitious, well-developed, and dichotomously branched. The stem is erect or creeping, green, delicate, and covered by single-veined scaly leaves. At the base, leafless stems are developed; after reaching the soil, they produce adventitious roots known as rhizophores. Leaves are small, green, single-veined, and spirally arranged in two or four rows. At the base, a thin membranous structure known as a ligule is present, which is of two types: photosynthetic (microphyllous) and reproductive (sporophylls). Photosynthetic leaves are loosely arranged all over the stem, whereas sporophyll leaves are compactly arranged, producing a conelike structure at the tip known as a strobilus. Sporophyllous leaves are of two types: microsporophylls (which produce small spores) and megasporophylls (which produce large spores). In some species, microsporophylls are present towards the apex, and megaspores are present towards the base or on one side of the strobilus (Fraser-Jenkins, 2017).

Different species of *Selaginella* are used in traditional medicine in several countries to treat a variety of diseases or health problems, such as cancer, cardiovascular problems, diabetes, gastritis, hepatitis, skin diseases, and urinary tract infections. Extracts from some *Selaginella* spp. have shown activity as antinociceptive, anti-inflammatory, antimutagenic, antispasmodic, cytotoxic, immunostimulant, and RNA reverse transcriptase inhibitory agents. The genus *Selaginella* is a rich source of steroids, biflavonoids, alkaloids, secolignans, neo-lignans, and caffeoyl derivatives. Other compounds, such as alkaloidal glycosides, phenylpropanoids, and lignans, have also been reported in some *Selaginella* species. However, only a few studies have been performed on the bioactive components of this genus.

8.2 BIOCHEMICAL DIVERSITY IN THE GENUS *SELAGINELLA*

The various compounds isolated from different species of *Selaginella* have been identified and classified as alkaloids, benzenoids, carbohydrates, chromones, coumarins, flavonoids, lignans, oxygen heterocycles, phenylpropanoids, pigments, quinoids, and steroids (Adnan et al., 2021; Almeida et al., 2013; Chao et al., 1987; Shim et al., 2018; Weng and Noel, 2013).

8.2.1 ALKALOIDS

Many alkaloids were reported from the few species of *Selaginella*, which are cited in Table 8.1.

TABLE 8.1 Different Types of Alkaloids Present in the Genus *Selaginella*

Name of Alkaloid	*Selaginella* Species in Which Found	Sources
Hordenine	*Selaginella doederleinii* and *S. repanda*	Adnan et al. (2021); Almeida et al. (2013); Chao et al. (1987); Chen et al. (1995); Wang et al. (2009); Weng and Noel (2013); Zheng et al. (2004a, b)
Hordenine-[6-O-(4-hydroxy-cinnamoyl)-β-D-glucosyl]-(1,3)-α-L-rhamnoside		
Hordenine-O-α-L-rhamnopyranoside		
N-methyltyramine-O-α-L-rhamnoside		
Selaginellic acid	*Selaginella moellendorfii* and *S. repanda*	
5- Hydroxyselaginellic acid	*Selaginella moellendorfii*	
5- Hydroxy-N8,N8-dimethylpseudophrynaminol		
N- Selaginelloyl-L-phenylalanine		
Neoselaginellic acid		
N- -(5-Hydroxyneoselaginelloyl)-L-phenylalanine		
Adenosine	*Selaginella tamariscina*	
Guanosine		

8.2.2 BENZENOIDS

Benzenoids are also reported from a few species of *Selaginella* are mentioned in Table 8.2.

TABLE 8.2 Different Types Of Benzenoids Are Reported from *Selaginella* Species

Name of Benzenoids	Selaginella Species in Which Found	Sources
4-Hydroxy-benzoic acid	*S. pulvinata*	Adnan et al., 2021; Almeida et al., 2013; Chao et al., 1987; Chen et al., 1995; Wang et al., 2009; Weng and Noel, 2013; Zheng et al., 2004a, b
Arbutin	*Selaginella tamariscina*	
Vanillic acid		
Syringic acid		

8.2.3 CARBOHYDRATES

The discovered carbohydrates from different species of *Selaginella* are listed in Table 8.3.

TABLE 8.3 Different Therapeutic Carbohydrates Extracted from *Selaginella* Species

Name of Carbohydrates	*Selaginella* Species in Which Found	Sources
Selaginose	*S. adunca, S. asperula, S. epirrhizos, S. galeotti, S. geniculata, S. kraussiana, S. marginata, S. parkeri, S. plumosa, S. sanguinolenta, S. stellata, S. sulcata*	Adnan et al., 2021; Almeida et al., 2013; Fischer and Kandler, 1975; Moore et al., 1993; Weng and Noel, 2013; Zheng et al., 2001
2-Carboxy-arabinitol	*S. mertensii*	
Mycose	*S. pulvinata*	

8.2.4 CHROMONE

From a single species, *Selaginella uncinata* from which three types of chromones are reported (Table 8.4).

TABLE 8.4 Different Types of Chromones Extracted from *Selaginella* Species

Name of Chromones	*Selaginella* Species in Which Found	Sources
8-Methyl-eugenitol	*S. uncinata*	Adnan et al., 2021; Almeida et al., 2013; Ma et al., 2001, 2002, 2003; Shim et al., 2018; Weng and Noel, 2013
Uncinoside A		
Uncinoside B		

8.2.5 LIGNANS

The common lignans found in *Selaginella* species are given in Table 8.5.

TABLE 8.5 Different Types of Lignans Extracted from *Selaginella* Species

Name of Lignans	*Selaginella* Species in Which Found	Sources
5-Acethyl-dihydro-2-(3,'5'-dimethoxy-4'-hydroxy-phenyl)-7-methoxybenzofuran	*S. tamariscina*	Adnan et al., 2021; Almeida et al., 2013; Feng et al., 2011; Shim et al., 2018; Weng and Noel, 2013
Syringaresinol		

TABLE 8.5 *(Continued)*

Name of Lignans	*Selaginella* Species in Which Found	Sources
Tamariscinoside B		
Tamariscinoside C		
(–)-Lirioresinol A	*S. doederleinii*	
(–)-Lirioresinol B		
(+)-Matairesinol		
(–)-Nortracheloside		
(–)-Nortracheloside	*S. moellendorfii*	
Matairesinol		
Moellenoside B		
Pinoresinol diglucoside	*S. sinensis*	
Sinesiol A		

8.2.6 FLAVONOIDS

In the genus *Selaginella,* different species possess a variety of flavonoids are given in Table 8.6.

TABLE 8.6 Different Types of Flavonoids Extracted from *Selaginella* Species

Name of Flavonoids	*Selaginella* Species in Which Found	Sources
Amentoflavone	*S. braunii, S. davidii, S. delicatula, S. denticulata, S. kraussiana, S. moellendorfii, S. pulvinata, S. rupestris, S. sanguinolenta, S. selaginoides, S. sinensis, S. stauntoniana, S. tamariscina, S. uncinata, S. willdenowii*	Adnan et al., 2021; Cao et al., 2010a, b; Chanravarthy et al., 1981; Huneck and Khaidav, 1985; Kunert et al., 2008; Lee et al., 1992; Lin et al., 2000; Lopez-Saez et al., 1994a, b; Ma et al., 2001; Qasim et al., 1985; Shim et al., 2018; Sun et al., 1997
2,3-Dihydroamentoflavone	*S. bryopteris*	
Lanaroflavone		
Tetrahydro-amentoflavone		
Amentoflavone-7,4,7,4-tetramethylether	*S. moellendorfii*	
Apigenin-7-O-β-neohesperidoside		
Kayaflavone		
Podocarpusflavone A		
Chrysoeriol		

TABLE 8.6 *(Continued)*

Name of Flavonoids	*Selaginella* Species in Which Found	Sources
4,'7"-Di-O-methyl-amentoflavone	*S. sinensis, S. willdenowii*	
Apigenin	*S. doederleinii*	
6-(2-Hydroxy-5-acetylphenyl)-apigenin	*S. tamariscina*	
2,'8"-Biapigenin		
6-(5-Carboxyl-2- methoxyphenyl)-apigenin	*S. uncinata*	
2,"3"-Dihydro-3,'3"'-biapigenin	*S. labordei*	
Bilobetin	*S. willdenowii*	
Chamaecyparin	*S. difusa, S. jungermannioides, S. stellata*	
Cryptomerin B	*S. denticulata, S. tamariscina*	
Delicaflavone	*S. delicatula*	
Genistin	*S. sinensis*	
Ginkgetin	*S. moellendorfii, S. sinensis*	
Heveaflavone	*S. doederleinii, S. lepidophylla*	
Robustaflavone	*S. delicatula, S. denticulata, S. lepidophylla, S. selaginoides, S. sinensis, S. willdenowii*	

8.2.7 POLYPHENOLS

These are widely distributed in plants and have gained much attention due to their antimutagenic, antitumor, antioxidant, and free radical scavenging abilities, which potentially have beneficial implications for human health. Selaginellin is a group of polyphenols with a chemical scaffold only found to date in the *Selaginella* genus. Selaginellin, the first member of this compound class, was identified in *S. sinensis* and was isolated as a racemic mixture, containing a p-quinone methide unit and an alkynylphenol moiety (Zhang et al., 2007). At present, 14 compounds structurally related to selaginellin have been identified from different *Selaginella* species (Weng and Noel, 2013). Pharmacological studies demonstrate that selaginellin confers protective effects on differentiated neuronal cells cultured *in vitro* under different apoptotic conditions, making selaginellins interesting targets for exploring new neuroprotective agents (Weng and Noel, 2013). The different selaginellins reported from *Selaginella* species are given in Table 8.7.

TABLE 8.7 Different Types of Polyphenols Extracted from *Selaginella* Species

Name of Polyphenols	*Selaginella* Species in Which Found	Sources
Selaginellin	*S. sinensis*	Adnan et al., 2021; Cao et al., 2010a, b; Cheng et al., 2008; Shim et al., 2018; Tan et al., 2009, Xu et al., 2011a, b, c; Zheng et al., 2011
Selaginellins A	*S. tamariscina*	
Selaginellins B		
Selaginellins C	*S. pulvinata*	
Selaginellins D		
Selaginellins E		
Selaginellins F		
Selaginellins G		
Selaginellins H		

8.2.8 TERPENOIDS

Two monoterpenes (two isoprene units – 10 carbons) and five sesquiterpenes (three isoprene units – 15 carbons) have been found in *Selaginella* (Wang et al., 2011; Li et al., 2012) and other plant lineages. Monoterpenes and sesquiterpenes constitute an important class of volatile, semi-volatile, and non-volatile hydrocarbons and biosynthetically elaborated compounds produced by plants. Different species of *Selaginella* synthesize various types of terpenoids (Table 8.8).

TABLE 8.8 Different Types of Terpenoids Extracted from *Selaginella* Species

Name of Terpenoids	*Selaginella* Species in Which Found	Sources
Linalool	*S. moellendorffii, S. tamariscina, S. delicatula*	Adnan et al., 2021; Hui et al., 2005; Li et al., 2012; Shim et al., 2018; Tan et al., 2004; Wang et al., 2011; Yang et al., 2010.
2,7-dimethyl-8-hydroxyocta-4,6-dienoic acid 8-O-β- D-glucopyranoside		
Cedrol		
(+)-Germacrene D		
(−)-β-elemene		
β-sitosterol		
3β,16(-dihydroxy-5,17β-cholestan-21-carboxylic acid	*S. tamariscina*	
3β-(3-hydroxybutyroxy)-16α-hydroxy-5α, 17β-cholestan-21-carboxylic acid	*S. tamariscina*	
Glycyrrhetinic acid	*S. delicatula*	
Pulvinatadione	*S. pulvinata*	
pentacyclic friedelin	*S. delicatula*	

8.2.9 STEROIDS

Many steroids have been isolated effectively from the plants, among which, β-sitosterol is the common one that is, omnipresence. In *Selaginella*, different steroids reported from different species are listed in Table 8.9.

TABLE 8.9 Different Types of Terpenoids Extracted from *Selaginella* Species

Name of Terpenoids	*Selaginella* Species in Which Found	Sources
β-sitosterol	*S. moellendorffii, S. bryopteris S. lepidophylla and S. doederleinii*	Adnan et al., 2021; Hui et al., 2005; Li et al., 2012; Tan et al., 2004; Wang et al., 2011; Yang et al., 2010
2-dehydrocampesterol	*S. doederleinii*	
24α-ethyl-cholest-5-en-3β-ol	*S. doederleinii*	
24α-methyl-cholest-5-en-3β-ol	*S. doederleinii*	
24β-methyl-cholest-5-en-3β-ol,	*S. doederleinii*	
24α-ethyl-cholesta-5,22-dien-3β-ol	*S. doederleinii*	
3β-16α-dihydroxy-(5α)-cholestan-21-oic acid	*S. pulvinata*	

8.3 PHARMACOLOGY OF *SELAGINELLA* SPECIES

Selaginella species have various bioactive molecules such as flavonoids, biflavonoids, triterpenes, steroids, saponins, tannins and other secondary metabolites. Hence, it attracted the special attention of researchers worldwide to study the antibacterial, antifungal, anticancer, and anti-inflammatory activities. The present review aims to summarize the literature on the pharmacological ability of plant extracts of *Selaginella* spp.

8.3.1 ANTIBACTERIAL ACTIVITY

The crude ethanol and aqueous extract of *Selaginella bisulcata* have antibacterial activity. Both the crude extracts showed bacterial lysis against *Bacillus subtilis* AR-2 (Gr +ve) and *Escherichia coli* XL1-Blue (Ganguly, 2019). The antibacterial property of *S. bisulcata* might be attributed to the presence of a number of phenolic compounds. Verma et al. (2015) studied the antibacterial activity from crude extracts of *S. bryopteris* in acetone and n-hexane fractions, and isolated compounds were screened against 14 different bacteria. No antimicrobial activity was observed in the n-hexane extract; however, the acetone extract of *Selaginella bryopteris* showed activity against all the microbes tested in this study. The maximum effect of the acetone extract was observed against *K. pneumoniae*, *S. aureus*, and *C. krusei*. The antibacterial activity of *S. bryopteris* leaf extract on *Staphylococcus aureus* and *Bacillus* sp. (isolated from mastitis-affected cow milk) was studied by Singh et al. (2018). *S. bryopteris* leaf methanolic extracts (50 mg) showed promising antibacterial activity against *S. aureus* and *Bacillus* sp. However, the maximum antibacterial activity was found against *S. aureus* than the *Bacillus* sp. The antibacterial activity of *Selaginella convoluta* from different fractions of hexane, chloroform, and ethyl acetate was studied by Macedo et al. (2018). The fractions were tested against nine different bacterial strains, and they observed that the ethyl acetate fraction has a low MIC value for all the strains. Among these, very low MIC values were observed for *Enterococcus faecalis*, *Escherichia coli*, and *Serratia marcescens*. Hence, it could be considered a promising antibacterial agent for these bacteria. The crude extract of *S. repanda* was obtained through the microdilution method. It was used to study the antagonistic potential against pathogenic bacterial strains *S. flexneri, S. aureus*, *E. coli*, and *P. aeruginosa* by Adnan et al. (2021). Selaginellin derivatives were isolated from the whole plant crude ethyl acetate fractions of *Selaginella pulvinata* and exhibited significant antibacterial activity against *Staphylococcus aureus*. The study revealed that the crude extract of *S. repanda* showed significant antagonistic activity against all four tested bacterial strains. A novel compound, isocryptomerin, was isolated from *Selaginella tamariscina.* It has antibacterial and synergistic activities against Gram-positive and Gram-negative bacterial strains, including clinical isolates of antibiotic-resistant species such as methicillin-resistant *Staphylococcus aureus* (MRSA). Wang et al. (2018) studied the chemical composition of the volatile oils of three species of *Selaginella*, viz. *S. moellendorfii, S. helvetica*, and *S. delicatula*. The main constituents of the essential oils were phytoene, 2-pentyl furan, and copaene. Their antibacterial activity was evaluated against three Gram-positive bacteria (*Staphylococcus aureus*, *Diplococcus pneumoniae*, and *Corynebacterium pyogenes*) and three Gram-negative bacteria (*Escherichia coli*, *Pseudomonas aeruginosa*, and *Proteus bacillus* vulgaris) by the agar disc diffusion method. The results of the antibacterial test indicated that the volatile oils of three species of *Selaginella* showed good antibacterial activity, especially against *Corynebacterium pyogenes*.

8.3.2 ANTIFUNGAL ACTIVITY

Candidiasis is a fungal infection caused by *Candida albicans*. This infection commonly affects the skin, oral mucosa, vagina, and gastrointestinal tract. However, excessive use of azole antifungicides to treat *Candida albicans* infections can lead to the development of resistance. Therefore, various pteridologists carried out research on candidiasis by using different extracts of *Selaginella* species. Preliminary phytochemical and antimicrobial studies on a spike-moss, *Selaginella inaequalifolia*, were conducted by Irudayaraj et al. (2010). The results of antifungal studies showed that *Candida albicans* is resistant to ethanol extract but susceptible to petroleum ether extract. The petroleum ether extract showed maximum inhibition with a 45 mm inhibition zone in *C. albicans*. Cao et al. (2010a, b) studied the antimicrobial activity of selaginellin derivatives of *S. pulvinata* from the ethyl acetate fraction. Six selaginellin derivatives isolated from the ethyl acetate fraction of the whole plant of *Selaginella pulvinata* showed antifungal activities against *Candida albicans*. Warella et al. (2023) studied the antifungal activity of *Selaginella plana* ethanol extract against *Candida albicans*. The results demonstrated that *Selaginella plana* ethanol fraction inhibited the growth of *Candida albicans* when administered starting at a concentration of 12.5%; however, the antifungal potential of *Selaginella plana* extract that induced cell death was only observed at a concentration of 100%. The fungicidal activity was exclusively identified in the undiluted, pure extract. The inhibitory and cytotoxic effects of *Selaginella plana* on *Candida albicans* cells were attributed to the presence of bioactive compounds in *Selaginella plana*, including flavonoids, tannins, terpenoids, and saponins. The antifungal effect of amentoflavone derived from *Selaginella tamariscina* has been reported by Jung et al. (2006).

8.3.3 ANTI-INFLAMMATORY ACTIVITY

Inflammation is a defensive and crucial mechanism for preserving the organism's integrity in the process of physical, chemical, and infectious damage. The response of inflammation is often observed in severe lesions because of erroneously damaging normal tissue. At inflammatory sites, the shifting of cells is dependent on the release of vasoactive and chemotactic factors. It increases local blood flow and microvascular permeability, promoting the migration of leukocytes from the intravascular space into the tissues. *Selaginella convoluta* has been widely used as folk medicine in various regions. The study reported the anti-inflammatory effects of *S. convoluta* fractions employing various experimental models by Macedo et al. (2018). The major natural phyto-compounds in the genus *Selaginella* are flavonoid dimers (biflavonoids), and these biflavonoids isolated from *S. tamariscina* show anti-inflammatory activity on lipopolysaccharide (LPS)-induced murine macrophages and colon epithelial cells (Shim et al., 2018). Silver nanoparticles were isolated from an aqueous fraction of *Selaginella myosurus* by Kedi et al. (2018), and these particles were tested for anti-inflammatory activity. The two different models, namely albumin denaturation and carrageenan-induced rat hind paw edema, were used to assess the anti-inflammatory potential of isolated nanoparticles. The study revealed that the silver nanoparticles may act as reducing or inhibiting agents on the release of acute inflammatory mediators. Hence, this work demonstrated that silver nanoparticles mediated by *Selaginella myosurus* could be considered a potential source for anti-inflammatory drugs. The 'Sanjeevni' (*Selaginella bryopteris* L.) has promising anti-inflammatory phytomedicine for the treatment of patients with inflammatory skin diseases (Paswan et al., 2017). The mechanisms underlying the anti-inflammatory effects of topically administered aqueous, polar, and non-polar methanolic fractions (10 mg/20 μL) of *Selaginella bryopteris* were studied. *Selaginella bryopteris* produced no mortalities when administered orally at doses from 250 to 2000 mg/kg bw. Interestingly, treatment with the non-polar methanolic fraction (10 mg/20 μL) showed a significant reduction in redness (2.4 ± 0.5), edema (30.4 ± 1), and effectively reduced the LPO level (32.3 ± 3.3).

8.3.4 ANTI-CANCER PROPERTY

The ethanolic fraction of *Selaginella tamariscina* exhibited antimetastatic activity on lung cancer cell lines *in vivo* and *in vitro* (Kim et al., 2020). In metastasis, epithelial-mesenchymal transition (EMT) is a well-known and important step and thus can be a key target in cancer treatment. The aqueous fraction of *S. tamariscina* was tested to study its ability to inhibit migration and invasion of A459 and HT29 cancer cell lines. It was confirmed that aqueous extracts of *S. tamariscina* inhibited EMT in human cancer cell lines.

8.3.5 ANTI-DIABETIC PROPERTY

The anti-diabetic effects of the total flavonoids of *Selaginella tamariscina* (Beauv.) Spring were evaluated by Zheng et al. (2011). The total flavonoids of *S. tamariscina* have anti-diabetic activity, as shown by the decrease in serum levels of fasting blood glucose (FBG), glycosylated hemoglobin A1C (HbA1c), triglycerides (TG), total cholesterol (TC), free fatty acids (FFA), low-density lipoprotein cholesterol (LDL-C), and glucagon, as well as increased serum levels of high-density lipoprotein cholesterol (HDL-C), insulin, and C-peptide.

8.3.6 ANTI-ALZHEIMER ACTIVITY

Selagintriflavonoid was isolated from *Selaginella doederleinii* and has the strongest inhibition; therefore, selagintriflavonoid acts as a promising compound for treating Alzheimer's disease (Zou et al., 2017).

8.4 CONCLUSION

The crude extract from different species of *Selaginella* has revealed various *in vitro* bioactivities, viz., antimicrobial, antiviral, antidiabetic, antimutagenic, anti-inflammatory, antinociceptive, antispasmodic, anticancer, and anti-Alzheimer. The genus *Selaginella* is a rich source of bioactive compounds. Different types of biochemical constituents, including alkaloids, benzenoids, carbohydrates, chromones, terpenoids, flavonoids, and lignans, were isolated from various species of *Selaginella*. Many of the species have been exploited as ethnomedicine around the globe to cure fever, jaundice, hepatic disorders, cardiac diseases, diarrhea, sore throat, and cough, remove blood stasis, and stop external bleeding after trauma and separation of the umbilical cord. Though various biochemicals have been isolated from *Selaginella* species, flavonoids have been recognized as the most active component in the genus. The biochemical constituents have been reported from only a few species; other species have not been explored; hence, more research is needed to explore all natural product constituents and their bioactivities in the different species of *Selaginella*.

ACKNOWLEDGMENT

The authors are thankful to the Head of the Department of Botany for providing the facilities to carry out the work.

KEYWORDS

- **alkaloids**
- **antimicrobial**
- **antioxidant**
- **flavonoids**
- **phenolics**
- ***Selaginella***

REFERENCES

Adnan, M., Siddiqui, A. J., Jamal, A., Hamadou, W. S., Awadelkareem, A. M., Sachidanandan, M., & Patel, M. (2021). Evidence-based medicinal potential and possible role of *Selaginella* in the prevention of modern chronic diseases: Ethnopharmacological and ethnobotanical perspective. *Records of Natural Products, 15*(5), 330–355.

Almeida, J. R. G. S., Sa, P. G. S., Macedo, L. A. R. O., Filho, J. A. S., de Oliveira, V. R. D., & Filho, J. M. B. (2013). Phytochemistry of the genus *Selaginella* (Selaginellaceae). *Journal of Medicinal Plants Research, 7*, 1858–1868.

Cao, Y., Chen, J. J., Tan, N. H., Oberer, L., Wagner, T., Wu, Y. P., Zeng, G. Z., Yan, H., & Wang, Q. (2010b). Antimicrobial selaginellin derivatives from *Selaginella pulvinata. Bioorganic & Medicinal Chemistry Letters, 20*(8), 2456–2460.

Cao, Y., Tan, N. H., Chen, J. J., Zeng, G. Z., Ma, Y. B., Wu, Y. P., Yan, H., Yang, J., Lu, L. F., & Wang, Q. (2010a). Bioactive flavones and biflavones from *Selaginella moellendorfii* Hieron. *Fitoterapia, 81*(4), 253–258.

Chanravarthy, B. K., Rao, Y. V., Gambhir, S. S., & Gode, K. D. (1981). Isolation of amentoflavone from *Selaginella rupestris* and its pharmacological activity on the central nervous system, smooth muscles and isolated frog heart preparations. *Planta Medica, 43*(1), 64–70.

Chao, L. R., Seguin, E., Tillequin, F. O., & Koch, M. (1987). New alkaloid glycosides from *Selaginella doederleinii. Journal of Natural Products, 50*, 422–426.

Chen, P., Sun, J. Y., Xie, N. G., & Shi, Y. G. (1995). Chemical constituents of daeycai (*Selaginella doederleinii*). *Zhong Cao Yao, 26*, 397–399.

Cheng, X. L., Ma, S. C., Yu, J. D., Yang, S. Y., Xiao, X. Y., Hu, J. Y., Lu, Y., Shaw, P. C., But, P. P., & Lin, R. C. (2008). Selaginellin A and B, two novel natural pigments isolated from *Selaginella tamariscina*. *Chemical and Pharmaceutical Bulletin, 56*(7), 982–984.

Feng, W. S., Li, K. K., & Zheng, X. K. (2011). A new norlignan lignanoside from *Selaginella moellendorfii* Hieron. *Acta Pharmaceutica Sinica B, 1*, 36–39.

Fischer, M., & Kandler, M. (1975). Identifizierung von Selaginose und deren Verbreitung in der Gattung *Selaginella*. *Phytochemistry, 14*(12), 2629–2633.

Fraser-Jenkins, C. R., Gandhi, K. N., Kholia, B. S., & Benniamin, A. (2017). *An annotated checklist of Indian pteridophytes Part-1 (Lycopodiaceae to Thelypteridaceae)*. Bishen Singh Mahendra Pal Singh.

Ganguly, G. (2019). Phytochemistry and antibacterial activity of crude extracts and extracted phenols of *Selaginella bisulcata* Spring. *International Research Journal of Basic and Applied Sciences, 4*, 76–89.

Hui, R. H., Hou, D. Y., Li, T. C., Guo, H., & Liu, X. Y. (2005). Study on terpene constituents of herb *Selaginella* from Qianshan. *Journal of Anshan Normal University, 7*, 31–32.

Huneck, S., & Khaidav, T. (1985). Amentoflavone from *Selaginella sanguinolenta*. *Pharmazie, 40*(6), 431.

Irudayaraj, V., Janaky, M., Johnson, M., & Selvan, N. (2010). Preliminary phytochemical and antimicrobial studies on a spike-moss *Selaginella inaequalifolia* (Hook. & Grev.) Spring. *Asian Pacific Journal of Tropical Medicine, 3*(12), 957–960.

Jung, H. J., Sung, W. S., Yeo, S. H., Kim, H. S., Lee, I. S., Woo, E. R., & Lee, D. G. (2006). Antifungal effect of amentoflavone derived from *Selaginella tamariscina*. *Archives of Pharmacal Research, 29*, 746–751.

Kedi, P. B., Meva, F. E., Kotsedi, L., Nguemfo, E. L., Zangueu, C. B., Ntoumba, A. A., Mohamed, H. E. A., Dongmo, A. B. D., & Maaza, M. (2018). Eco-friendly synthesis, characterization, *in vitro* and *in vivo* anti-inflammatory activity of silver nanoparticle-mediated *Selaginella myosurus* aqueous extract. *International Journal of Nanomedicine, 13*, 8537.

Kim, G. L., Jang, E. H., Lee, D. E., Bang, C., Kang, H., Kim, S., Yoon, S. Y., Lee, D. H., Na, J. H., Lee, S., & Kim, J. H. (2004). Amentoflavone, active compound of *Selaginella tamariscina*, inhibits *in vitro* and *in vivo* TGF-β-induced metastasis of human cancer cells. *Archives of Biochemistry and Biophysics, 687*, 108384.

Kunert, O., Swamy, R. C., Kaiser, M., Presser, A., Buzzi, S., Rao, A. V. N. A., & Schühly, W. (2008). Antiplasmodial and leishmanicidal activity of biflavonoids from Indian *Selaginella bryopteris*. *Phytochemistry Letters, 1*, 171–174.

Lee, C. W., Choi, H. J., Kim, H. S., Kim, D. H., Chang, I. S., Moon, H. T., Lee, S. Y., Oh, W. K., & Woo, E. R. (2008). Biflavonoids isolated from *Selaginella tamariscina* regulate the expression of matrix metalloproteinase in human skin fibroblasts. *Bioorganic & Medicinal Chemistry, 16*(2), 732–738.

Li, G., Kollner, T. G., Yin, Y., Jiang, Y., Chen, H., Xu, Y., Gershenzon, J., Pichersky, E., & Feng, C. (2012). Nonseed plant *Selaginella moellendorffii* has both seed plant and microbial types of terpene synthases. *Proceedings of the National Academy of Sciences of the United States of America, 109*, 14711–14715.

Lin, L. C., Kuo, Y. C., & Chou, C. J. (2000). Cytotoxic biflavonoids from *Selaginella delicatula*. *Journal of Natural Products, 63*(5), 627–630.

Lopez-Saez, J. A., Perez-Alonso, M. J., & Negueruela, A. V. (1994a). The biflavonoids of *Selaginella selaginoides*. *Zeitschrift für Naturforschung C – Journal, 49*(3/4), 265–266.

Lopez-Saez, J. A., Perez-Alonso, M. J., & Negueruela, A. V. (1994b). Biflavonoids of *Selaginella denticulata* growing in Spain. *Zeitschrift für Naturforschung C – Journal, 49*(3, 4), 267–270.

Ma, L. Y., Ma, S. C., Wei, F., Lin, R. C., But, P. P. H., & Lee, S. F. (2003). Uncinoside A and B, two new antiviral chromone glycosides from *Selaginella uncinata*. *Chemical and Pharmaceutical Bulletin, 51*(11), 1264–1267.

Ma, L. Y., Wei, F., Ma, S. C., & Lin, R. C. (2002). Two new chromone glycosides from *Selaginella uncinata*. *Chinese Chemical Letters, 13*(8), 748–751.

Ma, S. C., But, P. P., Ooi, V. E. C., He, Y. H., Lee, S. H. S., Lee, S. F., & Lin, R. C. (2001). Antiviral amentoflavone from *Selaginella sinensis*. *Biological and Pharmaceutical Bulletin, 24*(3), 311, 312.

Macedo, L. A. R. O., de Oliveira Júnior, R. G., Souza, G. R., de Oliveira, A. P., de Lavor, E. M., Silva, M. G., Pacheco, A. G. M., de Menezes, I. R. A., Coutinho, H. D. M., Pessoa, C. D. O., da Costa, M. P., & Almeida, J. R. G. S. (2017). Chemical composition, antioxidant and antibacterial activities and evaluation of cytotoxicity of the fractions obtained from *Selaginella convoluta* (Arn.) Spring (Selaginellaceae). *Pharmaceutical Biotechnology, 32*, 506–512.

Moore, B. D., Isidoro, E., & Seemann, J. R. (1993). Distribution of 2-carboxyarabinitol among plants. *Phytochemistry, 34*(3), 703–707.

Paswan, S. K., Gautam, A., Verma, P., Rao, C. V., Sidhu, O. P., Singh, A. P., & Srivastava, S. (2017). The Indian magical herb 'Sanjeevni' (*Selaginella bryopteris* L.) – A promising anti-inflammatory phytomedicine for the treatment of patients with inflammatory skin diseases. *Journal of Pharmacopuncture, 20*(2), 93–99.

Patil, S. M., Patil, M. N., & Patil, S. S. (2020). Diversity and ethnomedicinal pteridophytes of Kolhapur district, Maharashtra. *Indian Fern Journal, 37*(1, 2), 226–238.

PPG-I. (2016). A community-derived classification for extant lycophytes and ferns. *Journal of Systematics and Evolution, 54*(6), 563–603.

Qasim, M. A., Roy, S. K., Kamil, M., & Ilyas, M. (1985). Phenolic constituents of Selaginellaceae. *Indian Journal of Chemistry, 24B*(2), 220.

Shim, S. Y., Lee, S. G., & Lee, M. (2018). Biflavonoids isolated from *Selaginella tamariscina* and their anti-inflammatory activities via ERK 1/2 signaling. *Molecules, 23*, 926.

Singh, R. S., Thakur, D., Jha, V. K., Kumar, U., Pal, U. K., Singh, P. K., & Fulzele, D. P. (2018). Antibacterial activity of *Selaginella bryopteris* fronds extract on bacteria isolated from mastitic milk. *International Journal of Current Microbiology and Applied Sciences, 7*, 4711–4715.

Sun, C. M., Syu, W. J., Huang, Y. T., Chen, C. C., & Ou, J. C. (1997). Selective cytotoxicity of ginkgetin from *Selaginella moellendorfii*. *Journal of Natural Products, 60*(4), 382–384.

Tan, G. S., Chen, L. Z., Xu, K. P., Zheng, Q. C., Xu, Z., Huang, Z. H., Shu, J. H., & Deng, T. (2004). Study on the chemical constituents of *Selaginella pulvinata* Maxim. *Chinese Journal of Organic Chemistry, 24*, 1082–1085.

Tan, G. S., Xu, K. P., Li, F. S., Wang, C. J., Li, T. Y., Hu, C. P., Shen, J., Zhou, Y. J., & Li, Y. (2009). Selaginellin C, a new natural pigment from *Selaginella pulvinata* Maxim (Hook et Grev.). *Journal of Asian Natural Products Research, 11*, 1001–1004.

Verma, M., Gangwar, M., Sahai, M., Nath, G., & Singh, T. D. (2015). Antimicrobial activity of phytochemicals isolated from *Selaginella bryopteris*. *Chemical and Natural Compounds, 51*(2), 341–345.

Wang, G., Ma, X., Li, D., Jiang, Y., Liang, Q., & Hui, C. (2018). Chemical composition of essential volatile oils of *Selaginella* spp. and their antibacterial activity. *Bangladesh Journal of Botany, 47*(3), 719–726.

Wang, H. S., Sun, L., Wang, Y. H., Shi, Y. N., Tang, G. H., Zhao, F. W., Niu, H. M., Long, C. L., & Li, L. (2011). Carboxymethyl flavonoids and a monoterpene glucoside from *Selaginella moellendorffii*. *Archives of Pharmacal Research, 34*, 1283–1288.

Wang, Y. H., Long, C. L., Yang, F. M., Wang, X., Sun, Q. Y., Wang, H. S., Shi, Y. N., & Tang, G. H. (2009). Pyrrolidinoindoline alkaloids from *Selaginella moellendorfii*. *Journal of Natural Products, 72*(6), 1151–1154.

Warella, J. C., Rahma, K., Widodo, A. D. W., & Setiabudi, R. J. (2023). Antifungal activity of *Selaginella plana* (Desv. ex Poir.) Hieron extract against *Candida albicans in vitro*. *Folia Medica Indonesiana, 59*(3), 295–301.

Weng, J. K., & Noel, J. P. (2013). Chemodiversity in *Selaginella*: A reference system for parallel and convergent metabolic evolution in terrestrial plants. *Frontiers in Plant Science, 19*, 1–17.

Xu, K. P., Zou, H., Liu, G. R., Long, H. P., Li, J., Li, F., Zou, Z., Kuang, J., Xie, X., & Tan, G. (2011b). Selaginellin M, a new selaginellin derivative from *Selaginella pulvinata*. *Journal of Asian Natural Products Research, 13*, 1051–1055.

Xu, K., Zou, H., Li, F., Xiang, H., Zou, Z., Long, H., Li, J., Kuang, J., Xie, X., & Tan, G. (2011a). Two new selaginellin derivatives from *Selaginella tamariscina* (Beauv.) Spring. *Journal of Asian Natural Products Research, 13*, 356–360.

Xu, K., Zou, H., Tan, Q., Li, F., Liu, J., Xiang, H., Zou, Z., Long, H., Li, Y., & Tan, G. (2011c). Selaginellins I and J, two new alkynyl phenols, from *Selaginella tamariscina* (Beauv.) Spring. *Journal of Asian Natural Products Research, 13*, 93–96.

Yang, J. W., Pokharel, Y. R., Kim, M. R., Woo, E. R., Choi, H. K., & Kang, K. W. (2006). Inhibition of inducible nitric oxide synthase by sumaflavone isolated from *Selaginella tamariscina*. *Journal of Ethnopharmacology, 105*(1, 2), 107–113.

Zhang, L., Liang, Y., Wei, X., & Cheng, D. (2007). A new unusual natural pigment from *Selaginella sinensis* and its noticeable physicochemical properties. *Journal of Organic Chemistry, 72*, 3921–3924.

Zheng, X. K., Bi, Y. F., Feng, W. S., Wang, J. F., & Niu, J. Z. (2004a). Study on chemical constituents of *Selaginella tamariscina* (Beauv.) Spring. *Yao Xue Xue Bao, 39*(4), 266–268.

Zheng, X. K., Shi, S. P., Bi, Y. F., Feng, W. S., Wang, J. F., & Niu, J. Z. (2004b). The isolation and identification of a new lignanoside from *Selaginella tamariscina* (Beauv.) Spring. *Yao Xue Xue Bao, 39*(9), 719–721.

Zheng, X. K., Zhang, L., Wang, W. W., Wu, Y. Y., Zhang, Q. B., & Feng, W. B. (2011). Anti-diabetic activity and potential mechanism of total flavonoids of *Selaginella tamariscina* (Beauv.) Spring in rats induced by high-fat diet and low dose STZ. *Journal of Ethnopharmacology, 137*, 662–668.

Zheng, X., Du, J., Xu, Y., Zhu, B., & Liao, D. (2007). A new steroid from *Selaginella pulvinata*. *Fitoterapia, 78*(7–8), 598–599.

Zheng, X., Liao, D. F., Zhu, B. Y., & Tuo, Q. H. (2001). Study on chemical constituents of *Selaginella pulvinata*. *Zhong Cao Yao, 32*, 17–18.

Zou, Z., Xu, P., Zhang, G., Cheng, F., Chen, K., Li, J., Zhu, W., Cao, D., Xu, K., & Tan, G. (2017). Selagintriflavonoids with BACE1 inhibitory activity from the fern *Selaginella doederleinii*. *Phytochemistry, 134*, 114–121.

CHAPTER 9

A Comprehensive Review of the Pharmacological Activities of *Dicranopteris linearis* (Burm.f.) Underw.: A Medicinal Fern with Diverse Therapeutic Potential

PULICHERLAYUGANDHAR,[1] PASUPULETISIVARAMAKRISHNA,[2,3,*] and BHASHA SHANMUGAM[3]

[1]*Survey of Medicinal Plants Unit, Regional Ayurveda Research Institute, Itanagar, Arunachal Pradesh, India*

[2]*Department of Botany, Sri Venkateswara University, Tirupati, Andhra Pradesh, India*

[3]*Department of Botany, Government College for Men, Kadapa, Andhra Pradesh, India*

[4]*Department of Education in Science and Mathematics (DESM), North East Regional Institute of Education, NCERT, Shillong, Meghalaya, India*

*Corresponding authors

ABSTRACT

This review explores the therapeutic potential of *Dicranopteris linearis* (Burm. f.) Undrew., a pteridophytic plant known for its adaptability and rich phytochemical content. Different parts of the plant have been used traditionally for treating wounds, fever, and inflammatory conditions. *D. linearis* has been found to possess a wide range of pharmacological

Bioactives and Pharmacology of Pteridophytes. T. Pullaiah (Ed.)

DOI: 10.1201/9781779642738-9

activities, including anthelmintic, antibacterial, anti-inflammatory, antinociceptive, antioxidant, antipyretic, chemopreventive, cytotoxic, gastroprotective, hepatoprotective, urease inhibitory, and wound healing activities. Phytochemical analysis revealed essential bioactive compounds, such as flavonoids, phenolics, and terpenoids, which contribute to these activities. The review highlights the strong antioxidant capacity of extracted plant fractions, the enhanced antimicrobial effects, and the extracts' minimal cytotoxicity, indicating safety and therapeutic potential. The significant findings include the plant's comparable efficacy to standard analgesics in pain management, its anti-inflammatory properties validated in a rat model, and its ability to promote wound healing through enhanced cell migration. Overall, *D. linearis* offers promise as a natural source for novel therapeutic agents, with further research needed to isolate and characterize its bioactive compounds for clinical applications. The plant's extensive medicinal properties make it a valuable candidate for future pharmaceutical development.

9.1 INTRODUCTION

Medicinal plants are crucial for human health in the current scenario as they offer natural and effective alternatives for treating various ailments and contribute to discovering new pharmaceuticals (Sandhya et al., 2006). People rely on medicinal plants for healthcare through traditional medicine, modern pharmaceuticals, and preventive health measures. These plants can be used directly or in extracted forms due to their diverse secondary metabolites responsible for their therapeutic properties (Savithramma et al., 2011). Medicinal plants often possess a wide range of phytochemical properties that are highly significant in human medicine and veterinary science. Natural products derived from these plants have been pivotal in developing new drugs, playing a key role in treating and preventing diseases. Numerous plants have been demonstrated to be effective against various health conditions and have undergone extensive screening to identify their therapeutic compounds (Guan and He, 2015). In developing countries, about 80% of the population relies on medicinal plants for their primary healthcare needs due to accessibility, affordability, and cultural practices. The resurgence of public interest in plant-based medicine, coupled with the rapid growth of the pharmaceutical industry in recent years, has significantly increased the demand for medicinal plants. This trend is underscored by the fact that approximately 80% of traditional medicines use plant extracts, highlighting their enduring role in modern healthcare and their potential to drive the development of new therapeutic agents. Many compounds used in contemporary drugs are derived from medicinal plants, often featuring complex structures that enhance their medicinal efficacy (Savithramma et al., 2016).

Dicranopteris linearis, commonly known as Old World forked fern or uluhe, is a notable pteridophytic plant with significant medicinal value. The plant belongs to the family Gleicheniaceae, and this fern is widely distributed across tropical and subtropical regions, including Southeast Asia, India, China, Japan, the Pacific islands, and parts of Africa and Madagascar. This fern is highly adaptable and can be found in diverse habitats such as lowland rainforests, open woodlands, and disturbed areas like roadsides and clearings. It often forms dense, extensive ground cover, especially in areas with abundant sunlight and sufficient moisture. This fern is renowned for its distinctive morphology, characterized by forked, wiry fronds that spread extensively, creating an intricate network. The resilience and adaptability of *D. linearis* enable it to flourish in a variety of environmental conditions, contributing to its widespread presence.

9.2 TRADITIONAL/ETHNOMEDICINAL USES

The fern is commonly used to treat inflammatory conditions. Its fronds are often prepared as poultices or decoctions to reduce swelling, pain-related maladies, and alleviate pain (Vasuda, 1999). In parts of the Kumaun Himalaya, plant extracts have been used to treat sterility in women (Hussaini et al., 2012). In some cultures, the plant is used to treat intestinal worms and stomach ulcers (Srivastava, 2007). Decoctions made from the fronds are consumed to lower fevers and treat symptoms of colds and flu (Zakaria et al., 2008). The decoction of the whole plant has been used to treat throat pain, and the extract from fronds is used for asthma by the tribal people of Palani Hills (Kodaikanal), Western Ghats, South India (Sathiyaraj et al., 2015). Extract from this plant is applied to treat various skin ailments, such as external wounds, ulcers, abscesses, boils, and sores (Lai et al., 2015). The plant's leaves are applied to wounds and cuts to promote healing and prevent infections (Singh, 2018).

Recent scientific research has highlighted several pharmacological activities of *D. linearis*, validating many of its traditional uses. The fern exhibits antimicrobial effects against a range of bacterial pathogens. This activity supports its traditional use in treating infections and wounds (Thomas et al., 2007). The plant has shown potential in protecting the liver from damage caused by toxins and oxidative stress, indicating its use in liver health and detoxification (Ismail et al., 2014). Preliminary studies suggest that certain compounds isolated from *D. linearis* may have anticancer properties, inhibiting the growth of cancer cells and inducing apoptosis (Baharuddin et al., 2018). Studies have shown that extracts from *D. linearis* possess significant anti-inflammatory properties. The presence of flavonoids and other phenolic compounds contributes to its ability to reduce inflammation and alleviate pain (Zakaria et al., 2019). The plant is rich in antioxidants, which help to neutralize free radicals and protect cells from oxidative stress. This property is beneficial in preventing various chronic diseases and promoting overall health (Zakaria et al., 2020a). As interest in plant-based medicine continues to grow, *D. linearis* stands out as a valuable species with promising applications in health and wellness. Its role in traditional medicine, coupled with ongoing research into its bioactive compounds, highlights this fern's importance in ethnobotanical practices and contemporary therapeutic development.

9.3 PHYTOCHEMICAL ANALYSIS

A series of studies have been carried out to analyze the phytochemical constituents of *D. linearis*, uncovering a diverse range of bioactive compounds with potential therapeutic applications. These investigations include qualitative and quantitative analysis of phytochemicals, identification of bioactive compounds through GC-MS analysis, separation of compounds using various chromatographic techniques, structural elucidation using Nuclear Magnetic Resonance (NMR) spectroscopy, evaluation of its ability to disrupt bacterial biofilms, and assessment of alpha-glucosidase enzyme inhibition for the development of antidiabetic drugs. These studies include Rajesh et al. (2016), which aimed to analyze the qualitative and quantitative analysis of phytochemicals and determine the bioactive compounds in the extracts of *D. linearis*. The qualitative phytochemical analysis of aqueous, acetone, chloroform, ethanol, and petroleum ether extracts revealed that the acetone extract showed strong positivity for 12 phytoconstituents, except for anthocyanin, compared to other solvent extracts. Quantitative phytochemical analysis indicated a significant presence of terpenoids (97.0±1.15 mg/g), tannins (30.8±0.44 mg tannic acid equivalents/g), phenols (28.6±0.33 mg gallic acid equivalents/g), and flavonoids (8.50±0.29 mg quercetin equivalents/g), in descending order of concentration. Kale (2015) focused on analyzing the bioactive components of the whole plant extract of *D. linearis* using Gas Chromatography-Mass Spectrometry (GC-MS). The study aimed to identify and understand the medicinally important compounds present in this plant, known for its therapeutic uses. The GC-MS analysis revealed three main bioactive compounds in the ethanol extract of *D. linearis*: Nonane, 1,1-diethoxy; n-Hexadecanoic acid; and Hexadecanoic acid, ethyl ester. These compounds were identified based on their molecular structure, mass, and fragments. Notably, n-Hexadecanoic acid (palmitic acid) was recognized as a prevalent compound with various beneficial properties, including antioxidant and antibacterial activities. The study concluded that the ethanol extract of *D. linearis* contains several bioactive compounds with potential medicinal value, particularly for their antioxidant and antimicrobial properties.

Anh et al. (2022) investigated the chemical constituents of the leaves of *D. linearis* by focusing on extracting and isolating compounds from the leaf material using various chromatographic techniques. The study successfully isolated and identified four compounds: three flavonols, including isoquercetin, quercetin, kaempferol, and a sterol, stigmast-5,22-dien-3β-ol-3-O-β-D-glucopyranoside. The structures of these compounds were elucidated using NMR spectroscopy. The research highlights the potential of *D. linearis* as a source of bioactive compounds with possible therapeutic applications, given the known biological activities of the isolated compounds. Mawang et al. (2017) investigated the bioactive components of *D. linearis* leaves, focusing on their potential to disrupt biofilms formed by *Staphylococcus aureus*. The leaves were extracted using hexane (HEX), dichloromethane, ethyl acetate, and methanol. Further purification and bioassay-guided fractionation were performed to determine the active compound responsible for biofilm disruption. The results showed that the methanol extract was the only fraction that exhibited antimicrobial activity, while the hexane fraction demonstrated biofilm disruption ability. α-Tocopherol, isolated from the active fraction, was found to disrupt the biofilm matrix rather than directly affecting the bacterial cells. The study concluded that α-tocopherol is a promising bioactive compound with the potential to disrupt *S. aureus* biofilms. This suggests its potential application in treating biofilm-associated infections. Duong et al. (2023) focused on isolating active compounds from *D. linearis* leaves. The leaves were collected, dried, and ground into a fine powder, with 5.0 kg of the powder exhaustively extracted using methanol, resulting in 280 g of crude extract. This extract was partitioned using n-hexane, n-hexane:EtOAc (1:1, v/v), and EtOAc, yielding four fractions: H (27.44 g), HEA (34.79 g), EA (23.00 g), and MeOH (60.03 g). The HEA fraction was further purified through silica gel column chromatography, leading to the isolation of compounds DL1 (40.0 mg), DL3 (4.6 mg), DL8 (3.2 mg), DL9 (2.7 mg), DL4 (15.1 mg), DL5 (2.6 mg), and DL6 (2.6 mg). Similarly, fractionation and purification of the EA extract yielded additional compounds, including DL13 (12.0 mg), DL11 (40.0 mg), DL12 (16.0 mg), DL10 (22.0 mg), DL7 (5.1 mg), and DL2 (8.3 mg). The isolated 13 compounds were then evaluated for their alpha-glucosidase inhibitory activity *in vitro* using the alpha-glucosidase enzyme isolated from *Saccharomyces cerevisiae* and the substrate compound 4-nitrophenyl β-D-glucopyranoside, with acarbose as a positive control. The results revealed that compounds DL3, DL5-DL8, and DL10-DL13 showed potent inhibitory effects with IC50 values ranging from 67.1 to 282.1 μM. Molecular docking studies were performed on compounds DL3, DL5, and DL8 to gain insights into the inhibitory mechanisms. The study isolated a series of compounds from *D. linearis* leaves and identified several of these compounds as potent alpha-glucosidase inhibitors. The findings highlight the potential of *D. linearis* as a valuable natural source for developing new antidiabetic agents, warranting further research into the therapeutic applications of these compounds.

The extensive tropical distribution of *D. linearis* and its deep-rooted history in traditional and folk medicine have sparked significant interest among scholars worldwide. This fascination has led to a surge in research to evaluate plant pharmacological activities. Researchers have employed a wide range of methodologies, including *in vitro* experiments to assess the direct effects of plant extracts on cell lines, *in vivo* studies to observe the impact of these extracts on animal models like mice and rats, and phytochemical analyses to identify the active compounds responsible for

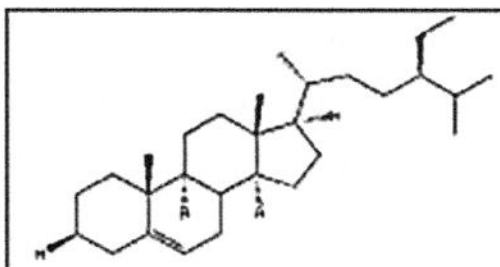 **β-Sitosterol (DL1)**	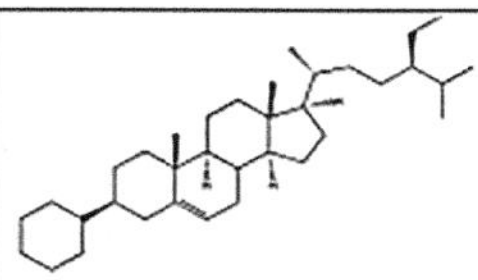 **Daucosterol (DL2)**	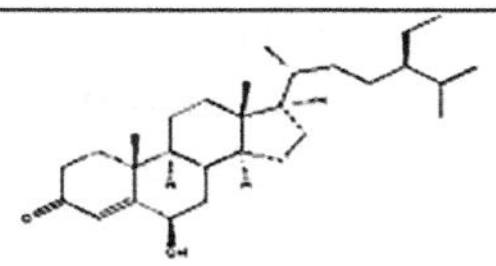 **6-hydroxystigma-4-en-3-one (DL3)**
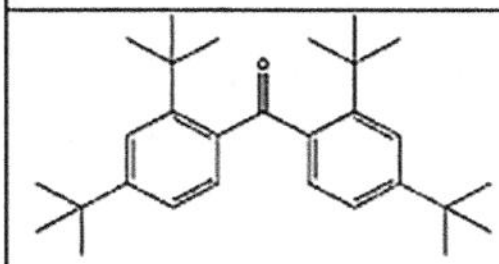 **2,2',4,4'-tetra-*tert*-butylbenzophenone (DL4)**	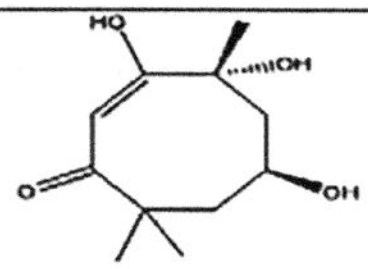 **Chakyunglupulin B(DL5)**	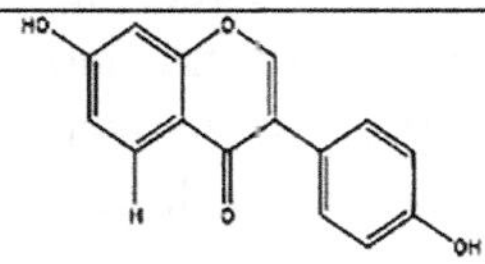 **Daidzein (DL6)**
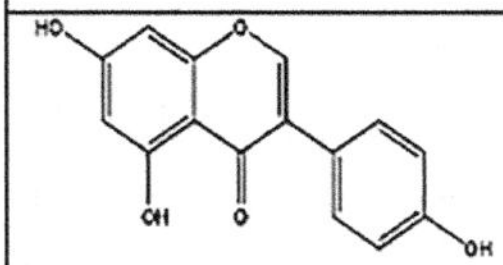 **Genistein (DL7)**	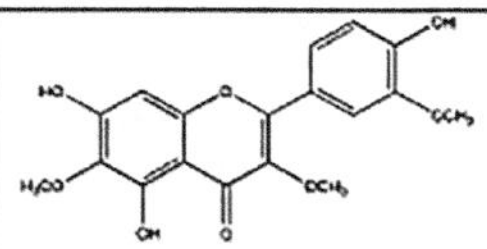 **Jaceidin(DL8)**	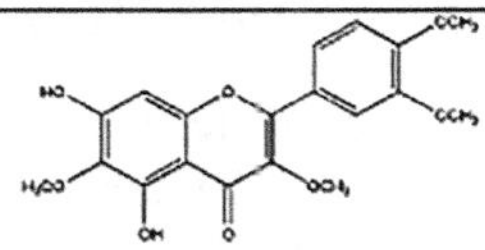 **Bonanzin (DL9)**
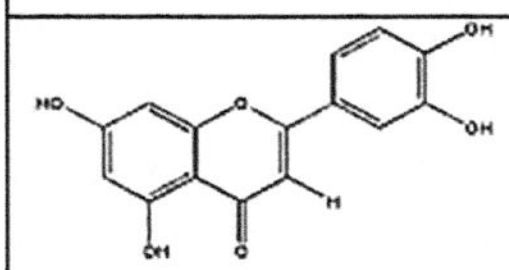 **Luteolin (DL10)**	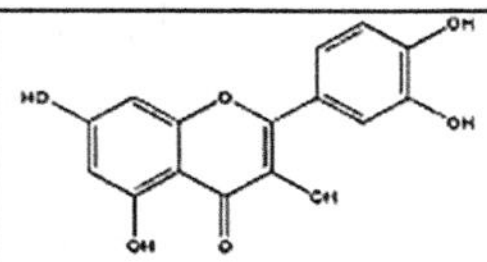 **Quercetin (DL11)**	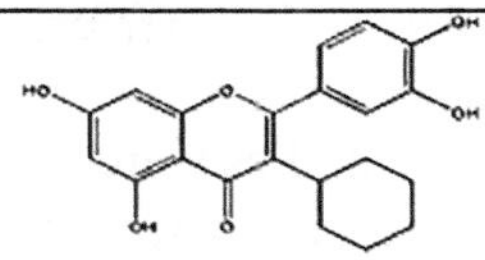 **Isoquercetin (DL12)**
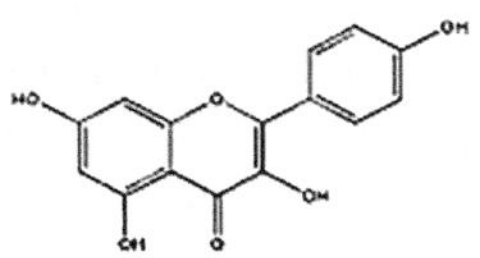 **Kaempferol (DL13)**	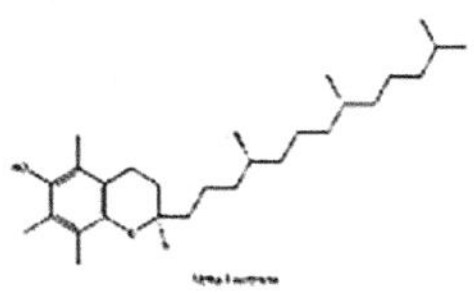 **α-Tocopherol**	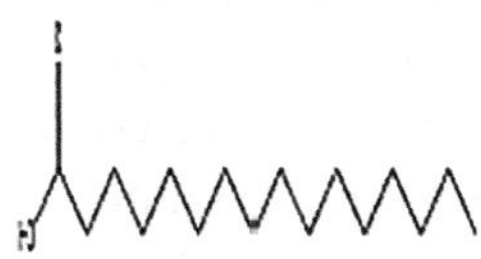 **Palmitic acid**

Source: Anh et al. (2022); Duong et al. (2023); Mawang et al. (2017).

the observed pharmacological effects. The findings from these studies have validated the traditional uses of *D. linearis* and opened new avenues for drug development. By isolating and characterizing the plant's active constituents, scientists are working towards creating novel pharmaceuticals that could offer effective treatments for various health conditions.

Scientists from various disciplines and regions have embarked on studies to unravel the therapeutic potential of *D. linearis*, focusing on its diverse biological activities. The review has highlighted the plant's significant biological activities, such as its anthelmintic, antibacterial, anti-inflammatory, antinociceptive, antioxidant, antipyretic, chemopreventive, cytotoxic, gastroprotective, hepatoprotective, urease inhibitory, and wound healing activities, showcasing its potential therapeutic benefits. This review aims to summarize and elaborate on these pharmacological activities, providing a comprehensive overview of the current state of research and its implications for future drug development.

9.4 PHARMACOLOGICAL ACTIVITIES

9.4.1 ANTIBACTERIAL ACTIVITY

Antibacterial activity refers to the ability of a substance to inhibit the growth or kill bacteria. This property is crucial in developing treatments for bacterial infections, which can cause a wide range of diseases in humans and animals. Substances with antibacterial activity can disrupt the bacterial cell wall, interfere with DNA replication, or inhibit other essential bacterial functions, leading to the death or inhibition of bacterial growth. Several studies were conducted to evaluate the efficacy of various extracts of *D. linearis*. These studies aimed to validate the quantitative assessment of the antibacterial potency of these plant extracts.

Kumarpal (2013) investigated the antibacterial properties of various parts of *Dicranopteris linearis*. Fresh rhizome, rachis, and frond specimens were collected, dried at 40°C, and powdered. Extracts were prepared by soaking 5 grams of each powdered sample in ethanol, followed by 24 hours of shaking at 37°C. The extracts were then filtered and tested for antibacterial activity against *Escherichia coli* (gram-negative) and *Bacillus megaterium* (gram-positive) using the disc diffusion method. Sterile distilled water was used as a control, and the antibacterial efficacy was determined by measuring the inhibition zones after 24 hours of incubation at 37°C. The study found that the rhizome extract exhibited the highest antibacterial activity against both gram-negative and gram-positive bacteria, followed by the frond extract, which also showed good effectiveness. The rachis extract was less effective in comparison. These findings suggest that the rhizome and frond extracts of *D. linearis* possess significant antibacterial properties, particularly against gram-negative bacteria, indicating their potential use as natural antimicrobial agents. The study supports further research into isolating and characterizing the active compounds in these extracts, which could contribute to developing new antibacterial treatments.

Thomas et al. (2007) explored the antimicrobial properties of *D. linearis* rhizome and petiole. Fresh specimens were dried, powdered, and extracted using Soxhlet extraction with petroleum ether, acetone, methanol, and water, resulting in extracts with yields of 0.34%, 5.2%, 3.1%, and 0.4%, respectively. The study tested these extracts against *Micrococcus luteus* (MTCC 6164), *Escherichia coli* (MTCC 443), *Staphylococcus aureus* subsp. *aureus* (MTCC 96), and *Pseudomonas aeruginosa* (MTCC 741). Bacterial cultures were prepared, and antibacterial activity was evaluated using the disc diffusion method with Mueller Hinton Agar. A solvent mixture of DMSO and sterilized distilled water was used as a control, and test materials were dissolved in 10% DMSO to make a 50 mg/ml stock solution. Each disc was loaded with 20 µL of the stock solution, dried, and placed on the agar surface. Plates were incubated at 37°C for 24 hours, and inhibition zones were measured. The Minimum Inhibitory Concentration (MIC) and Minimum Bactericidal Concentration (MBC) were also determined. Results showed that the acetone extract exhibited the highest antibacterial activity, followed by the methanol extract, while water and petroleum ether extracts were ineffective. The acetone extract was effective against two gram-positive strains (*M. luteus* and *S. aureus*) and one gram-negative strain (*P. aeruginosa*), whereas the methanol extract was only effective against gram-positive strains. Gram-positive bacteria were more susceptible to the extracts, with *M. luteus* being the most susceptible and *E. coli* the least. The acetone extract's inhibition zones against *M. luteus*, *S. aureus*, and *P. aeruginosa* were comparable to those of streptomycin, although the zone for *E. coli* was about half that of streptomycin. The acetone extract had significant antibacterial activity, particularly against *S. aureus*, with an MIC of 8 mg/ml and an MBC of 16 mg/ml for *M. luteus*, and an MIC and MBC of 16 mg/ml for *S. aureus*. These findings suggest that the acetone extract of *D. linearis* could be a valuable source of natural antibacterial compounds, warranting further investigation into its potential therapeutic applications. Continuing the above study, Thomas et al. (2023) explored the antibacterial properties of *D. linearis* frond extracts. Fronds were washed, shade-dried, powdered, and extracted using Soxhlet extraction with petroleum ether, ethyl acetate, methanol, and distilled water. The extracts were concentrated using a rotary evaporator and tested against bacterial strains including *Shigella flexneri*, *Pseudomonas aeruginosa*, *Serratia marcescens*, *Escherichia coli*, and *Salmonella typhi murium* using the agar disc diffusion method. Standard antibiotics Amoxyclav and Methicillin served as positive controls, while the pure solvents were used as negative controls. The study found that petroleum ether and aqueous extracts were ineffective against all tested bacterial strains. In contrast, the ethyl acetate and methanol extracts showed significant antibacterial activity, with the methanol extract being particularly effective against *P. aeruginosa* (16 ± 1 mm inhibition zone), followed by *S. flexneri, S. marcescens*, and *E. coli*. The solvents used as negative controls showed no inhibitory action, while Amoxyclav exhibited a zone of inhibition (ZOI), and methicillin showed no response. These findings suggest that the antibacterial compounds in *D. linearis* are more effectively extracted using methanol and ethyl acetate, highlighting their potential as natural antibacterial agents. The study underscores the need for further investigation into the active components and mechanisms of action of these extracts. Ahmad et al. (2023) investigated the antimicrobial activity of encapsulated *D. linearis* extracts. The crude extract (CE) was partitioned into petroleum ether, chloroform, and aqueous extracts, which were then filtered and evaporated to a constant weight. These extracts were encapsulated using solvent displacement dialysis with either ethyl cellulose (EC) or methylcellulose (MC) and freeze-dried for preservation. The encapsulated extracts were tested for antimicrobial activity against *E. coli* (ATCC 25922), *B. subtilis* (CDR), and *S. flexneri* (ATCC 12022) using the broth dilution technique. The growth media's pH was adjusted to 4, simulating an acidic environment, and the bacterial cultures were incubated with the encapsulated extracts at 37°C for 24 hours. Antimicrobial activity was assessed by measuring turbidity at 600 nm using a UV-Vis spectrophotometer. The study found that the MCCH (methyl cellulose chloroform) encapsulated extract exhibited excellent antibacterial activity against both Gram-negative bacteria (*E. coli* and *S. flexneri*) and Gram-positive bacteria (*B. subtilis*). In contrast, petroleum ether encapsulated extracts were the least effective. MCCH showed the highest antimicrobial activity, with bacterial counts of 2.42×10^7 CFU/mL for *E. coli*, 1.25×10^7 CFU/mL for *S. flexneri*, and 5.21×10^7 CFU/mL for *B. subtilis*. The findings suggest that MC encapsulation not only preserves but enhances the bioactive properties of *D. linearis* extracts, making them potent antibacterial agents. This method shows promise for developing effective antimicrobial treatments, particularly for use in acidic environments like the human stomach.

These studies on *D. linearis* have consistently demonstrated significant antibacterial properties, particularly against both Gram-positive and Gram-negative bacteria. Kumarpal (2013) found that the rhizome and frond extracts were highly effective, with the rhizome showing the highest activity. Thomas et al. (2007) observed that acetone extracts were most potent, especially against *S. aureus*, and methanol extracts were effective against Gram-positive strains. Further research by Thomas et al. (2023) indicated that methanol and ethyl acetate extracts of *D. linearis* fronds exhibited significant inhibitory action against pathogenic strains like *P. aeruginosa*, *S. flexneri*, *S. marcescens*, and *E. coli*. Additionally, Ahmad et al. (2023) demonstrated that encapsulated extracts, particularly those using methyl cellulose (MCCH), showed excellent antibacterial activity against both Gram-negative and Gram-positive bacteria. These findings underscore the potential of *D. linearis* as a source of natural antibacterial compounds. Future studies should focus on isolating and characterizing the active compounds responsible for these effects and investigating their mechanisms of action and potential therapeutic applications.

9.4.2 ANTHELMINTIC ACTIVITY

Anthelmintic activity refers to the ability of a substance to expel or destroy parasitic worms within a host organism. Researchers typically use *in vitro* and *in vivo* assays to evaluate the activity of medicinal plant extracts. *In vitro* assays involve exposing parasitic worms to different concentrations of plant extract to determine the minimum effective dose that can kill or inhibit the worms. To assess the anthelmintic potential of *D. linearis*, Kalpanadevi et al. (2016a) extracted 1 g of dried rhizome powder using various solvents, including ethanol, acetone, chloroform, aqueous solution, and petroleum ether. The extracts were prepared using an Ultra-Turrax mixer, soaked overnight, filtered, and evaporated to a constant weight. Adult live *Gastrothylax crumenifer* parasites were collected from infected sheep and goats and incubated with different concentrations of *D. linearis* extracts in Hedon-Fleig (H-F) solution. Oxyclozanide (1%)

served as the standard control, and H-F salt solution was the negative control. The motility of the flukes was monitored at intervals of 0, 10, 15, and 30 minutes, and 1 hour using a dissection microscope, with motility scores assigned based on the movement observed. The study revealed that the relative motility (RM) values decreased with increased incubation time for both the standard control and all tested concentrations of *D. linearis* extract. Higher concentrations of the extract (3–5 mg/mL) killed all worms within 10 minutes, similar to the standard control, while lower concentrations (1–2 mg/mL) achieved the same effect within an hour. Morphological studies using stereo-zoom and light microscopy, along with histopathological and scanning electron microscope examinations, revealed significant degenerative changes in oral suckers, teguments, and reproductive organs at higher extract concentrations. Lower concentrations caused milder effects. Severe tegmental damage, including blebbing and degeneration, was observed, particularly at higher concentrations. The study concluded that *D. linearis* exhibits substantial anthelmintic activity against *G. crumenifer*, with higher extract concentrations showing rapid and potent effects comparable to the standard control, oxyclozanide. These findings suggest the potential of *D. linearis* as a natural anthelmintic agent.

9.4.3 ANTI-INFLAMMATORY ACTIVITY

Anti-inflammatory activity refers to the ability of a substance to reduce inflammation in the body. Zakaria et al. (2006) investigated the anti-inflammatory activity of *D. linearis* leaves using *in vitro* assays. The leaves were extracted with chloroform, and the extract was diluted in DMSO to prepare concentrations of 10, 100, and 200 mg/kg. Distilled water and acetylsalicylic acid (ASA) were used as controls. The study found that the chloroform extract of *D. linearis* leaves (CEDL) exhibited significant anti-inflammatory activity, comparable to or greater than 100 mg/kg ASA, particularly in reducing carrageenan-induced paw edema. The effect was noticeable in the first 6 hours post-administration, suggesting that CEDL's anti-inflammatory properties may be linked to its ability to inhibit pro-inflammatory mediators like prostaglandins. This supports the traditional use of *D. linearis* in treating conditions such as ulcers and asthma in Papua New Guinea and India.

In another study, Zakaria et al. (2008) investigated the anti-inflammatory effects of *D. linearis* leaves, which were extracted with distilled water and freeze-dried to yield a crude extract (AEDL). Various doses of AEDL, ranging from 13.2 to 1320 mg/kg, were tested using the carrageenan-induced paw edema method in rats. ASA and distilled water served as controls. The study revealed that AEDL exhibited significant, dose-independent anti-inflammatory activity, particularly at 66.0 and 660.0 mg/kg doses, with effects lasting 2–6 hours before fading. The findings suggest that *D. linearis* has promising anti-inflammatory properties, comparable to those of ASA, supporting its potential therapeutic use.

However, in another study, Zakaria et al. (2019) extracted *D. linearis* leaves by soaking 160 g of leaf powder in methanol for 72 hours. The methanolic extract (MEDL) was further partitioned into petroleum ether (PEDL), ethyl acetate (EADL), and an aqueous partition (AQDL). The AQDL extract was then assessed for anti-inflammatory activity using lipoxygenase (LOX) and xanthine oxidase (XO) assays. In the LOX assay, AQDL showed a minimal inhibitory effect (<20% inhibition) at a concentration of 100 μg/mL, indicating limited potential to reduce LOX-mediated inflammation. In the XO assay, AQDL failed to inhibit XO-mediated activity. These results suggest that while AQDL exhibits some impact on LOX, its overall anti-inflammatory potential is limited and may require further investigation to determine its efficacy.

In continuation of the above study, Zakaria et al. (2020a) processed the methanolic extract of *D. linearis* (MEDL) to obtain an aqueous partition (EADL). The process involved dissolving MEDL in methanol, mixing it with distilled water and petroleum ether, and then with ethyl acetate. After multiple extractions, the final aqueous residue was freeze-dried. The anti-inflammatory potential of EADL was evaluated using lipoxygenase (LOX) and xanthine oxidase (XO) assays. In the LOX assay, EADL showed approximately 26% inhibition at 100 mg/mL, while in the XO assay, it exhibited only 1.5% inhibition. These results indicate that EADL has modest anti-inflammatory activity, suggesting it could be a valuable candidate for further development as a natural anti-inflammatory agent.

The studies conducted by Zakaria et al. (2006, 2008, 2019, 2020a) on the anti-inflammatory activity of *D. linearis* extracts provide compelling evidence supporting the traditional use of this plant in treating inflammatory conditions. The various extraction methods and assays employed across these studies demonstrated that different plant partitions exhibit significant anti-inflammatory effects, particularly the chloroform and ethyl acetate extracts. These effects were dose-independent in some cases and were comparable to, or even exceeded, the activity of standard anti-inflammatory drugs like ASA within specific time frames. However, the limited inhibitory effects observed in the lipoxygenase (LOX) and xanthine oxidase (XO) assays, particularly in the aqueous extract (AQDL), indicate that the anti-inflammatory potential of *D. linearis* may vary depending on the extraction method and the specific bioactive compounds present. These findings suggest that while *D. linearis* holds promise as a natural anti-inflammatory agent, further research is needed to isolate and identify the specific compounds responsible for these effects, optimize extraction methods, and fully understand the mechanisms involved.

9.4.4 ANTINOCICEPTIVE ACTIVITY

Antinociceptive activity refers to the ability of a substance to reduce or block the sensation of pain without affecting consciousness. Nociception is the sensory process that signals potential harm to the body, leading to the perception of pain. Antinociceptive agents work by interrupting this process, providing relief from pain. Evaluating antinociceptive activity using medicinal plant extracts is crucial for discovering new, natural pain relief options. This research supports the development of alternative pain management strategies and validates and integrates traditional medicine practices into contemporary healthcare. To carry out this activity, Zakaria et al. (2006) extracted *Dicranopteris linearis* leaves using chloroform (1:20 w/v) over 72 hours. The resulting supernatant was air-dried, dissolved in dimethyl sulfoxide (DMSO) to prepare a 200 mg/kg stock solution, which was then diluted to 20 mg/kg for antinociceptive testing. Male Balb-C mice were divided into groups and treated with DMSO, acetylsalicylic acid (ASA, 100 mg/kg), or chloroform extract of *Dicranopteris linearis* (CEDL) at doses of 10, 100, or 200 mg/kg for abdominal constriction and hot plate tests. For the formalin test, rats received similar treatments. CEDL demonstrated significant dose-dependent antinociceptive effects, with 20 mg/kg showing efficacy comparable to ASA, reducing abdominal constrictions by 50%. Higher doses (100 and 200 mg/kg) were even more effective, reducing constrictions by about tenfold. CEDL was effective in both early and late phases of the formalin test, outperforming 5 mg/kg morphine. In the hot plate test, CEDL showed sustained dose-dependent activity, with 100 and 200 mg/kg effective throughout the experiment, while the 20 mg/kg dose was effective for 4 hours. In the same way, Zakaria et al. (2008) prepared an aqueous extract of *Dicranopteris linearis* (AEDL) by soaking dried leaves in distilled water (1:20 w/v) for 72 hours, filtering, and freeze-drying the supernatant. From this, solutions of AEDL at 10%, 50%, and 100% concentrations were made, along with higher doses of 660 and 1320 mg/kg. Male BALB/c mice and Sprague-Dawley rats were used for antinociceptive testing, with treatments including AEDL, ASA, and morphine. Mice underwent abdominal constriction and hot-plate tests, while rats were tested using the formalin test. In the abdominal constriction test, AEDL showed significant dose-dependent antinociceptive effects, with the highest dose (1320 mg/kg) causing complete inhibition of constrictions. In the hot-plate test, AEDL demonstrated significant, dose-independent activity that lasted throughout the experiment, outperforming morphine in terms of duration. In the formalin test, AEDL exhibited significant antinociceptive effects in both the early and late phases of pain, with notable activity at doses between 66.0 and 660 mg/kg. The 1320 mg/kg dose was excluded from analysis due to lethality, but AEDL effectively reduced pain responses in the abdominal constriction, hot plate, and formalin tests. The findings suggest that AEDL has potent antinociceptive activity through both peripheral and central mechanisms, blocking nociceptive effects from chemical and thermal stimuli. Continuing the above study, Zakaria et al. (2012) investigated the antinociceptive activity of distilled water extracts of *D. linearis* leaves, prepared at doses of 300, 500, and 1000 mg/kg. They used 264 male Balb-C mice, divided into 11 groups, which received subcutaneous injections of the extracts, distilled water (as a vehicle), or ASA, as a positive control). 30 minutes later, the mice were tested using the acetic acid-induced abdominal constriction method. The number of constrictions was counted over 25 minutes. To explore the mechanisms involved, the study included 13 additional groups pre-treated with various antagonists or blockers, including naloxone, atropine, and yohimbine. Significant antinociceptive activity was observed with all extract doses, showing dose-dependent effects. Pretreatment with naloxone, atropine, yohimbine, and theophylline significantly reversed the antinociceptive effects. The results suggest that *D. linearis* exhibits antinociceptive activity, likely through opioid, muscarinic, $\alpha 2$-adrenergic, and adenosine receptors. This indicates its potential as a natural pain-relief agent and a candidate for future drug development.

In another study, Zakaria et al. (2020b) prepared a methanol extract of *D. linearis* leaves by soaking the powdered leaves in methanol (1:20 w/v) for 72 hours. After filtering and evaporating the supernatant at 40°C under reduced pressure, a crude methanol extract (MEDL) was obtained.

Male ICR mice (25–35 g; 5–7 weeks old) were used to assess MEDL's antinociceptive activity. For the abdominal constriction test, mice ($n = 6$) received oral treatments of MEDL (25, 150, 300 mg/kg), vehicle (2% DMSO), or acetylsalicylic acid (ASA; 100 mg/kg) before intraperitoneal administration of 0.6% acetic acid. The number of abdominal constrictions was counted to determine antinociceptive activity. In the hot plate test, mice ($n = 6$) were pretreated with MEDL (25, 150, 300 mg/kg), morphine (MOR; 5 mg/kg), or vehicle, and latency to nociceptive response was measured. In the formalin test, mice ($n = 6$) received oral treatments before a 5% formalin injection, and the time spent licking the injected paw was recorded in the early (0–5 minutes) and late (15–30 minutes) phases. To ensure MEDL did not cause muscle relaxation or sedation, a rotarod test was conducted on mice treated with MEDL (300 mg/kg), diazepam (DZP; 4 mg/kg), or vehicle. MEDL showed significant antinociceptive activity in the abdominal constriction test, comparable to ASA at higher doses. In the hot plate test, MEDL (150, 300 mg/kg) significantly increased response latency, with 300 mg/kg showing effects up to 120 minutes. MEDL also reduced response times in the formalin test, with dose-dependent effects in the late phase. The rotarod test indicated that MEDL (300 mg/kg) did not impair motor coordination, unlike DZP. Overall, MEDL demonstrates significant antinociceptive activity and potential as an effective pain reliever. Hossin et al. (2021) investigated the antinociceptive activity of *D. linearis* using a 96% ethanol extract from leaf powder. The extract, obtained by soaking 600 g of leaf powder in 3 L of ethanol for 72 hours, was concentrated and yielded about 3 g of crude extract, stored as *Dicranopteris linearis* Ethanolic Leaf Extract (DLET). Swiss albino mice (24–30 g, 45 days old) were divided into five groups ($n = 6$) for testing: control (vehicle), positive controls (Morphine 5 mg/kg, ASA 100 mg/kg), and DLET groups (100, 200, 300 mg/kg). To assess central antinociceptive activity, methods included tail pressure, tail flick, tail immersion, and hot plate tests. Mice underwent these tests before and after drug administration to measure pain responses. The tail pressure test used a metal clip to induce mechanical pain, while the tail flick and immersion tests applied radiant heat and hot water, respectively. The hot plate test assessed response to thermal stimuli. For peripheral antinociceptive evaluation, acetic acid-induced abdominal constriction and formalin-induced paw licking tests were used. Acute toxicity was tested by administering high doses of DLET (100, 250, 500, 1000 mg/kg), revealing no significant adverse effects except indigestion at the highest dose, leading to the use of lower doses in main experiments. The obtained results showed that DLET exhibited dose-dependent antinociceptive effects. In the tail pressure test, DLET (400 mg/kg) had increasing efficacy up to 120 minutes, while morphine was most effective at 60 minutes. In the tail flick and immersion tests, DLET showed increasing analgesia over time but was less effective than morphine. The hot plate test showed that morphine had sustained analgesia, while DLET's effects were moderate. In the abdominal constriction test, DLET (400 mg/kg) showed moderate pain relief, and in the formalin test, DLET (200, 400 mg/kg) reduced pain in the late phase. Overall, DLET demonstrated moderate antinociceptive activity in both central and peripheral models, suggesting potential for pain relief. Future studies should focus on optimal dosing, solvent selection, and phytochemical analysis.

The studies conducted by Zakaria et al. (2006, 2008, 2012, 2020b) and Hossin et al. (2021) have systematically evaluated the antinociceptive activity of various extracts of *D. linearis* using different pain assessment models. The findings from these studies demonstrate that *D. linearis* possesses significant antinociceptive properties, which were effective in both peripheral and central pain models. The extracts, whether prepared using chloroform, aqueous, methanol, or ethanol, consistently showed dose-dependent antinociceptive effects, often comparable to or superior to standard analgesics like ASA and morphine. The results highlight the potential of *D. linearis* as a natural source of pain-relieving agents, indicating its relevance in traditional medicine and its promise for future pharmacological applications. The studies suggest that the antinociceptive activity of *D. linearis* may involve multiple mechanisms, including opioid receptor activation and inhibition of inflammatory mediators. However, further research is necessary to optimize extraction methods, identify the specific bioactive compounds responsible for these effects, and determine safe and effective dosing regimens. The findings of these studies lay a strong foundation for the potential development of *D. linearis* into a viable natural alternative for pain management.

9.4.5 ANTIOXIDANT ACTIVITY

Antioxidant activity refers to the ability of certain substances to counteract the damaging effects of free radicals or prevent oxidative stress in the body.

Antioxidants are compounds that can safely interact with free radicals and terminate the chain reaction before vital molecules are damaged. They act as scavengers, helping to prevent cell and tissue damage and reducing the risk of inflammation, cancer, heart disease, and other conditions. Many medicinal plants have been used in traditional medicine systems for centuries. Evaluating their antioxidant activity can provide scientific validation for their traditional uses. To evaluate the antioxidant activity of medicinal plant extracts, researchers have used various *in vitro* assays to determine the antioxidant potential. The obtained results from these assays can guide the selection of plants or plant compounds for further research and potential therapeutic use.

Zakaria et al. (2011) evaluated the antioxidant potential of *D. linearis* extracts. Leaves were air-dried, powdered, and soaked in distilled water, chloroform, and methanol (1:20 w/v) for 72 hours. The aqueous extract (AEDL) was freeze-dried, while chloroform (CEDL) and methanol (MEDL) extracts were concentrated under reduced pressure. AEDL was dissolved in water, and CEDL/MEDL in DMSO at 100 μg/ml. The antioxidant activity was assessed using DPPH radical scavenging and superoxide anion radical scavenging assays. For the DPPH assay, samples in ethanol were reacted with 200 mM DPPH in methanol at 37°C for 30 minutes, and the absorbance at 520 nm was measured. The superoxide anion scavenging activity was determined by assessing the reduction of nitro blue tetrazolium (NBT) following a reaction with xanthine/xanthine oxidase. Results showed that MEDL demonstrated the highest antioxidant activity, with 37.8% to 99.7% inhibition in the DPPH test and 84.2% to 99.7% scavenging in the superoxide test, outperforming AEDL and CEDL. These findings highlight MEDL's significant antioxidant potential.

In another study, Kamisan et al. (2014) investigated the antioxidant activity of *D. linearis* leaves. The leaves were shade-dried for a week and ground into powder. A total of 160 g of the powder was soaked in methanol (1:20 w/v) for 72 hours, then filtered and evaporated under reduced pressure, yielding 48.4 g (30.3%) of crude extract (MEDL). The antioxidant potential of MEDL was evaluated using three assays: DPPH radical scavenging, superoxide anion radical scavenging, and Oxygen Radical Absorbance Capacity (ORAC). For the DPPH assay, 40 μl of a 1 mg/ml MEDL solution was mixed with 50 μl of 1 mM DPPH solution and 150 μl ethanol. Absorbance was measured at 520 nm after 30 minutes. For the superoxide anion assay, superoxide radicals were generated using a PMS-NADH system, and MEDL (25–50 μg) was tested for its ability to reduce NBT. Absorbance was measured at 560 nm with L-ascorbic acid as a control. In the ORAC assay, a fluorescein solution and MEDL were incubated with AAPH to generate peroxyl radicals. Fluorescence intensity was measured at 485 nm excitation and 520 nm emission over 3 hours. Results showed that MEDL exhibited significant antioxidant activity at a concentration of 200 μg/ml. It achieved a DPPH radical scavenging activity of 98.94 ± 1.14%, comparable to ascorbic acid. The superoxide anion scavenging activity was 93.2 ± 1.81%, and the ORAC value was 24,272.50 ± 2056.53 μM Trolox Equivalent (TE)/100 g. These findings indicate that MEDL possesses potent antioxidant activity, highlighting its potential as a natural antioxidant source for combating oxidative stress-related conditions.

Ponnusamy et al. (2015) evaluated the antioxidant activity of *D. linearis* leaf extracts. Freeze-dried leaf powder was extracted using methanol with ultrasound assistance for 20 minutes, repeated twice. The combined extracts were filtered, evaporated, and partitioned between ethyl acetate and water. The ethyl acetate fraction (5 g) was chromatographed on silica gel with a chloroform-MeOH gradient, isolating seven subfractions: sF1 (0.01 g), sF2 (0.27 g), sF3 (1.31 g), sF4 (0.95 g), sF5 (0.74 g), sF6 (0.25 g), and sF7 (0.06 g). Antioxidant activities were assessed using DPPH radical scavenging and Ferric Reducing Antioxidant Potential (FRAP) assays. For the DPPH assay, 10 μl of each subfraction (1 mg/ml) was mixed with 190 μl of 0.2 mM DPPH solution, incubated for 30 minutes in the dark, and absorbance was measured at 517 nm. For the FRAP assay, 150 μl of the FRAP reagent was added to 20 μl of sample (1 mg/ml), incubated at 37°C for 30 minutes, and absorbance was measured at 593 nm. Both assays were quantified using Trolox calibration curves. The results showed that the methanol extract's subfractions had enhanced antioxidant activity compared to the crude extract. Notably, sF5 exhibited the strongest DPPH scavenging effect, followed by sF7 and sF4, showing 50–70% activity relative to Trolox. In the FRAP assay, sF5 also displayed the highest antioxidant effect, with sF4 showing greater activity than sF7. These three subfractions demonstrated ferric ion reductive capacities of 25% to 40% of Trolox, while other subfractions had lower activity.

In a study, an *in vitro* free radical scavenging assay was performed using 2,2-diphenyl-1-picrylhydrazyl (DPPH) to evaluate antioxidant activity (Rajesh et al., 2016). Approximately 1 g of dried powder from the whole plant of *D. linearis* was extracted with 20 mL each of 75% ethanol, acetone, chloroform, aqueous, and petroleum ether using an ultra-turrax mixer at 13,000 rpm for 1 minute, followed by overnight soaking at room temperature. The mixture was then filtered through Whatman No. 1 paper using a Buchner funnel, evaporated under vacuum using a rotavapor at 40°C to a constant weight, and dissolved in the respective solvents. The dissolving rate of the crude extracts was approximately 100%, and the solutions were stored at 18°C until use. For the DPPH assay, 100 μl of each extract was mixed with 2.7 ml of methanol, followed by 200 μl of 0.1% methanolic DPPH. The mixture was incubated in darkness for 30 minutes. A blank sample with methanol and DPPH served as a control. Absorption readings were taken every 5 minutes using a UV double-beam spectrophotometer at 517 nm. Antioxidant activity was compared to a 0.16% BHT standard. The results indicated that the acetone extract exhibited the highest antioxidant potential at 96.0%, while the petroleum ether extract showed the lowest activity at 52.7%. Kalpanadevi et al. (2016b) assessed the antioxidant potential of *D. linearis* using extracts from dried plant powder. The powder was extracted with 20 mL of 75% ethanol, acetone, chloroform, aqueous solution, and petroleum ether for 1 minute with an Ultra-Turrax mixer at 13,000 rpm. After soaking overnight at room temperature, the mixtures were filtered through Whatman No. 1 paper and evaporated under vacuum at 40°C until a constant weight was achieved. The dried extracts were dissolved in their respective solvents, reaching an approximate 100% dissolution rate, and stored at 18°C for further use. Antioxidant activity was evaluated using the 2,2-diphenyl-1-picrylhydrazyl (DPPH) free radical assay. Extracts (100 μL) were mixed with 2.7 mL methanol and 200 μL of 0.1% methanolic DPPH, and incubated in the dark for 30 minutes. Absorption maxima were measured at 517 nm using a UV spectrophotometer, with a control blank prepared with methanol and DPPH. The antioxidant activity was compared with a synthetic standard (0.16% BHT), and the experiment was performed in triplicates. Results indicated that the acetone extract exhibited the highest antioxidant potential, with a 96% inhibition rate, while the petroleum ether extract showed the lowest activity at 52.7%. This study underscores the significance of solvent choice in optimizing antioxidant extraction from *D. linearis*, which is crucial for developing antioxidant-rich health and wellness products.

Zakaria et al. (2019) investigated the antioxidant potential of *D. linearis*. The dried leaves were ground into a coarse powder and extracted with absolute methanol (1:20 w/v) for 72 hours, repeating the process three times to prepare the methanolic extract (MEDL). About 20 g of MEDL was mixed with 1000 mL methanol and 200 mL distilled water, then partitioned using petroleum ether (PE) and ethyl acetate (EA). The PE fraction (PEDL) and EA fraction (EADL) were separated, filtered, and concentrated, while the remaining aqueous solution was freeze-dried to obtain the aqueous partition (AQDL). The antioxidant potential of AQDL was assessed through DPPH, superoxide anion radical scavenging, and oxygen radical absorbance capacity (ORAC) assays. For the DPPH assay, AQDL was mixed with DPPH and ethanol, and absorbance was measured at 520 nm after 30 minutes. In the superoxide anion radical scavenging assay, superoxide radicals were generated and measured by the reduction of nitroblue tetrazolium (NBT) at 560 nm. The ORAC test involved reacting AQDL with a peroxyl-radical generator and measuring fluorescence every minute for 3 hours using a BMG Omega Fluostar Fluorescence Spectrophotometer. Results showed that AQDL had strong antioxidant properties, with DPPH and superoxide radical scavenging activities ranging from 75% to 85%, compared to 200 μg/mL ascorbic acid and SOD standards. AQDL also demonstrated a high ORAC value (>15,000 μmol TE/100 g), comparable to 200 μg/mL Trolox. These findings suggest that AQDL is a potent antioxidant and could be valuable for developing health and wellness products. To evaluate the antioxidant potential, Zakaria et al. (2020b) took dried *D. linearis* leaves at 40°C for two weeks and ground them into powder. They extracted 160 g of the powder with absolute methanol (1:2 w/v) for 72 hours, repeating this process twice more. The supernatants were filtered through cloth, cotton wool, and Whatman No. 1 filter paper, then evaporated under reduced pressure at 40°C, resulting in a methanolic extract (MEDL). This was concentrated, and 2 g of the extract was partitioned sequentially with petroleum ether and ethyl acetate. The final residue was freeze-dried and designated as the ethyl acetate partition (EADL). The antioxidant potential of EADL was evaluated using DPPH, superoxide anion, and Oxygen Radical Absorbance Capacity (ORAC) assays. For the DPPH assay, 50 μL of EADL (1 mg/mL) was mixed with 50 μL of 1 mM DPPH in ethanol and 150 μL of absolute ethanol, and the absorbance was measured at 520 nm after 30 minutes. In the superoxide anion assay, the mixture included Tris-HCl buffer, NBT, NADH, and

EADL, and the absorbance was recorded at 560 nm after 5 minutes. For the ORAC test, sodium fluorescein and EADL (200 μg/mL) were incubated with AAPH, and fluorescence intensity was measured every minute for 3 hours using a spectrophotometer. Results showed EADL achieved 94% DPPH radical scavenging activity at 200 μg/mL, with an IC50 value of 76 μg/mL, and 93% superoxide scavenging activity. The ORAC assay indicated EADL had an antioxidant capacity of approximately 555 μmol Trolox Equivalent (TE) per gram of dry extract. These results highlight EADL's strong antioxidant properties, suggesting its potential in health and therapeutic applications.

In the same way, Zakaria et al. (2021) extracted *D. linearis* leaf powder using chloroform (1:20 w/v) over 72 hours, repeating the process twice more. The combined extract was concentrated at 40°C to yield the chloroform extract (CEDL), which was formulated into doses of 50, 250, and 500 mg/kg using 8% Tween 80. To assess its antioxidant potential, the DPPH radical scavenging assay was conducted, measuring absorbance at 520 nm to determine the percentage inhibition (I%) and effective concentration (EC50). The superoxide anion radical scavenging assay involved reacting the extract with superoxides generated from NBT solution, xanthine, and EDTA, with absorbance recorded at 560 nm. The ORAC assay mixed the extract with a phosphate buffer solution (PBS), followed by sodium fluorescein and AAPH, and fluorescence was monitored over two hours. The DPPH assay revealed minimal antioxidant activity for CEDL at 200 μg/ml, with about 15% radical scavenging activity, and no IC50 value was observed. Similarly, the superoxide anion assay showed less than 30% radical scavenging activity at the same concentration. However, CEDL demonstrated substantial antioxidant capacity in the ORAC assay, with an ORAC value of approximately 47,000 μmol TE/100 g dry weight. These results indicate that while CEDL shows limited antioxidant activity in the DPPH and superoxide anion assays, it exhibits significant antioxidant potential in the ORAC assay, highlighting the importance of using multiple assays to thoroughly evaluate the antioxidant properties of plant extracts.

In another study, Thomas et al. (2023) investigated the antioxidant potential of *D. linearis* by preparing methanol extracts of the fronds using the Soxhlet extraction method. About 30 grams of shade-dried, powdered fronds were extracted with 150 ml of methanol over 8–12 hours. The extracts were then concentrated under vacuum using a rotary evaporator. The antioxidant activity was assessed using the DPPH radical scavenging assay, where 1 ml of varying concentrations of the extract was mixed with 1 ml of DPPH solution (0.1 mM) and incubated for 30 minutes in the dark. Absorbance was measured at 517 nm, with ascorbic acid as the reference standard. The study found that the DPPH scavenging activity of the methanol extract increased with concentration. At the lowest concentration (62.5 μg/ml), the extract showed 24.55% scavenging activity, which significantly rose to 98.83% at the highest concentration (1000 μg/ml). These findings suggest that *D. linearis* is a potent source of natural antioxidants, capable of effectively combating oxidative stress and related diseases. Ahmad et al. (2023) evaluated the antioxidant activity of encapsulated *D. linearis* extracts using the DPPH radical scavenging assay. A mixture of 200 μL of the encapsulated extracts and 200 μL of 60 μM DPPH was incubated at room temperature in the dark for 30 minutes. The absorbance was measured at 517 nm, with ascorbic acid serving as the positive control. Among the encapsulated extracts, methylcellulose chloroform (MCCH) showed the highest scavenging activity at 86.40%, followed by ethyl cellulose encapsulated chloroform (ECCH) at 83.2%, and ethyl methyl cellulose encapsulated chloroform (ECMCCH) at 78.60%. The study highlighted methyl cellulose (MC) as an effective encapsulation material for preserving and enhancing the antioxidant properties of *D. linearis* chloroform extracts. The high scavenging activities of MC encapsulated extracts, particularly MCCH, suggest that the bioactive properties of the extracts were well-maintained post-encapsulation. This method also effectively released polyphenols from the MC polymer matrix, indicating that MC microencapsulation could be a promising approach for improving the stability and bioavailability of antioxidants. These findings support the potential use of MCCH as a stable and effective antioxidant agent for various applications. The antioxidant activities of *D. linearis* were explored through various extraction methods and assays in multiple studies, demonstrating significant variations based on the extraction solvent and methodology used. Across these studies, methanolic extracts (MEDL) consistently exhibited the highest antioxidant potential, as evidenced by their superior DPPH and superoxide anion radical scavenging activities and high Oxygen Radical Absorbance Capacity (ORAC) values. Notably, the subfractions derived from MEDL also showed enhanced antioxidant properties, with specific subfractions like sF5 demonstrating strong activity comparable to known standards such as Trolox and ascorbic acid. The effectiveness of different extracts varied, with aqueous and ethyl acetate partitions also displaying noteworthy antioxidant capacities, though generally lower than those of the methanolic extracts. Conversely, chloroform extracts (CEDL) exhibited minimal activity in DPPH and superoxide anion scavenging assays but showed significant ORAC values, highlighting the importance of assay selection in evaluating antioxidant properties. Overall, these findings underscore the potential of *D. linearis* as a valuable source of natural antioxidants, mainly through methanolic extraction. The diversity in antioxidant activity among different extracts suggests that *D. linearis* could be harnessed for various applications in health and therapeutic interventions, depending on the desired antioxidant profile. Further studies could focus on isolating and characterizing the specific bioactive compounds responsible for these activities, paving the way for their potential use in combating oxidative stress-related conditions.

9.4.6 ANTIPYRETIC ACTIVITY

Antipyretic activity refers to the ability of a substance to reduce fever. Fever is a typical physiological response to infection or illness, often caused by the release of pyrogens (fever-inducing substances) that affect the hypothalamus in the brain, leading to an increase in body temperature. Antipyretic agents work by inhibiting the synthesis of prostaglandins, particularly prostaglandin E2 (PGE2), which elevates the set point of body temperature in the hypothalamus. Many traditional medicinal systems, such as Ayurveda and Traditional Chinese Medicine, have long used plants to treat fever, which are often more accessible and sustainable, especially in regions with limited access to modern pharmaceuticals. Evaluating their antipyretic activity can support the development of affordable and locally available remedies. Zakaria et al. (2008) evaluated the antipyretic activity of *D. linearis* by preparing an aqueous extract (AEDL) from air-dried leaves. The extract was tested at doses of 13.2, 66.0, 132.0, 660, and 1320 mg/kg, with ASA and distilled water (dH_2O) serving as controls. Pyrexia was induced in test subjects using a Brewer's yeast (BY) suspension, and rectal temperatures were measured at various intervals up to 8 hours post-treatment. The study found that AEDL exhibited significant, dose-independent antipyretic activity. Specifically, the 13.2 and 66.0 mg/kg doses effectively reduced fever for up to 6 hours, while the 132.0 mg/kg dose significantly lowered body temperature within 1 hour of administration. All AEDL doses normalized rectal temperatures rapidly, comparable to the effect of ASA. Notably, the extract alone did not affect normal body temperature, indicating its targeted antipyretic effect. The findings suggest that AEDL has a potent antipyretic effect, supporting its traditional use in managing fever. The rapid normalization of body temperature by AEDL highlights its potential as a natural antipyretic agent. This study underscores the need for further research to explore the mechanisms behind AEDL's antipyretic properties and its therapeutic potential in fever treatment.

9.4.7 CHEMOPREVENTIVE ACTIVITY

Chemopreventive activity refers to the ability of a substance to prevent, inhibit, or delay the development of cancer. This activity involves blocking or reversing the process of carcinogenesis. Chemoprevention aims to intervene before cancer fully develops, potentially reducing the incidence and severity of cancer in high-risk populations. Evaluating chemopreventive activity using medicinal plant extracts is valuable in cancer prevention research, offering the potential for developing safe and effective natural therapies. Rodzi et al. (2013) evaluated the chemopreventive potential of *D. linearis* methanolic extract (MEDL) in an *in vivo* study using female ICR mice. The study involved soaking powdered *D. linearis* leaves in methanol, followed by filtration, concentration, and storage of the extract for further experiments. 48 mice were divided into six groups: three treatment groups receiving 30, 100, and 300 mg/kg of MEDL, a positive control group (10 mg/kg curcumin), a negative control group, and a vehicle control group (acetone). The experiment employed a two-stage carcinogenesis protocol using 7,12-dimethylbenz[a]anthracene (DMBA) as the carcinogen, followed by repeated applications of croton oil for 15 weeks. Tumor development was monitored weekly, including parameters such as body weight, latency period, tumor incidence, burden, and volume. After 15 weeks, the mice were sacrificed, and skin samples were collected for histopathological analysis. Results showed that tumors in the carcinogen control group were most numerous and largest, while no tumors were observed in the vehicle control group. Tumor number and size decreased with increasing MEDL doses, with the 300 mg/kg dose showing the most significant reduction, almost eliminating tumors. Tumor incidence was highest in the carcinogen control group (87.5%) and lowest in the 300 mg/kg MEDL group (12.5%). The latency period for tumor appearance was extended to 15 weeks in the high-dose MEDL group, compared to 6 weeks in the carcinogen control group. Histopathological analysis revealed severe skin abnormalities in the carcinogen control group, while the high-dose MEDL group showed minimal epidermal changes. The study concluded that higher doses of MEDL significantly reduced tumor number, size, and incidence, delayed tumor onset, and resulted in minimal skin abnormalities. These findings suggest that MEDL has a dose-dependent protective effect against carcinogen-induced tumor promotion, indicating its potential as a therapeutic agent in cancer prevention.

9.4.8 CYTOTOXIC ACTIVITY

Cytotoxic activity refers to the ability of a substance to cause damage to cells, leading to cell dysfunction or death. This property is particularly significant in cancer treatment, where the goal is to eliminate malignant cells without affecting healthy ones. Cytotoxicity is typically assessed by measuring cell viability, proliferation, or survival reduction after exposure to a test substance. Medicinal plants are rich sources of bioactive compounds that may offer novel mechanisms of action against cancer cells. An ideal cytotoxic agent should target cancer cells specifically while sparing normal cells. Evaluating plant extracts helps to identify those with selective cytotoxicity. Zakaria et al. (2011) evaluated the cytotoxic activity of *D. linearis* by air-drying its leaves, grinding them into powder, and extracting them using distilled water, chloroform, and methanol. The resulting crude extracts, labeled as AEDL, CEDL, and MEDL, were stored at 4°C and dissolved in appropriate solvents for testing. The MTT assay was employed to assess the antiproliferative effects of these extracts on various cancer cell lines (MCF-7, HeLa, HL-60, K-562, HT-29, 3T3, and MDA-MB-231) cultured in RPMI 1640 medium with fetal bovine serum and antibiotics. Cells were plated in 96-well plates, and the extracts were added at concentrations of 12.5, 25, 50, and 100 μg/ml, followed by a 72-hour incubation. Absorbance was measured at 550 nm to determine cell viability, with cytotoxicity expressed as the IC50 value, indicating the concentration required to inhibit 50% of cell growth. Results showed that DMSO alone had no antiproliferative effects, and none of the extracts were cytotoxic to normal 3T3 cell lines. However, CEDL and MEDL exhibited selective antiproliferative activity against several cancer cell lines. CEDL and MEDL showed IC50 values of 92 and 99 μg/ml, respectively, against MCF-7 cells; 23 and 64 μg/ml against HeLa cells; and 27 and 9 μg/ml against HL-60 cells. For the K-562 cell line, the IC50 values were 88 and 67 µg/ml, respectively. MEDL alone inhibited MDA-MB-231 cells with an IC50 of 40 µg/ml, while none of the extracts were effective against HT-29 cells. This study highlights the selective antiproliferative potential of CEDL and MEDL extracts against specific cancer cell lines, suggesting their promise as therapeutic agents. Further research is needed to optimize their efficacy and explore their underlying mechanisms.

In another study, Thomas et al. (2023) extracted fronds of *D. linearis* with methanol to know the cytotoxicity against rat skeletal muscle cell line (L6 cell line) using the MTT assay. L6 cells were seeded in 96-well plates and incubated under controlled conditions. The test samples were prepared in DMEM media, sterilized, and added to the cultured cells at varying concentrations (6.25 to 100 µg/mL). After 24 hours of incubation, the media was replaced with an MTT solution to develop formazan crystals, followed by DMSO addition to dissolve the crystals. The absorbance was measured at 570 nm. The results showed minimal cytotoxicity, with cell viability remaining above 94% even at the highest concentration of 100 µg/mL. This suggests that the *D. linearis* extracts are well-tolerated by L6 cells, indicating low toxicity and the potential for safe use in therapeutic applications. However, further research is necessary to confirm these findings across other cell lines and *in vivo* models to fully establish the safety and efficacy of *D. linearis* extracts for broader applications. Baharuddin et al. (2018) explored the anticancer potential of methanol (MEDL) and petroleum ether (PEEDL) extracts from *D. linearis* leaves, focusing on their ability to inhibit cancer cell proliferation by inducing S-phase arrest and apoptosis. For this, leaves were collected, dried, and ground into a powder, which was then soaked in methanol and petroleum ether for extraction. The methanol extract yielded 3.52 g (MEDL), while the petroleum ether extract yielded 0.15 g (PEEDL). Stock solutions were prepared at a concentration of 100 mg/mL and further diluted for testing against various cancer cell lines, including breast (MCF-7 and MDA-MB-231), cervical (HeLa), colon (HT-29), liver (HepG2), and lung (A549) cells. The extracts were also tested on mouse fibroblast cells (3T3) to evaluate selectivity. The study found that MEDL significantly inhibited the growth of MDA-MB-231 breast cancer cells with an IC50 value of 22.4 mg/mL, whereas PEEDL showed no cytotoxic effects on the tested cancer cell lines. MEDL induced S-phase cell cycle arrest and apoptosis in MDA-MB-231 cells, as confirmed by morphological analysis, acridine orange-propidium iodide staining, and flow cytometry. Early apoptosis was further validated using annexin V-FITC and PI staining, revealing an increase in apoptotic cells from 15.07% after 24 hours to 18.24% after 48 hours of treatment. The findings suggest that methanol extract from *D. linearis* leaves could be a promising anticancer agent, particularly against MDA-MB-231 breast cancer cells, due to its ability to induce S-phase arrest and apoptosis. This study supports further investigation into MEDL as a targeted cancer therapy.

The studies on *D. linearis* leaves have provided valuable insights into the cytotoxic activity of its extracts against various cancer cell lines. The findings indicate that the chloroform and methanol extracts (CEDL and MEDL) exhibit selective antiproliferative effects on specific cancer cell lines, such as MCF-7, HeLa, HL-60, K-562, and MDA-MB-231, without significant toxicity to normal cells. The methanol extract, in particular, has shown promise by inducing S-phase arrest and apoptosis in MDA-MB-231 breast cancer cells, suggesting a potential mechanism for its anticancer activity. These results warrant further research to optimize these extracts' efficacy and explore their therapeutic potential in cancer treatment. Additionally, the minimal cytotoxicity observed in the rat skeletal muscle cell line (L6) by the *D. linearis* extracts supports their safety and potential for use in various applications. Overall, the cytotoxic activity of *D. linearis* extracts represents a significant area of research that could lead to the development of new anticancer agents from natural sources.

9.4.9 GASTROPROTECTIVE EFFECT

The gastroprotective effect refers to the ability of a substance to safeguard the stomach lining from damage caused by various factors such as gastric acid, nonsteroidal anti-inflammatory drugs, alcohol, or stress. This protective action is crucial in preventing conditions like gastritis, peptic ulcers, and gastric bleeding. Many medicinal plants have been traditionally used to treat gastrointestinal disorders. Evaluating their gastroprotective effects can provide scientific validation for these traditional uses and identify new natural remedies for gastrointestinal protection. Researchers use various experimental models to assess the gastroprotective effect of medicinal plant extracts, including *in vitro* studies and *in vivo* animal models. A study by Hussaini et al. (2012) explored the gastroprotective effects of *Dicranopteris linearis* ethanolic leaf extract (DLELE) against ethanol-induced gastric mucosal injury in rats. To prepare the extract, 100 grams of dried leaf powder was soaked in 500 ml of 95% ethanol for three days, filtered, and concentrated using a rotary evaporator. The resulting extract was dissolved in a 0.5% carboxymethyl cellulose (CMC) solution and administered orally to rats at doses of 250 and 500 mg/kg. The study involved healthy male Sprague Dawley rats, divided into four groups of six. Gastric ulcers were induced by administering absolute ethanol via orogastric intubation. Control groups received either a vehicle (CMC) or a standard treatment with omeprazole (20 mg/kg), while experimental groups received DLELE at the specified doses. One hour post-treatment, the rats were euthanized, and their stomachs were examined for mucus production, gastric juice pH, ulcer formation, and histological changes. Results showed that pretreatment with DLELE significantly decreased gastric acidity and increased mucus production compared to the ulcer control group. The extract notably reduced gastric ulcer areas, with rats treated with 500 mg/kg DLELE showing marked flattening of gastric mucosal folds. DLELE's protective effects were dose-dependent and comparable to omeprazole. Histological analysis confirmed these findings, revealing less mucosal damage, reduced ulcer areas, and minimal inflammation in DLELE-treated rats. In conclusion, DLELE demonstrated significant gastroprotective effects, reducing gastric acidity, increasing mucus production, and providing substantial protection against ethanol-induced gastric ulcers. Its efficacy was comparable to the standard drug omeprazole, indicating DLELE's potential as a therapeutic agent for preventing and treating gastric ulcers.

9.4.10 HEPATOPROTECTIVE ACTIVITY

Hepatoprotective activity refers to the ability of a substance to protect the liver from damage caused by various factors such as toxins, drugs, or diseases. The liver metabolizes drugs, detoxifies harmful substances, and synthesizes essential proteins. Liver damage can lead to serious health issues, including liver failure, jaundice, and cirrhosis. Evaluating the hepatoprotective activity of medicinal plant extracts can help identify natural substances that support liver health and prevent liver damage. Researchers use various experimental models to assess the hepatoprotective activity of medicinal plant extracts, including *in vitro* studies and *in vivo* animal models. Din et al. (2012) studied the hepatoprotective effects of *Dicranopteris linearis* chloroform leaf extract (CEDL) on liver toxicity induced by paracetamol (PCM) and carbon tetrachloride (CCl_4) in rats. The leaves were dried, powdered, and soaked in chloroform, yielding approximately 2.8 grams of dried extract. Adult male Sprague-Dawley rats were used for the study, and CEDL was administered orally in doses of 50, 250, and 500 mg/kg over seven days. Silymarin (200 mg/kg) was used as a reference drug. An acute toxicity study with a single dose of 5000 mg/kg CEDL showed no toxicity or mortality. For the hepatoprotective assay, liver toxicity was induced by administering PCM (3 g/kg) or CCl_4 (1.5 ml/kg). After inducer administration, blood was collected for biochemical analysis, and livers were examined histologically. Results showed that PCM and CCl_4 significantly increased liver enzyme levels (AST and ALT), indicating liver damage. However, CEDL significantly reduced these levels in a dose-dependent manner, suggesting hepatoprotective effects. Histopathological analysis revealed that PCM and CCl_4 caused severe liver damage, including necrosis, leukocyte infiltration, hemorrhaging, and steatosis. Pretreatment with CEDL reduced these effects, with the highest dose (500 mg/kg) restoring liver morphology to near-normal conditions. In conclusion, CEDL demonstrated significant hepatoprotective effects against PCM and CCl_4-induced liver toxicity in rats. The extract was safe at high doses and effectively reduced elevated liver enzyme levels, preserving liver architecture and reducing damage. These findings suggest that CEDL could be a promising natural agent for liver protection, warranting further research for therapeutic use.

Kamisan et al. (2014) evaluated the hepatoprotective effects of methanolic extract of *Dicranopteris linearis* (MEDL) using male Sprague Dawley rats and male ICR mice. The study comprised six groups of six rats each, with Group I receiving 10% DMSO for 7 days followed by 50% olive oil on the seventh day, serving as a control. Group II received 10% DMSO for 7 days followed by CCl_4 on the seventh day to induce liver damage. Group III was treated with 200 mg/kg silymarin for 7 days, then CCl_4 on the seventh day. Groups IV, V, and VI were given 50 mg/kg, 250 mg/kg, and 500 mg/kg of MEDL, respectively, for 7 days, followed by CCl_4 on the seventh day. On the seventh day, all groups except Group I received 1 ml/kg of 50% CCl_4. After 48 hours, the rats were euthanized, and blood was collected for biochemical analysis of ALT, ALP, and AST levels. Livers were also examined for biochemical and histopathological analysis. Plasma samples were analyzed with a Hitachi 902 Automatic Chemical

Analyzer, and liver tissues were prepared for histopathology. Results showed that the CCl_4-treated group (Group II) had significantly elevated ALT, AST, and ALP levels, indicating liver damage. MEDL treatment at all doses reduced these enzyme levels in a dose-dependent manner, demonstrating hepatoprotective effects. The silymarin-treated group also showed a significant reduction in enzyme levels. Histopathological examination supported these findings, showing severe liver damage in the CCl_4-treated group and considerable improvement in the MEDL-treated groups. The silymarin-treated group exhibited liver architecture similar to the control group. In conclusion, MEDL demonstrated significant hepatoprotective activity against CCl_4-induced liver damage, effectively lowering liver enzyme levels and improving liver histology in a dose-dependent manner. This suggests MEDL's potential as a natural hepatoprotective agent for preventing and treating liver injury. Ismail et al. (2014) investigated the hepatoprotective effects of an aqueous extract of *Dicranopteris linearis* leaves (DLAE) against liver toxicity induced by PCM and CCl_4 in rats. The leaves were dried, powdered, and soaked in distilled water for 72 hours to prepare the extract, which was then freeze-dried, yielding 15.7 g from 80 g of powder. Adult male Sprague-Dawley rats, weighing 180–200 g, were used in the study. An acute toxicity test was first conducted by administering a single oral dose of 5000 mg/kg of DLAE, which showed no signs of toxicity or mortality, and the rats behaved normally. For the hepatoprotective assay, DLAE was administered orally at doses of 50, 250, and 500 mg/kg before inducing liver toxicity with PCM (3 g/kg) and CCl_4 (0.15 ml/kg). The results were compared to a vehicle group (dH_2O) and a standard drug group (200 mg/kg silymarin). The PCM and CCl_4 treatments significantly increased liver enzyme levels (ALP, ALT, and AST), indicating liver damage. However, pretreatment with DLAE significantly reduced these enzyme levels in a dose-dependent manner, demonstrating its hepatoprotective effects. Silymarin, used as a positive control, also significantly reduced enzyme levels. Histopathological analysis of liver tissues confirmed these findings. The livers of PCM and CCl_4-induced rats pretreated with dH_2O showed severe damage, including necrosis, leukocyte infiltration, and steatosis. Conversely, pretreatment with silymarin and DLAE, at all tested doses, effectively reversed these hepatotoxic effects, restoring liver architecture toward normal conditions. In conclusion, DLAE exhibited significant hepatoprotective activity, reducing elevated liver enzyme levels and restoring liver structure in rats subjected to PCM and CCl_4-induced toxicity. These findings suggest DLAE's potential as a natural hepatoprotective agent, warranting further research for its development as a treatment for liver toxicity.

Zakaria et al. (2017) studied the hepatoprotective effects of the MEDL against acetaminophen (APAP)-induced liver toxicity in rats. The extract was prepared by soaking 160 g of dried leaves in methanol, yielding 48.4 g of dried extract. To assess acute toxicity, a single 5000 mg/kg dose of MEDL was administered to mice, with observations over 14 days showing no toxic effects or mortality. Hematological and biochemical analyses revealed no significant differences in most parameters between the MEDL-treated and control groups, except for minor variations in MCV, WBC, bilirubin, and creatinine levels, which remained within normal ranges. Histological examination showed no pathological changes in vital organs. For the hepatoprotective study, six groups of rats were treated with varying doses of MEDL, a control solution, or silymarin, followed by APAP administration. MEDL pre-treatment significantly reduced elevated liver enzyme levels (ALT, AST) and improved antioxidant enzyme activities (SOD, CAT), indicating its hepatoprotective effects. Histological analysis of the liver tissue confirmed that MEDL mitigated the damage caused by APAP, preserving liver architecture, particularly at higher doses (250 and 500 mg/kg), comparable to silymarin. *In vitro* anti-inflammatory tests revealed that MEDL exhibited low inhibitory activity against lipoxygenase (LOX) and no activity against xanthine oxidase (XO). Phytochemical analysis identified several bioactive compounds in MEDL, including apigenin-7-O-glucoside, ferulic acid hexose, catechin, rutin, and gallic acid, known for their antioxidant and anti-inflammatory properties. GC-MS analysis further identified 48 volatile compounds, such as triphenylphosphine oxide and methyl-9,12,15-octadecatrienoate, which contribute to the extract's biological activities. The study concluded that MEDL effectively protects against APAP-induced liver intoxication, likely by activating the endogenous antioxidant system and through its anti-inflammatory properties. The presence of various bioactive compounds in MEDL suggests that they may act synergistically to enhance its hepatoprotective potential, making it a promising natural remedy for liver toxicity. Zakaria et al. (2019) investigated the hepatoprotective activity of *Dicranopteris linearis* by preparing a methanolic extract (MEDL) from the dried leaves. The extract was further partitioned into petroleum ether (PEDL), ethyl acetate (EADL), and aqueous (AQDL) fractions.

Male Sprague Dawley rats were divided into six groups to assess the effects of these extracts. The groups included a normal control, a hepatotoxic control, a positive control treated with silymarin, and three test groups receiving different doses of AQDL (50, 250, or 500 mg/kg). These treatments were administered daily for seven days, followed by an acetaminophen (PCM) dose to induce liver toxicity in all but the normal control group. After 48 hours, the rats were euthanized, and their blood and liver samples were collected for analysis. The study measured liver function markers such as ALT, AST, ALP, and total bilirubin, along with antioxidant enzyme activities (SOD, CAT) and malondialdehyde (MDA) levels in liver tissues. Histopathological analysis was also conducted to assess liver damage. The results showed that AQDL significantly reduced liver damage induced by PCM, as evidenced by lower liver weight and liver-to-body weight ratios compared to the untreated hepatotoxic group. AQDL treatment also led to significant decreases in serum ALT, AST, ALP, and total bilirubin levels, indicating improved liver function. Additionally, AQDL restored the activities of SOD and CAT enzymes, though it did not significantly reduce elevated MDA levels. Histopathological examination revealed that AQDL pretreatment preserved liver architecture and reduced the extent of liver damage compared to PCM-treated rats without AQDL. The study concluded that AQDL from *D. linearis* exhibits potent hepatoprotective effects, suggesting its potential therapeutic application for liver diseases and warranting further research.

In the same way, Zakaria et al. (2020a) evaluated the hepatoprotective activity of *Dicranopteris linearis* leaves by preparing a methanolic extract (MEDL), which was further partitioned into ethyl acetate (EADL). Male Sprague Dawley rats were divided into six groups for the study. Group 1 received a vehicle treatment, while Group 2 was given acetaminophen (PCM) to induce liver toxicity. Group 3 received silymarin as a positive control, and Groups 4, 5, and 6 were pre-treated with different doses of EADL (50, 250, or 500 mg/kg) before PCM administration. The study found that PCM administration significantly increased liver weight (LW) and liver-to-body weight ratios, indicating hepatotoxicity. However, pre-treatment with EADL at all doses significantly reduced these increases, similar to the effects seen with silymarin, suggesting that EADL has protective effects against PCM-induced liver damage. Additionally, PCM significantly elevated liver enzymes (ALT, AST, ALP) and total bilirubin (TB) levels, indicating liver injury. EADL pre-treatment at all doses significantly attenuated these elevations, demonstrating its ability to mitigate liver damage. Furthermore, EADL enhanced the activities of antioxidant enzymes CAT and SOD in the liver, which were reduced by PCM intoxication, indicating its role in activating the body's antioxidant defense system. Histopathological analysis supported these findings, showing that EADL pre-treatment preserved liver cellular architecture, reducing necrosis and lymphocyte infiltration compared to the PCM-intoxicated group. Overall, the study highlights EADL's potential as a hepatoprotective agent against chemical-induced liver injury, showing its effectiveness in reducing liver damage and enhancing antioxidant defenses.

Continuing the above study, Zakaria et al. (2021) conducted a study to evaluate the hepatoprotective activity of *D. linearis* leaf extract. The leaves were extracted with chloroform, and the resulting extract (CEDL) was prepared in doses of 50, 250, and 500 mg/kg using 8% Tween 80. Adult male Sprague-Dawley rats were divided into six groups, with Group 1 serving as the normal control and Group 2 as the negative control. Group 3 received 50 mg/kg NAC as a positive control, while Groups 4–6 received different doses of CEDL. On the seventh day, all groups except Group 1 were administered acetaminophen (PCM) to induce liver toxicity. After 48 hours, blood was collected for biochemical analysis, and the rats were euthanized for liver histopathology. The study found that PCM significantly increased liver weight and liver-to-body weight ratio, indicating hepatotoxicity. However, pretreatment with CEDL at 250 and 500 mg/kg, as well as with NAC, significantly reduced these increases. Serum levels of ALT and AST, which were elevated in PCM-intoxicated rats, were also significantly reduced by CEDL or NAC pretreatment. Histopathological analysis revealed severe liver damage in PCM-treated rats, including necrosis and inflammation, which was mitigated by higher doses of CEDL or NAC, preserving normal liver structure. Moreover, PCM administration significantly decreased the activities of antioxidant enzymes CAT and SOD in the liver, a marker of oxidative stress. Pretreatment with higher doses of CEDL or NAC restored these enzyme activities, suggesting that CEDL's hepatoprotective effects are likely due to its antioxidative properties. Overall, CEDL effectively protected against PCM-induced liver damage by reducing liver weight, normalizing enzyme levels, and preserving liver architecture, making it a potential candidate for treating drug-induced liver injury. In summary, the studies collectively underscore the significant hepatoprotective potential of various extracts from *D. linearis* leaves, with promising implications for liver health. The research highlights the efficacy of chloroform, methanol, and aqueous extracts of *D. linearis* in mitigating liver damage induced by toxic agents such as PCM and CCl_4. Across different studies, these extracts exhibited consistent hepatoprotective effects, evidenced by reduced liver enzyme levels, preservation of liver architecture, and improved biochemical and histopathological parameters. Overall, the various extracts of *D. linearis* consistently demonstrated hepatoprotective properties through biochemical and histopathological improvements. The absence of acute toxicity at high doses further supports the safety profile of these extracts. The ability of these extracts to restore normal liver function and architecture, alongside their impact on liver enzyme levels, suggests that *D. linearis* holds promise as a natural agent for liver protection and warrants further research for potential therapeutic applications.

9.4.11 UREASE INHIBITORY ACTIVITY

Urease inhibitory activity refers to the ability of a substance to inhibit the enzyme urease. Urease is an enzyme that catalyzes the hydrolysis of urea into ammonia and carbon dioxide. This reaction is crucial for nitrogen metabolism in various organisms. The evaluation of urease inhibitory activity typically involves measuring how much a plant extract can inhibit urease activity in a controlled setting. This is often done using *in vitro* assays where the plant extract is mixed with urease and urea, and the amount of ammonia produced is measured. It helps identify plants that can be used to manage conditions related to elevated urease activity and explore their broader applications in medicine. Ahmad et al. (2023) investigated the urease inhibitory activity of *D. linearis* extracts, both in crude and encapsulated forms. To assess this activity, urease was prepared in a phosphate buffer, and the phenol-hypochlorite reaction was used to estimate the liberated ammonia. In the experiment, urease was combined with different extracts and incubated, with clarithromycin serving as the positive control. The reaction was halted with sulfuric acid, followed by the addition of sodium nitroprusside and sodium hypochlorite solutions. Absorbance was measured at 625 nm to determine the inhibition of urease activity. The study revealed that the encapsulated extracts, particularly ECCH and MCCH, exhibited significant urease inhibitory activity, with inhibition percentages of 51.0 ± 4.9% and 50.0 ± 0.21%, respectively. The

ECMCCH extract showed the lowest inhibition at 25.8 ± 3.27%. These results suggest that the ethyl cellulose (EC) encapsulated extracts were more effective, highlighting the superior properties of EC as an encapsulation material due to its permeability, biocompatibility, and non-toxic nature. The findings underscore the potential of EC and MC encapsulations in offering gastroprotective and urease-inhibitory activities, making them promising non-antibiotic therapeutic agents for treating *H. pylori*-related gastrointestinal conditions. The encapsulated CH extracts, in particular, show promise for further development as therapeutic agents, providing new avenues for addressing gastrointestinal issues caused by urease-producing bacteria. The study emphasizes the importance of effective encapsulation techniques to enhance the bioavailability and therapeutic potential of plant-derived bioactive compounds.

9.4.12 WOUND HEALING ACTIVITY

Wound healing is a complex and dynamic process involving several stages, including hemostasis, inflammation, proliferation, and remodeling, to restore the integrity of damaged tissues. Wound healing activity refers to the ability of a substance to promote or enhance this healing process. Medicinal plants have been used for centuries to treat wounds and promote healing. Researchers use various experimental models to assess the wound healing activity of medicinal plant extracts, including *in vitro* cell culture assays and *in vivo* animal models. Ponnusamy et al. (2015) evaluated the wound-healing potential of *D. linearis* leaf extracts using an *in vivo* model. The leaves were freeze-dried, powdered, and extracted with methanol (MeOH) through ultrasound-assisted extraction. The process was repeated, and the combined extracts were filtered and evaporated under reduced pressure. The resulting extract was partitioned between ethyl acetate (EtOAc) and water, and the EtOAc fraction was chromatographed on silica gel, yielding seven subfractions (sF1-sF7). Subfraction sF5, in particular, showed significant antioxidant activity and was further tested for wound-healing efficacy. Hs27 cells were cultured and subjected to a scratch assay, where cells were treated with sF5 (10 mg/ml) in triplicate. The treatment accelerated cell migration, with visible results within 8 hours and complete wound closure by 24 hours, compared to controls, which only achieved 50% wound closure by the same time. These findings indicate that sF5 significantly enhances cell migration, promoting rapid wound healing. The study underscores the potential of the sF5 fraction from *D. linearis* leaves as a promising candidate for developing wound-healing therapies. The results were obtained through a meticulous extraction and chromatographic process, suggesting that sF5 could be valuable in creating effective wound-healing treatments.

9.5 CONCLUSION

Dicranopteris linearis, a fern with a rich history in traditional medicine and widespread distribution in tropical and subtropical regions, has been revealed as a potential source of bioactive compounds with significant pharmacological activities. This fern, known for its diverse traditional uses, has demonstrated a wide range of therapeutic properties. Recent scientific research supports the traditional applications of *D. linearis*, highlighting its efficacy in treating inflammatory conditions, infections, and various other health ailments. The plant's richness in secondary metabolites, such as flavonoids, phenolic compounds, and terpenoids, is pivotal to its diverse pharmacological activities, including antibacterial, anti-inflammatory, antioxidant, antipyretic, cytotoxic, and wound-healing properties. Phytochemical analyses have revealed the presence of numerous bioactive compounds, some of which have shown promising results in inhibiting bacterial growth, protecting the liver, and exhibiting potential anticancer properties. Additionally, compounds with significant antioxidant and antimicrobial activities further validate the traditional medicinal uses of *D. linearis*. Research on this plant underscores the importance of medicinal plants in modern healthcare, not only for their direct therapeutic benefits but also for their potential in drug development. For instance, the plant's ability to disrupt bacterial biofilms and inhibit alpha-glucosidase points to its potential in treating biofilm-associated infections and managing diabetes. As the interest in plant-based medicine continues to grow, *D. linearis* stands out as a valuable species with promising applications in health and wellness. Future research should focus on isolating and characterizing its bioactive compounds to develop novel pharmaceuticals. The comprehensive evaluation of *D. linearis* pharmacological activities provides a solid foundation for further studies to harness its full therapeutic potential. Finally, it is concluded that the present reviewed plant holds significant promise in the field of medicinal plants, offering natural and effective alternatives for treating various ailments and contributing to the discovery of new pharmaceuticals. Its extensive traditional uses, backed by scientific research, make it a noteworthy candidate for future drug development and therapeutic applications.

KEYWORDS

- **bioactive compounds**
- **gastroprotective effect**
- **histopathological analysis**
- **liver metabolizes drugs**
- **medicinal fern**
- **pharmacological activities**
- **phytochemical analysis**

REFERENCES

Ahmad, A. A., Kasim, K. F., Gopinath, S. C., Anbu, P., & Sofian-Seng, N. S. (2023). Encapsulation of *Dicranopteris linearis* extract using cellulose microparticles for antiulcer medication. *International Journal of Biological Macromolecules, 253*, 126795. https://doi.org/10.1016/j.ijbiomac.2023.126795.

Anh, N. T., Thu, N. T. H., Van, N. N. K., & Huy, D. T. (2022). Study on chemical constituents of the leaves of *Dicranopteris linearis* (Burm. f.) Underw. *Ho Chi Minh City University of Education Journal of Science, 19*(10), 1593–1598.

Baharuddin, A. A., Roosli, R. A. J., Zakaria, Z. A., & Md. Tohid, S. F. (2018). *Dicranopteris linearis* extract inhibits the proliferation of human breast cancer cell line (MDA-MB-231) via induction of S-phase arrest and apoptosis. *Pharmaceutical Biology, 56*(1), 422–432.

Din, S. S., Mamat, S. S., Ismail, N. A., Zainulddin, W. N. W., Zabidi, Z., Yahya, F., Kamisan, F. H., Mohtarrudin, N., Othman, F., Suhaili, Z., & Zakaria, Z. A. (2012). Effect of chloroform extract of traditional *Dicranopteris linearis* leaves against paracetamol- and CCl_4-induced liver toxicity in rats. *Cell Medicine, 2*(4), 1–5.

Duong, T. H., Vu, Y. T., Long, N. P., Phan, N. H. N., Pham, N. K. T., Sichaem, J., Kieu, N. K. D., Duong, C. B., Nguyen, T. T., Dang, V. S., & Nguyen, H. T. (2023). Bioactive-guided phytochemical investigations, *in vitro* and in silico alpha-glucosidase inhibition of two Vietnamese medicinal plants *Dicranopteris linearis* and *Psychotria adenophylla. Pharmaceuticals, 16*(9), 1–18.

Guan, Y. S., & He, Q. (2015). Plants consumption and liver health. *Evidence-Based Complementary and Alternative Medicine, 2015*(1), 1–10.

Hossin, A., Jabed, M. S. U., Tushar, R. R., Nawrin, K., & Billah, M. M. (2021). Central and peripheral antinociceptive activity of *D. linearis* ethanolic leaf extract. *Pharmaceutical and Pharmaceutical Sciences Discovery, 1*(2), 74–81.

Hussaini, J., Othman, N. A., Abdulla, M. A., Majid, N. A., Faroq, H. M., & Ismail, S. (2012). Gastroprotective effects of *Dicranopteris linearis* leaf extract against ethanol-induced gastric mucosal injury in rats. *Scientific Research and Essays, 7*(18), 1761–1767.

Ismail, N. A., Shamsahal, D. N. S., Mamat, S. S., Zabidi, Z., Wan Zainulddin, W. N., Kamisan, F. H., Yahya, F., Mohtarrudin, N., Mohd Desa, M. N., & Zakaria, Z. A. (2014). Effect of aqueous extract of *Dicranopteris linearis* leaves against paracetamol and carbon tetrachloride-induced liver toxicity in rats. *Pakistan Journal of Pharmaceutical Sciences, 27*(4), 831–835.

Kale, M. (2015). Analysis of bioactive components of *Dicranopteris linearis* (Burm. f.) Underwood whole plant ethanol extract by GC-MS. *Bulletin of Pure and Applied Sciences, Botany, 34*(1, 2), 9–12.

Kalpanadevi, R., Subramani, V., Annamalai, P., Nakulan, V., & Narayanaperumal, J. (2016b). Phytochemical analysis, *in vitro* antioxidant potential and gas chromatography–mass spectrometry studies of *Dicranopteris linearis*. *Asian Journal of Pharmaceutical and Clinical Research, 9*(2), 220–225.

Kalpanadevi, R., Vasantha, S., Panneerselvam, A., Rajesh, N. V., Jeyathilakan, N., & Jeeva, S. (2016a). *In vitro* study of trematodicidal action of *Dicranopteris linearis* (Burm. f.) Underw. extracts against *Gastrothylax crumenifer*. *Biomedicine & Pharmacotherapy, 84*, 2042–2053.

Kamisan, F. H., Yahya, F., Mamat, S. S., Fauzi, F. K., Mohtarrudin, N., TehLay, K., Salleh, M. Z., Hussain, M. K., & Zakaria, Z. (2014). Effect of methanol extract of *Dicranopteris linearis* against carbon tetrachloride-induced acute liver injury in rats. *BMC Complementary and Alternative Medicine, 14*, 1–10.

Kumarpal, S. (2013). Study of activity of some medicinal ferns of Darjeeling. *International Journal of Scientific Research Publications, 3*(8), 1–4.

Lai, C. S., Ponnusamy, Y., Lim, G. L., & Ramanathan, S. (2021). Antibacterial, antibiofilm and antibiotic-potentiating effects of a polyphenol-rich fraction of *Dicranopteris linearis* (Burm. f.) Underw. *Journal of Herbal Medicine, 25*, 10049.

Mawang, C. I., Lim, Y. Y., Ong, K. S., Muhamad, A., & Lee, S. M. (2017). Identification of α-tocopherol as a bioactive component of *Dicranopteris linearis* with disrupting property against preformed biofilm of *Staphylococcus aureus*. *Journal of Applied Microbiology, 123*(5), 1148–1159.

Ponnusamy, Y., Chear, N. J. Y., Ramanathan, S., & Lai, C. S. (2015). Polyphenols rich fraction of *Dicranopteris linearis* promotes fibroblast cell migration and proliferation *in vitro*. *Journal of Ethnopharmacology, 168*, 305–314.

Rajesh, K. D., Vasantha, S., Panneerselvam, A., & Rajesh, N. V. (2016). Phytochemical analysis, *in vitro* antioxidant potential and gas chromatography–mass spectrometry studies of *Dicranopteris linearis*. *Asian Journal of Pharmaceutical and Clinical Research, 9*(2), 1–6.

Rodzi, R., Cheah, Y. L., Ooi, K. K., Othman, F., Mohtarrudin, N., Tohid, S. F., Suhaili, Z., & Zakaria, Z. A. (2013). Chemopreventive potential of methanol extract of *Dicranopteris linearis* leaf on DMBA/croton oil-induced mouse skin carcinogenesis. *African Journal of Pharmacy and Pharmacology, 7*, 2484–2498.

Sandhya, B., Thomas, S., Isabel, W., & Shenbagarathai, R. (2006). Ethnomedicinal plants used by the Valaiyan community of Piranmalai Hills (reserved forest), Tamil Nadu, India – A pilot study. *African Journal of Traditional, Complementary and Alternative Medicines, 3*(1), 101–114.

Sathiyaraj, G., Muthukumar, T., & Ravindran, K. C. (2015). Ethnomedicinal importance of fern and fern allies traditionally used by tribal people of Palani Hills (Kodaikanal), Western Ghats, South India. *Journal of Medicinal Herbs and Ethnomedicine, 1*, 4–9.

Savithramma, N., Linga, R. M., & Suhrulatha, D. (2011). Screening of medicinal plants for secondary metabolites. *Middle-East Journal of Scientific Research, 8*(3), 579–584.

Savithramma, N., Yugandhar, P., Prasad, K. S., Ankanna, S., & Chetty, K. M. (2016). Ethnomedicinal studies on plants used by Yanadi tribe of Chandragiri Reserve Forest area, Chittoor District, Andhra Pradesh, India. *Journal of Intercultural Ethnopharmacology, 5*(1), 49–56.

Singh, A. V. (2018). Ethnobotany of ferns of Lakhimpur, Assam, India. *International Research Journal of Biological Sciences, 7*(7), 27–34.

Srivastava, K. (2007). Ethnobotanical studies of some important ferns. *Ethnobotanical Leaflets, 11*, 164–172.

Thomas, A. P., Sindu, N., & Dileesh, S. (2023). Phytochemical and bioactive efficiency of *Dicranopteris linearis*. *Materials Today: Proceedings*. https://doi.org/10.1016/j.matpr.2023.12.008.

Thomas, T., Kurup, G. M., & Sadasivan, C. (2007). Antibacterial activity of *Dicranopteris linearis* under *in vitro* conditions. *STAR: International Journal (Sciences), 1*(2), 191–195.

Vasudeva, S. M. (1999). Economic importance of pteridophytes. *Indian Fern Journal, 16*, 130–152.

Zakaria, Z. A., Abdul Ghani, Z. D. F., Raden Mohd. Nor, R. N. S., Gopalan, H. K., Sulaiman, M. R., Mat Jais, A. M., Somchit, M. N., Kader, A. A., & Ripin, J. (2008). Antinociceptive, anti-inflammatory, and antipyretic properties of an aqueous extract of *Dicranopteris linearis* leaves in experimental animal models. *Journal of Natural Medicines, 62*, 179–187.

Zakaria, Z. A., Ghani, Z. D. F. A., Nor, R. N. S. R. M., Gopalan, H. K., & Sulaiman, M. R. (2006). Antinociceptive and anti-inflammatory activities of *Dicranopteris linearis* leaves chloroform extract in experimental animals. *Yakugaku Zasshi, 126*(11), 1197–1203.

Zakaria, Z. A., Kamisan, F. H., & Kek, T. L., Salleh, M. Z. (2020a). Hepatoprotective and antioxidant activities of *Dicranopteris linearis* leaf extract against paracetamol-induced liver intoxication in rats. *Pharmaceutical Biology, 58*(1), 478–489.

Zakaria, Z. A., Kamisan, F. H., Mohd. Nasir, N., Teh, L. K., & Salleh, M. Z. (2019). Aqueous partition of methanolic extract of *Dicranopteris linearis* leaves protects against liver damage induced by paracetamol. *Nutrients, 11*(12), 1–19.

Zakaria, Z. A., Kamisan, F. H., Omar, M. H., Mahmood, N. D., Othman, F., Abdul, H. S. S., & Abdullah, M. N. H. (2017). Methanol extract of *Dicranopteris linearis* L. leaves impedes acetaminophen-induced liver intoxication partly by enhancing the endogenous antioxidant system. *BMC Complementary and Alternative Medicine, 17*, 1–14.

Zakaria, Z. A., Mohamed, A. M., Jamil, N. M., Rofiee, M. S., Somchit, M. N., Zuraini, A., Arifah, A. K., & Sulaiman, M. R. (2011). *In vitro* cytotoxic and antioxidant properties of the aqueous, chloroform and methanol extracts of *Dicranopteris linearis* leaves. *African Journal of Biotechnology, 10*(2), 273–282.

Zakaria, Z. A., Roosli, R. A. J., Marmaya, N. H., Omar, M. H., Basir, R., & Somchit, M. N. (2020b). Methanol extract of *Dicranopteris linearis* leaves attenuates pain via the modulation of opioid/NO-mediated pathway. *Biomolecules, 10*(2), 1–23.

Zakaria, Z. A., Sahmat, A., Azmi, A. H., Nur Zainol, A. S., Omar, M. H., Balan, T., Sulistyorini, L., Azizah, R., & Abdullah, M. N. H. (2021). Polyphenolics and triterpenes presence in chloroform extract of *Dicranopteris linearis* leaves attenuated paracetamol-induced liver intoxication in rat. *BMC Complementary Medicine and Therapies, 21*, 1–15.

Zakaria, Z. A., Sodri, N. H., Hassan, H., Anuar, K., & Abdullah, F. C. (2012). Effects of various receptor antagonists on the peripheral antinociceptive activity of aqueous extracts of *Dicranopteris linearis*, *Melastoma malabathricum*, and *Bauhinia purpurea* leaves in mice. *TANG [Human Medicine], 2*(4), 1–6.

CHAPTER 10

An Overview of the Bioactives and Pharmacology of *Pityrogramma* Species

SUNIL KUMAR

Department of Chemistry, Sant Kavi Baba Baijnath Government P.G. College, Harakh, Barabanki, Uttar Pradesh, India

*Corresponding authors

ABSTRACT

Genus *Pityrogramma* is known as the gold- and silver-backed ferns belonging to the family Pteridaceae. Some species of the genus are grown as ornamental plants in tropical Asia and used as traditional medicines. 89 compounds, including flavonoids and phenolic acids, were summarized here from different species of *Pityrogramma*. The extracts, fractions, and their compounds from the genus *Pityrogramma* were tested for antimicrobial, mucolytic, anticancer, pesticidal, anti-leishmania, and immunostimulant activities. The ethanolic extracts of *P. calomelanos* were found to have significant effects in most of the studied activities. Flavonoids are the major class of bioactive compounds in the genus.

10.1 INTRODUCTION

Genus *Pityrogramma,* which belongs to the highly morphologically diverse family Pteridaceae, is a genus of ferns commonly known as "goldfern" or "false maidenhair fern," consisting of about 17 tropical species (Zhang et al., 2017). Some species of this genus are cultivated in greenhouses for the yellow or white farina found on the lower surfaces of leaves. They are often grown as ornamental plants in gardens or as houseplants. The plants of this

Bioactives and Pharmacology of Pteridophytes. T. Pullaiah (Ed.)

DOI: 10.1201/9781779642738-10

genus are also used in traditional medicines. They are native to tropical and subtropical regions and are prized for their attractive fronds and ease of care (Windham and Yatskievych, 2003). One well-known species is *P. calomelanos*, which is commonly cultivated for its graceful appearance and tolerance of a wide range of growing conditions. The different extracts and bioactive compounds of *Pityrogramma* species have been reported for antimicrobial, mucolytic, anticancer activity/cytotoxicity, pesticidal, anti-leishmania, and immunostimulant activities (Souza et al., 2013a, 2013b). The objective of the present chapter is to summarize the bioactives and biological activities of *Pityrogramma* species.

10.2 BIOACTIVES

The structures of phytochemicals isolated and characterized in different parts of *P. calomelanos, P. chrysophylla, P. pallida, P. tartarea, P. tartarea, P. triangular, P. triangularis* and *P. trifoliata* are shown in Figures 10.1–10.4 and Table 10.1. Flavonoids are the major class of compounds including chalcones (1–23), flavanon (24–38), flavonol (39–73, 80–81), flavone (74–79, 82–83), phenolic acids (84–86) and others (87–89). The structure of all compounds were elucidated using mass spectrometry (MS), nuclear magnetic resonance (NMR), infrared (IR), ultraviolet-visible (UV-Vis), X-ray Crystallography. The literature survey showed that the flavonoids are major class of compound (Table 10.1).

TABLE 10.1 Distribution of Phytochemicals in *Pityrogramma* Species

Sl. No.	Phytochemicals	Species	References
	Chalcone		
1.	2,'6' Dihydroxy-4'-methoxychalcone	*P. trifoliata, P. austroamericana*	Rodriguez et al. (2022)
2.	2,'6' Dihydroxy-4,4'-methoxychalcone	*P. trifoliata, P. tartarea*	Rodriguez et al. (2022)
3.	2,'6'-Dihydroxy-4'-methoxy-5'-methylchalcone	*P. triangularis, P. austroamericana*	Wollenweber et al. (1985)
4.	2,'6'-Dihydroxy-4'-methoxy-3'-methylchalcone	*P. triangularis*	Wollenweber et al. (1985)
5.	Neosakuranetin	*P. calomelanos*	Hitz et al. (1982)
6.	Flavokawin B	*P. triangularis* var. *pallida.*	Wollenweber et al. (1981a)

TABLE 10.1 *(Continued)*

Sl. No.	Phytochemicals	Species	References
7.	Aurentiacin	*P. triangularis* var. *pallida.*	Wollenweber et al. (1981a)
8.	T-l	*P. trifoliata*	Dietz et al. (1980)
9.	T-3 (a)	*P. trifoliata*	Dietz et al. (1980)
10.	T-3 (b)	*P. trifoliata*	Dietz et al. (1980)
11.	T-3 (c)	*P. trifoliata*	Dietz et al. (1980)
12.	5,7,4'-Triisopropyloxy-8-cinnamoyl-4-phenyldihydrocoumarin	*P. calomelanos* var. *aureoflava*	Iinuma et al. (1986)
13.	β-(5,7.4'-trihydroxy-8-yl)-β-phenylpropionic acid	*P. calomelanos* var. *aureoflava*	Iinuma et al. (1986)
14.	Ceroptene	*Pityrogramma triangularis*	Markham et al. (1985)
15.	2,'6'-Dihydroxy-4'-methoxydihydrochalcone	*P. chrysophylla, P. tartarea* and *P. calomelanos*	Nilsson (1961); Star and Mabry (1971)
16.	2,'6,'4-Trihydroxy, 4'-methoxy-dihydrochalcone (asebogenin)	*P. calomelanos*	Hitz et al. (1982)
17.	2,'6,'4-trihydroxy, 4'-methoxy 3'-methyl dihydrochalcone	*P. triangularis* var. *viscosa*	Wollenweber et al. (1979)
18.	2,'6'-Dihydroxy-4,4'-dimethoxydihydrochalcone	*P. tartarea, P. calomelanos,* and *P. chrysophylla*	Nilsson (1961); Star and Mabry (1971)
19.	2,'6'-Dihydroxy-3,4,4'-trimethoxydihydrochalcone	*P. tartarea*	Iinuma et al. (1994)
20.	Compound D-l ((phenylpropionyl)-5,7-dihydroxydihydroneoflavone)	*P. trifoliata, P. tartarea* and *P. calomelanos*	Asai et al. (1991), Iinuma et al. (1994); Rodriguez et al. (2022)
21.	Calomelanol A	*P. calomelanos*	Asai et al. (1991); Rodriguez et al. (2022)
22.	Calomelanol B	*P. calomelanos*	Asai et al. (1991)
23.	Calomelanol C	*P. calomelanos*	Asai et al. (1991)
	Flavonoid		
24.	5-Hydroxy,7-methoxyflavanone	*P. trifoliata*	Rodriguez et al. (2022)
25	5-Hydroxy,4,'7-dimethoxyflavanone	*P. trifoliata,* and *P. calomelanos*	Rodriguez et al. (2022)
26.	Strobopinin	*P. pallida*	Wollenweber et al. (1981b)

TABLE 10.1 *(Continued)*

Sl. No.	Phytochemicals	Species	References
27.	Cryptostrobin	*P. pallida*	Wollenweber et al. (1981b)
28.	5-Methoxy-7-hydroxy-8-*C*-methyl-flavanone	*P. pallida*	Markham et al. (1987)
29	Comptonin (5-methoxy-7-hydroxy-6-methyl-flavanone	*P. pallida*	Markham et al. (1987)
30.	Desmethoxymatteucinol	*P. pallida*	Wollenweber et al. (1981b)
31.	5,7-Dimethoxy-8-*C*-methylflavanone	*P. pallida*	Markham et al. (1987)
32.	(2*S*)-5,7-Dihydroxy-4'-methoxy-6,8-dlmethylflavanone	*P. triangularis*	Star et al. (1978)
33.	Calomelonol G	*P. calomelanos*	Iinuma et al. (1992)
34.	Caluntelanol H	*P. calomelanos*	Iinuma et al. (1992)
35.	Calamelanol I	*P. calomelanos*	Iinuma et al. (1992)
36.	Calomelanol J	*P. calomelanos*	Iinuma et al. (1992)
37.	Trifoliatol	*P. chrysophylla*	Nilsson (1961)
38.	Compound T-2	*P. trifoliata*	Dietz et al. (1980)
39.	Kaempferol	*P. calomelanos*	de Souza et al. (2012)
40.	6-Hydroxykaempferol	*P. triangularis*	Star et al. (1975)
41.	Kaempferol 7-methyl ether (rhamnocitrin)	*P. tartarea* and *P. calomelanos*	Star and Mabry (1971)
42.	Kaempferol-7-O-methyl ether 3-O-rhamnoside	*P. triangularis*	Star et al. (1975)
43.	Kaempferol-7-*O*-methyl ether-3-*O*-glucoside	*P. triangularis*	Star et al. (1975)
44.	Quercetin	*P. calomelanos*	de Souza et al. (2012)
45.	Quercetin-3-*O*-glucoside	*P. triangularis*	Star et al. (1975)
46.	Quercetin-3-*O*-rhamnoside	*P. triangularis*	Star et al. (1975)
47.	Rutin	*P. calomelanos*	de Souza et al. (2012)
48.	Quercetagetin	*P. triangularis*	Star et al. (1975)
49.	Quercetagetin 3'-methylether 7-*O*-glucose	*P. triangularis*	Star et al. (1975)
50.	Gossypetin	*P. triangularis*	Star et al. (1975)
51.	Gossypetin 3,7-dimethyl ether	*P. triangularis*	Star et al. (1975)
52.	Gossypetin 3,7,3'-trimethyl ether	*P. triangularis*	Star et al. (1975)
53.	Gossypitrin	*P. triangularis*	Star et al. (1975)

TABLE 10.1 *(Continued)*

Sl. No.	Phytochemicals	Species	References
54.	Gossypin	*P. triangularis*	Star et al. (1975)
55.	Pityrogrammin	*P. triangularis*	Star et al. (1975)
56.	Herbacetin	*P. triangularis*	Star et al. (1975)
57	Herbacetin 3,7-dimethyl ether	*P. triangularis*	Star et al. (1975)
58.	Herbacetin 8,4'-dimethyl ether	*P. triangularis*	Star et al. (1975)
59.	Herbacetin 3,8- dlmethyl ether	*P. triangularis*	Wollenweber et al. (1985)
60.	Herbacetin 3,8,4'-trimethyl ether	*P. triangularis*	Wollenweber et al. (1985)
61.	4'-Methoxy-3,5,7-trihydroxyflavone	*P. triangularis*	Star et al. (1975)
62.	6-Methyl-8-methoxy-3,5,7-trihydroxyflavone	*P. triangularis*	Star et al. (1975)
63.	3,5-Dihydroxy-7,4'-dimethoxyflavone	*P. triangularis*	Star et al. (1975)
64.	5,8-Dihydroxy-3,7 dimethoxyflavone	*P. triangularis*	Wollenweber et al. (1985)
65.	3,5,7-Trihydroxy-8-methylflavone	*P. triangularis*	Wollenweber et al. (1985)
66.	5,7-Dihydroxy-3-methoxy-8-methylflavone	*P. triangularis*	Wollenweber et al. (1985)
67.	3,5-Dihydroxy-7-methoxy-8-methylflavone	*P. triangularis*	Wollenweber et al. (1985)
68.	5,8-Dihydroxy-3,7-dimethoxy-6-methylflavone	*P. triangularis*	Wollenweber et al. (1985)
69.	3,5,8-Trihydroxy-7-methoxy-6-methylflavone	*P. triangularis*	Wollenweber et al. (1985)
70.	3,5,7- Trihydroxy-6,8-dimethylflavone	*P. triangularis*	Wollenweber et al. (1985)
71.	5,7-Dihydroxy-3-methoxy-6,8-dimethylflavone	*P. triangularis*	Wollenweber et al. (1985)
72.	3,5-Dihydroxy-7-methoxy-6,8-dimethylflavone	*P. triangularis*	Wollenweber et al. (1985)
73.	5,7,4'-Trihydroxy-3-methoxy-6,8-dimethylflavone	*P. triangularis*	Wollenweber et al. (1985)
74.	Apigenin 7-methyl ether (genkwanin)	*P. tartarea* and *P. calomelanos*	Star and Mabry (1971)
75.	Norwogonin	*P. triangularis*	Star et al. (1975)

TABLE 10.1 *(Continued)*

Sl. No.	Phytochemicals	Species	References
76.	Baicalein	*P. triangularis*	Star et al. (1975)
77.	5-Hydroxy-7- methoxy-6-methylflavanone	*P. triangularis*	Wollenweber et al. (1985)
78.	5-Hydroxy-7- methoxy-8-methylflavanone	*P. triangularis*	Wollenweber et al. (1985)
79.	Scutellarin	*P. triangularis*	Star et al. (1975)
80.	Calomelanols D	*P. tartarea*	Iinuma et al. (1994)
81.	Compound D-2/a	*P. calomelanos*	Iinuma et al. (1993)
82.	Colomelonol F	*P. calomelanos*	Iinuma et al. (1992)
83.	Compound D-2/b	*P. calomelanos*	Iinuma et al. (1993)
a.	**Other**		
84.	Gallic acid	*P. calomelanos*	de Souza et al. (2012)
85.	Caffeic acid	*P. calomelanos*	de Souza et al. (2012)
86.	Chlorogenic acid	*P. calomelanos*	de Souza et al. (2012)
87.	Isoceroptene	*P. triangularis*	Markham et al. (1985)
88.	Calomelanolactone	*P. cabmelanos*	Bardouille et al. (1978)
89.	Pterosin Z	*P. cabmelanos*	Bardouille et al. (1978)

Chalcones consist of two or three aromatic rings linked by a three-carbon α, β-unsaturated carbonyl system. All flavonoids typically consist of a 15-carbon skeleton (C6-C3-C6), with two aromatic rings (A and B rings) connected by a three-carbon bridge (C-ring). The basic flavonoid structure can be modified by variations in the substitution pattern, degree of hydroxylation, and presence of glycosylation or other functional groups. The several functional groups, including phenolic hydroxyl groups (-OH), contribute to their antioxidant properties, as well as carbonyl groups (C=O) and double bonds (C=C) within the rings. These functional groups can undergo various chemical reactions, such as oxidation, reduction, and conjugation. Compounds 24–38 often exist as stereoisomers due to the presence of chiral centers at C-2 in their structures. The arrangement of substituents around these chiral centers gives rise to different stereoisomeric forms, such as cis/trans isomers and enantiomers, which can exhibit distinct biological activities. Some of these special chalcones (8–13) consist of three and four rings (Dietz et al., 1980). Similarly, flavonoids (33–38) have five rings (Dietz et al., 1980; Iinuma et al., 1992) (Figures 10.1–10.4).

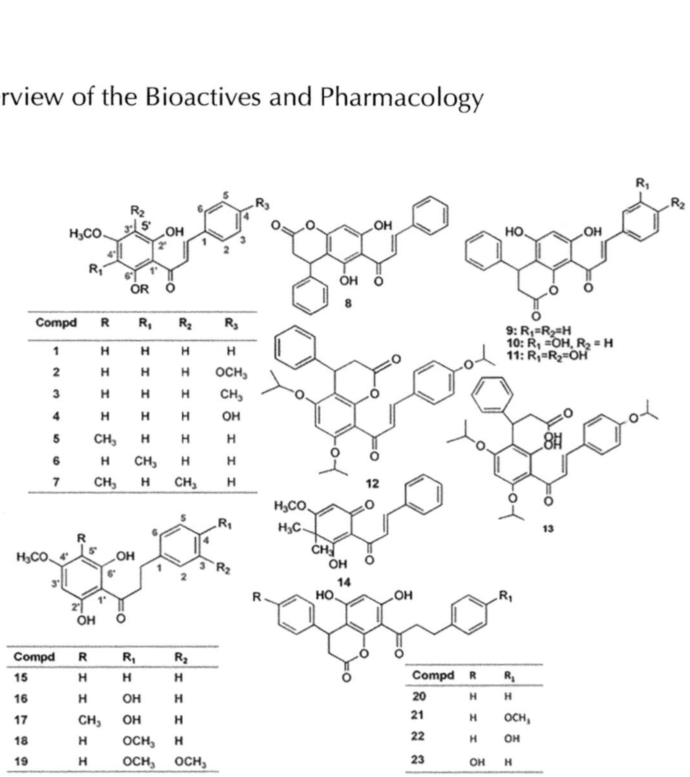

Compd	R	R_1	R_2	R_3
1	H	H	H	H
2	H	H	H	OCH_3
3	H	H	H	CH_3
4	H	H	H	OH
5	CH_3	H	H	H
6	H	CH_3	H	H
7	CH_3	H	CH_3	H

Compd	R	R_1	R_2
15	H	H	H
16	H	OH	H
17	CH_3	OH	H
18	H	OCH_3	H
19	H	OCH_3	OCH_3

Compd	R	R_1
20	H	H
21	H	OCH_3
22	H	OH
23	OH	H

FIGURE 10.1 Structure of chalcones (**1–23**) from different species of *Fityrogramma.*

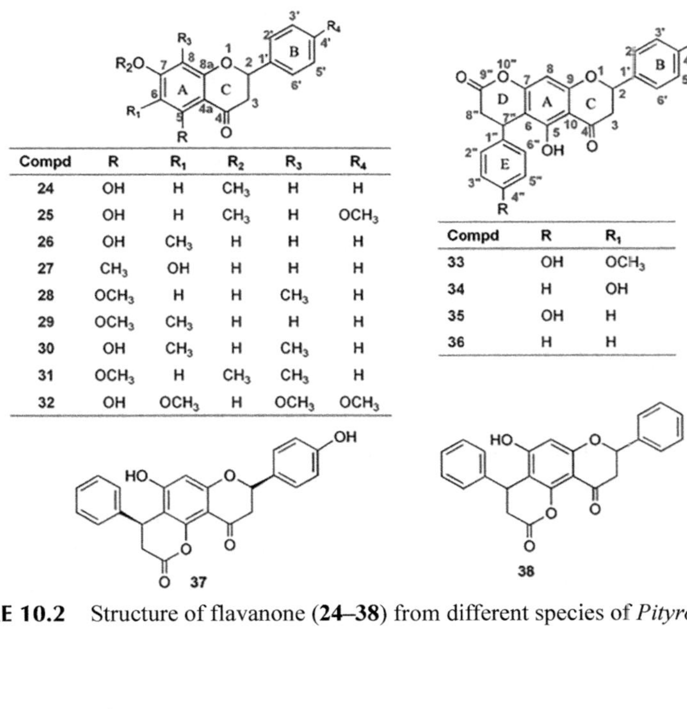

Compd	R	R_1	R_2	R_3	R_4
24	OH	H	CH_3	H	H
25	OH	H	CH_3	H	OCH_3
26	OH	CH_3	H	H	H
27	CH_3	OH	H	H	H
28	OCH_3	H	H	CH_3	H
29	OCH_3	CH_3	H	H	H
30	OH	CH_3	H	CH_3	H
31	OCH_3	H	CH_3	CH_3	H
32	OH	OCH_3	H	OCH_3	OCH_3

Compd	R	R_1
33	OH	OCH_3
34	H	OH
35	OH	H
36	H	H

FIGURE 10.2 Structure of flavanone (**24–38**) from different species of *Pityrogramma.*

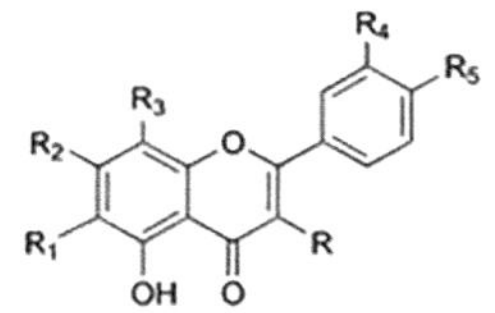

Compd	R	R1	R2	R3	R4	R5
39	OH	H	OH	H	H	OH
40	OH	OH	OH	H	H	OH
41	OH	H	OCH_3	H	H	OH
42	ORh	H	OCH_3	H	H	OH
43	OG	H	OCH_3	H	H	OH
44	OH	H	OH	H	OH	OH
45	OG	H	OH	H	OH	OH
46	ORh	H	OH	H	OH	OH
47	ORu	H	OH	H	OH	OH
48	OH	OH	OH	H	OH	OH
49	OH	OH	OG	H	OCH_3	OH
50	OH	H	OH	OH	OH	OH
51	OCH_3	H	OCH_3	OH	OH	OH
52	OCH_3	H	OH	OH	OCH_3	OH
53	OH	H	OG	OH	OH	OH
54	OH	H	OH	OG	OH	OH
55	OH	CH_3	OH	OCH_3	H	H
56	OH	H	OH	OH	H	OH
57	OCH_3	H	OCH_3	OH	H	OH
58	OH	H	OH	OCH_3	H	OCH_3
59	OCH_3	H	OH	OCH_3	H	OH
60	OCH_3	H	OH	OCH_3	H	OCH_3
61	OH	H	OH	H	H	OCH_3
62	OH	CH_3	OH	OCH_3	H	H
63	OH	H	OCH_3	H	H	OCH_3

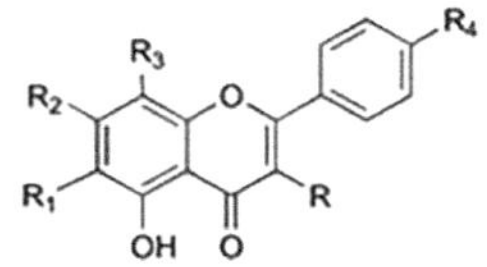

Compd	R	R1	R2	R3	R4
64	OCH_3	H	OCH_3	OH	H
65	OH	H	OH	CH_3	H
66	OCH_3	H	OH	CH_3	H
67	OH	H	OCH_3	CH_3	H
68	OCH_3	CH_3	OCH_3	OH	H
69	OH	CH_3	OCH_3	OH	H
70	OH	CH_3	OH	CH_3	H
71	OCH_3	CH_3	OH	CH_3	H
72	OH	CH_3	OCH_3	CH_3	H
73	OCH_3	CH_3	OH	CH_3	OH
74	H	H	OH	H	OH
75	H	H	OH	OH	H
76	H	OH	OH	H	H
77	H	CH_3	OCH_3	H	H
78	H	H	OCH_3	CH_3	H
79	H	OH	OGA	H	OH

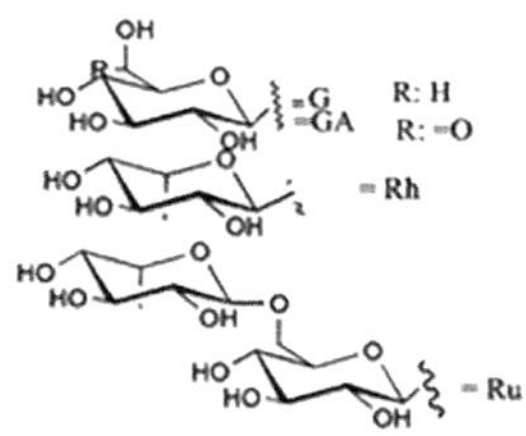

FIGURE 10.3 Structure of flavonols (**39–73**) and flavons (**74–79**) from different species of *Pityrogramma.*

10.3 PHARMACOLOGY

The species of *Pityrogramma* has possessed many biological activities such as antimicrobial, mucolytic, anticancer, pesticidal, anti-leishmania, and immunostimulant activities of different extracts. The ethanolic extract of *P. calomelanos*is most commonly reported for many activities due to its high content of flavonoids. Some of the bioactive compounds isolated from the genus *Pityrogramma* have also been reported for their activities.

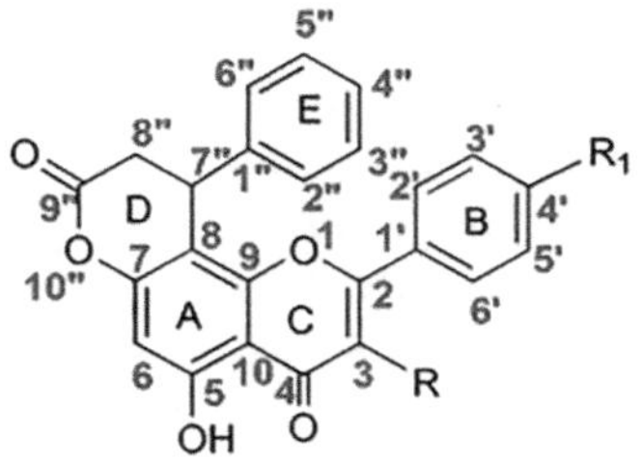

Compd	R	R_1
80	OH	OH
81	OH	H
82	H	OH
83	H	H

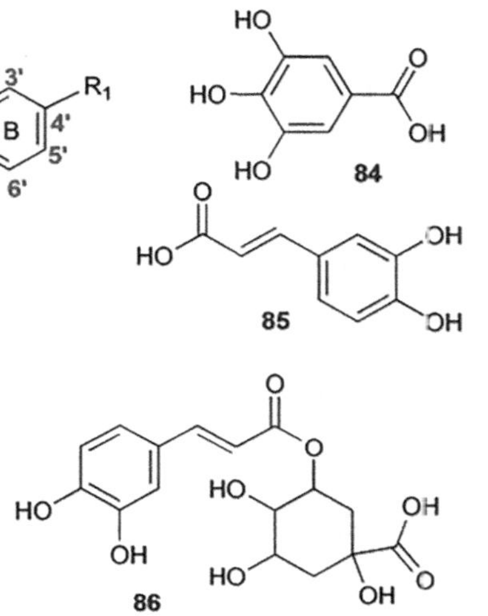

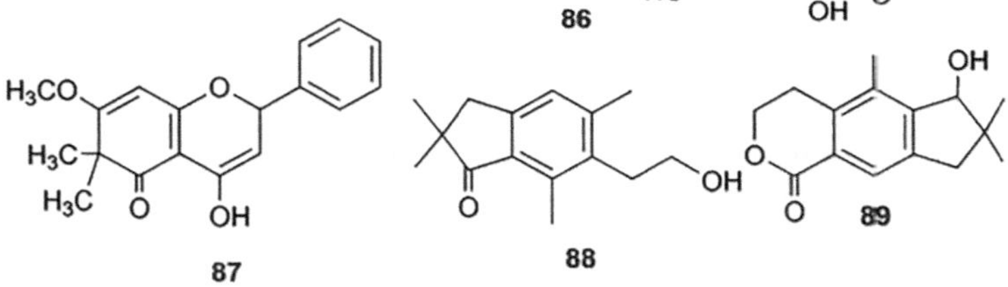

FIGURE 10.4 Structure of flavones (**80–83**) phenolic acids (**84–86**) and others (**87–89**) from different species of *Pityrogramma.*

10.3.1 ANTIMICROBIAL ACTIVITY

The ethanolic extract of the leaves of *P. calomelanos* and its combination with fluconazole were studied against *Candida albicans* and *Candida tropicalis* strains. The ethanolic extract showed minimum inhibitory concentrations (MIC) at 8192 mg/mL, indicating that it has weak antifungal activity. The potential of activity against these strains increased as the combination of the extracts with fluconazole increased (Freitas et al., 2017). The ethanolic extract of *P. calomelanos* exhibited significant potential against *S. aureus* (MIC/8 = 128 µg/mL) among the tested strains: *Candida albicans, C. krusei, C. tropicalis, Escherichia coli,* and *Pseudomonas aeruginosa* (Souza et al., 2012). The ethyl acetate fraction and hexane fraction of *P. calomelanos* were determined at MIC ≥1024 µg/mL. Both fractions had more potential against *E. coli.* The ethyl acetate fraction showed a significant effect against *S. aureus* (Souza et al., 2013a).

The yellow and white farina of *P. trifoliata*, as well as compounds 1–2, 5, 8–10, 15, 18, 19, 20, 24, and 25, were tested in the range from 250–7.8 μg/mL against *C. albicans, Cryptococcus neoformans, Trichophyton rubrum, Trichophyton mentagrophytes,* and *Microsporum gypseum* using the microbroth dilution method. Compounds 20 and 24 were notably active against both strains of *Trichophyton* with MICs and MFCs between 31.2 and 7.8 μg/mL. Compounds 1, 2, and 5 showed moderate activity with MICs and MFCs between 31.2 and 125 μg/mL, while the remaining compounds were inactive. Both yellow and white farina showed significant activity against *C. neoformans* (MIC = MFC = 15.8 μg/mL and MIC = MFC = 31.2 μg/mL, respectively) (Rodriguez et al., 2022).

10.3.2 MUCOLYTIC ACTIVITY

Mucolytic activity refers to the ability of certain substances to break down or thin mucus, making it easier to clear from the respiratory tract. Mucus plays a crucial role in protecting the respiratory system by trapping foreign particles, bacteria, and viruses, but excessive or thick mucus can impair breathing and lead to respiratory problems. Compound 44, isolated from the acetone extract of *P. calomelanos,* significantly reduced the viscosity of cow intestinal mucus. The compound showed a concentration of 0.4% mucolytic activity, which is equivalent to 0.1% acetylcysteine as a standard drug using the ANOVA test, followed by the LSD test (Sutoyo et al., 2019).

10.3.3 ANTICANCER ACTIVITY/CYTOTOXICITY

The ethanolic extract and fractions of *P. calomelanos* showed significant cytotoxic activity using fibroblasts NCTC929. The hexane and methanolic fractions did not have a cytotoxic effect (Souza et al., 2013a). An ethanolic extract and hexane fraction of *P. calomelanos* were studied against *Trypanosoma cruzi* (CL-B5 clone). The effective concentration capable of killing 50% of parasites (EC50) was 55.26 mg/mL and 73.57 mg/mL for the ethanolic extract and hexane fraction, respectively (de Souza et al., 2012). The anticancer activity of dihydrochalcone, isolated from *P. calomelanos*, was studied *in vitro* for apospory and apogamy. Half-strength Murashige and Skoog (MS) medium with 3.33 μM N6-benzyladenine (BA) and 2.32 μM kinetin (Kn) showed earlier development and produced higher numbers of aposporous gametophytes than half-strength MS basal medium. Crozier explants developed higher numbers (mean value 29.2) of gametophytes but were slower than frond explants (mean value 23.2). The gametophytes originated from the epidermal hairs progressed from uniseriate filamentous to cordate through bi-, tri-, and multiseriate and spatulate stages with the development of antheridia. Reduction in the nutrient and sucrose concentrations in the media favored apogamy. Sucrose-free 1/10 strength MS medium and agar plates developed a mean of 30.4 and 29.9 sporophytes, respectively, in the light. The greenhouse-established gametophytes developed sporophytes. The established sporophytes *ex vitro* showed a 95% survival rate. The established protocol accomplishes apogamy and apospory in *P. calomelanos*, and the aposporous gametophytes can be used for genetic transformation and development of transgenic *P. calomelanos* (Martin et al., 2006). The toxicity of compound 15 (which is isolated from *P. calomelanos*) was tested against *Artemia salina* in the BSLT assay with an LC50 of 3.957 μg/mL using Brine Shrimp Lethality. Therefore, compound 15 was found to be highly toxic and had potential as an anticancer agent (Najihah et al., 2018). Compounds 15, 39, and 44 were isolated from *P. calomelanos* and showed cytotoxic activity against murine leukemia P-388 cells with IC50 values of 1.6, 14.1, and 6.4 μg/mL, respectively. Thus, compounds 15 and 44 strongly inhibited murine leukemia P-388 cells (IC50 $<$ 10 μg/mL), while compound 39 showed moderate activity against murine leukemia P-388 cells ($10 \leq$ IC50 < 100 μg/mL) (Suyatno et al., 2014).

10.3.4 PESTICIDAL ACTIVITY

Pesticidal activity refers to the ability of a substance to kill, repel, or otherwise control pests such as insects, weeds, fungi, and other organisms that are harmful to crops, livestock, humans, or the environment. The phytoecdysteroid fraction of *P. calomelanos* showed maximum *Spodoptera litura* larvae mortality (68%) at 2000 ppm (LC50 = 1473 ppm; $F = 48.65$; $p = 0.049$). The reduction of pupation, pupal weight, and adult emergence was observed at higher concentrations. It could be due to the interaction of the phytoecdysteroid fraction with the insect endocrine system. Therefore, the biopesticide could be a better alternative to commercial chemical insecticides under the IPM program (Jayaseeli et al., 2022).

10.3.5 ANTI-LEISHMANIA ACTIVITY

Anti-leishmania activity refers to the ability of a substance to inhibit the growth or kill leishmania parasites, the causative agents of leishmaniasis. The ethanolic extract and its fractions of *P. calomelanos* were studied *in vitro* for leishmanicidal activity against promastigote forms of *Leishmania brasiliensis*. The results indicated a significant leishmanicidal potential of the ethanolic extract and the ethyl-acetate fraction (a concentration of 500 mg/mL inhibited 100% of the parasite cells). The hexane and methanol fractions did not show leishmanicidal activity (Souza et al., 2013b).

10.3.6 IMMUNOSTIMULANT ACTIVITY

Kaempferol (39) was isolated from the acetone extract of *P. calomelanos*. It showed immunostimulant activity with a phagocytosis index of 1.846 based on the carbon clearance method (Sutoyo et al., 2018).

KEYWORDS

- **anticancer activity**
- **anti-leishmania**
- **flavonoids**
- **mucolytic activity**
- **pharmacological activities**
- **photochemistry**
- **x-ray crystallography**

REFERENCES

Asai, F., Iinuma, M., Tanaka, T., & Mizuno, M. (1991). Complex flavonoids in farinose exudate from *Pityrogramma calomelanos*. *Phytochemistry, 30*, 3091–3093.

Bardouille, V., Mootoo, B. S., Hirotsu, K., & Clardy, J. (1978). Sesquiterpenes from *Pityrogramma calomelanos*. *Phytochemistry, 17*, 275–277.

De Souza, T. M., Braga, M. F., Saraiva, R. A., Nogara, P. A., Bueno, D. C., Boligon, A. A., Fone, M. L., da Rocha, J. B., Rolon, M., Vega, C., & de Arias, A. R. (2012). Cytotoxic and tripanocide activities of *Pityrogramma calomelanos* (L.) Link. *American Fern Journal, 102*, 198–207.

Dietz, V. H., Wollenweber, E., Favre-Bonvin, J., & Gómez, P. L. D. (1980). A novel class of complex flavonoids from the frond exudate of *Pityrogramma trifoliata*. *Zeitschrift für Naturforschung C, 35*, 36–40.

Freitas, M. A., Santos, A. T., Machado, A. J., Silva, A. R. P., Campina, F. F., Costa, M. S., Martins, G. M., Morais-Braga, M. F. B., Tintino, S. R., Menezes, I. R., & Ribeiro-Filho, J. (2017). Fern extracts potentiate fluconazole activity and inhibit morphological changes in *Candida* species. *Asian Pacific Journal of Tropical Biomedicine, 7*, 1025–1030.

Hitz, C., Mann, K., & Wollenweber, E. (1982). New flavonoids from the farina of *Pityrogramma* species. *Zeitschrift für Naturforschung C, 37*, 337–339.

Iinuma, M., Hamada, K., Mizuno, M., Asai, F., & Wollenweber, E. (1986). Complex flavonoids from *Pityrogramma* frond exudates: Synthesis of two flavones with C–C-linked dihydrocinnamoyl substituents. *Zeitschrift für Naturforschung C, 41*, 681–683.

Iinuma, M., Tanaka, T., & Asai, F. (1994). Flavonoids in frond exudates of *Pityrogramma tartarea*. *Phytochemistry, 36*, 941–943.

Iinuma, M., Tanaka, T., Asai, F., Miyauchi, K. I., & Wollenwebert, E. (1993). Spectral characters of a complex flavonoid isolated from the farinose exudate of *Pityrogramma calomelanos*. *Phytochemistry, 33*, 1247–1248.

Iinuma, M., Tanaka, T., Takenaka, M., Mizuno, M., & Asai, F. (1992). Five complex flavonoids in the farinose exudate of *Pityrogramma calomelanos*. *Phytochemistry, 31*, 2487–2490.

Jayaseeli, J. P. R., Selvaraj, P. S., John, J. S., & Velankanny, M. (2022). Phytochemical screening and pesticidal efficacy of phytoecdysteroid fraction of a fern *Pityrogramma calomelanos* (L.) Link against *Spodoptera litura* (Fabricius). *Journal of Biopesticides, 15*, 73–81.

Markham, K. R., Vilain, C., Wollenweber, E., Dietz, V. H., & Schilling, G. (1985). Isoceroptene, a novel polyphenol from *Pityrogramma triangularis*. *Zeitschrift für Naturforschung C, 40*, 317–320.

Markham, K. R., Wollenweber, E., & Schilling, G. (1987). Structure revision for two C-methyl flavanones from *Pityrogramma pallida*. *Journal of Plant Physiology, 131*, 45–48.

Martin, K. P., Sini, S., Zhang, C. L., Slater, A., & Madhusoodanan, P. V. (2006). Efficient induction of apospory and apogamy *in vitro* in silver fern (*Pityrogramma calomelanos* L.). *Plant Cell Reports, 25*, 1300–1307.

Najihah, N., Sutoyo, S., & Ismono, I. (2018). Toxicity assay of phenolic compound isolated from the dichloromethane extract of Silver Fern (*Pityrogramma calomelanos*). In *Seminar Nasional Kimia – National Seminar on Chemistry (SNK 2018)* (pp. 63–66). Atlantis Press.

Nilsson, M. (1961). Dihydrochalcones from the fronds of *Pityrogramma chrysophylla* var. *marginata* Domin. *Acta Chemica Scandinavica, 15*, 154–158.

Rodriguez, A. M., Derita, M., Cartagena, E., Hernández, M. A., & Bardón, A. (2022). Complex flavonoids of *Pityrogramma trifoliata*: Absolute configuration, antifungal effects, and localization in the plant. *Natural Product Communications, 17*, 1–10.

Souza, T. M., Morais-Braga, M. F. B., Saraiva, A. Á. F., Rolón, M., Vega, C., de Arias, A. R., Costa, J. G. M., Menezes, I. R. A., & Coutinho, H. D. (2013b). Evaluation of the

anti-Leishmania activity of ethanol extract and fractions of the leaves from *Pityrogramma calomelanos* (L.) Link. *Natural Product Research, 27*, 992–996.
Souza, T. M., Morais-Braga, M. F., Costa, J. G., Saraiva, A. A., & Coutinho, H. D. (2012). Enhancement of antimicrobial activity of antibiotics and antifungals by the use of natural products from *Pityrogramma calomelanos* (L.) Link. *Archives of Biological Sciences, 64*, 43–48.
Souza, T. M., Morais-Braga, M. F., Costa, J. G., Saraiva, A. A., Lima, M. A., & Coutinho, H. D. (2013a). Herbs in association with drugs: Enhancement of the aminoglycoside-antibiotic activity by *Pityrogramma calomelanos* (L.) Link. *Journal of Young Pharmacists, 5*, 188–190.
Star, A. E., & Mabry, T. J. (1971). Flavonoid frond exudates from two Jamaican ferns, *Pityrogramma tartarea* and *P. calomelanos*. *Phytochemistry, 10*, 2817–2818.
Star, A. E., Mabry, T. J., & Smith, D. M. (1978). Triangularin, a new chalcone from *Pityrogramma triangularis*. *Phytochemistry, 17*, 586, 587.
Star, A. E., Rösler, H., Mabry, T. J., & Smith, D. M. (1975). Flavonoid and ceroptin pigments from frond exudates of *Pityrogramma triangularis*. *Phytochemistry, 14*, 2275–2278.
Sutoyo, S., Ismono, I., & Mitarlis, M. (2018). Immunostimulant activity of flavonoid isolated from the acetone extract of silver fern (*Pityrogramma calomelanos*). In *Seminar Nasional Kimia – National Seminar on Chemistry (SNK 2018)* (pp. 86–88). Atlantis Press.
Sutoyo, S., Tukiran, T., Hidajati, N., & Kumalasari, N. I. (2019). Mucolytic activity of flavonol isolated from the silver fern (*Pityrogramma calomelanos*). In *National Seminar on Chemistry 2019 (SNK-19)* (pp. 8–10). Atlantis Press.
Suyatno, N. H., Umami, K., & Sari, I. P. (2014). Chemical constituents of Indonesian silver fern (*Pityrogramma calomelanos*) and their cytotoxicity against murine leukemia P-388 cells. In *Proceeding of International Conference on Research, Implementation and Education of Mathematics and Sciences* (Vol. 1, pp. 229–236).
Windham, M. D., & Yatskievych, G. (2003). Chromosome studies of cheilanthoid ferns (Pteridaceae: Cheilanthoideae) from the western United States and Mexico. *American Journal of Botany, 90*(12), 1788–1800.
Wollenweber, E., Dietz, V. H., Schilling, G., Favre-Bonvin, J., & Smith, D. M. (1985). Flavonoids from chemotypes of the goldback fern, *Pityrogramma triangularis*. *Phytochemistry, 24*, 965–971.
Wollenweber, E., Dietz, V. H., Smith, D. M., & Seigler, D. S. (1979). A novel C-methylated dihydrochalcone from *Pityrogramma triangularis* var. *viscosa*. *Zeitschrift für Naturforschung C, 34*, 876, 877.
Wollenweber, E., Rehse, C., & Dietz, V. H. (1981a). The occurrence of aurentiacin and flavokawin B on *Pityrogramma triangularis* var. *pallida* and *Didymocarpus* species. *Phytochemistry, 20*, 1167, 1168.
Wollenweber, E., Walter, J., & Schilling, G. (1981b). New flavanones and chalcones from the farinose frond exudate of *Pityrogramma pallida*. *Zeitschrift für Pflanzenphysiologie, 104*, 161–168.
Zhang, L., Zhou, X. M., Lu, N. T., & Zhang, L. B. (2017). Phylogeny of the fern subfamily *Pteridoideae* (*Pteridaceae*; *Pteridophyta*), with the description of a new genus: *Gastoniella*. *Molecular Phylogenetics and Evolution, 109*, 59–72.

CHAPTER 11

A Comprehensive Review of the Ethnomedicinal Properties, Bioactive Compounds, and Pharmacological Activities of *Cyrtomium fortunei* J. Sm.

SANTOSHKUMAR JAYAGOUDAR,[1,*] SAVALIRAM G. GHANE,[2] and PRADEEP BHAT[3]

[1]*Department of Botany, G. S. S. College and Rani Channamma University P. G. Centre, Belagavi, Karnataka, India*

[2]*Department of Botany, Plant Physiology Laboratory, Shivaji University, Vidyanagar, Kolhapur, Maharashtra, India*

[3]*ICMR–National Institute of Traditional Medicine, Belagavi, Karnataka, India*

*Corresponding authors

ABSTRACT

The medicinally important fern, that is, *Cyrtomium fortunei* J. Sm., belongs to the family Polypodiaceae. The plant grows perennially in the temperate biome and is native to the Eastern Himalaya, Korea, Indo-China, Japan, and Taiwan. The plant is used in Korean traditional medicine as an antitoxic and hemostatic agent. In Chinese herbal medicine, it is used in the treatment of many viral diseases. Powdered rhizomes and whole plant extracts are used as vermifuge agents for humans. The plant is used to treat a range of diseases, including influenza, cancer, pharyngitis, severe acute respiratory syndrome (SARS), and migraine. Phytochemical screening revealed the presence of protocatechaldehyde, woodwardin-säure methylester,

Bioactives and Pharmacology of Pteridophytes. T. Pullaiah (Ed.)

DOI: 10.1201/9781779642738-11

physcion, ursolic acid, betulin, woodwardinic acid, sutchuenoside A, kaempferol-3, (-)-epicatechin, (+)-catechin hydrate, kaempferol, crassirhizomoside-A, and phellopterin compounds from different extracts of the plant. Due to the presence of potent metabolites, the species shows promising pharmacological activities such as cholinesterase inhibition, antioxidant, depigmentation, anti-enterovirus, antitumor, and anti-inflammatory activities.

11.1 INTRODUCTION

Cyrtomium fortunei J. Sm. belongs to the family Polypodiaceae. *Phanerophlebia fortunei* (J.Sm.) Copel., *Polystichum falcatum* var. *fortunei* (J.Sm.) Matsum., *Polystichum fortunei* (J.Sm.) Nakai, *Aspidium falcatum* var. *fortunei* (J.Sm.) Makino, *Cyrtomium shandongense* J.X. Li, *Cyrtomium recurvum* Ching and Shing, *Cyrtomium confertiserratum* J.X. Li, H.S. Kung and Xiao J. Li, and *Cyrtomium reflexosquamatum* J.X. Li and F.Q. Zhou are synonyms of this species (POWO, 2024).

Average plant height is 25–50 cm, densely covered with brown scales. Stipe 12–26 cm, stramineous, lower part dense brown. Lamina oblong-lanceolate, base slightly contracted, unipinnate-imparipinnate, apex obtuse; rachis sparse lanceolate. Lateral pinnae 7–16 (up to 29) pairs, alternate, with short stalk, lanceolate, more or less falcate; margins cuneate, entire or sometimes serrulate, apex acuminate or rarely caudate; terminal pinna ovate-lanceolate, sometimes lower portion with 1 or 2 lobes, papery, glabrous on both surfaces; sori throughout the abaxial surface of pinnae; indusia grayish, margins entire. The plant is perennial and grows naturally in the temperate biome, distributed to the Eastern Himalaya, Korea, Indo-China, Japan, and Taiwan, and grows primarily (POWO, 2024). The proper use of crude plant extracts was in the age-old Indian traditional system of medicine 'Ayurveda' and 'Traditional Chinese Medicine' for thousands of years. Since ancient times, most of the ferns and their allies have been reported to provide many health benefits. These plants were used in several foods, teas, and drugs. The freshwater fern-weeds are used in the preparation of life-saving drugs (Baskaran et al., 2018; Goswami et al., 2016). The plant is used in Korean traditional medicine for its reputed effects as a counter toxin and hemostatic agent (Choi, 2013).

In Chinese herbal medicine, it is used in the treatment of many viral diseases. Powdered rhizomes and whole plant extracts are used as vermifuge agents for humans. The plant is used to treat several other ailments, viz. severe acute respiratory syndrome (SARS), influenza, cancer, acute and chronic pharyngitis, and migraine. Also, the herb is used to treat other respiratory diseases believed to be caused by coronavirus 14 strains (Yang et al., 2013a).

11.2 BIOACTIVES

The comprehensive and systematic studies on the phytochemicals and their bioactivity of *Cyrtomium fortunei* are limited (Yang et al., 2013a, b). The dried rhizomes (10 kg) were suspended in water and extracted (10 L × 3 times) with petroleum ether, ethyl acetate, 80% ethanol, and n-BuOH, respectively. The extracts were dried under vacuum, and 23 compounds were obtained from the fractions – Protocatechaldehyde, woodwardin saure methyl ester, physcion, ursolic acid, sitost-4-en-3-one, betulin, 3,'4,'5'-Trihydroxy-3,7-dimethoxyflavone, woodwardinic acid, sutchuenoside A, kaempferol-3,7-O-α-L-dirhamnoside, (-)-epicatechin, (+)-catechin hydrate, kaempferol, crassirhizomoside-A, and kaempferol-3-O-(3-O-acetyl-α-L-rhamnopyranoside) (Yang et al., 2013a). The water extract of the whole plant yielded five compounds, such as 6'-methylglucuronide-5-hydroxy-chromone, ethyl α-D-glucopyranoside, neoechinulin A, 9,12,13-trihydroxyoctadeca-10(E),15(Z)-dienoic acid, and phellopterin (Yang et al., 2013b). The chemical composition of essential oil (EO) from the fresh aerial parts (leaves) was determined by Yang et al. (2015) through gas chromatography-mass spectroscopy (GC–MS) instrumentation. In total, 26 compounds representing 93.19% of the oil composition were isolated. The compounds detected in EO are trans-1,4-dimethylcyclohexane, octane, 5-nonadecen-1-ol, n-hexadecanoic acid, 9Z,12Z-octadecadienoic acid, oleic acid, octadecanoic acid, octadecanal, 1,21-docosadiene, 1-nonadecene, 1-nonadecanol, squalene, 1,19-eicosadiene, cyclotetracosane, 1-hentetracontanol, 1,19-eicosadiene, 1-heptacosanol, 17-pentatriacontene, campesterol, 1,30-triacontanediol, 7-oxo-5a-cholestane-3b-ol acetate, diploptene, 22,23-dihydrostigmasterol, β-amyrin, and 3-methylheneicosane, stigmast-4-en-3-one (Figure 11.1).

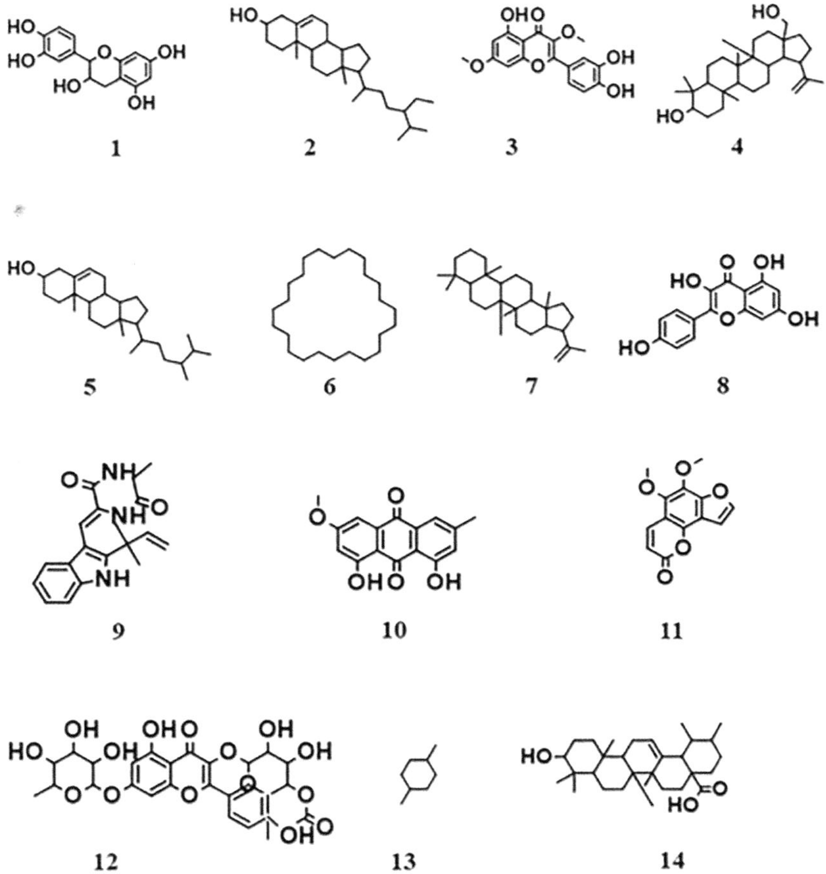

FIGURE 11.1 Structures of (−)-epicatechin (1), 22,23-Dihydrostigmasterol (2), 3,'4,'5-trihydroxy-3,7-dimethoxyflavone (3), Botulin (4), Campesterol (5), Cyclotetracosane (6), Diploptene (7), Kaempferol (8), Neoechinulin A (9), Physcion (10), Pimpinellin (11), Sutchuenoside A (12), Trans-1,4-dimethylcyclohexane (13), Ursolic acid (14).

11.3 PHARMACOLOGY

11.3.1 CHOLINESTERASE ACTIVITY

The cholinesterase activity of *C. fortunei* leaf extract performed through Ellman's method showed 100.70% relative enzyme activity (Chen et al., 2013).

11.3.2 DEPIGMENTATIONAL ACTIVITY

The inhibitory effect of the root extract on melanin production was evaluated by assessing its tyrosinase inhibitory effects. This extract exhibited depigmenting ability on UV-induced hyperpigmentation in the skin of brown guinea pigs. The dried root (100 g) was ground and extracted with methanol at 50°C. The filtered extract was evaporated in a vacuum at 45°C. The root extract exhibited melan-a and mushroom tyrosinase activity by 49% and 52%, respectively, at 500 mg/ml (Choi, 2013).

11.3.3 ANTI-ENTEROVIRUS ACTIVITY

The study aimed to search for new antiviral agents, specifically anti-enterovirus agents from traditional Chinese medicine (Guo et al., 2006). The aqueous extracts of more than 100 traditional medicinal plants in China were evaluated for MTT assay followed by *in vitro* anti-Coxsackie virus activity against different strains. Among the 16 selected plants, *Sargentodoxa cuneata, Sophora tonkinensis, Paeonia veitchii, Spatholobus suberectus,* and *Cyrtomium fortunei* exhibited significant anti-enterovirus activity against Coxsackie virus B5, including Echo virus 9 and 29, followed by Polio virus I. Cytotoxicity assay demonstrated that all the tested extracts had higher CC50 values compared to their EC50 values. The EC50 index of *C. fortunei* against Coxsackie virus B3 was 64.9 mg/L, followed by Coxsackie virus B5 (66.2 mg/L), Echo virus 9 (EC50 – 13.6 mg/L), Echo virus 29 (EC50 – 44.2 mg/L), and Polio virus I (EC50 – 52.8 mg/L) (Guo et al., 2006).

11.3.4 ANTIOXIDANT ACTIVITY

Antioxidant activity of the extract and its solvent fractions (n-hexane, 85% aqueous methanol, n-butanol, and water) of the whole plant was evaluated (Kim et al., 2021). Antioxidative ability was measured by DPPH, intracellular reactive oxygen species (ROS), peroxynitrite scavenging, DNA oxidation inhibition, and ferric reducing antioxidant power (FRAP). The extracts such as 85% aqueous-MeOH and n-BuOH fractions showed significant radical scavenging activities in HT-1080 cells. In another study, crude petroleum ether, ethyl acetate, and n-BuOH extracts were studied on different cell lines such as PC3, A375, BGC-823, and Bcap-37. Ethyl acetate and n-BuOH extracts exhibited higher DPPH free radical scavenging effects (96.7 ± 3.5% and 96.9 ± 3.2%, respectively, at 100 μg/mL concentration) (Yang et al., 2013c).

11.3.5 ANTI-INFLAMMATORY ACTIVITY

The anti-inflammatory activity of the extract and its solvent fractions (n-hexane, 85% aqueous methanol, n-butanol, and water) of the whole plant was analyzed by the inhibitory effect against mRNA expression of pro-inflammatory factors (COX-2, NO, IL-1β, and IL-6) in Raw 264.7 macrophages. The crude extract showed inhibition at 100 μg/ml concentration. The fractions of n-BuOH and 85% aqueous MeOH fraction greatly suppressed the expression levels of IL-1β and IL-6 at 100 μg/ml concentration (Kim et al., 2021).

11.3.6 ANTITUMOR ACTIVITY

The antitumor activities (*in vitro*) of crude ethyl acetate, petroleum ether, and n-BuOH extracts were performed against the cell lines viz. Bcap-37, BGC-823, A375, and PC3 through MTT assay, using adriamycin as a positive standard and culture medium with DMSO (0.1%) as a negative control (Yang et al., 2013c). Ethyl acetate and n-BuOH extracts, followed by four fractions (Fr.3, Fr.5, Fr.6, and Fr.12) exhibited antiproliferative activities against all the cell lines. Among all, fraction 5 demonstrated inhibitory activity of 82.4%, 81.2%, 75.1%, and 80.9% at 100 μg/mL each.

KEYWORDS

- **anti-inflammatory activity**
- **antioxidant activity**
- **antitumor activity**
- **fern**
- **hyperpigmentation**
- **phytochemical screening**
- **polypodiaceae**
- **reactive oxygen species**

REFERENCES

Baskaran, X. R., Geo Vigila, A. V., Zhang, S. Z., Feng, S. X., & Liao, W. B. (2018). A review of the use of pteridophytes for treating human ailments. *Journal of Zhejiang University–Science B (Biomedicine & Biotechnology), 19*(2), 85–119.

Chen, K. J., Zhao, G., Wu, S. C., Zhuang, Q. C., & Qian, F. (2013). Preliminary researches on possibility of several kinds of plants including *Dracaena angustifolia* used to relieve the damages caused by organic phosphorous compounds. *Advanced Materials Research, 815*, 358–361. https://doi.org/10.4028/www.scientific.net/AMR.815.358.

Choi, S. Y. (2013). Inhibitory effects of *Cyrtomium fortunei* J. Smith root extract on melanogenesis. *Pharmacognosy Magazine, 9*(35), 227–230. https://doi.org/10.4103/0973-1296.113272.

Goswami, H. K., Sen, K., & Mukhopadhyay, R. (2016). Pteridophytes: Evolutionary boon as medicinal plants. *Plant Genetic Resources, 14*(4), 328–355. https://doi.org/10.1017/S1479262116000290.

Guo, J. P., Pang, J., Wang, X. W., Shen, Z. Q., Jin, M., & Li, J. W. (2006). *In vitro* screening of traditionally used medicinal plants in China against enteroviruses. *World Journal of Gastroenterology, 12*(25), 4078–4081.

Kim, H., Kim, H., Kong, C. S., Lee, B. H., Sim, H. B., & Seo, Y. (2021). Antioxidant and anti-inflammatory activities of the halophyte *Cyrtomium falcatum. Ocean and Polar Research, 43*(3), 113–126.

POWO (Plants of the World Online). (2024). Facilitated by the Royal Botanic Gardens, Kew. http://www.plantsoftheworldonline.org/ (accessed on 17th November 2025).

Yang, S. J., Liang, N., Hu, D. Y., Xiang, H. M., Xue, W., & Yang, S. (2013c). Antitumor and antioxidant activities of extracts from *Cyrtomium fortumei* (J.) Smith. *Asian Journal of Chemistry, 25*(14), 7761–7764.

Yang, S. J., Liu, M. C., Liang, N., Xiang, H. M., & Yang, S. (2013b). Chemical constituents of *Cyrtomium fortumei* (J.) Smith. *Natural Product Research, 27*(21), 2066–2068. https://doi.org/10.1080/14786419.2013.824442.

Yang, S., Liu, M., Liang, N., Zhao, Q., Zhang, Y., Xue, W., & Yang, S. (2013a). Discovery and antitumor activities of constituents from *Cyrtomium fortumei* (J.) Smith rhizomes. *Chemistry Central Journal, 7*(1), 24.

Yang, S., Liu, M., Zhao, Q., Zhao, H., Xue, W., & Yang, S. (2015). Antiproliferative and apoptosis-inducing effect of essential oil extracted from *Cyrtomium fortumei* (J.) Smith leaves. *Medicinal Chemistry Research, 24*, 1644–1652.

CHAPTER 12

Phytochemical and Pharmacological Profile of *Actiniopteris radiata* Link (Pteridaceae)

K. RAJA KULLAYISWAMY* and N. SAROJINI DEVI

Centre of Excellence for Biodiversity and Climate Resilience, Dharmavana Nature Ark, Hyderabad, Telangana, India

*Corresponding author

ABSTRACT

Actiniopteris radiata, commonly known as the Peacock Fern, is a notable species among ferns, thriving among rocks in both dry and wet areas of tropical and subtropical regions, including Africa, Asia, and parts of Australia. It is a small, erect, and rather stiff fern that contains various bioactive compounds such as quercetin-3-rutinoside (rutin), hentriacontane, hentriacontanol, β-sitosterol palmitate, β-sitosterol, β-sitosterol-D-glucoside, an unidentified glucoside, glucose, and fructose. *A. radiata* exhibits a range of medicinal properties, including antibiotic, antidiabetic, antifungal, antioxidant, and antifertility activities, marking it as a key medicinal plant. However, further experimental studies are necessary to isolate and identify single active molecules from this group of compounds to better understand its pharmacological potential.

12.1 INTRODUCTION

Ferns are a valuable source of medication for various ailments, despite their immense medicinal potential being underutilized compared to angiosperms (Benniamin, 2011). The medicinal properties of certain

Bioactives and Pharmacology of Pteridophytes. T. Pullaiah (Ed.)

DOI: 10.1201/9781779642738-12

ferns are documented in ancient literature by Theophrastus, Sushruta, Charaka, Dioscorides, and others. Some ferns used as drugs are also listed in the pharmacopoeias of different countries (Benniamin, 2011). Ethnic communities worldwide employ ferns to treat ailments such as dysentery, malaria, stomach ache, urinary disorders, and burns. Ferns are incorporated into Ayurvedic, Homeopathic, and Unani medicines for various treatments. Recently, numerous phytochemical and pharmacological studies on ferns have been conducted, revealing information about the bioactive components of important medicinal ferns (Benniamin, 2011). *Actiniopteris radiata* is one notable species among all ferns. It grows among rocks in dry and wet areas of tropical and subtropical regions, including Africa, Asia, and parts of Australia. Commonly known as the Peacock Fern, it is a small, delicate fern characterized by its fan-shaped fronds that resemble the spread feathers of a peacock's tail. It shows many medicinal properties like antimicrobial, anti-inflammatory, antioxidant, and is used for wound healing, gastrointestinal and urinary disorders, fever, and malaria. Phytochemical and pharmacological studies on this species have identified several bioactive components, including flavonoids, tannins, saponins, and phenolic compounds. These studies have further validated its traditional uses and highlighted its potential in modern medicine (Yaseen et al., 2023). *Actiniopteris radiata* (Sw.) Link synonyms include *Asplenium radiatum* Sw., *Acropteris dichotoma* (Kuhn) Farw., and *Actiniopteris dichotoma* Kuhn. Common names of this species are English – Peacock's tail; Sanskrit – Mayurishikha; Telugu – Nemalipinchamu; Hindi – Morpankhi; Marathi – Mapursika; Tamil – Mayilatumshikhai; Malayalam – Nanmukhappullu. This species is used as an anthelmintic, as an alternative in prolonged malarial fevers, and as an astringent to arrest hemorrhages (Puri, 1970).

Actiniopteris radiata is a small, erect, rather stiff fern, 15–20 cm high, resembling a miniature fan palm. The rhizome is erect and clothed with scales; fronds are tufted and 5–20 cm tall. The stipes have a few persistent scales, and the lamina is palmate and fan-like, comprising many dichotomous segments, coriaceous, with edges that inflex on drying and an apex that is toothed. The fertile segments are longer than the sterile ones, and the sori are copious, covering the whole lower surface. This plant is a xerophyte and grows in the crevices of rocks and walls of old buildings. The mature diploid sporophyte is the most recognizable stage of the life cycle of *Actiniopteris radiata*. The leaves, known as fronds, are multi-pinnate. The short stem is entirely underground with a rhizome from which roots emerge. A cross-section of the leaf reveals the epidermis, cortex, and a cylinder with vascular bundles. The plant *Actiniopteris radiata* is used as a styptic, anthelmintic (Nadkarni, 2005), and dietary supplement in cases of malnutrition (Naik et al., 2015), and is also used to treat acute headaches, dysentery, and intestinal worms when a paste with sugar is taken twice daily for three days (Chopra, 1994; Pushpalata et al., 1990). The paste is also used as an aphrodisiac and tonic. In syphilis, a paste with sugar candy serves as a cooling agent. To cure rickets, two fronds paste is given to children twice daily. The whole plant paste mixed with cow's milk treats piles and leucorrhoea, and milk is given twice daily for 2–3 days. In epilepsy, a paste with sugar candy is administered. Leaf paste with honey treats leucorrhoea, taken twice daily. Soaked plants in water, taken in the morning, help control blood pressure and tuberculosis. Dried plant powder, taken once daily for four days, treats cough. A paste of 5–6 leaves with cow's milk regulates menstrual periods, while plant ash with milk is given after menses for conception. Conversely, a paste of 8–10 leaves with thin curd is used for birth control. A decoction of leaves is also used for tuberculosis (Thulsi Rao et al., 2007).

Sikarwar et al. (2006) have documented various medicinal uses of *Actiniopteris radiata*, which include the application of a paste made from the whole plant (including roots) on cuts and wounds; administration of a mixture of 20 grams of the plant paste with sugar to eradicate intestinal worms, taken morning and evening; plant paste mixed with sugar, administered twice daily to enhance semen quantity and as a tonic to boost potency; use of plant paste with sugar candy as a cooling agent for syphilis; daily consumption of a paste made from two fronds of *Actiniopteris radiata*, twice a day, to treat rickets in children; preparation of a paste by soaking the whole plant in water, filtering it, mixing with cow's milk, and administering it every morning for 21 days for irregular menses, piles, and leucorrhoea treatment. Plant paste is mixed with cow's milk, given twice daily for two to three days, to alleviate vomiting and diarrhea. Other ethnomedicinal uses include plant paste with sugar candy administered for treating epilepsy. Smoking dried fronds of *Actiniopteris radiata* by the Sahariya tribe to treat sore throat. Administration of plant paste with honey twice daily to women for treating leucorrhoea. Application of the whole plant paste for gynecological complaints, headache, and dysentery, particularly in Rajasthan (Jain, 1990).

12.2 BIOACTIVES

Preliminary phytochemical studies by Hungund and Pathak (1971) revealed that the roots and leaves of *Actiniopteris radiata* contain alkaloids. A preliminary study by Manonmani and Sara (2013) showed the presence of secondary metabolites in *Actiniopteris radiata*. In the leaves and rhizomes, they identified sugars, alkaloids, phenolic compounds, and flavonoids, while in the rhizomes alone, they found anthraquinones and amino acids. Bhambie et al. (1972) reported that the ethanolic extract of the powdered sporophyte contains Quercetin-3-rutinoside (rutin). Taneja and Tiwari (1974) reported the presence of hentriacontane, hentriacontanol, β-sitosterol palmitate, β-sitosterol, β-sitosterol-D-glucoside, an unidentified glucoside, glucose, and fructose. Hemalatha et al. (2012) also reported the same compounds as above from the whole plant of *Actiniopteris radiata* along with structures. Those compounds are hentriacontane ($C_{31}H_{64}$ [CH_3-$(CH_2)_{29}$-CH_3]), hentriacontanol (known as 1-hentriacontanol $C_{31}H_{64}O$ [CH_3-$(CH_2)_{29}$-CH_2-OH]), β-sitosterol palmitate ($C_{45}H_{78}O_2$), β-sitosterol ($C_{29}H_{50}O$), β-sitosterol-D-glucoside ($C_{35}H_{60}O_6$), and quercetin-3-rutinoside ($C_{27}H_{30}O_{16}$). These compounds are identified from the Mayurishikha plant using IR, HNMR, CNMR, and direct comparison with authentic samples. The ethanol extract was diluted with absolute ethanol and tested, revealing the presence of 3-hydroxyflavones. Paper chromatography identified a single flavone, Quercetin-3-rutinoside (Rutin). Harborne (1967) found a high percentage of phenolic compounds in *Actiniopteris radiata*, making ferns resistant to fungi and microorganisms. This plant species is known for its anthelmintic and styptic properties due to the presence of Quercetin-3-rutinoside, also found in other ferns like *Pteridium aquilinum* and *Dryopteris oligophlebis*. Quercetin-3-rutinoside reduces blood capillary fragility, acts as a herbal remedy and antioxidant, relaxes smooth muscles, is used in multivitamin preparations, and has antibiotic properties by inhibiting the germination of *Puccinia graminis* f. *tritici* uredospores (Bhambie et al., 1972).

Sankar and Venkateswarlu (2020) reported 11 phytochemical compounds in various solvent extracts of *A. radiata*: phytosterols, terpenoids, glycosides, saponins, flavonoids, tannins, carbohydrates, alkaloids, amino acids, oils, and phenols. They estimated phenolic and flavonoid contents, finding phenolic content in hexane extract at 3.02 ± 0.6 mg/g, ethyl acetate at 18.1 ± 0.2, and hydro-alcoholic at 24.28 ± 0.3.

Flavonoid content was 6.32 ± 0.52, 21.05 ± 0.38, and 22.68 = 0.6, respectively, for the same extracts (Figures 12.1–12.3).

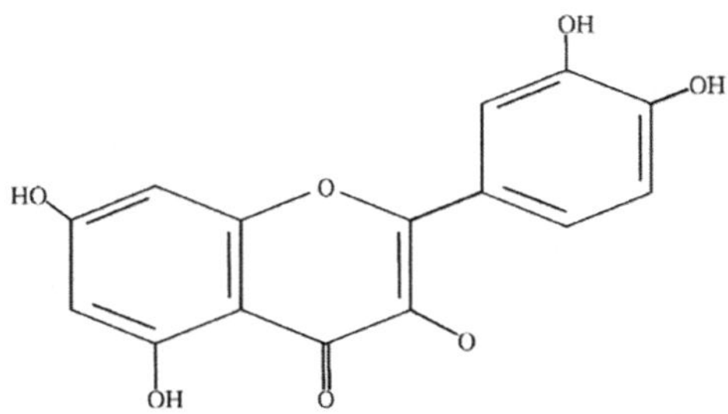

FIGURE 12.1 Structure of rutinose.

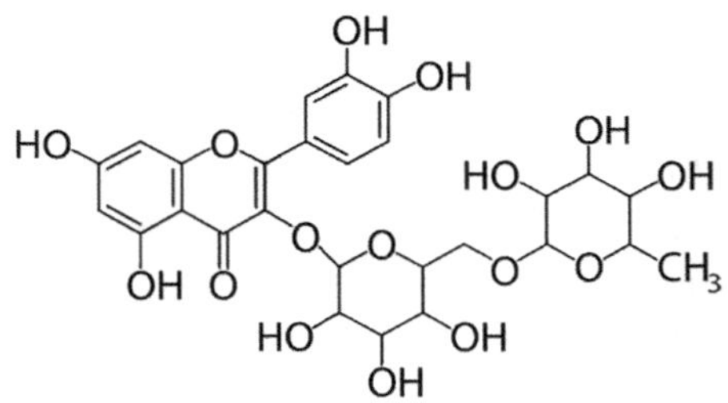

FIGURE 12.2 Rutin (quercetin-3-rutinoside).

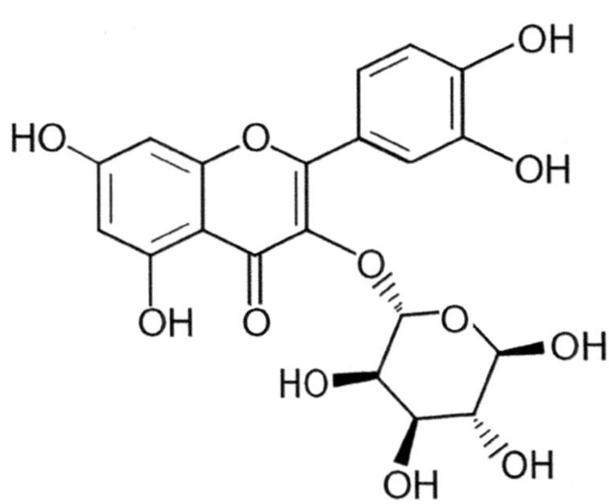

FIGURE 12.3 Quercitrin.

12.3 PHARMACOLOGICAL ACTIVITIES

12.3.1 LIPID PEROXIDATION ASSAY (LPOA)

Manjunath et al. (2011) induced lipid peroxidation (LPO) in human red blood cells (RBC) using the Fe^{2+}-ascorbate system and measured it as thiobarbituric acid reacting substances (TBARS) following Buege and Aust's method (1978). They conducted the reaction using RBCs (10^8 cells/mL) in Tris-HCl buffer (20 mM, pH 7.0) with $CuCl_2$ (2 mM), ascorbic acid (10 mM), and hexane, chloroform, ethyl acetate, ethanol, and aqueous extracts of *A. radiata* (200–1000 μg/mL) in a final volume of 1 mL. The mixture was incubated at 37°C for 1 hour. Lipid peroxidation was assessed by measuring malondialdehyde (MDA) equivalents using the TBA-TCA method (0.375% w/v TBA, 15% w/v TCA, and 0.25N HCl). The results showed that at a 200 μg/mL concentration, n-Hexane, chloroform, ethyl acetate, ethanol, and aqueous extracts of *A. radiata*, as well as ascorbic acid, inhibited lipid peroxidation by 34.66%, 26.86%, 31.53%, 38.53%, 28.9%, and 40.24%, respectively. Among the extracts, the ethanol extract of *A. radiata* exhibited the highest anti-lipid peroxidation activity. The anti-LPO activity ranking from highest to lowest was L-ascorbic acid > ethanol extract of *A. radiata* > n-Hexane extract > ethyl acetate extract > aqueous extract > chloroform extract.

12.3.2 ANTIOXIDANT ACTIVITY

Manjunath et al. (2011) performed the DPPH activity of *A. radiata*. The plant extracts and ascorbic acid were assessed for their free radical scavenging activity. Results indicated that the DPPH radical scavenging efficacy of n-Hexane, chloroform, ethyl acetate, ethanol, and aqueous extracts of *A. radiata,* as well as ascorbic acid, followed this order of potency: L-ascorbic acid > ethanol extract of *A. radiata* > n-Hexane extract > ethyl acetate extract > aqueous extract > chloroform extract. The percentage inhibitions were 73.2%, 62.63%, 58.3%, 55.1%, 51.3%, and 48.66%, respectively, at a concentration of 1 mg/mL. Paulsamy et al. (2013) also investigated the DPPH scavenging activity. They utilized methanol extracts of samples at concentrations ranging from 50 to 250 μg/mL. *Actiniopteris radiata* extract demonstrated a concentration-dependent increase in free radical scavenging activity. The IC50 value for *A. radiata* extract was determined to be 93.48, indicating its efficacy in scavenging

DPPH radicals. This activity was compared with the standard antioxidant, BHT (IC50 = 22.83), which also exhibited significant DPPH radical scavenging activity in the study.

The DPPH radical scavenging activity was also measured by Sankar and Venkateswarlu (2020). The results showed DPPH free radical IC50 values for the n-Hexane extract (HE) 225 μg, the ethyl acetate extract (EAE) 256 μg, and the hydro-alcoholic extract (HAE) 145 μg, compared with ascorbic acid (AA) standard at 95 μg. The inhibition percentages of the same extracts at different concentrations were as follows: At 20 μg concentration: HE: 7.33%, EAE: 6.33%, HAE: 10.67%, AA: 16.33%; At 40 μg concentration: HE: 15.00%, EAE: 12.00%, HAE: 23.33%, AA: 30.67%; At 80 μg concentration: HE: 20.67%, EAE: 23.33%, HAE: 38.67%, AA: 46.33%; At 160 μg concentration: HE: 39.67%, EAE: 40.11%, HAE: 52.33%, AA: 66.00%; At 320 μg concentration: HE: 65.33%, EAE: 56.00%, HAE: 71.00%, AA: 82.33% (Sankar and Venkateswarlu, 2020).

Antioxidant activity was evaluated using the ABTS•+ assay. In this study, Paulsamy et al. (2013) determined the total antioxidant activity of methanolic extracts from *Actiniopteris radiata*, finding it to exhibit the highest activity at 2523.11 μmol/g. ABTS•+ radicals were generated by incubating ABTS with potassium persulfate. Compounds within the extracts that inhibit potassium persulfate activity can reduce ABTS•+ production, influencing the assay's outcome.

Manjunath et al. (2011) evaluated the hydroxyl radical scavenging activity of *A. radiata*. The results showed that all solvent extracts of *A. radiata* demonstrated the ability to protect 2-deoxy-D-ribose from oxidative degradation by scavenging hydroxyl radicals in a concentration-dependent manner. Ascorbic acid exhibited the highest hydroxyl radical scavenging activity, followed by the ethanol extract of *A. radiata* (EEAR), n-Hexane extract (HEAR), ethyl acetate extract (EaEAR), aqueous extract (AEAR), and chloroform extract (ChEAR).

The scavenging activity of *A. radiata* extracts on hydroxyl radicals was measured by Sankar and Venkateswarlu (2020) using the established method. The results showed hydroxyl free radical IC50 values as follows: n-Hexane extract (HE) was not determined, ethyl acetate extract (EAE) 155 μg, and hydro-alcoholic extract (HAE) 128 μg, compared with ascorbic acid (AA) standard at 103 μg. The inhibition percentages of the same extracts at different concentrations were: At 20 μg concentration: HE: 4.00%, EAE: 8.33%, HAE: 13.00%, AA: 14.67%; At 40 μg concentration:

HE: 9.00%, EAE: 17.00%, HAE: 24.00%, AA: 29.33%; At 80 μg concentration: HE: 18.33%, EAE: 28.00%, HAE: 38.00%, AA: 44.67%; At 160 μg concentration: HE: 34.33%, EAE: 51.00%, HAE: 57.67%, AA: 62.67%; At 320 μg concentration: HE: 45.67%, EAE: 70.67%, HAE: 79.67%, AA: 78.00% (Sankar and Venkateswarlu, 2020).

The ferrous ion chelating (FIC) assay was conducted by Paulsamy et al. (2013) using whole plant methanolic extracts from the two study species. In this assay, extract samples (250 μL) were used. The reaction was initiated by adding 5 mmol/L ferrozine (0.2 mL). The absorbance of the resulting solution was then measured spectrophotometrically at 562 nm. EDTA was used as the standard to evaluate the chelating activity of the extracts, with results expressed as mg EDTA equivalent/g extract. The superoxide scavenging activity of the *Actiniopteris radiata* extracts was evaluated by Sankar and Venkateswarlu (2020). The results showed superoxide free radical IC50 values for the n-Hexane extract (HE) 281 μg/ml, the ethyl acetate extract (EAE) 140 μg/ml, and the hydro-alcoholic extract (HAE) 125 μg/ml. The inhibition percentages of the same extracts at different concentrations were: At 20 μg/ml concentration: HE: 6.00%, EAE: 10.33%, HAE: 12.33%, AA: 18.33%; At 40 μg/ml concentration: HE: 11.33%, EAE: 18.67%, HAE: 23.67%, AA: 35.33%; At 80 μg/ml concentration: HE: 19.67%, EAE: 33.67%, HAE: 39.33%, AA: 54.33%; At 160 μg/ml concentration: HE: 36.00%, EAE: 55.67%, HAE: 59.67%, AA: 69.33%; At 320 μg/ml concentration: HE: 55.33%, EAE: 69.67%, HAE: 80.33%, AA: 85.67% (Sankar and Venkateswarlu, 2020).

12.3.3 HEPATOPROTECTIVE ACTIVITY

Manjunath et al. (2014) investigated the hepatoprotective activity of *Actiniopteris radiata* using CCl_4-induced toxicity in rats. They used the ethanolic extract of *Actiniopteris radiata* (EEAR) on CCl_4-induced oxidative stress in male Wistar albino rats. EEAR was administered for 8 consecutive weeks to the rats. Group I served as the control, Group II as the toxin control (30% CCl_4), and Groups III and IV received EEAR at doses of 250 mg/kg and 500 mg/kg, respectively. The antioxidant status in the liver was assessed by measuring the activities of antioxidant enzymes, including superoxide dismutase (SOD), catalase (CAT), and glutathione peroxidase (GPx), as well as the levels of lipid peroxidation (LPO) and reduced glutathione (GSH). The results showed that CCl_4 induced oxidative stress, evidenced by an increase in LPO and a reduction in SOD, CAT, GPx, and GSH levels. Treatment with EEAR (250 and 500 mg/kg) mitigated the CCl_4-induced oxidative stress. Analysis of the isozyme patterns of these antioxidant enzymes revealed variations in SOD2, CAT, GPx2, and GPx3 in CCl_4-treated rats, which were normalized after EEAR treatment.

The extracts of *A. radiata* were found to be safe in toxicity studies, showing no mortality or adverse physiological and psychological conditions at a dose of 2000 mg/kg b.w. in animals under 24–72h observation (Sankar and Venkateswarlu, 2020). Following this safety confirmation, the extracts were evaluated for their hepatoprotective activity using paracetamol-induced liver toxicity at doses of 100 mg/kg, 200 mg/kg, and 400 mg/kg b.w. The extracts at these concentrations exhibited antioxidant activity and dose-dependent hepatoprotective activity. They effectively reduced elevated liver function parameters such as AST (SGOT), ALT (SGPT), ALP, and total bilirubin levels ($P<0.05$). Sankar and Venkateswarlu (2020) selected six groups of rats for this hepatoprotective activity experiment. Group I rats, treated with vehicle, showed no significant changes. Group II rats, treated with paracetamol, showed significant changes in biomarker enzyme levels. Group III rats, administered paracetamol (200 mg/kg b.w., s.c.) and Liv 52 (25 mg/kg per day, p.o.), showed significant improvements in biomarker enzyme levels compared to Group II rats. The percentage protection offered by Liv 52 against changes in aspartate aminotransferase (serum glutamic-oxaloacetic transaminase (AST (SGOT)), alanine amino-transferase (serum glutamic-pyruvic transaminase (ALT (SGPT)), alkaline phosphatase (ALP), and total bilirubin levels. Group IV, V, and VI rats were administered n-Hexane extract (HE), ethyl acetate extract (EAE), and hydro-alcoholic extract (HAE) at doses of 100 mg/kg, 200 mg/kg, and 400 mg/kg respectively, resulting in changes in biomarker enzyme levels.

The results revealed that Group I (Vehicle) had no significant protection, and hence no protection percentages are reported for AST (SGOT), ALT (SGPT), ALP, and Total Bilirubin. Group II (Paracetamol, 200 mg/kg) showed significant results in biomarker enzymes AST (SGOT), ALT (SGPT), ALP, and Total Bilirubin. Group III (Paracetamol, 200 mg/kg + Liv 52, 25 mg/kg) showed 93.58% protection for AST (SGOT), 92.83% protection for ALT (SGPT), 94.67% protection for ALP, and 93.57% protection for Total Bilirubin. Group IV (Extract HE, 100 mg/kg) had 11.59% protection for AST (SGOT), 11.81% protection for ALT (SGPT),

13.08% protection for ALP, and 13.02% protection for Total Bilirubin. Group V (Extract HE, 200 mg/kg) had 27.71% protection for AST (SGOT), 27.00% protection for ALT (SGPT), 26.27% protection for ALP, and 26.49% protection for Total Bilirubin. Group VI (Extract HE, 400 mg/kg) showed 49.87% protection for AST (SGOT), 50.63% protection for ALT (SGPT), 47.89% protection for ALP, and 47.63% protection for Total Bilirubin. For extract EAE groups, Group IV (100 mg/kg) had 23.30% protection for AST (SGOT), 23.21% protection for ALT (SGPT), 23.63% protection for ALP, and 24.81% protection for Total Bilirubin. Group V (200 mg/kg) had 41.56% protection for AST (SGOT), 41.07% protection for ALT (SGPT), 41.93% protection for ALP, and 44.56% protection for Total Bilirubin. Group VI (400 mg/kg) showed 75.82% protection for AST (SGOT), 72.43% protection for ALT (SGPT), 73.95% protection for ALP, and 74.27% protection for Total Bilirubin. For extract HAE groups, Group IV (100 mg/kg) had 32.62% protection for AST (SGOT), 31.22% protection for ALT (SGPT), 30.33% protection for ALP, and 31.55% protection for Total Bilirubin. Group V (200 mg/kg) had 53.02% protection for AST (SGOT), 54.12% protection for ALT (SGPT), 53.48% protection for ALP, and 53.14% protection for Total Bilirubin. Group VI (400 mg/kg) showed 82.24% protection for AST (SGOT), 82.14% protection for ALT (SGPT), 84.18% protection for ALP, and 82.85% protection for Total Bilirubin. The percentage (%) protection of *Actiniopteris radiata* extracts on enzyme levels at different doses is as follows. The Control group (Group I) received no drug doses (0 mg/kg b.w) and had the following results: AST (84.33 ± 1.45 U/l), ALT (51.50 ± 0.99 U/l), ALP (215.00 ± 1.46 U/l), and Total Bilirubin (0.25 ± 0.01 mg/dl). The Paracetamol group (Group II) received a dose of 200 mg/kg b.w and had the following results: AST (216.67 ± 1.65 U/l), ALT (170.00 ± 1.93 U/l), ALP (531.00 ± 2.78 U/l), and Total Bilirubin (1.35 ± 0.02 mg/dl). The Paracetamol and Liv 52 group (Group III) received a dose of 25 mg/kg b.w and had the following results: AST (92.83 ± 1.08 U/l), ALT (60.00 ± 1.06 U/l), ALP (231.83 ± 2.01 U/l), and Total Bilirubin (0.34 ± 0.01 mg/dl). The HE groups received different doses. Group IV received 100 mg/kg b.w with the following results: AST (201.33 ± 1.58 U/l), ALT (156 ± 1.15 U/l), ALP (489.67 ± 1.16 U/l), and Total Bilirubin (1.21 ± 0.01 mg/dl). Group V received 200 mg/kg b.w with the following results: AST (180 ± 1.57 U/l), ALT (138 ± 0.97 U/l), ALP (448 ± 1.81 U/l), and Total Bilirubin (1.07 ± 0.01 mg/dl). Group VI received 400 mg/kg b.w with the following results: AST (150.67 ± 2.22

U/l), ALT (110 ± 1.63 U/l), ALP (379.67 ± 1.8 U/l), and Total Bilirubin (0.84 ± 0.01 mg/dl).

The EAE groups received different doses as well. Group IV received 100 mg/kg b.w with the following results: AST (185.83 ± 1.35 U/l), ALT (142.5 ± 1.12 U/l), ALP (456.33 ± 1.89 U/l), and Total Bilirubin (1.08 ± 0.01 mg/dl). Group V received 200 mg/kg b.w with the following results: AST (161.67 ± 1.05 U/l), ALT (121.33 ± 1.36 U/l), ALP (398.5 ± 1.65 U/l), and Total Bilirubin (0.87 ± 0.01 mg/dl). Group VI received 400 mg/kg b.w with the following results: AST (116.33 ± 1.73 U/l), ALT (84.17 ± 1.08 U/l), ALP (297.33 ± 1.20 U/l), and Total Bilirubin (0.55 ± 0.01 mg/dl). The HAE groups also received different doses. Group IV received 100 mg/kg b.w with the following results: AST (173.5 ± 1.38 U/l), ALT (133 ± 1.46 U/l), ALP (435.17 ± 1.51 U/l), and Total Bilirubin (1.01 ± 0.02 mg/dl). Group V received 200 mg/kg b.w with the following results: AST (146.5 ± 1.48 U/l), ALT (105.83 ± 1.01 U/l), ALP (362 ± 1.79 U/l), and Total Bilirubin (0.78 ± 0.01 mg/dl). Group VI received 400 mg/kg b.w with the following results: AST (107.83 ± 1.19 U/l), ALT (72.67 ± 1.15 U/l), ALP (265 ± 1.41 U/l), and Total Bilirubin (0.45 ± 0.01 mg/dl).

12.3.4 ANTI-DIABETIC ACTIVITY

Basha et al. (2013) evaluated the anti-diabetic properties of *Actiniopteris radiata* following Miller (1959) DNSA method. They prepared two sets of samples (AR-I & AR-II) with concentrations of 100, 250, and 500 µg. Notably, AR-II at a concentration of 500 µg showed significant inhibition of amylase activity, while the rest were moderately inhibited. In this experiment, acarbose was used as a standard inhibitor.

12.3.5 ANTI-FERTILITY VERSUS ANTI-INFERTILITY PROPERTIES

Jyothi et al. (2019) conducted an experiment on albino mice using the methanolic extract of *Actiniopteris radiata* against standard ethylene estradiol (0.468 mg/kg). They divided the mice into four groups: Group 1 (Control) received only distilled water and food, Group 2 (Standard) received ethylene estradiol at 0.468 mg/kg, Group 3 received the methanolic extract of *Actiniopteris radiata* (MEAR) at 175 mg/kg, and Group 4 received MEAR at 300 mg/kg. The results showed an intriguing contradiction: at a lower dose (175 mg/kg), the extract exhibited anti-fertility effects, while at a higher dose (300 mg/kg), this effect was not observed. These dual properties in the same plant enhanced the scope of further research to evaluate active bio-compounds that could help in laying down standards for the authentication of the original drug.

The only clinical study on humans was conducted by Dixit and Bhatt (1975) at the Research Institute (Ayurveda) in Jaipur, Rajasthan. They studied the antifertility effects of *Actiniopteris radiata* (Mayursikha) combined with *Ocimum americanum* (Nagad Babchi) in 10 volunteers. Administered after menstruation for three days, the combination prevented conception in 9 out of 10 cases, suggesting a non-toxic oral contraceptive. Sharma et al. (1999) studied the antifertility effects of a 50% ethanolic extract of *Actiniopteris radiata* in male rats with the combination of *Ocimum americanum*, showing reduced fertility due to suppressed sperm production and motility, which was reversible after 8–10 weeks.

12.3.6 ANTIHISTAMINIC AND ANTICHOLINERGIC ACTIVITY

Vivek and Gautam (2011) investigated the antihistaminic and anticholinergic activity of *Actiniopteris radiata* whole plant ethanolic extract (AREE – Test extract). In this study, they used five groups of rats: Group I (Control – Histamine & Acetylcholine), Group II (Vehicle – PEG-400, 5%/ml), Group III (Test extract – 2 mg/ml), Group IV (Test extract – 4 mg/ml), and Group V (Test extract – 10 mg/ml). Ileum contractions induced by the agonists were considered as 100%, and reductions induced by the extract were calculated. The percentage of ileum relaxation or contraction was expressed as mean ± SEM. The results showed that AREE had an antagonistic effect on histamine and acetylcholine-induced contractions. Significant effects were observed at doses of 2, 4, and 10 mg/ml for histamine, with relaxation percentages of 68.21 ± 3.872, 52.80 ± 3.142, and 46.52 ± 1.24, respectively. Similarly, the ethanol extract at 2, 4, and 10 mg/ml relaxed tissues precontracted by acetylcholine, with relaxation percentages of 64.22 ± 1.865, 56.98 ± 2.881, and 42.40 ± 1.254, respectively.

12.3.7 THE ANALGESIC ACTIVITY

Jyothi et al. (2018) investigated the analgesic activity of the MEAR using the acetic acid-induced writhing model in albino mice. Group I served as the control, Group II was treated with Pentazocin at 5 mg/kg, Group III received MEAR at 175 mg/kg, and Group IV received MEAR at 300 mg/kg. In the acetic acid-induced writhing model, the extract demonstrated significant inhibition of writhing responses, showing 80.93% and 93.33% inhibition at 175 mg/kg and 300 mg/kg, respectively, compared to the control.

12.3.8 ANTIMICROBIAL ACTIVITY:

Manjunath et al. (2008) evaluated the antibacterial activity of *Actiniopteris radiata* using the agar well diffusion bioassay. The standard drugs used for comparison with the plant extract were ciprofloxacin and chloramphenicol. Minimum Inhibitory Concentration (MIC) and Minimum Bacterial Concentration (MBC) were determined for the extracts that showed total growth inhibition according to the described protocol. Ethanol, n-hexane, chloroform, and ethyl acetate extracts of *A. radiata* were tested for antibacterial activity against selected bacteria, namely *Escherichia coli, Pseudomonas aeruginosa, Salmonella typhimurium, Staphylococcus aureus, Klebsiella pneumoniae*, and *Bacillus subtilis*. The highest zone of inhibition (ZOI) was achieved with ethanol and n-hexane extracts at a concentration of 50 mg/ml against all the tested bacterial strains. However, an average ZOI was observed with the ethyl acetate extract at both concentrations (25 and 50 mg/ml). The chloroform extract showed the least ZOI against Gram-negative bacteria at low concentrations (25 mg/ml). Nevertheless, higher concentrations of chloroform extract (50 mg/ml) exhibited distinct antibacterial properties against both Gram-positive and Gram-negative bacteria. Comparatively, MBC values were higher than the MIC values. The lowest MIC values were observed in ethanol and n-hexane extracts, moderate MIC values were obtained in the ethyl acetate extract, and higher values were found in the chloroform extract. The ethanol extract showed the lowest MIC and MBC values against *S. aureus* and the highest MIC values against *S. typhimurium*. The plant extracts were also screened for qualitative analysis. The chloroform extract revealed the presence of flavonoids and sterols, while the n-hexane extract contained wax, glycosides, and saponins. The ethyl acetate extract contained flavonoids and glycosides, and the ethanol extract contained flavonoids, glycosides, phenols, tannins, and alkaloids.

Naik and Jadge (2010a) tested the antibacterial activity of *Actiniopteris radiata* extracts (petroleum ether, benzene, chloroform, acetone, and ethanol) at concentrations of 150 mg/ml and 300 mg/ml against various Gram-positive and Gram-negative bacteria, including *E. coli, Shigella, Salmonella typhi, Pseudomonas aeruginosa, Vibrio cholerae, Bacillus subtilis, Klebsiella pneumoniae, Proteus vulgaris,* and *Staphylococcus aureus*. The ethanolic extract of *Actiniopteris radiata* at concentrations of 150 mg/ml and 300 mg/ml, as well as the aqueous extract at a concentration of 300 mg/ml, showed antibacterial activity against *S. typhi* and *E. coli.* No antibacterial activity was observed against the other tested Gram-positive and Gram-negative organisms. Kshirsagar and Mehta (1972) reported that the MEAR showed antibacterial activity against *Bacillus subtilis, Bacillus cereus, Bacillus megaterium, Staphylococcus aureus*, and *E. coli.*

12.3.9 ANTIFUNGAL ACTIVITY

Naik and Jadge (2010b) tested the antifungal activity of *Actiniopteris radiata* extracts against *Candida albicans, Aspergillus niger*, and *Mucor* using the cup-plate method at concentrations of 150 mg/ml and 300 mg/ml, with Griseofulvin as the standard drug. All extracts were found to be inactive against *Candida albicans, Aspergillus niger*, and *Mucor* when compared with Griseofulvin.

12.3.10 WOUND HEALING ACTIVITY

Yaseen et al. (2023) reported in his review that the ethanolic extract derived from *A. radiata* has potential in wound healing in rats. The study employed excision, incision, and dead space wound models in rats, with topical application, and compared the outcomes with a standard 5% w/w Povidone-iodine ointment. The results revealed that the alcoholic extract of *A. radiata* exhibited significant wound healing activity in both excision and incision wound models, comparable to that of the 5% w/w Povidone-iodine ointment. Additionally, the extract showed a dose-dependent effect on granulation tissue and hydroxyproline content, indicating its potential in promoting wound healing. Overall, these findings strongly suggest the promising wound healing activity of the alcoholic extract.

KEYWORDS

- ***Actiniopteris radiata***
- **ascorbic acid**
- **bioactives**
- **paper chromatography**
- **pharmacological importance**
- **pteridaceae**
- **quercetin-3-rutinoside**

REFERENCES

Basha, S. C., Sreenivasulu, V., & Pramod, N. (2013). Anti-diabetic activity of *Actiniopteris radiata* (Link). *Asian Journal of Research in Chemistry and Pharmaceutical Sciences, 1*(1), 1–6.

Benniamin, A. (2011). Medicinal ferns of North Eastern India with special reference to Arunachal Pradesh. *Indian Journal of Traditional Knowledge, 10*(3), 516–522.

Bhambie, S., Tiwari, H. P., & George, C. X. (1972). Chemical analysis of *Actiniopteris radiata* (Sw.) Link. *Current Science, 42*(21), 778.

Buege, J., & Aust, D. S. (1978). Microsomal lipid peroxidation. *Methods in Enzymology, 52*, 302–310. https://doi.org/10.1016/S0076-6879(78)52032-6.

Chopra, R. N. (1994). *Indigenous Drugs of India* (pp. 647–648). Academic Publisher.

Dixit, S. K., & Bhatt, G. K. (1975). Folk-lore studies on unknown indigenous antifertility drugs. *Journal of Research in Indian Medicine, 10*(3), 77–81.

Harborne, J. B. (1967). *Comparative Biochemistry of Flavonoids* (p. 383). Academic Press.

Hemalatha, K., Sunandana, R. N., & Shivaji, N. (2012). Phytochemical constituents of *Actiniopteris radiata* Linn. whole plant. *Annals of Phytomedicine, 1*(1), 34–38.

Hungund, B. L., & Pathak, C. H. (1971). A survey of plants in Gujarat, India for alkaloids, saponins and tannins. *USDA Forest Service, Northeastern Forest Experiment Station*, 1–9.

Jain, S. K. (1990). *Dictionary of Indian Folk Medicine and Ethnobotany* (pp. 1–350). Deep Publications.

Jyothi, C., Ali, M. L., & Pratima, M. (2018). Analgesic efficacy of methanolic extract of *Actiniopteris radiata* (Sw.) Link. in albino mice. *International Journal of Current Advanced Research, 7*(5), 12736–12738.

Jyothi, C., Arjun, N. S., Sundar, M., Ali, M. L., & Pratima, M. (2019). *In vivo* assessment of antifertility potential of pteridophytic plants: *Actiniopteris radiata* (Sw.) L. and *Selaginella bryopteris* (L.) Baker in Swiss albino mice. *Asian Journal of Pharmacy and Pharmacology, 5*(1), 152–158.

Kshirsagar, M. K., & Mehta, A. R. (1972). Survey of ferns in Gujarat state (India) for the presence of antibacterial substances. *Planta Medica, 22*, 215–219.

Manjunath, M., & Sharma, P. V. G. K. (2008). *In vitro* evaluation of antibacterial activity of *Actiniopteris radiata* (Sw.) Link. *Journal of Pharmacy and Chemistry, 2*(2), 112–117.

Manjunath, M., Lavanya, G., Sivajyothi, R., & Reddy, O. V. S. (2011). Antioxidant and radical scavenging activity of *Actiniopteris radiata* (Sw.) Link. *Asian Journal of Experimental Science, 25*(1), 73–80.

Manjunath, M., Lavanya, G., Sivajyothi, R., Jayakumar, T., Paresh, D., Kjell, M., & Vijayasarathi, R. O. (2014). Protective effect of *Actiniopteris radiata* (Sw.) Link. against CCl_4-induced oxidative stress in albino rats. *Journal of Ethnopharmacology, 153*(3), 744–752. https://doi.org/10.1016/j.jep.2014.03.040.

Manonmani, R., & Sara, S. C. (2013). Preliminary phytochemical analysis of *Actiniopteris radiata* (Swartz) Link. *International Research Journal of Pharmacy, 4*(6), 214–216.

Miller, G. L. (1959). Use of dinitrosalicylic acid reagent for determination of reducing sugar. *Analytical Chemistry, 31*(3), 426–428.

Nadkarni, K. M. (2005). *Indian Materia Medica* (Vol. 1, p. 38). Popular Prakashan Pvt. Ltd.

Naik, J. B., & Jadge, D. R. (2010a). Evaluation of analgesic activity of *Actiniopteris radiata* (Sw.) Link. *Journal of Pharmacy Research, 3*(7), 1556–1557.

Naik, J. B., & Jadge, D. R. (2010b). Antibacterial and antifungal activity of *Actiniopteris radiata* (Sw.) Link. *International Journal of PharmTech Research, 2*(3), 1751–1753.

Naik, J. B., Jadge, D. R., Deshpande, A. N., Gadgul, A. B., & Siddiqui, A. R. (2015). Pharmacological and phytochemical review on *Actiniopteris radiata. International Journal of Pharmaceutical Sciences and Research, 6*(4), 1373–1377.

Paulsamy, S., Moorthy, D., Nandakumar, K., & Saradha, M. (2013). Evaluation of *in vitro* antioxidant potential of methanolic extracts of the ferns, *Actiniopteris radiata* (Sw.) Link. and *Equisetum ramosissimum* Desf. *International Journal of Research and Development in Pharmacy and Life Sciences, 2*(3), 451–455.

Puri, H. S. (1970). Indian pteridophytes used in folk remedies. *American Fern Journal, 60*(4), 137, 138.

Pushpalata, H., Santha, T. R., Pattanshetty, J. K., & Holla, B. V. (1990). Folk medicines from some rural areas of Bangalore district. *Aryavaidyan, 12*, 215–219.

Sankar, K. G., & Venkateswarlu, B. S. (2020). Phytochemical analysis, antioxidant and hepatoprotective activity of *Actiniopteris radiata. International Journal of Pharmacy and Pharmaceutical Sciences, 12*(8), 69–74.

Sharma, A., Mathur, A., Verma, P., Joshi, S. C., & Dixit, V. P. (1999). Effects of *Actiniopteris dichotoma* (Sw.) on reproductive function of male rat. *Journal of Endocrinology and Reproduction, 3*(1, 2), 47–59.

Sikarwar, R. L. S., Anil, J., & Vivek, K. (2006). Medicinal uses of *Actiniopteris radiata. Journal of Tropical Medicinal Plants, 7*(2), 279–282.

Taneja, S. C., & Tiwari, H. P. (1974). Chemical constituents of *Actiniopteris radiata* (Sw.) Link. *Current Science, 43*(23), 749.

Thulsi, R. K., Reddy, K. N., Pattanaik, C., & Reddy, C. S. (2007). Ethnomedicinal importance of pteridophytes used by Chenchus of Nallamalais. *Research Journal of Medicinal Plants, 4*, 128–133.

Vivek, R. A., & Gautam, P. V. (2011). *In vitro* antihistaminic and anticholinergic activity of *Actiniopteris radiata* (Sw.) Link. on isolated rat ileum. *International Journal of PharmTech Research, 3*(1), 5–7.

Yaseen, M., Khannam, A., & Unnisa, M. (2023). *Actiniopteris radiata* (Link.): An updated critical review. *International Journal of Scientific Research, 12*(12), 215–221.

CHAPTER 13

Biomolecules and Therapeutics of *Salvinia natans* (L.) All.: A Review

SOMASUNDARAM RAMACHANDRAN* and
VEERAMANENI ALEKHYA

Department of Pharmacology, GIET School of Pharmacy, Chaitanya Knowledge City, Rajahmundry, Andhra Pradesh, India

*Corresponding author

ABSTRACT

Salvinia natans (L.) All. belongs to the family Salviniaceae. Though it is most prevalent in Africa, Asia, Central Europe, and South America. *S. natans* has 11 active chemicals, which can be divided into four categories: steroids, megastigmanes, lignans, phenanthrene, and derivatives of nepetolactone. The 70% ethanol extract of *S. natans* revealed no signs of triterpenoids or tannins, but it contained active components such as sugars, alkaloids, glycosides, phenols, flavonoids, and saponins. It was discovered to have biological activities such as antioxidant, hepatoprotective, and antibacterial properties. It was also discovered to be used in the treatment of diabetic foot ulcers.

13.1 INTRODUCTION

There are 12 species in the *Salvinia* genus, and they are all known to be distributed over the African continent. These species are members of the Salvinia family. *Salvinia natans* (L.) All. is the annual species among the other *Salvinia* species. The rhizome of the sporophytic plant is long, thin, and grows horizontally. The rhizome is entirely

Bioactives and Pharmacology of Pteridophytes. T. Pullaiah (Ed.)

DOI: 10.1201/9781779642738-13

covered in whorls of leaves. It is an annual floating water fern with a superficial moss-like appearance. Other names for it include floating fern, floating watermoss, floating moss, and water butterfly wings. It is found throughout the world, although it is most prevalent in Central Europe, Africa, South America, and Asia. To thrive, it requires a large amount of standing fresh water, sunlight, and humid air. The leaves have two shapes: oblong or hemispherical. They typically show up in groups of three. The third leaf is buried, while the other two are lateral and float on the surface of the water. Morphologically, the levitating and submerged leaves differ significantly. The long, filiform, delicately divided leaves dangle below the surface, and they resemble the roots.

13.2 BIOACTIVES

A phytochemical analysis of *Salvinia natans* revealed the presence of 11 compounds, which were further divided into 4 steroids (1–4), 3 megastigmanes (5–7), 2 lignans (8 & 9), 1 phenanthrene (10), and 1 nepetolactone derivative (11). The structures are elucidated using NMR spectroscopy and published data. All of these compounds were first isolated from *S. natans*, and the compounds (3–6 and 8–11), a total of eight, were first isolated from the *Salvinia* genus. The analysis demonstrated the importance of the isolates chemotaxonomically (Pham et al., 2022). The details of the bioactives found are 1) β-sitosterol, 2) stigmasterol, 3) Stigmast-4,22-dien-3,6-dione, 4) 4α-methyl-24-methylene-5α-cholestane-3β-ol, 5) 3S, 5R, 6S-trihydroxy-7-megastigmen-9-one, 6) (7E,3R,5S,6R)-3,6-dihydroxy-10-normegastigm-7-en-9-oic acid, 7) Dehydrovomifoliol, 8) (+)-lariciresinol, 9) (+)-1-hydroxysyringaresinol, 10) 9,10-dihydro-2,4,7-phenanthrenetriol, 11) {[(3S,4S,7R,8S)-1-oxo-3,7-dimethyl-4-(O-β-D-glucopyranosyl) oxy] hexa-hydro cyclopenta [c]-furan-3-yl}-methyl trans-4-hydroxy cinnamate (Pham et al., 2022). The preliminary phytochemical analysis of *S. natans* 70% ethanol extract showed the presence of active ingredients like sugars, alkaloids, glycosides, hydroxybenzenes, flavonoids, and saponins, but no traces of triterpenoids or tannins.

1

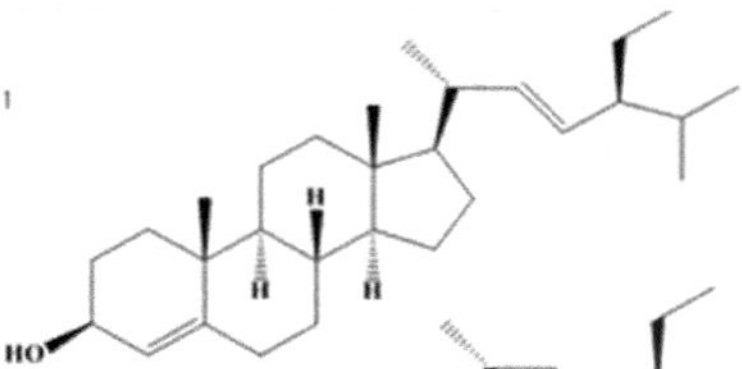

2

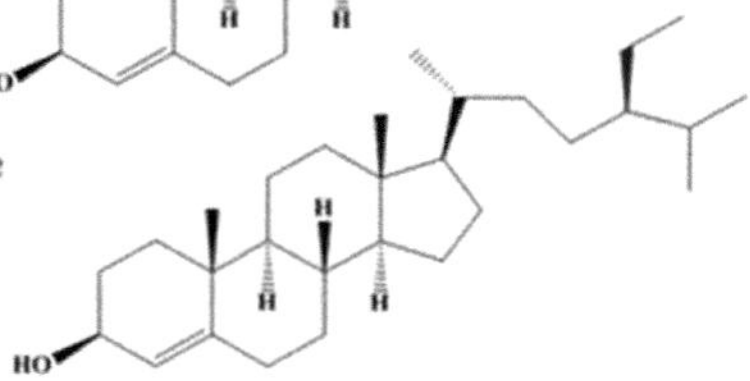

3

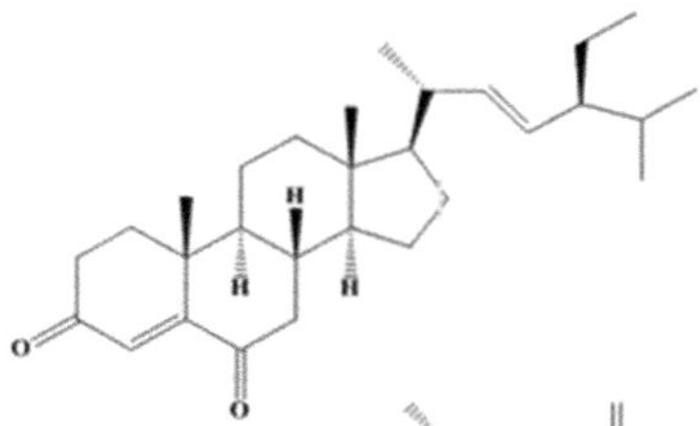

4

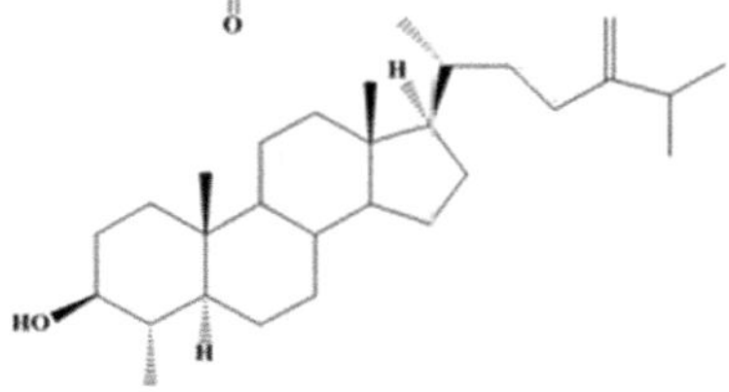

5

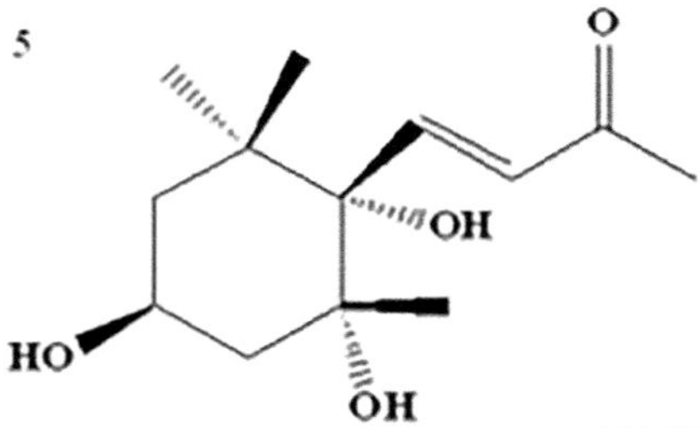

6

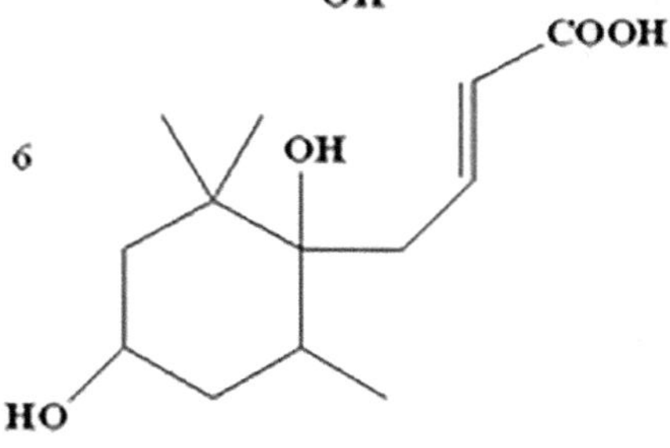

7

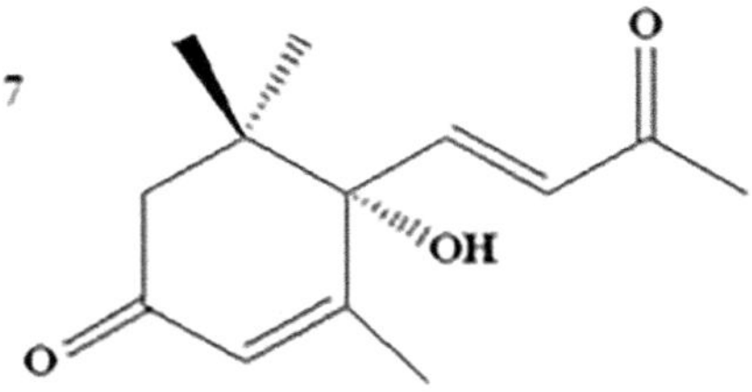

10

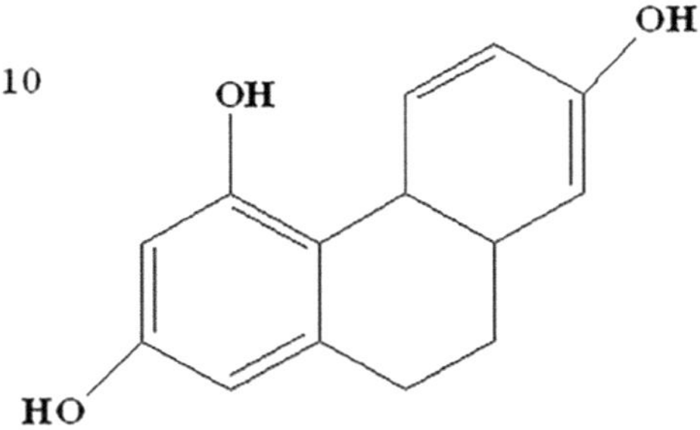

8

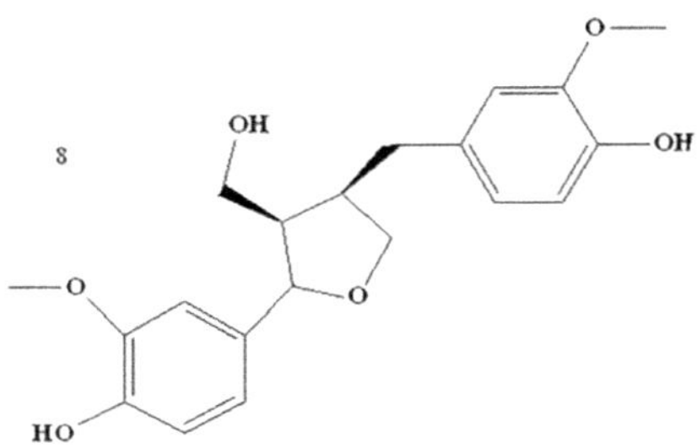

9

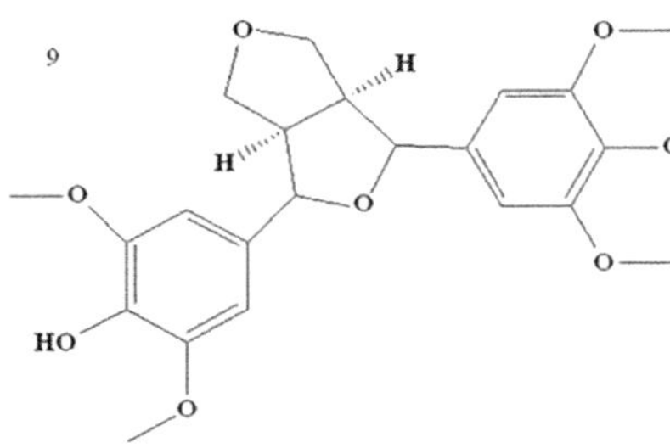

11

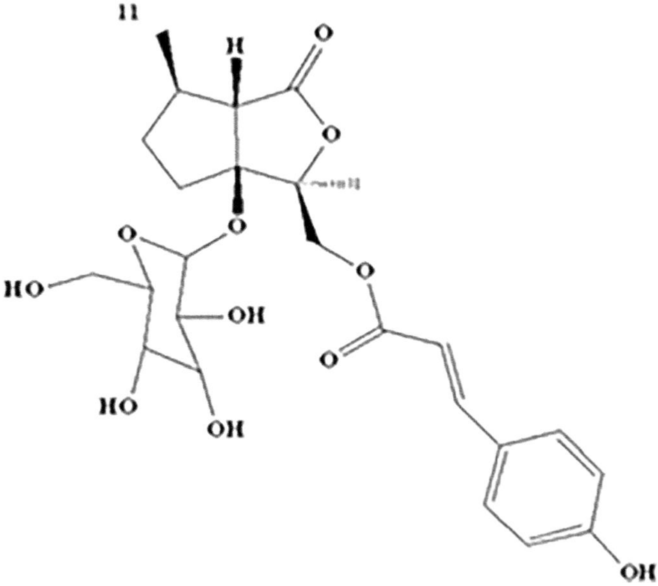

13.3 BIOLOGICAL ACTIVITIES

13.3.1 ANTIOXIDANT ACTIVITY

Several assays were used to establish whether *S. natans* had any active chemical compounds, and the findings indicate that the species has a high concentration of active components. Carbohydrates, flavonoids, phenols, alkaloids, glycosides, and saponins accounted for 70% of the phytochemical results of the ethanol extract; triterpenoids and tannins were not found. To evaluate the antioxidant capacity of *S. natans* extracts, the free radical scavenging assay (DPPH) was used, which is a fast method to show the radical scavenging activity (Knani et al., 2023). The results of the investigation showed that, at different doses, all extracts exhibited antioxidant activity by scavenging radicals (DPPH). From the above study, it is evident that the compounds that have been isolated from the extracts of *S. natans* possess antioxidant properties.

13.3.2 HEPATOPROTECTIVE ACTIVITY

Carbon tetrachloride (CCl_4) has been demonstrated to be hepatotoxic, serving as an animal model for liver damage caused by free radicals. Srilakshmi et al. (2020) investigated whether the new dibenzoyl glycoside derivative from *S. natans*, natansnin, might protect cells against oxidative stress and degeneration caused by CCl_4. Lipid peroxides and oxidized glutathione levels were dramatically elevated by CCl_4, while those of reduced glutathione, SOD, and CAT were significantly lowered. CCl_4 causes significant changes in histopathology and raises levels of apoptotic proteins. The inflammatory mediators iNOS and COX-2, as well as the apoptotic proteins like PARP, BAX, BID, Caspases-3, and cytochrome C, were significantly raised upon treatment with CCl_4 (Srilakshmi et al., 2020).

The most well-characterized system for xenobiotic-induced hepatotoxicity and for evaluating a drug's potential anti-hepatotoxic or hepatoprotective properties is liver injury caused by CCl_4. Natansnin pretreatment improved rats' acute liver injury caused by CCl_4, as shown by the oxidative, cellular degeneration, and cell death parameters previously described. In fact, natural compounds like natansnin have been shown to be effective in preventing hepatotoxicity caused by CCl_4 (Recknagel et al., 1989).

Treatment with natansnin significantly lowered the inflammatory mediator levels and CCl_4-induced apoptotic proteins. Further administration of natansnin significantly attenuated the apoptosis induced by CCl_4, as demonstrated by the reduction in TUNEL-positive cells. This study proved that natansnin protects rat liver tissue against oxidation, which can be correlated to its antioxidant properties as well.

13.3.3 ANTIBACTERIAL ACTIVITY

This study was carried out to find out how well the water plants, *Nymphaea alba* and *S. natans*, inhibited the growth of three dangerous bacteria: *Vibrio* sp., *Staphylococcus aureus,* and *Escherichia coli.* The bacteria were extracted from branches in the city of Al-Qurna that were connected to the Shatt Al-Arab River. It was assessed that the ethanol and acetone extracts of the entire plant might inhibit harmful bacteria (such as *E. coli, Vibrio* sp., and *S. aureus*). Each extract's zone of inhibition (ZOI) against the test bacteria was measured and contrasted with a conventional antibiotic disc. The ethanol and acetone extracts demonstrated a larger inhibitory zone against *Vibrio* sp., measuring 26.2 mm and 18.8 mm, respectively. The minimum inhibitory concentration values for acetone extracts were reported at low concentrations ranging from 20.0 mg/mL up to 160.0 mg/mL, in contrast to ethanol extracts, which were found to range from 10.0 mg/mL to 80.0 mg/mL (Maliki et al., 2017). Though the growth of antibiotic-resistant bacteria will restrict their efficiency, *S. natans* extracts may play a significant role in the future in treating infections of the skin, throat, urinary tract, and digestive system.

13.3.4 DIABETIC FOOT ULCER

At P Value <0.01, the Salvinia Flavonoid (SF) extract exhibited the maximum efficiency against all isolated pathogenic bacteria from the foot ulcers of diabetes patients, while the other extracts displayed variable efficacy against various bacterial species at different periods. The study's findings revealed that there was no inhibition/mm^2 in the hexane extract against the same isolated bacteria (Knani et al., 2023).

13.3.5 OTHER ACTIVITIES

Salvinia natans finds its application in general exhaustion, fever, eczema, and skin diseases. The herb extract exhibits antimicrobial and antioxidant activity (Ya et al., 2015).

KEYWORDS

- **antibacterial**
- **antioxidant**
- **floating watermoss**
- **hepatoprotective**
- **megastigmanes**
- **natansnin**
- **phenanthrene**
- **salvinia flavonoid**
- ***Salvinia natans***

REFERENCES

Knani, Z. M. I. A., Ashoor, A. S. A., & Hussein, A. N. (2023). Antioxidant activity of some *Salvinia natans* L. extracts and a study of their efficacy against isolated bacteria from diabetic foot ulcers. *Journal of Survey in Fisheries Sciences, 10*(1), 1954–1968.

Maliki, G. M. A., Khafaji, K. K. A., & Karim, R. M. (2017). Antibacterial activity of two water plants, *Nymphaea alba* and *Salvinia natans* leaves against pathogenic bacteria. *International Journal of Fisheries and Aquatic Studies, 5*(5), 353–355.

Pham, H. G., Tuan, H. M., Cao, T. Q., & Vu, N. K. (2022). Chemical constituents from *Salvinia natans* (L.) All. (Salviniaceae) and their chemotaxonomic significance. *Biochemical Systematics and Ecology, 102*(3), 104410.

Recknagel, R. O., Glende, E. A. Jr., Dolak, J. A., & Waller, R. L. (1989). Mechanisms of carbon tetrachloride toxicity. *Pharmacology & Therapeutics, 43*, 139–154.

Srilaxmi, P., Sareddy, G. R., Kavi, K. P. B., Setty, O. H., & Babu, P. P. (2010). Protective efficacy of natansnin, a dibenzoyl glycoside from *Salvinia natans* against CCl_4-induced oxidative stress and cellular degeneration in rat liver. *BMC Pharmacology, 10*(13). https://doi.org/10.1186/1471-2210-10-13 (accessed on 17th November 2025).

Ya, V. B. (2015). Aquatic plants of the Far East of Russia: A review on their use in medicine, pharmacological activity. *Bangladesh Journal of Medical Science, 14*(1).

CHAPTER 14

Bioactive Compounds and Pharmacological Activities of *Azolla pinnata* R.Br. (Water Fern)

M. INDIRA

Department of Biotechnology, Vignan's Foundation for Science, Technology and Research (Deemed to be University), Vadlamudi, Andhra Pradesh, India

ABSTRACT

Azolla pinnata R.Br., commonly known as water fern, is a remarkable aquatic plant recognized for its extensive array of bioactive compounds and diverse pharmacological activities. This aquatic fern is rich in secondary metabolites, including phenols, sugars, tannins, steroids, and anthraquinone glycosides. It also contains substantial amounts of proteins, vitamins, and minerals, making it a powerhouse of nutrition and therapeutic potential. *Azolla* exhibits a range of pharmacological activities such as antioxidant, anti-inflammatory, and antimicrobial properties. This chapter delves into the diverse bioactive constituents and pharmacological activities of *Azolla pinnata*, highlighting its potential applications in medicine.

14.1 INTRODUCTION

Azolla, a genus within the Salviniaceae family, is composed of small-leafed, free-floating aquatic ferns that have garnered extensive study due to their remarkable biological and ecological characteristics (Rashad, 2021). Native to the tropical, subtropical, and warm temperate regions of Africa, Asia, and the Americas, these ferns thrive in freshwater environments and

Bioactives and Pharmacology of Pteridophytes. T. Pullaiah (Ed.)

DOI: 10.1201/9781779642738-14

exhibit heterosporous spore production, a distinguished feature among ferns. The genus *Azolla* encompasses six species that are globally distributed and are categorized into two subgenera: *EuAzolla* and *Rhizosperma* (Lydia et al., 2023). The *EuAzolla* subgenus includes four species: *A. filiculoides, A. caroliniana, A. mexicana*, and *A. microphylla*, each adapted to different climatic and environmental conditions. The *Rhizosperma* subgenus comprises two species: *A. nilotica* and *A. pinnata*, which are similarly widespread across suitable habitats (Korsa et al., 2024).

Azolla pinnata R.Br. is a notable aquatic fern, measuring between 1–2.5 cm in diameter, characterized by its short, branched, floating stems with dangling roots (Farook et al., 2019). This species thrives in warm temperate and tropical waters, commonly found in ponds, ditches, and rice fields globally (Roy et al., 2016). Known by various names such as mosquito fern, duckweed fern, fairy moss, green gold mine, and water velvet, *A. pinnata* features leaves with a thick, green aerial dorsal lobe and a thin, colorless floating ventral lobe (Roy et al., 2018). *Azolla pinnata* exhibits a triangular or polygonal texture and floats on the water surface either individually or in dense mats (Rashad, 2021). Extensive research has highlighted its vast potential applications, including antimicrobial properties and antioxidant activity (Farook et al., 2019; Jacob et al., 2020; Ravi et al., 2020).

14.2 BIOACTIVE COMPOUNDS

Azolla pinnata spp. is a multifaceted plant with a rich phytochemical profile that supports a variety of health-promoting and environmental benefits (Rashad, 2021). The plant contains significant levels of crude protein, crude fat, and crude fiber, along with essential bioactive compounds such as alkaloids, terpenoids, flavonoids, tannins, polyphenols, saponins, triterpenoid derivatives, amino acids, fatty acids, and quinones (Prasad et al., 2014). These phytochemicals contribute to its diverse therapeutic and nutritional properties. *Azolla pinnata* is renowned for its exceptional nutritional profile, rich in beta-carotene, vitamins, essential amino acids, minerals, saponins, flavonoids, and high-quality protein (Rashad, 2021). It also contains substantial levels of calcium and other vital nutrients (Bhatt et al., 2020).

Additionally, *Azolla* harbors a variety of secondary metabolites, including several notable bioactive compounds. These include dodecanoic acid, tetradecanoic acid, hexadecanoic acid, 9-octadecenal, docosane, peonidine 3-O-glucoside, rutin, thiamine, vitexin, astragalin, hyperoside, quercetin, choline, and tamarixetin (Bhuvaneswari et al., 2022; Sanjay et al., 2021). Other important compounds found in *Azolla* are 3,7,11,15-tetramethyl-2-hexadecen-1-ol, loliolide, 6,7-dimethoxy-2-tetralone, 2-methoxy-5-[(E)-1-propenyl] phenol, cis-2-[2-(hydroxymethyl) cyclopentyl] ethanol, cycloheptane 1,3,5-tris(methylene), and 4-(6,6-dimethyl-2-methylenecyclohex-3-enylidene) pentan-2-ol (Ravi et al., 2018). These compounds contribute to *Azolla*'s bioactivity and potential therapeutic applications. Moreover, *Azolla* contains C-glycosylflavones such as saponarin (isovitexin 7-O-glucoside) and homoorientin (luteolin 6-C-glucoside), which are known for their antioxidant activity and health benefits (Teixeira et al., 2001). Fatty acids such as oleic acid, lignoceric acid, and palmitic acid are identified in *Azolla pinnata* extracts (Bhuvaneswari et al., 2022). The presence of phytochelatins and metallothioneins in *Azolla* demonstrates its remarkable capacity to sequester and concentrate heavy metals from its environment, highlighting its potential as an effective tool for bioremediation (Talebi et al., 2019). These secondary metabolites bestow *Azolla* with a multitude of health benefits such as antioxidant, anti-inflammatory, anticarcinogenic, antiviral, antidiabetic, antihypertensive, cardioprotective, neuroprotective, and hepatoprotective properties (Ravi et al., 2018) (Figure 14.1) (Table 14.1).

TABLE 14.1 Bioactive Compounds and Their Pharmacological Activities

Name of the Bioactive Compound	Molecular Formula	Phytochemical Class	Pharmacological Activity	References
Dodecanoic acid (Lauric acid)	$C_{12}H_{24}O_2$	Fatty acid	Antimicrobial activity	Nakatsuji et al. (2009)
Tetradecanoic acid (Myristic acid)	$C_{14}H_{28}O_2$	Fatty acid	Promote proliferation and differentiation of neural stem cells	Shang et al. (2022)
Hexadecanoic acid	$C_6H_{31}O_2^-$	Fatty acid	Anti-inflammatory	Aparna et al. (2012)
9-Octadecenal	$C_{18}H_{34}O$	Fatty aldehyde	Antimicrobial and anti-inflammatory	Hatami et al. (2016)
Docosane	$C_{22}H_{46}$	Straight chain alkane	Used as microcapsule	Fredi et al. (2019)
Peonidine 3-O-glucoside	$C_{22}H_{23}O_{11}^+$	Anthocyanin	Antioxidant and anti-inflammatory agent	Sari et al. (2019)
Rutin	$C_{27}H_{30}O_{16}$	Flavonol glycoside	Antioxidant and anti-inflammatory	Ganeshpurkar and Saluja (2017)

TABLE 14.1 *(Continued)*

Name of the Bioactive Compound	Molecular Formula	Phytochemical Class	Pharmacological Activity	References
Thiamine	$C_{12}H_{17}N_4OS^+$	Primary alcohol	Key role in intracellular glucose metabolism; essential for normal growth and development	Polegato et al. (2019
Vitexin	$C_{21}H_{20}O_{10}$	Flavone glycoside	Platelet aggregation inhibitor; antineoplastic agent	Aslam et al. (2015)
Astragalin	$C_{21}H_{20}O_{11}$	Monosaccharide derivative	Used for osteoporosis	Liu et al. (2019)
Hyperoside	$C_{21}H_{20}O_{12}$	Monosaccharide derivative	Hepatoprotective agent	Gajendar et al. (2023)
Quercetin	$C_{15}H_{10}O_7$	Flavonol	Anti-bacterial agent; antioxidant, and antineoplastic agent	Qi et al. (2022)
Choline	$C_5H_{14}NO^+$	Choline class	Neurotransmitter	Gibb (2017)
Tamarixetin	$C_{16}H_{12}O_7$	Mono methoxy flavone	Antioxidant; anti-inflammatory activity	Park et al. (2018)
3,7,11,15-Tetramethyl-2-hexadecen-1-ol (Phytol)	$C_{20}H_{40}O$	Acyclic diterpene alcohol	Precursor for the synthesis of vitamin E and vitamin K1.	Byju et al. (2013)
Loliolide	$C_{11}H_{16}O_3$	Benzofurans	Plant metabolite	
6,7-Dimethoxy-2-tetralone	$C_{12}H_{14}O_3$	Cyclic ketone	Precursor for the synthesis of dopaminergic compounds	Qandil et al. (1999)
2-methoxy-5-[(E)-1-propenyl) phenol], (Isochavibetol)	$C_{10}H_{12}O_2$	Essential oil	Antimicrobial activity	Nielsen et al. (2017)
Cis-2- [2- (hydroxy methyl) cyclo pentyl] ethanol	$C_8H_{16}O_2$	Alcoholic compound	Antibacterial agent	Manivel and Nawaz Khan (2009)
Cycloheptane 1,3,5-tris (methylene)-	$C_{10}H_{14}$	Cyclic hydrocarbon	Antifungal activity	Alkhafaji et al. (2013)
4-(6,6-dimethyl-2-methylene cyclohex-3-enylidene) pentan-2-ol	$C_{14}H_{22}O$	Terpenoid	Melamine and dye effects	Thangaraj et al. (2022)
Isovitexin 7-O-glucoside (saponarin)	$C_{27}H_{30}O_{15}$	C-glycosyl compound	Anti-inflammatory and anti-allergic effect	Min et al. (2021)
Luteolin 6-C-glucoside (homoorientin)	$C_{21}H_{20}O_{11}$	Flavone compound	Anticancer activity	Ye et al. (2016)

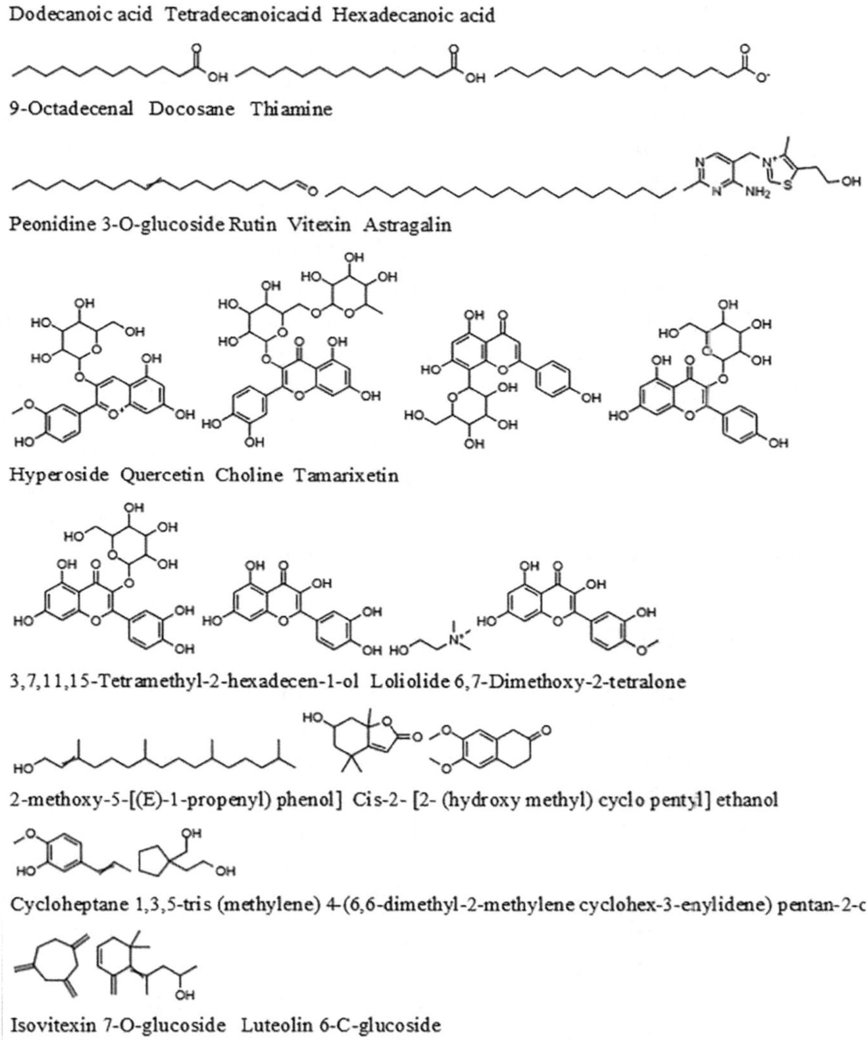

FIGURE 14.1 Chemical structures of bioactive compounds of *Azolla pinnata.*

14.3 PHARMACOLOGICAL ACTIVITIES

14.3.1 ANTIMICROBIAL ACTIVITY

Jacob et al. (2020) investigated the antimicrobial activity of *Azolla pinnata* extracts. The study revealed that the benzene extract exhibited significant inhibitory effects against *Staphylococcus* sp. and *Bacillus* sp., but showed less inhibition against *Pseudomonas* sp. In contrast, the ethanolic extract did not demonstrate any antimicrobial activity compared to the benzene extract. Bouattou and Ali-Nehari (2021) investigated the antimicrobial properties of the ethanolic extract of *Azolla pinnata* against *Staphylococcus aureus, Escherichia coli,* and the fungus *Fusarium oxysporum*. The results demonstrated that *Azolla pinnata* exhibits both antibacterial and antifungal activities. Farook et al. (2019) evaluated the antimicrobial activity of *Azolla pinnata* using various solvents. The extracts showed antimicrobial activity in the following order: acetone, benzene, ethanol, methanol, and water. The acetone extract of *Azolla pinnata* showed the highest zones of inhibition against *Staphylococcus aureus* (13 mm) and *Pseudomonas aeruginosa* (10 mm). The benzene extract showed zones of inhibition of 12 mm for *S. aureus* and 8 mm for *P. aeruginosa*. The ethanol extract had inhibition zones of 12 mm for *P. aeruginosa* and 9 mm for *S. aureus*. The methanol extract showed zones of 12 mm for *S. aureus* (12 mm) and 11 mm for *P. aeruginosa*. No inhibition zones were observed for the water extract.

14.3.2 ANTIOXIDANT ACTIVITY

Elrasoul et al. (2020) investigated the antioxidant activity of an ethanolic extract of *Azolla pinnata*. They analyzed the effects of lead acetate and *Azolla pinnata* extract on hepatic tissue lipid peroxidation and antioxidant defense system biomarkers in rats. Oral administration of lead acetate significantly increased malondialdehyde (MDA) levels compared to control rats. However, rats intoxicated with lead acetate and treated with *Azolla pinnata* extract showed a significant reduction in MDA content in hepatic tissues. Additionally, the supplemented lead acetate-intoxicated rats exhibited significantly elevated levels of glutathione (GSH), superoxide dismutase (SOD), and catalase activities in hepatic tissues compared to rats intoxicated with lead acetate alone. Qoms et al. (2024) investigated the antioxidant activity of *Azolla pinnata* fern protein hydrolysate (AFPH) generated using different enzymes – alcalase, flavourzyme, and papain – to achieve various degrees of hydrolysis (10% for partial hydrolysis, 20% for moderate hydrolysis, and 30% for extensive hydrolysis). The DPPH scavenging activity reached its maximum as hydrolysis progressed to the extensive rate, with no significant difference observed between the enzymes, ranging from 81.33% to 82.62%. For ABTS, the extensively hydrolyzed AFPH prepared using alcalase exhibited the highest activity at 97.48%, whereas the papain-derived hydrolysate showed the lowest activity at 84.95%. In contrast, extensive hydrolysis using papain resulted in the highest ferric reducing capacity at 978.83 μm $FeSO_4$/g, while alcalase and flavourzyme hydrolysates exhibited lower capacities at 837.58 μm $FeSO_4$/g and 785.50 μm $FeSO_4$/g, respectively.

14.3.3 ANTI-INFLAMMATORY ACTIVITY

Azolla pinnata has demonstrated promising anti-inflammatory activities in various studies. The plant's bioactive compounds, including flavonoids, phenolic acids, and polysaccharides, contribute significantly to its anti-inflammatory properties (Saxena et al., 2021). These compounds exhibit potent inhibitory effects on pro-inflammatory mediators and enzymes such as cyclooxygenase (COX) and lipoxygenase (LOX), which play critical roles in the inflammatory response (Abraham et al., 2015). Experimental studies have shown that the extracts from *Azolla pinnata* can significantly reduce inflammation in animal models. Elrasoul et al. (2020) demonstrated that the ethanolic extract of *Azolla pinnata* had a protective effect against lead-induced hepatotoxicity in rats. The study found that the extract reduced serum levels of pro-inflammatory cytokines such as IL-1β, IL-10, and TNF, which are responsible for inflammation. Furthermore, they found that the *Azolla pinnata* extract could mitigate the toxic effects of lead by restoring antioxidant enzyme activities and reducing oxidative stress.

14.3.4 NUTRITIONAL PROPERTIES

Singh et al. (2021) investigated the nutritional analysis of *Azolla pinnata* under controlled conditions. In this study, the crude protein (CP) content of *Azolla* was determined to be 23.74%, suggesting its potential as a natural protein source for livestock feed. This CP value aligns closely with the results from previous studies. However, variations in CP values have been reported, ranging from higher values of 27.09% and 32.05% to lower values of 20.02% and 17.59%. These differences may be attributed to the nutritional quality of the inputs used. The crude fiber content in this study was 14.25%, consistent with values reported in another research. The total ash content was found to be 15.05%, which is in close agreement with other findings in the literature.

In another study by Bhatt et al. (2020), it was found that the fresh biomass of *Azolla pinnata* consists of crude protein (13.1–28.1%), crude fiber (11.3–22.8%), neutral detergent fiber (NDF) (35.4–52.3%), acid detergent fiber (ADF) (24–38.9%), lignin (9.3–13.5%), ether extract (1.9–5.1%), ash (9.8–21.6%), starch (2.7–5.5%), and gross energy (17.0 MJ/kg dry matter (DM)). The minerals present include calcium (5.8–17 g/kg DM), phosphorus (0.3–15.5 g/kg DM), potassium (10.9–22.5 g/kg DM), sodium (2.8–12.5 g/kg DM), magnesium (3.9–6.1 g/kg DM), manganese (208–1429 mg/kg DM), zinc (11–77 mg/kg DM), copper (10–28 mg/kg DM), and iron (71–8200 mg/kg DM). The amino acids include alanine (5.3–7.4%), arginine (5.1–6.6%), aspartic acid (8.2–10.3%), cystine (0.7–2.3%), glutamic acid (11.6–13.5%), glycine (4.5–6.6%), histidine (1.6–2.4%), isoleucine (3.7–5.4%), leucine (7.0–9.2%), lysine (3.5–6.5%), methionine (1.2–1.9%), phenylalanine (5.2–5.6%), proline (3.5–6.8%), serine (3.9–5.6%), threonine (4.0–5.3%), tryptophan (1.5–2.0%), tyrosine (3.2–4.1%), and valine (3.8–6.8%).

14.3.5 ANTINUTRITIONAL PROPERTIES

Aquatic plants are highly valued in the culinary world for their broad potential applications as human food, animal feed, and bio-fertilizers (Prabakaran et al., 2022). *Azolla pinnata* extract was studied for its anti-nutritional activities. The concentration of harmful compounds, which determines a food's edibility, can vary significantly based on factors such as growing conditions, harvesting practices, processing methods, testing techniques, and the food's age at evaluation (Popova and Mihaylova, 2019). The concentration of anti-nutritional compounds in *Azolla pinnata* are as follows: saponin is the highest at $9.13 \pm 0.012\%$, followed by phytate at $2.534 \pm 0.08\%$, tannin at $0.3675 \pm 0.033\%$, oxalate at $0.0349 \pm 0.013\%$, and cyanide at $0.0045 \pm 0.0004\%$. The notably low levels of oxalate and hydrogen cyanide in *Azolla* species are remarkable. Consequently, incorporating this water fern into animal feed and its extract into the human diet is unlikely to pose significant health risks to ruminants consuming it (Chandrababu et al., 2024).

14.3.6 BIO-INSECTICIDE

Ravi et al. (2018) investigated the efficacy of *Azolla pinnata* plant extracts against *Aedes aegypti* and *Aedes albopictus* mosquitoes. In the adulticidal assay, mortality significantly increased with higher concentrations of *A. pinnata* extracts, exhibiting LC50 and LC95 values of 2572.45 ppm and 6100.74 ppm respectively against *A. aegypti* and 2329.34 ppm and 5315.86 ppm respectively against *A. albopictus*. The ovicidal test demonstrated 100% egg mortality for both species. During the oviposition deterrence test, neither *A. aegypti* nor *A. albopictus* laid eggs in the cups containing *A. pinnata* extract, preferring to oviposit in cups filled with water instead.

KEYWORDS

- **anti-inflammatory**
- **antimicrobial activity**
- **antioxidant**
- ***Azolla pinnata***
- **bioactive compounds**
- **bioinsecticide**
- **nutritional profile**

REFERENCES

Abraham, G., Yadav, R. K., & Kaushik, G. K. (2015). Antimicrobial activity and identification of potential antimicrobial compounds from aquatic pteridophyte, *Azolla microphylla* Kaulf. *Indian Journal of Experimental Biology, 53*, 232–235.

Alkhafaji, H. H., Altameme, H. J., & Alsharifi, S. M. (2022). Detection of bioactive chemical compounds in the methanolic extract of *Azolla filiculoides* Lamark fern by GC–MS technique. *Iraqi Journal of Agricultural Sciences, 53*(4), 922–930.

Aparna, V., Dileep, K. V., Mandal, P. K., Karthe, P., Sadasivan, C., & Haridas, M. (2012). Anti-inflammatory property of *n*-hexadecanoic acid: Structural evidence and kinetic assessment. *Chemical Biology & Drug Design, 80*(3), 434–439.

Aslam, M. S., Ahmad, M. S., & Mamat, A. S. (2015). Pharmacological potential of vitexin. *Indian Research Journal of Pharmacy and Sciences, 2*(2), 114–122.

Bhatt, N., Chandra, R., Kumar, S., Singh, K., & Pratap, N. (2020). Nutritive analysis of *Azolla pinnata* and its cultivation during winter season. *International Journal of Current Microbiology and Applied Sciences, 9*(3), 2012–2018.

Bhuvaneswari, S. S., Kumudha, D., Prabha, T., & Sivakumar, T. (2022). *Azolla pinnata* R. Br.: An aquatic macrophyte as a potential therapeutic candidate. *Annals of Phytomedicine, 11*(1), 133–141.

Bouattou, K., & Ali-Nehari, A. (2021). Antimicrobial activities of *Azolla pinnata* harvested from the North-West of Algeria. In *Euro-Mediterranean Conference for Environmental Integration* (pp. 199, 200).

Byju, K., Vasundhara, G., Anuradha, V., Nair, S. M., & Kumar, N. C. (2013). Presence of phytol, a precursor of vitamin E in *Chaetomorpha antinnina. MAPANA Journal of Sciences, 12*(2), 57–65.

Chandrababu, K. A., Parvathy, U., Meenu, B., & Binsi, P. K. (2024). Exploring the micro-nutritional and anti-nutritional aspects of *Azolla* for its application as functional food ingredient. *Asian Journal of Dairy and Food Research*, DR2101, 1–6.

Elrasoul, A. S. A., Mousa, A. A., Orabi, S. H., Mohamed, M. A. E. G., Gad-Allah, S. M., Almeer, R., & Eldaim, M. A. A. (2020). Antioxidant, anti-inflammatory, and anti-apoptotic effects of *Azolla pinnata* ethanolic extract against lead-induced hepatotoxicity in rats. *Antioxidants, 9*(10), 1014.

Farook, M. A., Mohamed, H. M., Kumar, G. S., Subash, S., Paranjothi, M., Naveez, V. M., & Ahmed, I. A. (2019). Phytochemical screening, antibacterial and antioxidant activity of *Azolla pinnata. International Journal of Research and Analytical Reviews, 6*(2), 240–247.

Fredi, G., Dire, S., Callone, E., Ceccato, R., Mondadori, F., Pegoretti, A. (2019). Docosane-organosilica microcapsules for structural composites with thermal energy storage/release capability. *Materials. 12*(8), 1286.

Gajender, Mazumder, A., Sharma, A., & Azad, M. A. (2023). A comprehensive review of the pharmacological importance of dietary flavonoids as hepatoprotective agents. *Evidence-Based Complementary and Alternative Medicine, 2023*, Article ID 4139117, 1–17.

Ganeshpurkar, A., & Saluja, A. K. (2017). The pharmacological potential of rutin. *Saudi Pharmaceutical Journal, 25*(2), 149–164.

Gibb, A. J. (2017). Choline and acetylcholine: What a difference an acetate makes! *Journal of Physiology, 595*(4), 1021.

Hatami, S., Mohamadi, S. A., & Yavarmanesh, M. (2016). Chemical composition and antibacterial activity of organic extra virgin olive oil from Iran. *Nutrition & Food Science, 46*(3), 388–395.

Jacob, M. M., Jom, M., Sherin, A., & Shahla, B. (2020). *Azolla pinnata*: Potential phytoremediation, antimicrobial, and antioxidant applications. *Proceedings International, 2*(1), 81.

Korsa, G., Alemu, D., Ayele, A. (2024). Azolla plant production and their potential applications. *International Journal of Agronomy. 2024*(1), 1716440.

Liu, L., Wang, D., Qin, Y., Xu, M., Zhou, L., Xu, W., & Li, D. (2019). Astragalin promotes osteoblastic differentiation in MC3T3-E1 cells and bone formation *in vivo. Frontiers in Endocrinology, 10*, 409564.

Lydia, N., Basamba, T. A., Nyakoojo, C., Mustafa, A. S., Saidi, N., Mutumba, G. M., & Ssenku, J. E. (2023). Ecological distribution and genetic diversity of *Azolla* in Uganda. *BMC Plant Biology, 23*(1), 131.

Manivel, P., & Nawaz, K. F. (2009). Synthesis of 2-(2-(hydroxymethyl) phenyl) ethanol derivatives as potential antibacterial agents. *Journal of the Chilean Chemical Society, 54*(2), 180–182.

Min, S. Y., Park, C. H., Yu, H. W., & Park, Y. J. (2021). Anti-inflammatory and anti-allergic effects of saponarin and its impact on signaling pathways of RAW 264.7, RBL-2H3, and HaCaT cells. *International Journal of Molecular Sciences, 22*(16), 8431.

Nakatsuji, T., Kao, M. C., Fang, J. Y., Zouboulis, C. C., Zhang, L., Gallo, R. L., & Huang, C. M. (2009). Antimicrobial property of lauric acid against *Propionibacterium acnes*: Its therapeutic potential for inflammatory acne vulgaris. *Journal of Investigative Dermatology, 129*(10), 2480–2488.

Nielsen, C. K., Kjems, J., Mygind, T., Snabe, T., Schwarz, K., Serfert, Y., & Meyer, R. L. (2017). Antimicrobial effect of emulsion-encapsulated isoeugenol against biofilms of food pathogens and spoilage bacteria. *International Journal of Food Microbiology, 242*, 7–12.

Park, H. J., Lee, S. J., Cho, J., Gharbi, A., Han, H. D., Kang, T. H., & Park, Y. M. (2018). Tamarixetin exhibits anti-inflammatory activity and prevents bacterial sepsis by increasing IL-10 production. *Journal of Natural Products, 81*(6), 1435–1443.

Polegato, B. F., Pereira, A. G., Azevedo, P. S., Costa, N. A., Zornoff, L. A., Paiva, S. A., & Minicucci, M. F. (2019). Role of thiamin in health and disease. *Nutrition in Clinical Practice, 34*(4), 558–564.

Popova, A., & Mihaylova, D. (2019). Anti-nutrients in plant-based foods: A review. *Open Biotechnology Journal, 13*(1), 68–76.

Prabakaran, S., Mohanraj, T., Arumugam, A., Sudalai, S. (2022). A state-of-the-art review on the environmental benefits and prospects of Azolla in biofuel, bioremediation and bio-fertilizer applications. *Industrial Crops and Products. 183*, 114942.

Prasad, R. M. V., Jagadeeswara, R. S., Jayalaxmi, P., & Kumar, D. S. (2014). A study on the nutritive value of *Azolla pinnata*. *Livestock Research International, 2*(1), 13–15.

Qandil, A. M., Miller, D. W., & Nichols, D. E. (1999). A practical and cost-effective synthesis of 6,7-dimethoxy-2-tetralone. *Synthesis, 12*, 2033–2035.

Qi, W., Qi, W., Xiong, D., & Long, M. (2022). Quercetin: Its antioxidant mechanism, antibacterial properties and potential application in prevention and control of toxipathy. *Molecules, 27*(19), 6545.

Qoms, M. S., Arulrajah, B., Ibadullah, W. Z. W., Ramli, N. S., Shamsudin, R., Chau, D. M., & Saari, N. (2024). Antihypertensive, antidiabetic, and antioxidant properties of novel *Azolla pinnata* fern protein hydrolysates: Inhibition mechanism, stability, profiling, and molecular docking. *Food and Bioprocess Technology, 18*, 1–18.

Rashad, S. (2021). An overview on the aquatic fern *Azolla* spp. as a sustainable source of nutrients and bioactive compounds with resourceful applications. *Egyptian Journal of Aquatic Biology and Fisheries, 25*(1), 775–782.

Ravi, R., Husna, Z. N. S., Rozhan, N. N., Nik Yusoff, N. R., Mat Rasat, M. S., Ahmad, M. I., & Mohd Amin, M. F. (2018). Evaluation of two different solvents for *Azolla pinnata* extracts on chemical compositions and larvicidal activity against *Aedes albopictus* (Diptera: Culicidae). *Journal of Chemistry*, 2018, 7453816. https://doi.org/10.1155/2018/7453816.

Ravi, R., Rajendran, D., Oh, W. D., Mat Rasat, M. S., Hamzah, Z., Ishak, I. H., & Mohd Amin, M. F. (2020). The potential use of *Azolla pinnata* as an alternative bio-insecticide. *Scientific Reports, 10*(1), 19245. https://doi.org/10.1038/s41598-020-76372-2.

Ravi, R., Zulkrnin, N. S. H., Rozhan, N. N., Nik Yusoff, N. R., Mat Rasat, M. S., Ahmad, M. I., & Amin, M. F. M. (2018). Chemical composition and larvicidal activities of *Azolla*

pinnata extracts against *Aedes* (Diptera: Culicidae). *PLOS ONE, 13*(11), e0206982. https://doi.org/10.1371/journal.pone.0206982.
Roy, D. C., & Pakhira, M. C., & Bera, S. (2016). A review on biology, cultivation and utilization of *Azolla*. *Advances in Life Sciences, 5*(1), 11–15.
Roy, D. C., Saha, A., & Biswas, S. (2018). *Azolla*: An unconventional forage crop. In *Forage Crops of the World, Volume II: Minor Forage Crops* (pp. 337–348).
Sanjay, V., Muruganandham, L., & Sabina, E. P. (2021). Identifying the potential activity of *Azolla pinnata* through *in vitro* assay and SEM. *Songklanakarin Journal of Science and Technology, 43*(5). https://doi.org/10.14456/sjst-psu.2021.65.
Sari, D. R. T., Cairns, J. R. K., Safitri, A., & Fatchiyah, F. (2019). Virtual prediction of the delphinidin-3-O-glucoside and peonidin-3-O-glucoside as anti-inflammatory agents of TNF-α signaling. *Acta Informatica Medica, 27*(3), 152–157. https://doi.org/10.5455/aim.2019.27.152–157.
Saxena, M. K., Singh, N., Kumar, S., Mp, D., & Datta, S. (2021). Potent pharmaceutical products from aquatic plants – Review. *Asian Journal of Pharmaceutical and Clinical Research, 14*(1), 48–63.
Shang, Q., Chen, G., Zhang, P., Zhao, W., Chen, H., Yu, D., & Ren, H. (2022). Myristic acid alleviates hippocampal aging correlated with GABAergic signaling. *Frontiers in Nutrition, 9*, 907526. https://doi.org/10.3389/fnut.2022.907526.
Singh, T. V., Prasad, R., Jayalaxmi, P., & Rani, M. S. (2021). A study on nutrient analysis of *Azolla pinnata*. *Indian Journal of Animal Production and Management, 37*(1–4), 102–106.
Talebi, M., Tabatabaei, B. E. S., & Akbarzadeh, H. (2019). Hyperaccumulation of Cu, Zn, Ni, and Cd in *Azolla* species inducing expression of metallothionein and phytochelatin synthase genes. *Chemosphere, 230*, 488–497. https://doi.org/10.1016/j.chemosphere.2019.05.082.
Teixeira, G., Carrapiço, F., & Gomes, E. T. (2001). C-glycosylflavones in the genus *Azolla*. *Plant Biosystems, 135*(2), 233–237. https://doi.org/10.1080/11263500112331351331.
Thangaraj, M., Saravana, B. P., Thanasekaran, J., Joen-Rong, S., Manubolu, M., & Pathakoti, K. (2022). Phytochemicals of algae, *Arthospira platensis* (spirulina), *Chlorella vulgaris* (chlorella), and *Azolla pinnata* (*Azolla*). *GSC Biological and Pharmaceutical Sciences, 19*(2), 23–43.
Ye, T., Su, J., Huang, C., Yu, D., Dai, S., Huang, X., & Zhou, M. (2016). Isoorientin induces apoptosis, decreases invasiveness, and downregulates VEGF secretion by activating AMPK signaling in pancreatic cancer cells. *OncoTargets and Therapy, 12*(9), 7481–7492. https://doi.org/10.2147/OTT.S111026.

CHAPTER 15

Bioactives and Pharmacology of *Huperzia serrata* (Thunb.) Trevis. (=*Lycopodium serratum* Thunb.) (Lycopodiaceae)

R. KRIPASREE,[1] SINJUMOL THOMAS,[1] and BINCE MANI[2,*]

[1]*Department of Botany, Carmel College (Autonomous), Mala, Thrissur, Kerala, India*

[2]*Department of Botany, St. Thomas College Palai (Autonomous), Kottayam, Kerala, India*

*Corresponding author

ABSTRACT

Huperzia serrata has a rich source of bioactive compounds with many neuropharmacological activities and therapeutical potential. The major bioactive phytoconstituents isolated and characterized from *H. serrata* are Huperzine A, Huperserine E, 12-deoxyhuperzine O, Lycopodine-6α,11α-diol, Serralongamines B and D, Serratenediol, Serratezomines A, B and C and Serratezomine D. It has various biological activities especially anti-neuroinflammatory effects, as well as significance in traditional methods to treat various neurological conditions, attributed to bioactive compounds like the alkaloid, Huperzine A. The medicine to treat Myasthenia gravis and Alzheimer's disease has emerged especially with the discovery of Huperzine A. The discovery of novel lycopodium-alkaloids from *H. serrata* emphasizes its diverse pharmacological applications.

Bioactives and Pharmacology of Pteridophytes. T. Pullaiah (Ed.)

DOI: 10.1201/9781779642738-15

15.1 INTRODUCTION

The pharmacological and phytochemical exploration and investigation of the family Lycopodiaceae began in the early 1990s, marking a key period during which researchers determined the AChE inhibitory activity and isolated several lycopodium alkaloids. Huperzine A (Hup A) extracted from *Huperzia serrata* is the major discovery among them having the greatest efficacy. A new wave of study on the potential of lycopodium alkaloids to cure Myasthenia gravis and Alzheimer's disease has emerged especially with the discovery of Hup A (Ma et al., 2004).

Huperzia serrata (Thunb.) Trevis. [Syn: *Huperzia selago* var. *serrata* (Thunb.) Love & D. Love, *Lycopodium serratum* Thunb., *Urostachys serratus* (Thunb.) Herter], commonly referred to as Chinese club moss is a subshrub and grows chiefly in temperate biomes. It belongs to the family Lycopodiaceae and native to the Russian Far East to northeast China and Japan, Hawaiian Island, Mexico, Cuba to Hispaniola (POWO, 2024). It is commonly seen in forests, shrublands, and roadsides. They are mostly terrestrial plants with erect or ascending, dichotomously branched stems having bulbils at the upper portion. Leaves are narrowly elliptical and arranged at right angle to the stem. Its sporophylls are homomorphic with tropophylls and have yellowish reniform sporangia on adaxial sides of the sporophylls (Zhang and Iwatsuki, 2013). *Huperzia serrata* is used as folk medicine, especially in China for many health issues such as confusion, strain, swelling, schizophrenia, and in improving blood circulation. The principal secondary metabolites found in *H. serrata* revealed the presence of alkaloids, diterpenes, and triterpenes (Cao et al., 2021).

The Lycopodiaceae family encompasses a diverse array of plant species having medicinal value. The diverse biochemical profiles among these plants underscore the medicinal promise, prompting further exploration of their pharmaceutical applications. *Huperzia serrata*, have been the subject of extensive scientific inquiry mainly because of their unique pharmacological activities (Patil, 2020).

15.2 BIOACTIVES

Huperzine A and B are the two lycopodium alkaloids isolated for the first time from *Huperzia serrata* (Liu et al., 1986). Later, *H. serrata* were scientifically explored for the isolation and characterization of various biochemically active compounds especially focused on its lycopodium alkaloids. Sano et al. (1970) elucidated the structure of two serratane type lycopodium alkaloids, tohogenol diacetate and tohogeninol triacetate isolated from *L. serratum.* It was also reported that dehydrating these compounds could yield serratenediol diacetate and serratriol triacetate respectively. Morita et al. (2000) isolated three lycopodium alkaloids of serratezomines A, B and C. The skeletal structures of these compounds are seco-serratinine-type, serratinine-type, and lycodoline-type respectively. A distinct cytotoxic activity was showed by serratezomines A and B against, Murine lymphoma L1210 cell lines and human epidermoid carcinoma KB cells.

Takayama et al. (2001) identified a lycopodium alkaloid, lycoposerramine-A having 1,2,4-oxadiazolidin-5-one residue in the compound. It was the first report that the lycopdium alkaloid had such residue in its molecule. By analyzing the spectroscopic and X-ray data, the structure of the alkaloid was established. In the next year, Takayama et al. (2002) isolated seven fawcettimine-related lycopodium alkaloids, lycoposerramines C, D, E, P, Q, S, and U and determined their structural details. Takayama et al. (2003) again discovered a lot of alkaloids including 10 novel alkaloids to the science including lycoposerramines F, G, H, I, J, K, L, M, N, and O, having structures related to lycopodine. The structure of these compounds was elucidated with the help of the data obtained from spectroscopic analysis and chemical transformations.

Tan et al. (2002) isolated two irregular fawcettimine-type lycopodium alkaloids, huperzine Q and N-oxy huperzine Q, from the $CHCl_3$ fraction of the whole plant of the *H. serrata.* Their structures were identified as 13-epi-13β,16-epoxydihydrofawcettimine and N-oxy13-epi-13β,16-epoxydihydrofawcettimine by using spectroscopic studies and X-ray crystallographic analysis.

Zhou et al. (2003a) were able to isolate and characterize seven novel serratene-type triterpenoids along with 15 known compounds from the whole plants of *H. serrata*. The structures of the new compounds were identified by spectroscopic analysis as 21α-hydroxyserrat-14-en-3β-yl p-dihydrocoumarate, 21α-hydroxyserrat-14-en-3β-yldihydrocaffeate, 21α-hydroxyserrat-14-en-3β-ylpropanedioic acid monoester, 3α,21α-dihydroxyserrat-14-en-24-oic acid, 16-oxo-3α,21β-dihydroxyserrat-14-en-24-al, 16-oxo-3α,21β-dihydroxyserrat-14-en-24-oic acid, and 16-oxo-21β-hydroxyserrat-14-en-3α-yl acetate. The known compounds characterized were 21β-hydroxyserrat-14-en3α-ol,

21α-hydroxyserrat-14-en-3β-ol, 21α -hydroxyserrat-14-en-3β-yl acetate, 3β-hydroxyserrat-14-en21-one, 3α,21β-dihydroxyserrat-14-en-16-one, 21β-hydroxyserrat-14-en-3β-ol, 3β-hydroxyserrat-14-en-21α-yl acetate, 21β-hydroxyserrat-14-en-3β-yl acetate, 3β,21α-dihydroxyserrat-14-en-24-ol, 3α,21β-dihydroxyserrat-14-ene-24,29- diol, 3β,21β -dihydroxyserrat-14-en-16-one, 3α,21β-dihydroxyserrat-14-en-24-ol, 3α,21β,24- trihydroxyserrat-14-en-16-one, 3β,21β-dihydroxyserrat-14-en-24-ol, and 3β,21β-dihydroxyserrat14-en-29-ol. In the same year, Zhou et al. (2003b) characterized three serratane-type epoxy triterpenoids from ethanol extract of the whole plant of *H. serrata* and defined its structures as 14β,15β-epoxy-3β-hydroxyserratan-21β-ol, 14β,15β-epoxy-3β-hydroxyserratan-21α-ol, and 14β,15β-epoxy-3β-hydroxyserratan-21α-ol-3β- O-acetate. In another work, Zhou et al. (2004) identified three new serratene type triterpenoids, 14β,15β-epoxyserratan-3β,21β,29-triol, serrat-14-en-3β,21β,29-triol and serrat-14-en-3α,21β,24,29-tetraol along with 13 other known serratane-type triterpenoids from whole plant extracts of *H. serrata* by spectroscopic analysis. The known phytoconstituents identified were serrat-14-en-3β,21β,24-triol, serrat-14-en-3α,21β,24-triol, 3α,21β -dihydroxyserrat-14-en-16-one, serrat-14-en-3β,21α,24-triol, 3α,21β,24-trihydroxyserrat-14-en-16-one, 3β,21β-dihydroxyserrat-14-en-16-one, serrat-14-en-3β,21β-diol, serrat-14-en-3α,21β-diol, serrat-14-en-3β,21α-diol, 21α-hydoxyserrat-14-en-3β-ylacetate, 3β-hydroxyserrat-14-en-21α-yl acetate, 3β-hydroxy serrat-14-en-21α-one and 21β-hydroxyserrat-14-en-3β-yl acetate.

Katakawa and co-workers (2004, 2006, 2007, 2009) discovered 13 different compounds from four experiments. In 2004, lycoposerramine-B was reported as a new oxime-containing lycopodium alkaloid. The studies like J-resolved HMBC spectroscopy revealed the structure of lycoposerramine-B and validated the synthesis of this compound from serratinine, a known alkaloid. Katakawa et al. (2006) found three nitrone-containing phlegmarine type lycopodium alkaloids such as lycoposerramines X, Y, and Z. The structural pattern of all these compounds along with other related lycopodium alkaloids such as huperzines J and K was also determined based on spectroscopic analysis. Five fawcettimine-related lycopodium alkaloids such as 11α-hydroxyfawcettidine, 2α,11α-dihydroxyfawcettidine, 8α,11α-dihydroxyfawcettidine, 2β-hydroxylycothunine and 8α-hydroxylycothunine were isolated in 2007 and their structure was elucidated based on spectroscopic data. Among these five alkaloids, 2α,11α-dihydroxyfawcettidine showed a significant acetylcholinesterase (AChE) inhibitory activity. In 2009, lycoposerramine-R was isolated and characterized as a novel alkaloid having a distinct tetracyclic ring structure. They have also isolated and identified three fawcettimine-related alkaloids, lycoposerramine-T, N-methyl lycoposerramine-T and N-formyl lycoposerramine-T in the same study.

The experiment conducted by Shigeyama et al. (2007) on *L. serratum* yielded two new phlegmarine-type alkaloids, lycoposerramines V and W. Based on the data of spectroscopic analyses and asymmetric total syntheses, the structures of these compounds with the absolute configuration were also established. Wang et al. (2007) examined the phyto-constitution of the petroleum ether/Me_2CO (50:1) fraction of *H. serrata*. It resulted in the large-scale isolation of huperzine A along with three new compounds, 2-chlorohuperzine E, huperzine E, and huperzine F as well as two known alkaloids, huperzines E and F from the soluble fraction of the residue. They also elucidated its structure by using spectroscopic analysis. The whole plant extraction experiment conducted by Gao et al. (2008) resulted in the isolation and characterization of two nitrone alkaloids, huperzines M and N as a colorless oil and yellowish oil respectively. Both are phlegmarine-type lycopodium alkaloids with a nitrone moiety. Their configurations were determined by optical rotation, CD, and NOESY-NMR data, while their structures were clarified based on spectral evidence.

Kubota et al. (2009) studied *L. serratum* var. *serratum* and yielded two alkaloids, serratezomines D and E. Serratezomine D is a lucidine-type alkaloid and serratezomine E is a phlegmarane-type alkaloid. The spectroscopic data analysis figured out the detailed structures and stereochemistry of both compounds. Among these, serratezomine D showed a significant inhibitory action against the enzyme acetylcholinesterase. Yang et al. (2010) identified a new lycopodane-type lycopodium alkaloid, 6α-hydroxy-5,15-oxide-lycopodane, and seven known alkaloids (huperzines E and F, huperzinine, 12-deoxyhuperzine O, phlegmariurine B, 8-deoxy-13-dehydroserratinine and N-methyl-huperzine B) from the whole plant extract of *H. serrata* and elucidated their structure by spectroscopic methods. Among them, 12- deoxyhuperzine O was identified as a naturally occurring alkaloid for the first time and has an antagonistic effect on the N-methyl-D-aspartate receptor.

Jiang et al. (2010) isolated and characterized two new phyto-compounds of lycopodine alkaloids such as (5β,6β,15α)-15-methyllycopodane-5,6-diol, and (12β)-12-hydroxyhuperzine G from *H. serrata*. Six known phyto-compounds, huperzine A, B, G, lycoposerramine D, phlegmariurine B, and (8β)-8-hydroxyphlegmariurine B were also eluted and characterized in the same experiment. Jiang et al. (2014a) reported a phlegmarine-type novel lycopodium alkaloid serralongamine A from the methanol extracts of the whole plants of *L. serratum* var. *longipetiolatum* as a colorless amorphous solid and the detailed structure of this compound was depicted by the analysis of spectroscopy. Additionally, known compounds such as lycoposerramine-V, huperzine A, and huperzine B were also isolated and characterized in the same study. Phytochemical analysis conducted by Jiang et al. (2014b) in *H. serrata* resulted in the isolation of four new 5-deoxyfawcettimine-related lycopodium alkaloids, huperserines A–D, and one new lycodine-type alkaloid, huperserine E. Based on spectroscopic data from 1D and 2D NMR techniques, their structures were clarified. The alkaloids of the lycopodine and flawcettimine types were characterized by the presence of a 5-carbonyl or 5-hydroxyl group. This was the first report of lycopodium alkaloids of the 5-deoxyfawcettimine class. The *in vitro* acetylcholinesterase (AChE) inhibitory activity assay revealed that huperserine E has modest anti- AChE activity.

Ying et al. (2014) isolated three new lycopodium alkaloids, huperserramines A–C, in addition to the 15 known ones, lycopodine-6α,11α-diol, lycoposerramine H, lycoposerramine I, lycopodine-6α-ol, lycoposerramine M, diphaladine A, lycoposerramine K, lycoposerramine W, huperzine M, luciduline, phlegmariuine N, huperzine A, huperzine B, lycodine, and lycoposerramine R. By spectroscopic techniques including 2D-NMR and MS studies, their structures were determined. The inhibitory effects of each isolate on α-glucosidase and acetylcholinesterase (AChE) were assessed and lycopodine-6α,11α-diol showed a more powerful α-glucosidase inhibitory action.

Three new compounds of lycopodium alkaloids, serralongamines B–D were isolated by Ishiuchi et al. (2016). By using spectroscopic data and by analysing the chemical transformation patterns, they elucidated the structure of these compounds. It was reported that the effect of these alkaloids on foam cell formation (accumulation of cholesterol ester) in macrophages found that serralongamines B and D have a significant inhibitory effect in cholesterol ester accumulation.

From the whole plant of *H. serrata*, four new lycodine-type lycopodium alkaloids, 1-methyllycodine, 8α-hydroxy-15,16-dehydro-desN-methyl-α-obscurine, N-methyl-16-hydroxyhuperzine B, and N-methylhuperzine along with 11 known analogues, 6β-hydroxyhuperzine A, huperzine A, huperzine B, casuarine B, huperzinine, lycodine, N-methyllycodine, carinatumine B, des-N-methyl-β-obscurine, N-demethylhuperzinine, and 16-hydroxyhuperzine B were isolated and characterized. The structures of compounds were clarified using NMR spectroscopy and mass spectrometry. 1-methyllycodine was the first alkaloid of the lycodine type to have a methyl group at C-1. Furthermore, the single-crystal X-ray crystallography data verified the structure of 6β-hydroxyhuperzine A, and the same work reported its ^{13}C NMR for the first time (Liu et al., 2017).

Deng et al. (2022) isolated five new fawcettimine-type lycopodium alkaloids, hupertimines A-E from the whole plant, in addition to the two known alkaloids, 8α-hydroxyphlegmariurine B and 8β-hydroxyphlegmariurine B. Using single-crystal X-ray diffraction analysis, HR-MS, IR, H, C, and 2D NMR, the structures of all the compounds were determined. Hupertimines B was an example of a lycopodium alkaloid with a 5/5/5/5/6 pentacyclic ring system and a 1-aza-7-oxabicycloheptane unit, whereas hupertimines A was an example of having a structure with an ether bond between C-5 and C-13. BACE1 inhibitory activity of all compounds were determined. Furthermore, accurate H- and C-NMR data of 8β-hydroxyphlegmariurine B was also reported.

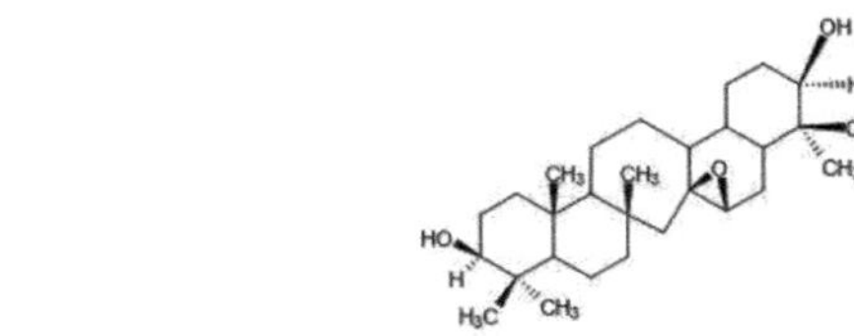

Huperzine A and B

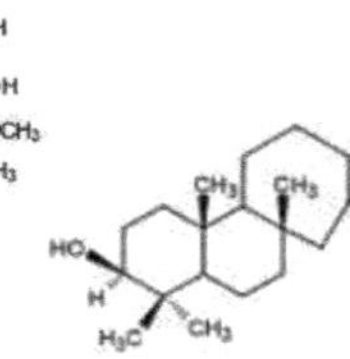

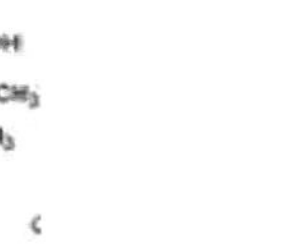

Huperzine Q N-oxyhuperzine Q

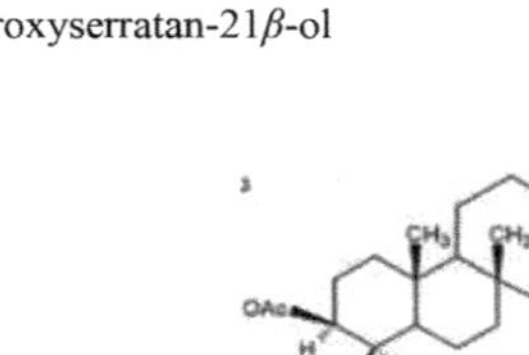

14*β*,15*β*-epoxy-3*β*-hydroxyserratan-21*β*-ol

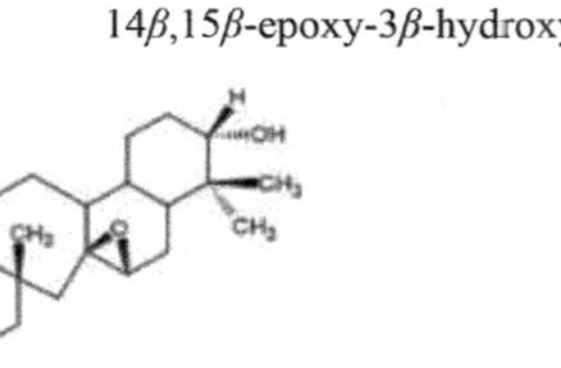

14*β*,15*β*-epoxy-3*β*-hydroxyserratan-21α-ol

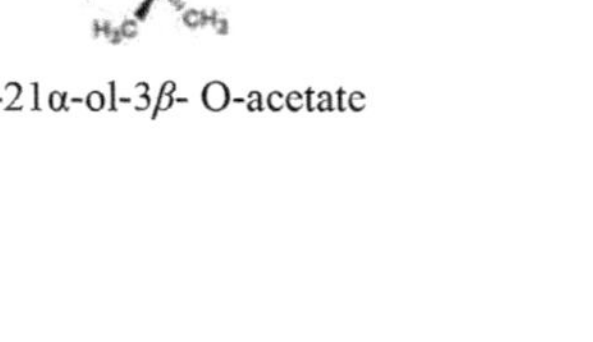

14*β*,15*β*-epoxy-3*β*-hydroxyserratan-21α-ol-3*β*- O-acetate

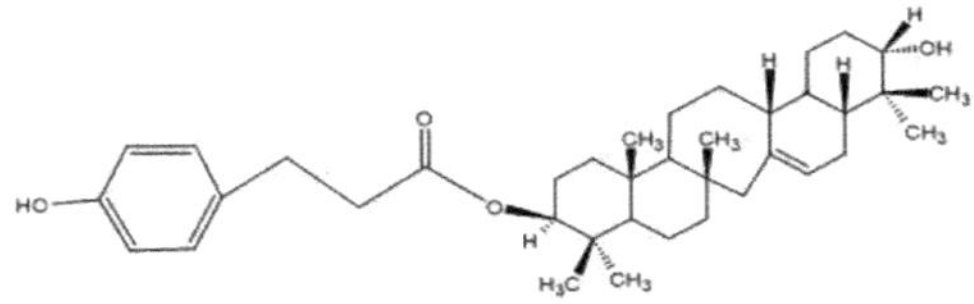

21α -hydroxyserrat-14-en-3*β*-yl*p*-dihydrocoumarate,

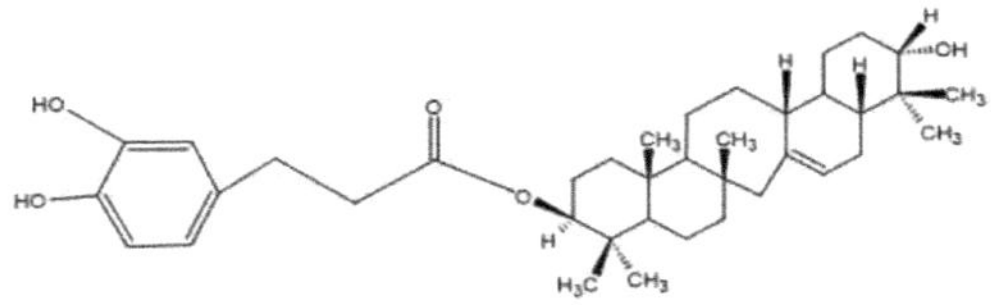

21α-hydroxyserrat-14-en-3*β*-yldihydrocaffeate

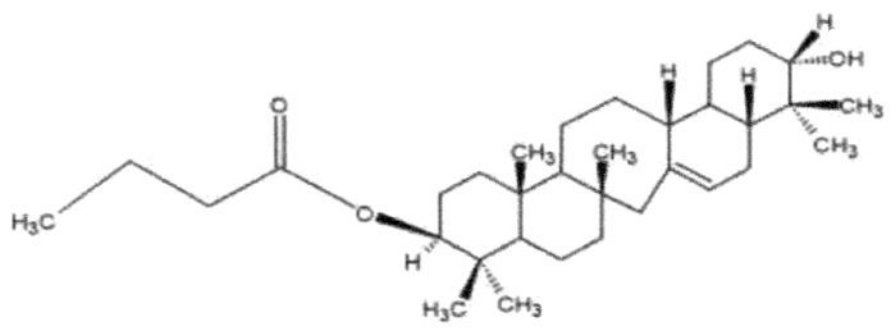

21α-hydroxyserrat-14-en-3*β*-ylpropanedioic acid monoester

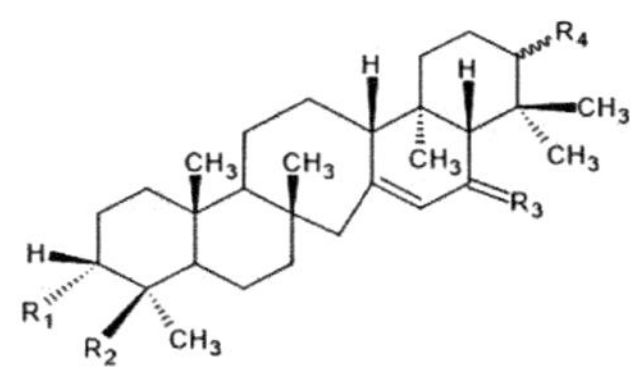

3α,21α-dihydroxyserrat-14-en-24-oic acid: R_1-OH; R_2-COOH; R_3- H_2; R_4- αOH

16-oxo-3α,21*β*-dihydroxyserrat-14-en-24-al: R_1-OH; R_2- CHO; R_3- O; R_4- *β*OH

16-oxo-3α,21*β*-dihydroxyserrat-14-en-24-oic acid:R_1-OH; R_2- COOH; R_3- O; R_4- *β*OH

16-oxo-21*β*-hydroxyserrat-14-en-3α-ylacetate:R_1-OAc; R_2- Me; R_3- O; R_4- *B*OH

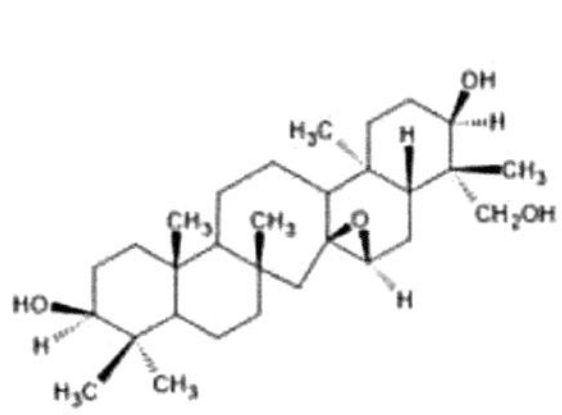

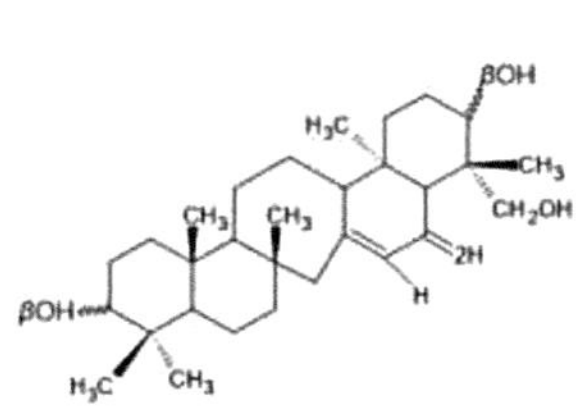

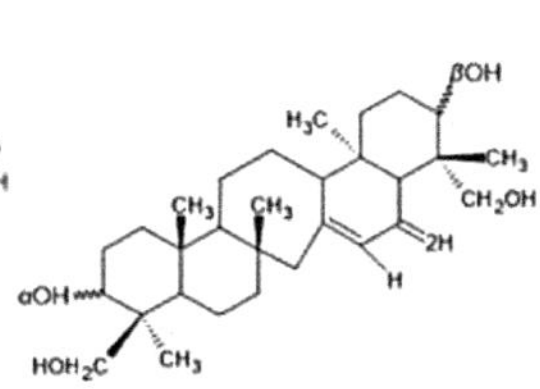

14β,15β-epoxyserratan-3β,21β,29-triol Serrat-14-en-3β,21β,29-triol Serrat-14-en-3α,21β,24,29-tetraol

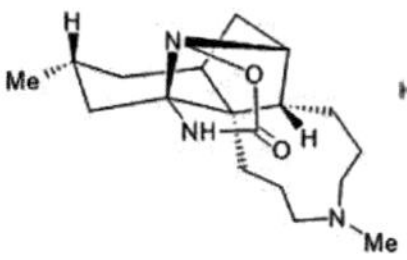

Lycoposerramine A

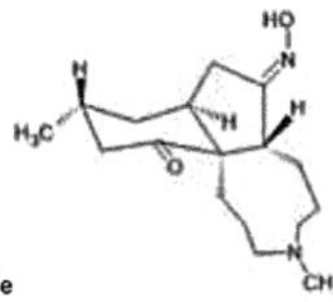

Lycoposerramine B

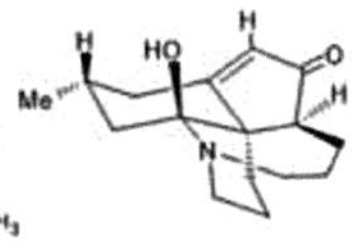

Lycoposerramine C

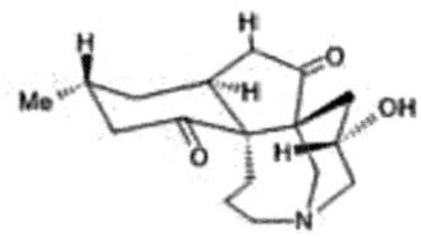

Lycoposerramine D

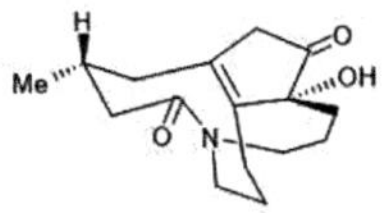

Lycoposerramine E

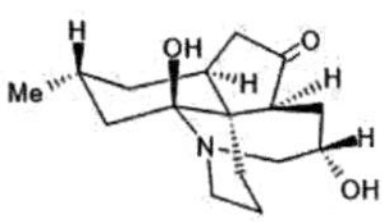

Lycoposerramine P

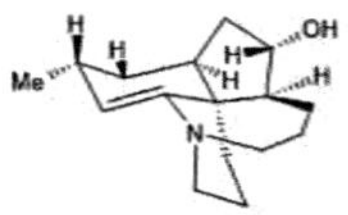

Lycoposerramine Q

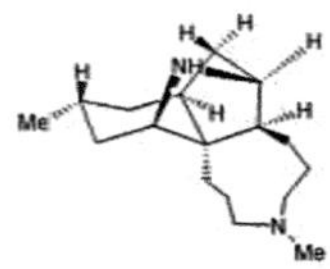

LycoposerramineS

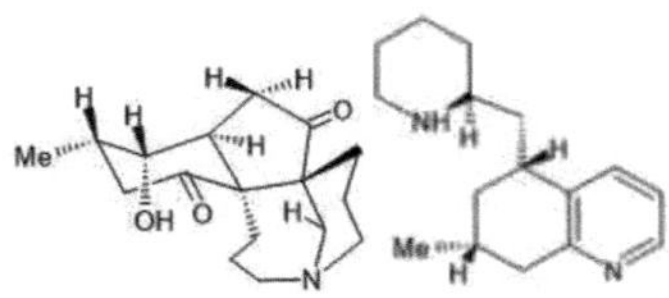

Lycoposerramine U

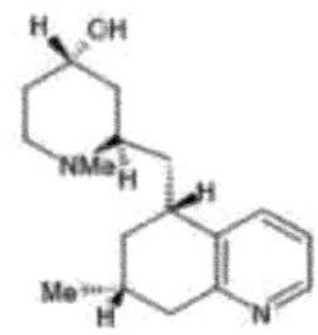

Lycoposerramines V and W

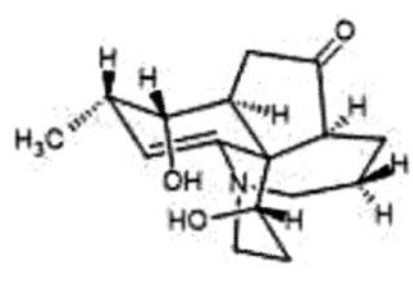

8α,11α-dihydroxyfawcettidine

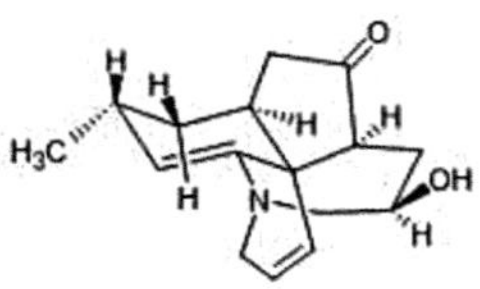

2β-hydroxylycothunine

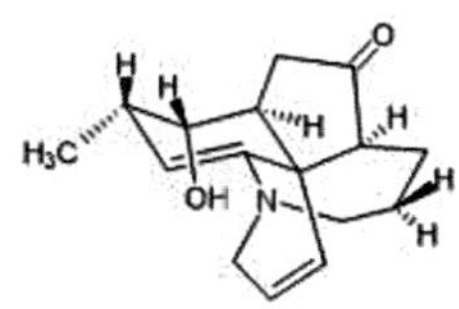

8α-hydroxylycothunine

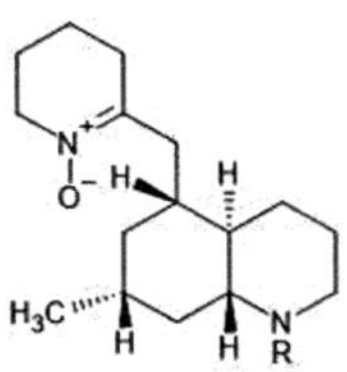

Lycoposerramine X: R=H
Huperzine J: R=CH3

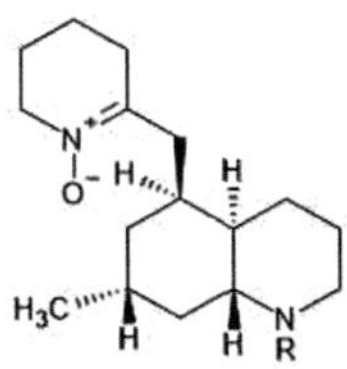

Lycoposerramine Y: R=CH_3
Huperzine K: R=H

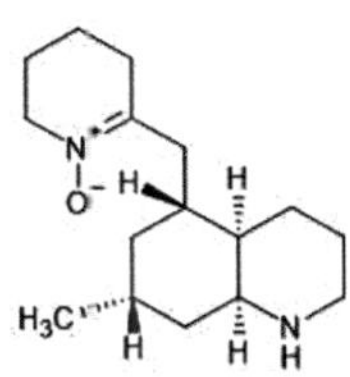

Lycoposerramine Z

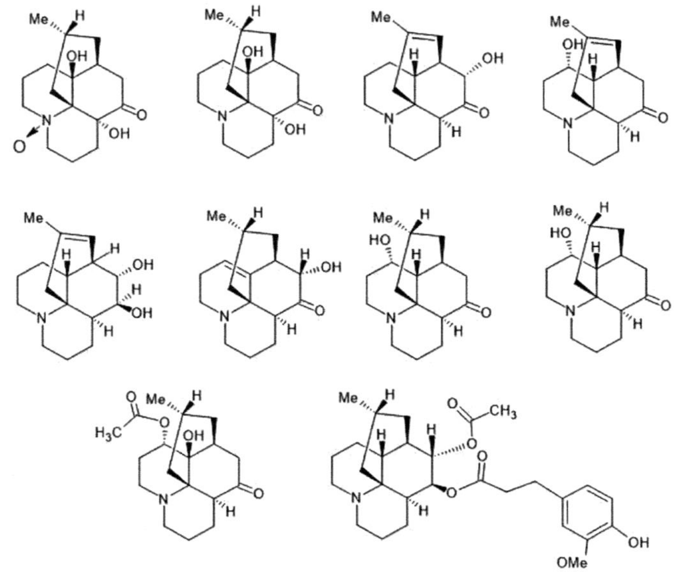

Lycoposerramines F-O

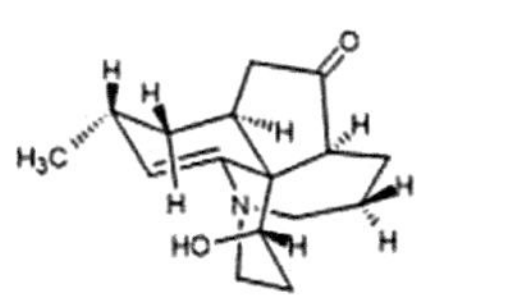

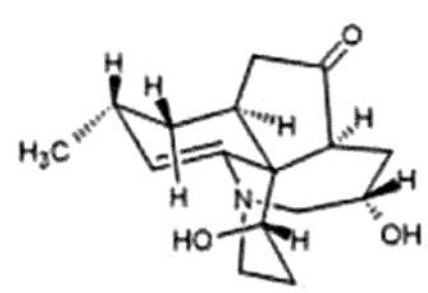

11α-hydroxyfawcettidine 2α,11α-dihydroxyfawcettidine

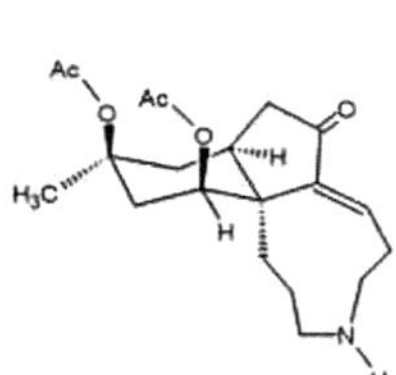

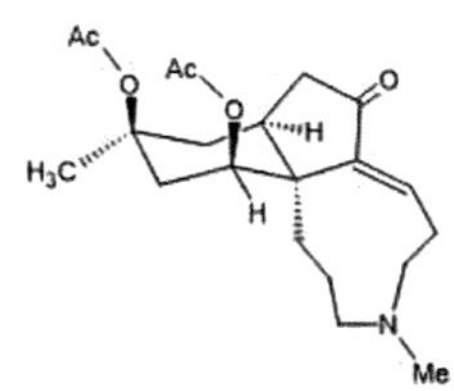

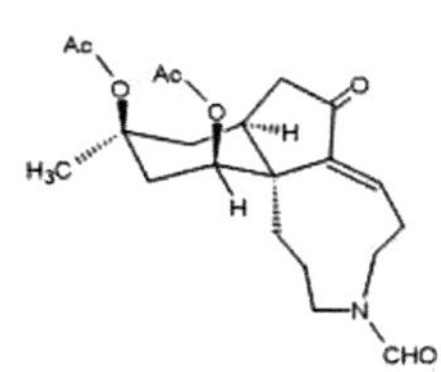

Lycoposerramine-T N-Methyllycoposerramine-T N-Formyllycoposerramine-T

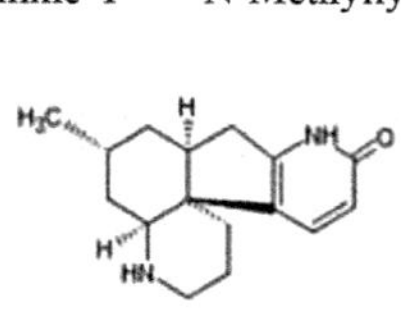

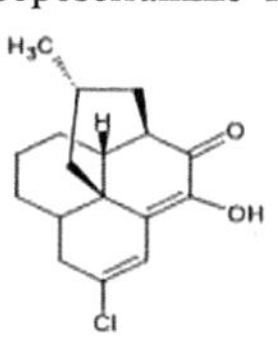

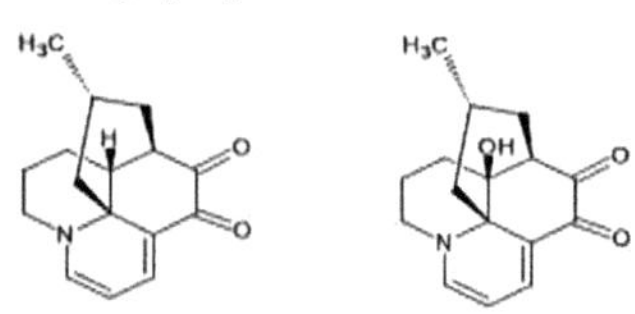

Lycoposerramine-R 2-chlorohuperzine E Huperzines E' and F'

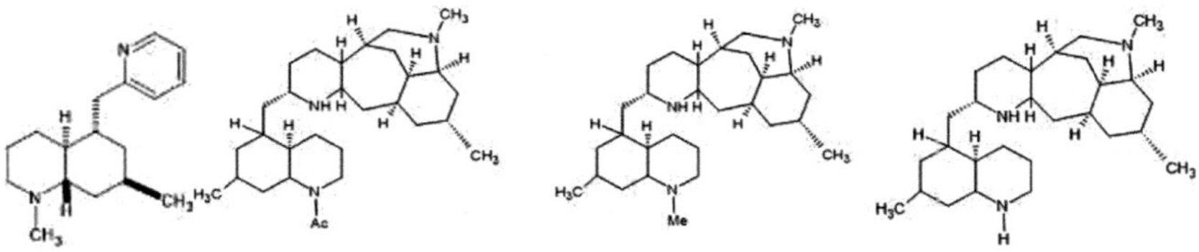

Serralongamines A–D

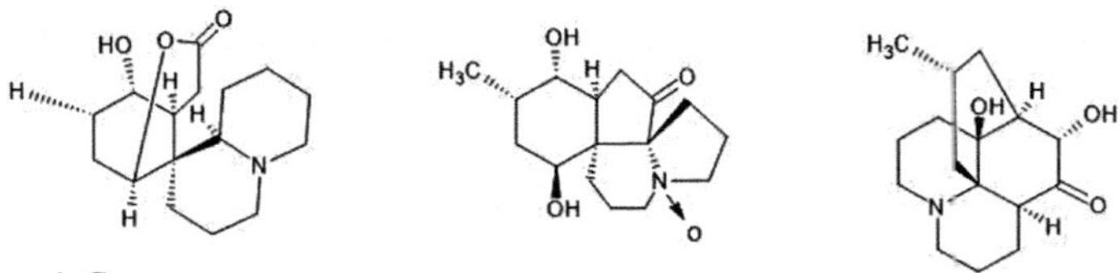

Serratezomines A-C

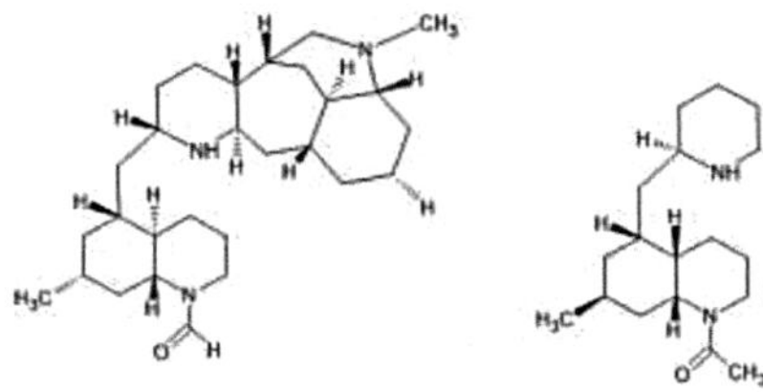

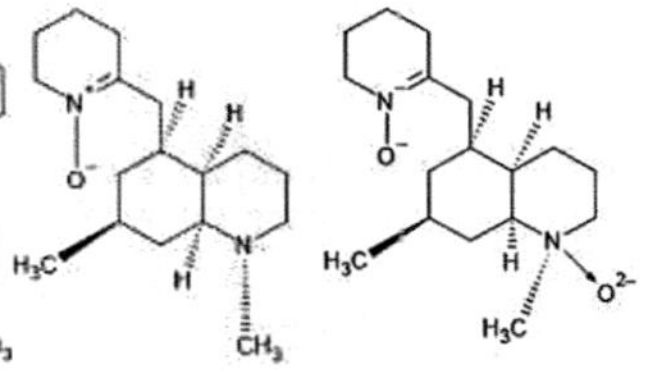

Serratezomines D and E

Huperzines M and N

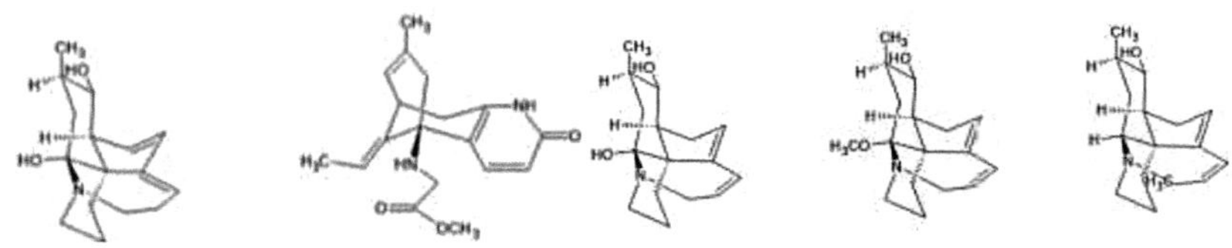

Huperserines A–E

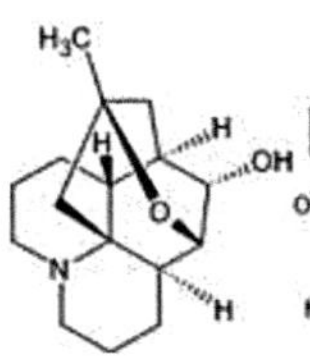

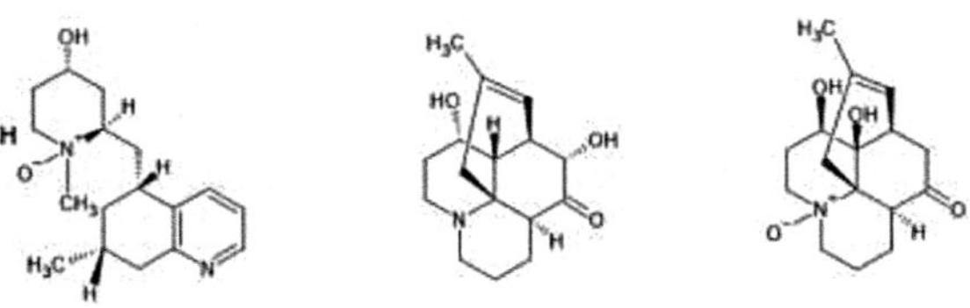

6α-hydroxy-5,15-oxide-lycopodane

Huperserramines A – C

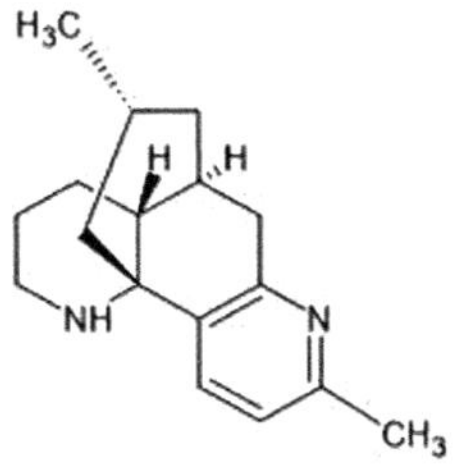

1-methyllycodine

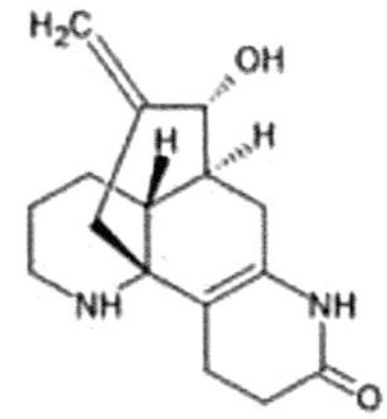

8α-hydroxy-15,16-dehydro-desN methyl-α-obscurin

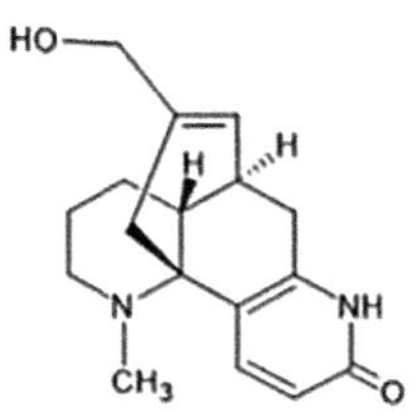

N-methyl-16hydroxyhuperzine

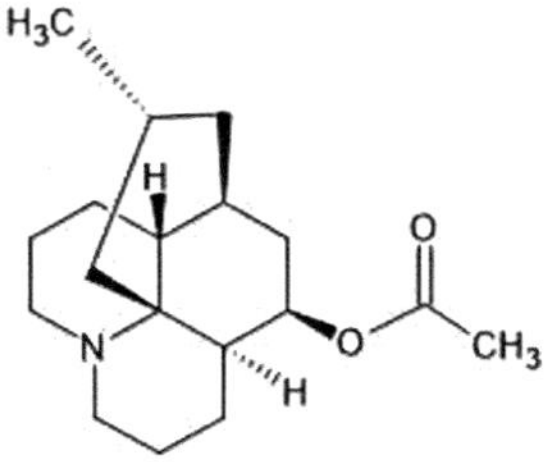

Lycopladine E

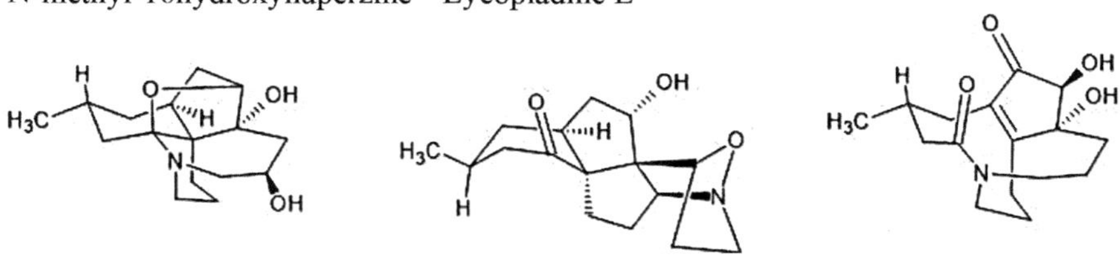

Hupertimines A-C

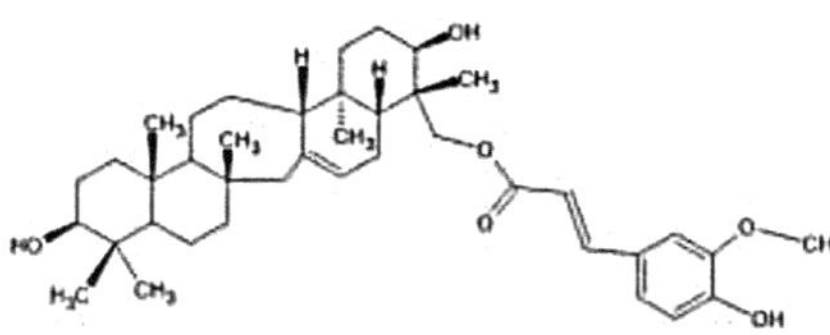

3β, 21β-diol-serrat-14-ene-29-ferulate

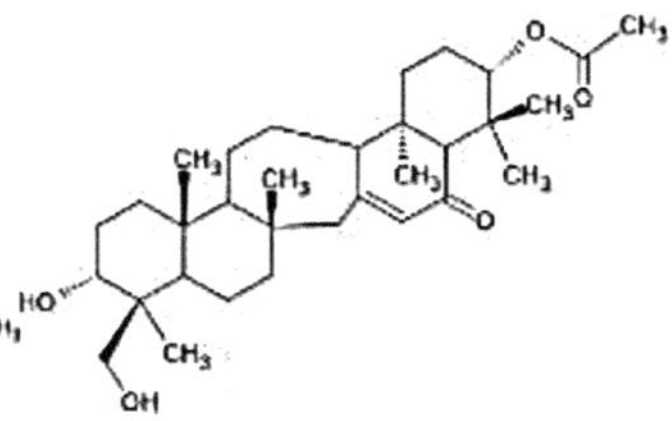

16-oxo-3α, 24-dihydroxy serrat-14-en-21α-yl acetate

15.3 PHARMACOLOGY

The pharmacological significance of *H. serrata* has been elucidated by different studies (Table 15.1) as follows:

15.3.1 ANTI-NEUROINFLAMMATORY EFFECTS

Dang et al. (2023) analyzed the anti-neuroinflammatory activity of alkaloid-enriched extract of *H. serrata* (HsAE) on BV-2 microglial cells stimulated by lipopolysaccharide (LPS). The LPS triggers BV-2 cells to produce high amounts of nitric oxide (NO). So, the effect of LPS-induced NO synthesis in BV-2 cells was compared by pre- or post-treatment of BV2- microglial cells with different concentrations of HsAE and huperzine A range from 25, 50, 100 and 150 μg/mL for 30 min before or after LPS (100 ng/mL) induction. N-monomethyl-L-arginine (L-NMMA), a nitric oxide synthase (NOS) was taken as positive control and Griess reagent was used to determine the concentration of NO. LPS-induced nitric oxide (NO) production was reduced by HsAE with half-maximal inhibitory concentration (IC_{50}) at pre- and post-treatment (99.79 and 92.40 mg/mL respectively). Furthermore, the IC_{50} value of huperzine A was similar to that of HsAE on NO secretion before (38.04 mg/mL) and after (40.23 mg/mL) treatment.

The activity of HsAE on pro-inflammatory mediators and pro-inflammatory cytokines such as prostaglandin E2 (PGE2), tumor necrosis factor-α (TNF-α), interleukin (IL)-6, and IL-1β was analysed by enzyme-linked immunosorbent assay. Complete inhibition of the secretion of PGE2, TNF-α, IL-6, and IL-1 β was shown during the pre-treatment with HsAE at concentrations of 50, 100, and 150 mg/mL compared to post-treatment with HsAE. It implies that prophylactic treatment is most preferable to inflammatory treatment. Microglial activation induces the release of pro-inflammatory mediators and cytokines, including PGE2, IL-6, IL-1 β and TNF-α. It would result in neuroinflammation due to pathogenesis and can worsen brain injury (Surace and Block, 2012). So, the result shows an effective therapeutic potentiality of *H. serrata* in neuroinflammatory treatment.

15.3.2 ANTIOXIDANT ACTIVITY

An efficient *in vitro* propagation system for *H. serrata* to produce Huperzine A (HupA) along with notable antioxidant potentiality was reported by Yang et al. (2021). The major objectives they investigated include, optimizing the sterilization methods for explants of *H. serrata*, analyzing its antioxidant activity, and determining HupA content *in vitro*. Different genotypes of wild *H. serrata* were used for propagation, and various hormone types and concentrations were tested to enhance growth rates. Evaluation of the HupA content of the *in vitro* cultured three genotypes by High-performance liquid chromatography (HPLC) resulted in significant production of 53.90–87.17 μg/g of HupA. The antioxidant capacity was assessed through the analysis of polyphenol content of *in vitro H. serrata* methanol extract (IVE) and wild *H. serrata* methanol extract (WHE) indicating the potential health benefits of the cultured plants. At a concentration of 0.25 mg/mL, the highest scavenging activity of IVE was 82.90 ± 0.56% for DPPH and 98.64 ± 1.00% for ABTS. However, the WHE exhibited a 78.50 ± 0.45% DPPH and 61.59 ± 0.68% ABTS radical scavenging activity. In the DPPH and ABTS scavenging assays, IC_{50} values of the IVE indicated that the total polyphenol content of WHE was 12.97 ± 0.81 mgGAE/g. IVE contains a total of 14.96 ± 0.65 mgGAE/g of polyphenols. The antioxidant analysis demonstrates that *H. serrata* has a potent antioxidant effect when cultured *in vitro*. The *in vitro H. serrata* extract could reduce Fe^{3+} and had significant scavenging activity on DPPH, ABTS, and OH^-(Yang et al., 2021).

Micro propagated plants comparatively have a higher amount of phytochemicals and show a remarkable antioxidant property than wild ones. This experiment showed beyond doubt that methanol extract of *in vitro* cultures of *H. serrata* had exclusively higher antioxidant activity compared to wild *H. serrata*. This study reveals a reliable antioxidant capacity of the callus of *in vitro H. serrata* along with culture protocol and a substantial basis to enhance the content of HupA.

15.3.3 ACETYLCHOLINESTERASE INHIBITORY ACTIVITY

Lycopodium alkaloids have the property to inhibit AChE. Huperzine A, which was first isolated from *H. serrata* exhibits a strong inhibitory effect against AChE and has been sold in China as a new drug in the treatment of Alzheimer's disease (Liu et al., 1986). After testing almost all the isolated lycopodium alkaloids for AChE inhibitory activities and using a screening method for AChE and butyrylcholinesterase activities, huperserine E from *H. serrata* showed AChE inhibitory activities (Zhu and Zhao, 2019). Huperserines A-E isolated from *H. serrata* were assayed for their AChE inhibitory activity by following the Ellman method of spectrophotometry. Huperzine A was taken as positive control, it has an IC_{50} value of 0.03 μM. Among all these tested compounds huperserine E showed AChE inhibitory activity at a moderate rate with an IC_{50} value of 6.71μM (Jiang et al., 2014b).

15.3.4 ANTI-CANCER ACTIVITY

Apoptosis is a physiological mechanism of programmed cell death to maintain homeostasis by destroying abnormal cells. Cancer can be effectively prevented by inducing apoptosis in the tumor cells. A wide range of plant-derived chemical compounds has been identified as apoptosis-inducing factors in several human cancer cell lines (Taraphdar et al., 2001).

Serratenediol, acompound isolated from *L. serratum* showed a significant apoptotic effect in the cultured HL-60 cell lines of human promyelocytic leukemia by regulating the ratio of Bax/Bcl-xL. Alcoholic extract of *L. serratum* (LSE) in various concentrations ranging from 6–100 μg/mL for 72 h shows a series of apoptotic events such as cell viability loss, morphological variations, and an increase in sub-G1 DNA content in cultured human promyelocytic leukemia HL-60 cells. The cytotoxicity assay of LSE and its derivatives was examined by using colorimetric MTT assay and showed a concentration-dependent growth inhibitory effect in tumour cell lines with a calculated IC_{50} value of 29.0 μg/mL. The number of cells having sub-G1 DNA content was determined by flow cytometry and confirmed the LSE-induced apoptosis in HL-60 cells. The biologically active compound, serratenediol (SE), was isolated and identified from the MC fraction of LSE by NMR spectroscopic data analysis with known triterpenes. SE showed an inhibitory activity of 81.8% with an IC_{50} value of 15.9 μM. The cell permeable DNA dye Hoechst 33342 was used to stain the HL-60 cells to evaluate the nuclear morphology under the fluorescent microscope and observed an extensive amount of apoptotic bodies in them. It indicates that LSE pretreatment has a significant effect on the apoptosis and inhibition of cell growth. Furthermore, the effects of SE on proteins related to apoptosis such as Bax, Bcl-xL, cytochrome C, caspase-3 and -9, and poly-ADP-ribose polymerase (PARP) were analyzed through western blotting. The anti-apoptotic protein, Bcl-xL showed a gradual decrease in the expression and the expression of Bax, a pro-apoptotic protein, had been elevated. Cytochrome C, which promotes mitochondrial apoptosis was prominently released into the cytoplasm. The active forms of caspase-9 and-3 were also elevated, and the PARP was cleaved. So, SE considerably has significant therapeutical potential in the treatment of cancer by regulating the Bax/Bcl-xL ratio in HL-60 cell lines, leading to apoptosis (Ham et al., 2012).

15.3.5 ANTI-DIABETIC ACTIVITY

Diabetes is a common medical condition marked by excessive plasma glucose levels. Major complications such as diabetic neuropathy, retinopathy, and cardiovascular diseases can be the most severe result of diabetes. The key enzyme α-Glucosidase is essential for the digestion of carbohydrates in the final step digestive process. α-glucosidase inhibitors will retard the release of d-glucose from dietary complex carbohydrates and consequently glucose absorption will be delayed, and ultimately leading to lower postprandial plasma glucose levels (Kumar et al., 2011).

Lycopodine-6α,11α-diol, lycoposerramine H, lycoposerramine I, lycopodine-6α-ol, lycoposerramine M, diphaladine A, lycoposerramine K, lycoposerramine W, huperzine M, luciduline, phlegmariuine N, huperzine A, huperzine B, lycodine, and lycoposerramine R obtained were examined for their AChE and α-glucosidase inhibitory activities (Ha et al., 2023). Other than huperzine A and B, none of the compounds showed evident AChE inhibition at 20μM concentration. However, a remarkable report they revealed is that lycopodine-6α,11α-diol, a novel chemical obtained in their previous work exhibited a more powerful and higher level of α-glucosidase inhibitory activity. The test was conducted against yeast α-glucosidase to assay the inhibitory activity. All the test was evaluated by taking acarbose as a positive control. The inhibitory activity showed IC_{50} value of 148.1 ± 5.5 μM compared to the IC_{50} value of 376.3 ± 2.7 μM for acarbose. It was the first report of α-glucosidase inhibitory activity of lycopodine-type alkaloids.

Two new serratene triterpenoids, 3β,21β-diol-serrat-14-ene-29-ferulate and 16-oxo-3α, 24-dihydroxy serrat-14-en-21α-yl acetate along with nine known analogs including seven serratenes and two serratanes and huperzine A were isolated from the methanolic extract of *H. searrata.* Among these compounds, serratene-type triterpenoids, serrat-14-en-3β,21β-diol, miyoshianol B, 16-oxo-3β,21β-dihydroxy-serrat-14-ene-24-ferulate, and 3-O-acetyltohogenol showed significant inhibitory activity against PTP1B (IC_{50} value of 13.50, 17.65, and 16.64 μM respectively), with promising therapeutical potential against type 2 diabetes. The stable binding of these compounds to the active site of PTP1B was confirmed by Molecular docking simulations (Ha et al., 2023) (Table 15.1).

TABLE 15.1 Major Phytochemicals and Their Bioactivities of *H. serrata*

Sl. No.	Phytochemical	Bioactivity	References
1.	Huperzine A	Anti-acetylcholinesterase and Antioxidant activity	Liu et al. (1986)
2.	12-deoxyhuperzine O	Potent inhibitory effect on the N methyl-D-aspartate (NMDA) receptor	Yang et al. (2010)
3.	Huperserine E	Modest anti-acetylcholinesterase (AChE) inhibitory activity.	Jiang et al. (2014b); Zhu and Zhao (2019)
4.	Lycopodine-6α,11α-diol	α-glucosidase inhibitory action	Ying et al. (2014)
5.	Serralongamines B and D	Prevention of early atherosclerotic lesions by targeting primarily foam cell formation (accumulation of cholesterol ester)	Ishiuchi et al. (2016)
6.	Serratenediol	Anti-cancer properties against HL-60 cell lines of human promyelocytic leukemia	Ham et al. (2012)
7.	Serratezomines A, B and C	Anti-cancer properties against murine lymphoma L1210 cells and human epidermoid carcinoma KB cells	Morita et al. (2000)
8.	Serratezomine D	Anti-acetylcholinesterase (AChE) inhibitory activity	Kubota et al. (2009)
9.	Total alkaloids (HsAE)	Neuroprotection against scopolamine-induced memory impairment connected to AChE depletion and oxidative stress	Dang et al. (2022)
10.	Serrat-14-en-3β,21β-diol,miyoshianol B (16-oxo-3β,21β-dihydroxy-serrat-14-ene-24-ferulate, 3-O-acetyltohogenol	Anti-diabetic activity by inhibitory action against PTP1B	Ha et al. (2023)

KEYWORDS

- **acetylcholinesterase**
- **diabetic neuropathy**
- **enzyme α-glucosidase**
- **enzyme-linked immunosorbent assay**
- **lycopodium alkaloids**
- **neuroinflammation**

REFERENCES

Cao, D., Sun, P., Bhowmick, S., Wei, Y., Guo, B., Wei, Y., Mur, L. A. J., & Sun, Z. (2021). Secondary metabolites of endophytic fungi isolated from *Huperzia serrata*. *Fitoterapia, 155*, 104970. https://doi.org/10.1016/j.fitote.2021.104970.

Dang, T. K., Hong, S. M., Dao, V. T., Nguyen, D. T., Nguyen, K. V., Nguyen, H. T., Ullah, S., Tran, H. T., & Kim, S. Y. (2022). Neuroprotective effects of total alkaloids fraction of *Huperzia serrata* on scopolamine-induced neurodegenerative animals. *Phytotherapy Research, 37*(1), 140–150. https://doi.org/10.1002/ptr.7602.

Dang, T. K., Hong, S. M., Dao, V. T., Tran, P. T. T., Tran, H. T., Do, G. H., Hai, T. N., Pham, H. T. N., & Kim, S. Y. (2023). Anti-neuroinflammatory effects of alkaloid-enriched extract from *Huperzia serrata* on lipopolysaccharide-stimulated BV-2 microglial cells. *Pharmaceutical Biology, 61*(1), 135–143. https://doi.org/10.1080/13880209.2022.2159450.

Deng, Z. T., Liu, Y. C., Zhu, Q. F., Jiang, S., Wu, X. D., & Zhao, Q. S. (2022). Hupertimines A–E, fawcettimine-type lycopodium alkaloids from *Huperzia serrata*. *Chemistry & Biodiversity, 19*(7), e202200454. https://doi.org/10.1002/cbdv.202200454.

Gao, W. Y., Li, Y. M., Jiang, S. H., & Zhu, D. Y. (2008). Two new nitrone alkaloids from *Huperzia serrata*. *Helvetica Chimica Acta, 91*(6), 1031–1035. https://doi.org/10.1002/hlca.200890110.

Ha, M. T., Le, T. T., Nguyen, V. T., Kim, J. A., Choi, J. S., & Min, B. S. (2023). Isolation and characterization of PTP1B inhibitory serratene-type triterpenoids from *Lycopodium serratum* Thunb. *Phytochemistry Letters, 57*, 231–238. https://doi.org/10.1016/j.phytol.2023.08.020.

Ham, Y. M., Yoon, W. J., Park, S. Y., Jung, Y. H., Kim, D., Jeon, Y. J., Wijesinghe, W. A. J. P., Kang, S. M., & Kim, K. N. (2012). Investigation of the component of *Lycopodium serratum* extract that inhibits proliferation and mediates apoptosis of human HL-60 leukemia cells. *Food and Chemical Toxicology, 50*(8), 2629–2634. https://doi.org/10.1016/j.fct.2012.05.019.

Ishiuchi, K., Jiang, W. P., Fujiwara, Y., Wu, J. B., & Kitanaka, S. (2016). Serralongamines B–D, three new lycopodium alkaloids from *Lycopodium serratum* var. *longipetiolatum*, and their inhibitory effects on foam cell formation in macrophages. *Bioorganic & Medicinal Chemistry Letters, 26*(11), 2636–2640. https://doi.org/10.1016/j.bmcl.2016.04.019.

Jiang, J., Liu, Y., Min, K., Jing, B., Wang, L., Zhang, Y., & Chen, Y. (2010). Two new lycopodine alkaloids from *Huperzia serrata. Helvetica Chimica Acta, 93*, 1187–1191. https://doi.org/10.1002/hlca.200900365.

Jiang, W. P., Ishiuchi, K., Wu, J. B., & Kitanaka, S. (2014a). Serralongamine A, a new lycopodium alkaloid from *Lycopodium serratum* var. *longipetiolatum. Heterocycles, 89*(3), 747–752. https://doi.org/10.3987/com-13-12928.

Jiang, W. W., Liu, F., Gao, X., He, J., Cheng, X., Peng, L. Y., Wu, X. D., & Zhao, Q. S. (2014b). Huperserines A–E, lycopodium alkaloids from *Huperzia serrata. Fitoterapia, 99*, 72–77. https://doi.org/10.1016/j.fitote.2014.09.005.

Katakawa, K., Kitajima, M., Aimi, N., Seki, H., Yamaguchi, K., Furihata, K., Harayama, T., & Takayama, H. (2004). Structure elucidation and synthesis of lycoposerramine-B, a novel oxime-containing lycopodium alkaloid from *Lycopodium serratum* Thunb. *Journal of Organic Chemistry, 70*(2), 658–663. https://doi.org/10.1021/jo0483825.

Katakawa, K., Kitajima, M., Yamaguchi, K., & Takayama, H. (2006). Three new phlegmarine-type lycopodium alkaloids, lycoposerramines-X, -Y and -Z, having a nitrone residue, from *Lycopodium serratum. Heterocycles, 69*, 223–229.

Katakawa, K., Kogure, N., Kitajima, M., & Takayama, H. (2009). A new lycopodium alkaloid, lycoposerramine-R, with a novel skeleton and three new fawcettimine-related alkaloids from *Lycopodium serratum. Helvetica Chimica Acta, 92*(3), 445–452. https://doi.org/10.1002/hlca.200800327.

Katakawa, K., Nozoe, A., Kogure, N., Kitajima, M., Hosokawa, M., & Takayama, H. (2007). Fawcettimine-related alkaloids from *Lycopodium serratum. Journal of Natural Products, 70*(6), 1024–1028. https://doi.org/10.1021/np0700568.

Kubota, T., Yahata, H., Yamamoto, S., Hayashi, S., Shibata, T., & Kobayashi, J. (2009). Serratezomines D and E, new lycopodium alkaloids from *Lycopodium serratum* var. *serratum. Bioorganic & Medicinal Chemistry Letters, 19*(13), 3577–3580. https://doi.org/10.1016/j.bmcl.2009.04.146.

Kumar, S., Narwal, S., Kumar, V., Prakashukne, O., Viswanathan, V., & Phadatare, A. G. (2011). Alpha-glucosidase inhibitors from plants: A natural approach to treat diabetes. *Pharmacognosy Reviews, 5*(9), 19–29. https://doi.org/10.4103/0973-7847.79096.

Liu, J., Zhu, Y., Yu, C., Zhou, Y. Z., Han, Y., Wu, F. W., & Qi, B. (1986). The structures of huperzine A and B, two new alkaloids exhibiting marked anticholinesterase activity. *Canadian Journal of Chemistry, 64*(4), 837–839. https://doi.org/10.1139/v86-137.

Liu, Y. C., Zhang, Z. J., Su, J., Peng, L. Y., Pan, L. T., Wu, X. D., & Zhao, Q. S. (2017). Lycodine-type lycopodium alkaloids from the whole plants of *Huperzia serrata. Natural Products and Bioprospecting, 7*(5), 405–411. https://doi.org/10.1007/s13659-017-0140-z.

Ma, X., & Gang, D. R. (2004). The lycopodium alkaloids. *Natural Product Reports, 21*(6), 752–772. https://doi.org/10.1039/B409720N.

Morita, H., Arisaka, M., Yoshida, N., & Kobayashi, J. (2000). Serratezomines A–C, new alkaloids from *Lycopodium serratum* var. *serratum. Journal of Organic Chemistry, 65*(19), 6241–6245. https://doi.org/10.1021/jo000661e.

Patil, S. P. (2020). Recently isolated lycodine-type lycopodium alkaloids and their total synthesis: A review. *Future Journal of Pharmaceutical Sciences, 6*(1), Article 108. https://doi.org/10.1186/s43094-020-00108-z.

POWO. (2024). *Huperzia serrata (Thunb.) Trevis*. Plants of the World Online. https://powo.science.kew.org/taxon/urn:lsid:ipni.org:names:20004840-1.

Sano, T., Tsuda, Y., & Inubushi, Y. (1970). Structures of tohogenol and tohogeninol. *Tetrahedron, 26*(12), 2981–2986. https://doi.org/10.1016/s0040-4020(01)92878-9.

Shigeyama, T., Katakawa, K., Kogure, N., Kitajima, M., & Takayama, H. (2007). Asymmetric total syntheses of two phlegmarine-type alkaloids, lycoposerramines-V and -W, newly isolated from *Lycopodium serratum*. *Organic Letters, 9*(20), 4069–4072. https://doi.org/10.1021/ol701871v.

Surace, M. J., & Block, M. L. (2012). Targeting microglia-mediated neurotoxicity: The potential of NOX2 inhibitors. *Cellular and Molecular Life Sciences, 69*(14), 2409–2427. https://doi.org/10.1007/s00018-012-1015-4.

Takayama, H., Katakawa, K., Kitajima, M., Seki, H., Yamaguchi, K., & Aimi, N. (2001). A new type of lycopodium alkaloid, lycoposerramine-A, from *Lycopodium serratum* Thunb. *Organic Letters, 3*(26), 4165–4167. https://doi.org/10.1021/ol0167762.

Takayama, H., Katakawa, K., Kitajima, M., Yamaguchi, K., & Aimi, N. (2002). Seven new lycopodium alkaloids, lycoposerramines-C, -D, -E, -P, -Q, -S, and -U, from *Lycopodium serratum* Thunb. *Tetrahedron Letters, 43*(46), 8307–8311. https://doi.org/10.1016/s0040-4039(02)02026-9.

Takayama, H., Katakawa, K., Kitajima, M., Yamaguchi, K., & Aimi, N. (2003). Ten new lycopodium alkaloids having the lycopodane skeleton isolated from *Lycopodium serratum* Thunb. *Chemical and Pharmaceutical Bulletin, 51*(10), 1163–1169. https://doi.org/10.1248/cpb.51.1163.

Tan, C. H., Ma, X. Q., Chen, G. F., & Zhu, D. Y. (2002). Two novel lycopodium alkaloids from *Huperzia serrata. Helvetica Chimica Acta, 85*(4), 1058–1061. https://doi.org/10.1002/1522-2675(200204)85:4<1058::AID-HLCA1058>3.0.CO;2-Q.

Taraphdar, A. K., Roy, M., & Bhattacharya, R. K. (2001). Natural products as inducers of apoptosis: Implication for cancer therapy and prevention. *Current Science, 80*(11), 1387–1396.

Wang, H. B., Tan, C. H., Tan, J. J., Gao, M. Y., Li, Y. M., Jiang, S. H., & Zhu, D. Y. (2007). Lycopodium alkaloids from *Huperzia serrata. Helvetica Chimica Acta, 90*(1), 153–157. https://doi.org/10.1002/hlca.200790009.

Yang, Y. F., Qu, S. J., Xiao, K., Jiang, S. H., Tan, J. J., Tan, C. H., & Zhu, D. Y. (2010). Lycopodium alkaloids from *Huperzia serrata*. *Journal of Asian Natural Products Research, 12*(11), 1005–1009. https://doi.org/10.1080/10286020.2010.522180.

Yang, Y., Dai, L., Wu, D., Dong, L., Tu, Y., Xie, J., & Luo, X. (2021). *In vitro* propagation, huperzine A content and antioxidant activity of three genotypic *Huperzia serrata*. *Plants, 10*(6), 1112–1112. https://doi.org/10.3390/plants10061112.

Ying, Y. M., Liu, X. S., Tong, C. P., Wang, J. W., Zhan, Z. J., & Shan, W. G. (2014). Lycopodium alkaloids from *Huperzia serrata*. *Helvetica Chimica Acta, 97*(10), 1433–1439. https://doi.org/10.1002/hlca.201400015.

Zhang, L. B., & Iwatsuki, K. (2013). Lycopodiaceae. In *Flora of China (Pteridophytes)* (Vol. 2, 3, pp. 13–34). Science Press. http://flora.huh.harvard.edu/china/mss/volume02/Flora_of_China_Volume_2_3_Lycopodiaceae.pdf.

Zhou, H., Jiang, S. H., Tan, C. H., Wang, B. D., & Zhu, D. Y. (2003b). New epoxyserratanes from *Huperzia serrata. Planta Medica, 69*(1), 91–94. https://doi.org/10.1055/s-2003-37033.

Zhou, H., Li, Y. S., Tong, X. T., Liu, H. Q., Jiang, S. H., & Zhu, D. Y. (2004). Serratane-type triterpenoids from *Huperzia serrata. Natural Product Research, 18*(5), 453–459. https://doi.org/10.1080/14786410310001643803.

Zhou, H., Tan, C. H., Jiang, S. H., & Zhu, D. Y. (2003a). Serratane-type triterpenoids from *Huperzia serrata. Journal of Natural Products, 66*(10), 1328–1332. https://doi.org/10.1021/np0301590.

Zhu, Q. F., & Zhao, Q. S. (2019). Chemical constituents and biological activities of lycophytes and ferns. *Chinese Journal of Natural Medicines, 17*(12), 887–891. https://doi.org/10.1016/s1875-5364(19)30108-6.

CHAPTER 16

Ethnomedicinal Importance, Active Bioconstituents, and Pharmacological Properties of Club Moss [*Lycopodiella cernua* (L.) Pic.Serm.]

PRADEEP BHAT,[1] SAVALIRAM G. GHANE,[2] and SANTOSHKUMAR JAYAGOUDAR[3,*]

[1]*ICMR–National Institute of Traditional Medicine, Belagavi, Karnataka, India*

[2]*Department of Botany, Plant Physiology Laboratory, Shivaji University, Vidyanagar, Kolhapur, Maharashtra, India*

[3]*Department of Botany, G. S. S. College and Rani Channamma University P. G. Centre, Belagavi, Karnataka, India*

*Corresponding authors

ABSTRACT

The traditionally important club moss (*Lycopodiella cernua* (L.) Pic.Serm.) of the Lycopodiaceae family is a perennial shrub that grows primarily in the wet tropical biome. The plant is widely used in Chinese, Iranian, and Indian traditional medicine to treat joint pain, contusions, rheumatism, tendons, the common cold, diarrhea, burns, trauma, wounds, and to stimulate blood circulation. Phytochemical investigation of a 95% ethanol extract of the whole plant recorded five serratene-type triterpenoids with four known analogues. Four Fawcett mine-type palhicerines A-D and two lycopodine-type (palhicerines E and F) alkaloids, together with 20 known compounds, were isolated. An aqueous MeOH (90%) extract of the whole plant was processed

Bioactives and Pharmacology of Pteridophytes. T. Pullaiah (Ed.)

DOI: 10.1201/9781779642738-16

to give the alkaloid-containing $CHCl_3$ fraction, from which 9 Fawcettimine-type, 14 lycopodine-type, and 3 lycodine-type alkaloids were isolated and characterized. The plant extracts and active fractions of the plant exhibit several pharmacological activities such as antihypertensive, anti-amnesic, neuroprotective, anti-inflammatory, antioxidant, and cytotoxic activities.

16.1 INTRODUCTION

Lycopodiella cernua (L.) Pic.Serm. belongs to the family Lycopodiaceae. *Lepidotis cernua* (L.) P.Beauv., *Lycopodium cernuum* L., *Palhinhaea cernua* (L.) Franco & Vasc., *Lycopodium heeschi* Müll.Hal., *Lycopodium hupeanum* Müll.Hal., *Lycopodium marianum* Willd., *Lycopodium polycephalum* Colenso, *Lycopodium secundum* Müll.Hal., *Lycopodium sikkimense* Müll. Hal., *Palhinhaea brevibracteata* (Alderw.) Holub, *Palhinhaea capillacea* (Willd. ex Spring) Holub, *Palhinhaea cernua* f. *ciliatomarginata* W.M.Chu, *Palhinhaea cernua* f. laxa (Blume) Satou, *Palhinhaea lufengensis* C.Y.Yang, *Palhinhaea mariana* (Willd.) Holub, *Palhinhaea polycephala* (Colenso) Holub, and so forth. are some of the synonyms of this plant (POWO, 2024).

Plants are terrestrial with a creeping primary stem, growing in open habitats or in shrubby vegetation. Secondary stems are erect and profusely branched, pinnately to dichotomously branched, with leaves abundant, spreading-falcate, linear, 2–4 x 0.2 mm, indistinctly spirally arranged, with sparse hairs 0.2–0.5 mm along decurrent leaf bases. Leaves are linear, 2.5 mm long, spirally arranged around the stem. Single sporangia are borne in strobilus on the adaxial surface; strobili are at the apex of secondary branches, small in size, sessile, pendent at the ends of branchlets, 0.5–1 cm x 2–3 mm (POWO, 2024).

Lycopodiella cernua is a widely distributed species recognized for several uses in traditional medicine. In China, it has been used since time immemorial as a traditional medicine to treat nephrolith, hepatitis, rheumatism, and whooping cough. It has also been demonstrated to be highly effective in attenuating joint pain (Hung et al., 2015). It is commonly known as ‘Pu-Di-Wu-Gong’ in Chinese, due to its appearance like a long procumbent centipede, which has been used to heal contusions, treat rheumatism, relax tendons, and stimulate blood circulation in traditional folk medicine (Tang et al., 2016). In Indian and Iranian traditional medicine, the plant is used in treating the common cold, diarrhea, burns, trauma, and wounds (Johnson et al., 2021).

16.2 BIOACTIVES

Lycopodium alkaloids are the characteristic metabolites of club mosses. Chemical differences exist in specimens of a single plant species collected from different geographic origins. A preliminary LC-MS detection and dereplication analysis of alkaloidal constituents of *L. cernuum* var. *sikkimense* collected from Fujian province resulted in the isolation and characterization of three undescribed *Lycopodium* alkaloids, lycocernuskines A–C, and six known cernuane-type compounds – cernuine, lycocernuine, 2α-hydroxycernuine, cernuine N-oxide, and lyco-cernuine N-oxide (Tang et al., 2019). Phytochemical investigation of a 95% ethanol extract of the whole plant recorded five serratene-type triterpenoids with four known analogues (Wei et al., 2018). Two *Lycopodium* alkaloids, acetyllycoposerramine-M and palcernine-A, were isolated from whole plant extracts together with 10 previously known compounds (Zhao et al., 2012).

Four fawcettimine-type palhicerines A-D and two lycopodine-type (palhicerines E and F) alkaloids, together with 20 known compounds, were isolated. An aqueous MeOH (90%) extract of the whole plant (10 kg) was processed to give the alkaloid-containing $CHCl_3$ fraction, from which 9 (nine) fawcettimine-type, 14 (fourteen) lycopodine-type, and 3 (three) lycodine-type alkaloids were isolated and characterized by their spectroscopic data and physicochemical properties and literature reports of the known structures. Compounds identified include lycopoclavamine-A, lycoposerramine-M, lycopodine, acetyllycoposerramine-M, acetylfawcettiine, dehydroisofawcettiine, clavolonine, diacetyllycofoline, 5-acetyllycofoline, anhydrolycodoline, gnidioidine, huperzine-E, 12-deoxyhuperzine-O, fawcettimine, phlegmariurine-B, palhinine-A, lyco-flexine N-oxide, lycodine, des-N-methyl-a-obscurine, and α-obscurine (Tang et al., 2016) (Figure 16.1).

16.3 PHARMACOLOGY

16.3.1 ANTIHYPERTENSIVE ACTIVITY

The hydroalcoholic extract of the aerial part of the plant was used to perform antihypertensive activity through *in vivo* studies, and its vascular effect was studied *in vitro* on the isolated aorta. Rats' normal blood pressure was 103.83 ± 3.7 / 82.13 ± 2 mmHg (systole/diastole), and it increased to 208 ± 2.2 / 150.06 ± 2.5 mmHg after 21 days of a high salt diet. They regained their normal blood pressure after 22 days

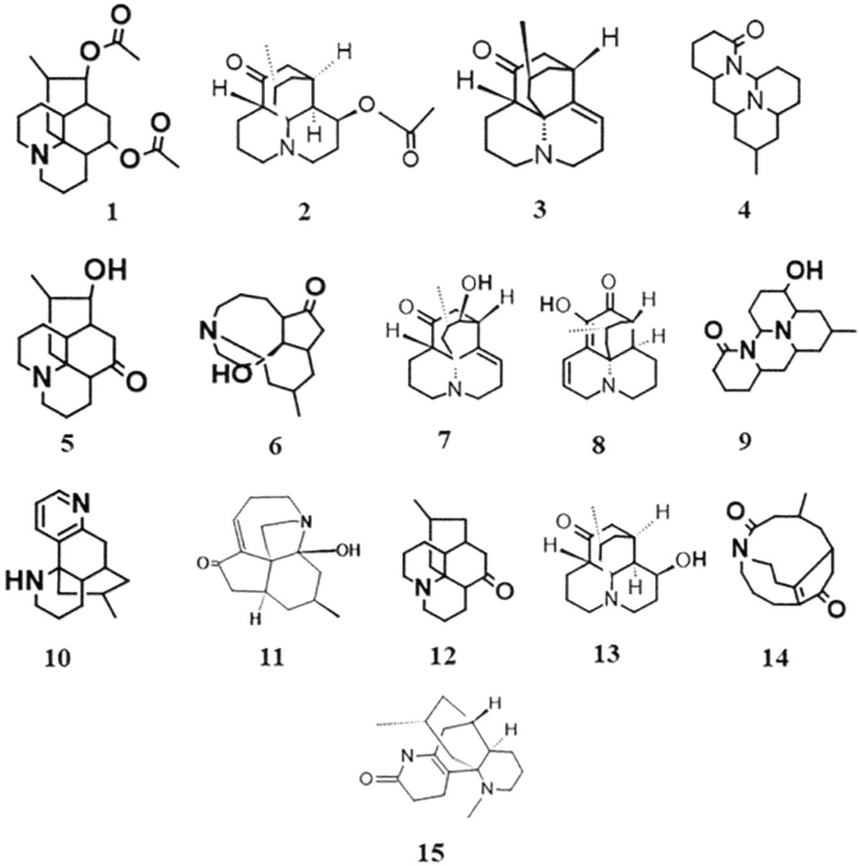

FIGURE 16.1 Structures of Acetylfawcettiine (1), Acetyllycoposerramine-M (2), Anhydrolycodoline (3), Cernuine (4), Clavolonine (5), Fawcettimine (6), Gnidioidine (7), Huperzine-E (8), Lycocernuine (9), Lycodine (10), lycopoclavamine-A (11), Lycopodine (12), lycoposerramine-M (13), phlegmariurine-B (14), α-obscurine (15).

when they were treated with the extract at 400, 200, and 100 mg/kg. The extract relaxed the isolated aorta contracted with norepinephrine, with an EC50 of 0.65 ± 0.02 mg/ml. The maximum of this contraction was 0.71 ± 0.05 g, which is considered as 100%. Injection in a cumulative manner of the extract into the bath significantly relaxed the isolated organ precontracted with norepinephrine (EC50: 0.65 ± 0.02 mg/ml) (Andriamaro et al., 2024).

16.3.2 ANTI-AMNESIC EFFECT

The whole plant is used in Vietnamese folk medicine for treating central nervous system (CNS) conditions. The alkaloid fraction from the methanol extract of the plant was evaluated for *in vitro* acetylcholinesterase (AChE) inhibitory activity in cognition-relevant brain areas of mice. In an *in vivo* study, the cognitive-enhancing effect of the methanol extract on amnesic mice induced by scopolamine was investigated by assessing a passive avoidance and a Morris water maze test. The methanol extract inhibited AChE activity in the mouse frontal cortex, hippocampus, and striatum with IC50 values of 26.7, 32.2, and 25.7 μg/mL, respectively. Administration of the methanol extract (10, 20, 50, and 100 mg/kg, p.o.) significantly reversed cognitive impairments in mice as measured by the passive avoidance test. Treatment with the methanol extract (50 mg/kg) reduced escape latencies in training trials and prolonged swimming times in the target quadrant during the probe trial in the water maze task (Chuong et al., 2014).

16.3.3 NEUROPROTECTIVE ACTIVITY

Palhinosides A–H flavone glucosidic truxinate esters, including β-truxinate and μ-truxinate forms, were isolated from the plant. The highly differentiated HT-22 cells were inoculated into 96-well plates at a density of 900 cells/well in medium containing 10% fetal bovine serum (FBS) and 1% double antibiotics (penicillin, streptomycin), then cultured in 5% CO_2 with a 100% humidified atmosphere at 37°C for 24 h. The test compounds and the positive control (Trolox) were prepared in serial dilutions from 0.6 to 15 μM in serum-free medium. Afterward, each well was supplemented with various concentrations of test compounds for 24 h before glutamate treatment. Cell viability was detected by a luminescent cell viability assay after the L-glutamate (50 mM) treatment for 24 h. The protective effects of compounds Palhinoside-D (4) and Palhinoside-E were better than that of Trolox at concentrations ranging from 1.3 to 15 μM, while compounds Palhinoside-F (6) and Palhinoside-G (7) showed effects similar to Trolox. The rest of the compounds exerted weak protective effects (Li et al., 2020).

16.3.4 ANTI-INFLAMMATORY ACTIVITY

The anti-inflammatory activity of the crude methanol extract of the whole plant was performed through cyclooxygenase (COX-2 and COX-1) enzyme inhibition assays at a final concentration of 100 ppm. Two trials

(4 replicate wells/trial) were conducted for the extract. In the cyclooxygenase inhibition assay, the methanolic extracts from the aerial parts of *L. cernua* exhibited high anti-inflammatory activity against COX-2 (74.78 ± 18.61%) compared to its activity against COX-1 (57.80 ± 3.46%). These values yield a ratio of 1.29 (Porquis et al., 2018).

16.3.5 ANTIOXIDANT ACTIVITY

The antioxidant activity of the whole plant extract was carried out through the DPPH assay. A final concentration of 0.33 mg/ml was added with 150 µl DPPH in a 96-well plate. Ascorbic acid at 0.4 mg/ml was used as a positive control. After 30 min of incubation, absorbance was read at 517 nm. The DPPH radical scavenging activity of the extract and the standard ascorbic acid were 19.13 ± 0.88% and 12.21 ± 0.9%, respectively (Porquis et al., 2018).

16.3.6 CYTOTOXICITY ACTIVITY

The MTT cytotoxicity assay of the crude methanol extract of the aerial and underground parts of the plant was performed using A549 and HEKn cells. It was found that both extracts were nontoxic to HEKn, with the aerial extract showing 54.75 ± 0.82 (200 ppm) and 95.43 ± 7.50 (20 ppm), and the underground extract showing 84.33 ± 0.83 (200 ppm) and 93.74 ± 32.18 (20 ppm). These extracts are also non-toxic against the A549 cell line (IC50 – 49.05 and 48.20 (µg/mL), respectively) (Porquis et al., 2018).

KEYWORDS

- **antihypertensive**
- **anti-inflammatory**
- **antioxidant**
- **crude methanol extract**
- **five serratene-type triterpenoids**
- **hydroalcoholic extract**
- **lycopodiaceae**
- **neuroprotective**

REFERENCES

Andriamaro, T. N., Ramanantsoa, N. J. P., Fidiniaina, R. J., Mohamad, C. C., & Rajaonarison, J. F. (2024). Antihypertensive activity of *Lycopodium cernuum* (Lycopodiaceae) hydroalcoholic extract on rats. *World Journal of Biology, Pharmacy and Health Sciences, 18*(2), 118–123.

Chuong, N. N., Trung, B. H., Luan, T. C., Hung, T. M., Dang, N. H., & Dat, N. T. (2014). Anti-amnesic effect of alkaloid fraction from *Lycopodiella cernua* (L.) Pic. Serm. on scopolamine-induced memory impairment in mice. *Neuroscience Letters, 575*, 42–46.

Hung, T. M., Lee, J. S., Chuong, N. N., Kim, J. A., Oh, S. H., Woo, M. H., Choi, J. S., & Min, B. S. (2015). Kinetics and molecular docking studies of cholinesterase inhibitors derived from water layer of *Lycopodiella cernua* (L.) Pic. Serm. (II). *Chemico-Biological Interactions*. https://doi.org/10.1016/j.cbi.2015.07.008 (accessed on 17th November 2025).

Johnson, M., Anwardeen, I., Periasamy, A., Siva, H. R., Dhinesh, K. M., Almeida, R. S., & Coutinho, H. D. M. (2021). Evaluation of phenolic constituents and toxicity of lycophytes and ferns of Shervarayan Hills aqueous extracts. *Chemistry Africa, 3*, 513–523.

Li, J., Tan, L. H., Zou, H., Zou, Z. X., Long, H. P., Wang, W. X., Xu, P. S., Liu, L. F., Xu, K. P., & Tan, G. S. (2020). Palhinosides A–H: Flavone glucosidic truxinate esters with neuroprotective activities from *Palhinhaea cernua*. *Journal of Natural Products, 83*(2), 216–222.

Porquis, H. C., Ang, A. M. G., Doblas, G. Z., Amoroso, V. B., Jacalan, D. R. Y., Batbatan, C. G., & Cruz, R. Y. D. (2018). Anti-inflammatory, antioxidant and cytotoxicity studies on *Lycopodiella cernua* (L.) J. Sm. in Bukidnon, Philippines. *Asian Journal of Biological and Life Sciences, 7*(2), 47–52.

POWO. (2024). *Plants of the World Online*. Facilitated by the Royal Botanic Gardens, Kew. http://www.plantsoftheworldonline.org/ (accessed on 17th November 2025).

Tang, Y., Li, N., Zou, Y., Ai, Y., Ma, G. L., Osman, E. E. A., Xiong, J., Li, J., Jin, Z. X., & Hu, J. F. (2019). LC-MS guided isolation and dereplication of lycopodium alkaloids from *Lycopodium cernuum* var. *sikkimense* of different geographical origins. *Phytochemistry, 160*, 25–30.

Tang, Y., Xiong, J., Zou, Y., Zhang, H. Y., & Hu, J. F. (2016). Palhicerines A–F, lycopodium alkaloids from the club moss *Palhinhaea cernua*. *Phytochemistry, 2016*, 1–10.

Wei, J. J., Wang, W. Q., Song, W. B., & Xuan, L. J. (2018). Serratene-type triterpenoids from *Palhinhaea cernua*. *Fitoterapia, 127*, 151–158.

Zhao, F. W., Sun, Q. Y., Yang, F. M., Luo, J. F., Hu, G. W., Liu, F., Wang, Y. H., & Long, C. L. (2012). Lycopodium alkaloids from *Palhinhaea cernua*. *Journal of the Brazilian Chemical Society, 23*(2), 349–354.

CHAPTER 17

Phytochemical and Pharmacological Activities of *Marsilea quadrifolia* L.

K. S. SHANTHI SREE,* A. SHOBHA RANI, P. SUJATHAMMA, and M. SWETHA PRIYA

Sri Padmavati Mahila Visvavidyalayam (Women's University), Tirupati, Andhra Pradesh, India

*Corresponding author

ABSTRACT

The well-known perennial aquatic plant *Marsilea quadrifolia* L. is found in marshy and wet areas and has long been utilized in Ayurvedic medicine. The concentration and physiologically active components of a plant are used to determine its therapeutic value. Plant-derived compounds are very important because they are used as pharmaceuticals, model molecules for drug manufacturing, or intermediates for synthetic pharmaceuticals. The current review highlights the bioactive compounds and biological activities of *Marsilea quadrifolia*.

17.1 INTRODUCTION

The compounds derived from plants have been used clinically for a long time as they are more accepted and patients have better tolerance (Tyler, 1999). The majority of phytochemicals found in medicinal plants are strong bioactive substances that can be utilized as building blocks to create effective medications (Kaufman et al., 1999). It is important to identify the chemical compounds present in local medicinal plants by herbalists, which are usually used for disease treatment (Sofowora, 1993).

Bioactives and Pharmacology of Pteridophytes. T. Pullaiah (Ed.)

DOI: 10.1201/9781779642738-17

Prominently known as European water clover or four-leaf clover in English, Niraaaikkirai, Aaraikeerai in Tamil, Neeraral in Malayalam, and Cauptiya, Sunsuniya in Hindi, *Marsilea quadrifolia* L. is a recognized pteridophytic herb that is a member of the Marsileaceae family. It is an aquatic fern with four divided leaves that resemble a trifolium or four-leaf clover. Leaflets are glaucous, the sporocarp is ellipsoid, dark brown in color, and present on stalks attached at the base of the petioles. Rooting can be seen at both nodes and internodes of the rhizome. Leaves can be found on land or in shallow water, floating or standing erect. Its petiole, which has a long stalk with four lobes like clovers, is held above the water or immersed in clayey, sandy substrates found in permanent lakes, seasonal ponds, and artificial habitats like fields and ditches. It can be found all over the world in both tropical and temperate climates. It grows along the banks of rivers and canals as well as in marshy areas (Sajem and Gosai, 2010). In submerged water, it is rooted at the bottom of the clayey soil (Marwat et al., 2007).

It is taken as food by many tribal communities such as Malasars, Kadar's, Malaimalasars, Pulaiyars, and Mudhuvars of Anamalais hills, which are present in the Western Ghats of Coimbatore district, Tamil Nadu, India, according to their seasonal availability (Ramachandran, 2007). In folklore, the whole plant juice with garlic is used to cure cough as well as convulsions of legs and muscles, and the whole plant's juice or paste relieves sleep disorders and hypertension when applied to the head externally. Two drops of juice from young leaves, extracted after crushing, can be given for migraine in the nostrils. The whole plant crushed with sugar candy or honey is used for curing diarrhea in infants (Sen and Behera, 2000). Leaf juice is used as a diuretic and febrifuge and is also used for abscesses and snake bites (Sajem and Gosai, 2010). The plant acts as an anti-inflammatory, depurative, and refrigerant (Sahu, 2011), and is also used to treat skin conditions, diarrhea, mental illnesses, diabetes, bronchitis, and cough. The significance of the plant's biological activities and bioactive substances is covered in this review.

17.2 BIOACTIVES

Preliminary phytochemical analysis of *Marsilea quadrifolia* crude extracts showed the presence of tannins, saponins, flavonoids, phenols, phytosterols, steroids, terpenoids, triterpenoids, alkaloids, tropane alkaloids, isoquinoline alkaloids, carbohydrates, glycosides, reducing sugars, and lipids, in varying concentrations (Gini and Jothi, 2013). The quantitative analysis of phenolic compounds, flavonoids, alkaloids, tannins, and saponins in the leaf and stem of *M. quadrifolia* revealed that the content of phenolic compounds was 8.34±0.92 mg/g and 7.31±0.46 mg/g, flavonoids was 7.46±0.64 mg/g and 6.45±0.68 mg/g, alkaloids was 6.12±0.51 mg/g and 5.89±0.61 mg/g, tannins was 6.58±0.72 mg/g and 6.07±0.56 mg/g, and saponins was 5.32±0.48 mg/g and 6.30±0.58 mg/g (Gopalakrishnan and Udayakumar, 2017).

The GC-MS analysis showed the presence of 39 phytocompounds in the methanolic extract of the leaf of *M. quadrifolia* including 9,12,15-octadecatrienoic acid, methyl ester (Z,Z,Z)-(12.96%); 4H-pyran-4-one, 2,3-dihydro-3,5-dihydroxy-6-methyl-(21.41%); 12,15-octadecatrien-1-ol (Z,Z,Z)-(3.54%); and n-hexadecanoic acid (17.47%); 2-furancarboxaldehyde, 5-(hydroxymethyl)-(9.39%). The methanolic extract of the stem of *M. quadrifolia* reported to have 29 bioactive compounds including 5-(hydroxymethyl)-(60.42%); 2-furancarboxaldehyde, 4H-pyran-4-one, 2; n-hexadecanoic acid-(6.00%); 6-octadecenoic acid (Z)-(2.69%); 3-dihydro-3, 5-dihydroxy-6-methyl-(13.88%); and furfural-(2.23%) (Gopalakrishnan and Udayakumar, 2014b).

Marsilea quadrifolia contains phytochemical compounds like β-sitosterol, methylamine, 3-hydroxy-triacontan-11-one, glucoflavones, marsilin (1-triacontanol-cerotate), hentriacotan-6-ol, marsileagenin A, C-glucosyl xanthones, and flavonol-O-mono and diglycoside (Khare, 2004).

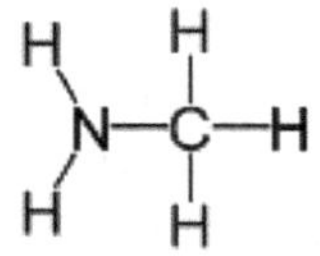

Methylamine

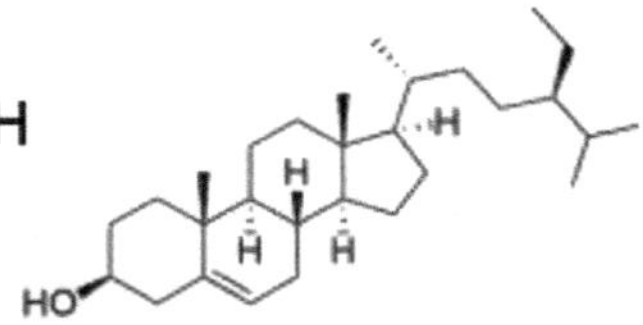

β-Sitosterol

17.3 PHARMACOLOGICAL REVIEW

17.3.1 ANTIMICROBIAL ACTIVITY

The antimicrobial activity was analyzed using different concentrations of leaf and stem extracts of *M. quadrifolia*, in which the best antimicrobial activity was observed in 5 mg of both leaf and stem extracts compared to the other concentrations of 2.5 and 3.75 mg. The zone of inhibition (ZOI) for both extracts ranges from 8 to 23 mm. The greatest ZOI for the aqueous leaf extract against *Streptococcus pyogenes* was 23 mm, whereas the ethanolic stem extract demonstrated a 21 mm ZOI against *Bacillus subtilis* (Gopalakrishnan and Udayakumar, 2014a). When compared to other solvent extracts, the antifungal activity of diethyl ether leaf extract demonstrated favorable results in all examined fungal strains. With the exception of 5 mg of methanolic extract, methanolic and aqueous leaf extracts displayed no inhibitory effect in all of the strains tested. Diethyl ether leaf extract (5 mg) demonstrated a maximal ZOI of up to 13 mm. A potent antifungal activity was observed in aqueous and diethyl ether stem extracts, showing a ZOI of 15 mm against *Aspergillus niger* The stem extract of diethyl ether also showed a maximum ZOI of 15 mm against *Trichoderma viride (*Gopalakrishnan and Udayakumar, 2014a). In another study of the antibacterial activity of *M. quadrifolia*, chloroform and ethyl acetate extracts showed a ZOI of 9 to 20 mm (Uma and Pravin, 2013).

17.3.2 ANTIOXIDANT ACTIVITY

Antioxidant activity of ethyl acetate, aqueous, and methanol extracts using the DPPH radical scavenging method exhibited potent activity in a concentration-dependent manner, with IC50 values of 10, 125, and 25 μg/ml, respectively, when compared with the standard BHT showing an IC50 value of 7.5 μg/ml (Ripa et al., 2009). The antioxidant activity of the plant extract using DPPH increased in a concentration-dependent manner with an IC50 value of 96.37 ± 3.62 μg/ml when compared with the standard, which had an IC50 of 16.59 ± 0.59 μg/ml. Total phenolic content was observed to be 165.75 ± 0.961 mg/g in GAE, and total antioxidant capacity was observed to be equivalent to ascorbic acid (224.90 ± 1.42 mg/g) (Zahan et al., 2011).

17.3.3 IN VITRO CYTOTOXIC ACTIVITY

The *in vitro* cytotoxic activity of the extracts of *M. quadrifolia* was studied using MCF-7 cells related to human breast cancer, revealing decreased cell viability and inhibition of cell growth in a dose-dependent manner. The IC50 value of the standard 5-fluorouracil, which is 9.3 μg/

ml, was compared to those of methanol, ethyl acetate, and aqueous extracts, which were 39.06, 47.82, and 187.5 μg/ml, respectively. This study indicated that ethyl acetate and methanol extracts of *M. quadrifolia* leaf can be used as potent medicinal drugs in the treatment of breast cancer (Ripa et al., 2009).

17.3.4 ANTI-STRESS ACTIVITY

The anti-stress activity of *M. quadrifolia* methanolic extract was carried out using physical stress models such as swimming endurance, post-swimming motor function test, toxic tolerance test, and restraint stress test in Swiss albino mice (18–25 g), by administering orally for 7 days at dosages of 200 and 400 mg/kg of body weight. In the anoxic tolerance test, *M. quadrifolia* ethanolic extracts significantly reduced the immobility time while increasing swimming endurance time, clonic convulsion timing, and post-motor function when compared with the control group, and behavioral and biochemical alterations in restraint stress were significantly reversed (Singh and Yadav, 2014).

17.3.5 HYPOGLYCEMIC ACTIVITY

Hypoglycemic activity was studied by inducing diabetes in rats using alloxan (110 mg/kg body weight). There was a considerable decrease in increased blood glucose levels in diabetic rats induced by alloxan. Plant extract (300 mg/kg) showed a 47.57% reduction in glucose levels in alloxan-induced rats, whereas it was a 44.38% reduction in metformin hydrochloride (reference drug @ 100 mg/kg) treated rats after 3 days.

17.3.6 ANTICONVULSANT ACTIVITY

The anticonvulsive potential of *M. quadrifolia* extracts was determined by adopting behavioral and electroencephalographic (EEG) analysis by inducing seizures in rats using pentylenetetrazole (PTZ) at a concentration of 60 mg/kg. The behavior and EEG were analyzed after 60 minutes of administration of *M. quadrifolia* extracts. Analyses were conducted on parameters such as EEG power, length of epileptic seizure, latency of seizure start, and seizure severity score. Both the water and ethanol extracts enhanced seizure latency, although the duration and seizure severity score of epileptic seizures decreased. EEG power analysis and recording both showed a decrease in seizure severity. It was demonstrated that the *M. quadrifolia* ethanol extract had superior effects over the water extract (Sahu et al., 2012).

17.3.7 ANTIANXIETY ACTIVITY

It was identified that *M. quadrifolia* was utilized by the locals and several traditional methods to treat anxiousness and related problems. To study the anti-anxiety activity, CNS effects were evaluated by taking Wistar rats and Swiss albino mice. Swimming and pole climbing tests were carried out with the plant extract at a dose of 500 mg/kg of body weight, which was able to reduce the spontaneous activity and exploratory behavioral patterns of the animals. This preliminary study indicates significant anti-anxiety activity of *M. quadrifolia* methanolic extract at a dosage of 500 mg/kg. This study supports the traditional medicinal usage of this plant against inflammation (Asham et al., 2013).

17.3.8 PSYCHOPHARMACOLOGICAL ACTIVITY

A hydroalcoholic extract of the entire *M. quadrifolia* plant was used to evaluate the psychopharmacological activities, such as behavior, exploratory behavior, muscle relaxant activity, and phenobarbitone-induced sleeping time. By using swimming and pole climbing tests, the plant extract demonstrated a significant reduction in spontaneous activity as well as a decrease in exploratory behavioral patterns, and traction testing revealed muscle relaxation (Reddy et al., 2011).

17.4 CONCLUSION

The results of GC-MS studies showed that the methanolic extract of *M. quadrifolia* possesses many pharmacologically important bioactive compounds. As *M. quadrifolia* is a traditionally important medicinal plant that is being used for many diseases, including diabetes, further studies can be carried out to isolate and identify the phytochemical compounds that can control these diseases, leading to drug development.

KEYWORDS

- **antimicrobial**
- **bioactive compounds**
- **drug production**
- **hypoglycemic activity**
- ***Marsilea quadrifolia***
- **pharmacology**
- **phenobarbitone-induced sleeping time**
- **synthetic pharmaceuticals**
- **triterpenoids**

REFERENCES

Asham, A., Tripathys, R., & Madhuri, Y. (2013). Preliminary screening of *Marsilea quadrifolia* extracts for their antianxiety potential. *International Journal of Pharmacy and Pharmaceutical Sciences, 5*(3), 137–139.

Gini, T. G., & Jothi, G. J. (2013). Preliminary phytochemical screening of whole plant extracts of *Peperomia pellucida* (Linn.) HBK (Piperaceae) and *Marsilea quadrifolia* Linn. (Marsileaceae). *International Journal of Pharmacognosy and Phytochemical Research, 5*(3), 200–214.

Gopalakrishnan, K., & Udayakumar, R. (2014a). Antimicrobial activity of *Marsilea quadrifolia* L. against some selected pathogenic microorganisms. *British Microbiology Research Journal, 4*(9), 1046–1056.

Gopalakrishnan, K., & Udayakumar, R. (2014b). GC-MS analysis of phytocompounds of leaf and stem of *Marsilea quadrifolia* (L.). *International Journal of Biochemistry Research & Review, 4*(6), 517–526. https://doi.org/10.9734/IJBCRR/2014/11350.

Gopalakrishnan, K., & Udayakumar, R. (2017). Phytochemical content of leaf and stem of *Marsilea quadrifolia* (L.). *Journal of Plant Science and Phytopathology, 1*, 26–37. https://doi.org/10.29328/journal.jpsp.1001003.

Kaufman, P. B., Cseke, L. J., Warber, S., Duke, J. A., & Brielmann, H. L. (1999). *Natural products from plants*. CRC Press.

Khare, C. P. (2004). *Encyclopedia of Indian Medicinal Plants: Rational Western Therapy, Ayurvedic and Other Traditional Usage*. Springer.

Marwat, S. K., Khan, M. A., Ahmad, M., Zafar, M., & Sultana, S. (2007). Aquatic plants of district Dera Ismail Khan, Pakistan. *Ethnobotanical Leaflets, 11*, 247–257.

Ramachandran, V. S. (2007). Wild edible plants of the Anamalais, Coimbatore district, Western Ghats, Tamil Nadu. *Indian Journal of Traditional Knowledge, 6*(1), 173–176.

Reddy, K. S., Reddy, C. S., & Ganapaty, S. (2011). Psychopharmacological studies of hydroalcoholic extract of whole plant of *Marsilea quadrifolia*. *Journal of Scientific Research, 4*(1), 279–285.

Ripa, F. A., Nahar, L., Haque, M., & Islam, M. M. (2009). Antibacterial, cytotoxic and antioxidant activity of crude extract of *Marsilea quadrifolia*. *European Journal of Scientific Research, 33*(1), 123–129.

Sahu, P. K. (2011). Plants used by Gond and Baiga women in ethnogynaecological disorders in Achanakmar Wildlife Sanctuary, Bilaspur, C.G. *International Journal of Pharmaceutical and Life Sciences, 2*(2), 559–561.

Sahu, S., Dutta, G., Mandal, N., Goswami, A. R., & Ghosh, T. (2012). Anticonvulsant effect of *Marsilea quadrifolia* Linn. on pentylenetetrazole-induced seizure: A behavioral and EEG study in rats. *Journal of Ethnopharmacology, 141*(1), 537–541.

Sajem, A. L., & Gosai, K. (2010). Ethnobotanical investigations among the Lushai tribes in North Cachar Hills district of Assam, Northeast India. *Indian Journal of Traditional Knowledge, 9*(1), 108–113.

Sen, S. K., & Behera, L. M. (2000). Ethnomedicinal plants used by the tribals of Bargarh district to cure diarrhea and dysentery. *Indian Journal of Traditional Knowledge, 7*(3), 425–428.

Singh, S., & Yadav, A. K. (2014). Protection of stress-induced behavioral and physiological alteration by *Marsilea quadrifolia* in rodents. *Journal of Chemical and Pharmaceutical Research, 6*(7), 2207–2217.

Sofowora, A. (1993). *Medicinal plants and traditional medicine in Africa* (p. 150). Spectrum Books.

Tyler, V. E. (1999). Phytomedicines: Back to the future. *Journal of Natural Products, 62*, 1589–1592.

Uma, R., & Pravin, B. (2013). *In vitro* cytotoxic activity of *Marsilea quadrifolia* Linn. of MCF-7 cells of human breast cancer. *International Research Journal of Medical Sciences, 1*(1), 10–13.

Zahan, R., Ripa, F. A., Alam, M. B., Haque, M. A., Mosaddik, M. A., & Nahar, L. (2011). Hypoglycemic and *in vitro* antioxidant effects of methanolic extract of *Marsilea quadrifolia* plant. *Pharmacognosy Journal, 3*(26), 86–92.

CHAPTER 18

Phytochemical and Pharmacological Activities of *Hemionitis arifolia* (Burm. f.) T. Moore

K. S. SHANTHI SREE,* A. SHOBHA RANI, M. SWETHA PRIYA, and G. SAVITHRI

Sri Padmavati Mahila Visvavidyalayam (Women's University), Tirupati, Andhra Pradesh, India

*Corresponding author

ABSTRACT

Hemionitis arifolia (Burm. f.) T. Moore is a terrestrial, rhizomatous fern belonging to the family Pteridaceae. It is distributed in India, Sri Lanka, Laos, and other countries of Southeast Asia. The fern has been evaluated for its antidiabetic, antibacterial, antifungal, antioxidant, and antistress activity. The presence of phytochemical constituents in *H. arifolia* supports its biological activity. The present review highlights the bioactive compounds and biological activities of *H. arifolia*.

18.1 INTRODUCTION

In India, we are fortunate to have a rich heritage of ancient medicinal systems that have been practiced for over 3000 years. One of the richest, most varied, and oldest cultural traditions is widely accepted in society and is well-known for using therapeutic herbs (Prashant et al., 2011). The use of plants and products derived from them as a source of medicine has become more popular in recent years. However, pteridophytes have been disregarded in pharmacogenetic and phytochemical research despite the

Bioactives and Pharmacology of Pteridophytes. T. Pullaiah (Ed.)

DOI: 10.1201/9781779642738-18

economic significance (Singh et al., 2010). Herbal medicine, sometimes referred to as having fewer side effects than synthetic medications, is more accessible, inexpensive, and culturally acceptable. Medicinal plants contain chemical compounds that can be used for treating illnesses or relieving pain.

Pteridophytes are important, although they have received little attention in phytochemical and antimicrobial research, despite the fact that several writers have examined their ethnobotanical value (Ahmad et al., 1998). Pteridophytes are among the earliest vascular plants and are commonly known as the "reptile group" of plants. Although they are less common than flowering plants, their uses are not well understood among indigenous groups. However, pteridophytes are essential to our planet's biodiversity. It is important to know about the potential of pteridophytes and evaluate their medicinal properties, which can help in the field of herbal medicine and further improve our knowledge of alternative remedies, in which *H. arifolia* is one of the ferns (Reddy et al., 2023).

Hemionitis arifolia is commonly called the heart fern and tongue fern. Synonyms of this species include *Hemionitis sagittata, Hemionitis cordata, Hemionitis cordifolia, Hemionitis toxotis, Gymnogramma arifolia, Asplenium arifolium*, and *Gymnogramma sagittata. H. arifolia* is a dwarf fern that looks attractive, and it is native to tropical Southeast Asia, India, Sri Lanka, Vietnam, Laos, and possibly China, Taiwan, and other nations in tropical areas. The fern exists as a terrestrial plant and as an epiphyte that grows on trees.

This plant is attractive with dark green cordate fronds, which grow on black stems; the fronds usually reach up to 5 to 7.5 cm in length. It is a small and shade-loving fern that grows well in low light intensity with high humidity. *H. arifolia* also grows in a shady area in an atrium or grows well as a terrarium plant. The fronds are leathery and heart-shaped, featuring tall and fertile frond leaves that can spread above the fern's foliage. The fern grows up to 15 to 20 cm in height.

18.2 BIOACTIVES

Bioactive compounds play a major role in maintaining human health. Previous reports show that green plants contain a wealth of efficient therapeutic agents that can be easily biodegradable, non-phytotoxic, and systematic. Pteridophytes have antimicrobial, antifungal, antiviral, anticancer, rheumatism, diabetic, inflammatory, fertility, diuretic, hepatoprotective, and sedative properties (Gayathiri et al., 2019). Preliminary phytochemical studies of *H. arifolia* methanolic extract possess various phytochemicals such as saponins, glycosides, phenols, flavonoids, tannins, carbohydrates, and steroids (Rakkimuthu et al., 2018; Reddy et al., 2023). These compounds often contribute to their medicinal properties and potential uses.

Gas chromatography mass spectrometry (GC-MS) analysis revealed the presence of nine compounds, and the phytoconstituents screened were benzamide, N-(4-chlorotetrahydro-3-thinyl)-3, 5-dimethoxy, sdioxide (31.52), Choestan-26-oic acid, 3,7,12-trihydroxy-, Androst-5-en-3-one, 19-acetoxy-4, 4-dimethyl-, oxine (26.14), retinoic acid, 5,8-epoxy-5,8-dihydro (24.80), pregnane-3,20-dione, 17,21-{(methylborylene)bis(oxy)}-, 5a- (23.76), N-(4-(chlorophenyl)isothiazol-5-yl)-2-methylpiperidin-2-imine (21.73), decanoic acid, 2-hexyl-, methyl ester (19.91), 4-Nitrobenzoic acid, 2,4,6-trichlorophenyl ester (18.70), and 4-(chloromethyl)-1-azabicyclo-(2.2.2)octane (18.26%). The nine compounds identified provide positive insights into anti-diabetic works, and a drug could be formulated in the future for diabetic conditions (Gayathiri et al., 2019).

Preliminary evaluation of phytochemicals showed that a range of extracts underwent qualitative chemical analysis. Petroleum ether extracts (PEE) included high concentrations of terpenoids and sterols. Low concentrations of terpenoids, sterols, and glycosides were detected in chloroform extract (CHE). High concentrations of phenolics, flavones, and flavonoids were found in ethyl acetate extract (EAE). Total ethanolic extract contains extremely low carbohydrates and a high concentration of phenolics, flavones, and flavonoids (TEE). Aqueous extracts (AQE) have minimal levels of glycosides, phenolics, flavones, and flavonoids, and carbohydrates, whereas large levels of saponins are present (Joy et al., 2017).

18.3 PHARMACOGNOSY

18.3.1 ANTIBACTERIAL ACTIVITY

A significant antimicrobial activity was observed in the vegetative and reproductive leaves of *H. arifolia* using agar diffusion and tube dilution assays against *E. coli* (MTCC-739) in both crude and alcoholic extracts.

Reproductive leaves also showed moderate antibacterial activity against *Bacillus subtilis* (MTCC-441) (Karmakar and Mukhopadhyay, 2011).

18.3.2 HYPOGLYCEMIC AND ANTIDIABETIC PROPERTIES

Antidiabetic evaluation of *H. arifolia* leaves was carried out using *in vitro* methods, which are caused by a combination of factors including glucose uptake facilitation in L-6 muscle cell lines and enzyme inhibitions such as DPP-IV and alpha-amylase inhibition. For humans and other animals to engage in their regular activities, the most crucial component needed is the ability of L-6 muscle cells to absorb glucose. This extract helps to facilitate it, and it's also harmless. As a result, this plant qualifies as an antidiabetic. Additional extract fractionation, component enrichment in each fraction, and targeted research will aid in the discovery of a better, more sustainable, non-toxic, and commercially feasible diabetic medication (Kumar et al., 2015).

18.3.3 ANTIOXIDANT ACTIVITY

The hydroalcoholic extract of *H. arifolia* revealed the presence of phenolic content of 31.78%, flavonoid content of 1.02%, and alkaloid content of 30.40% individually. Antioxidant activity of the ABTS assay, DPPH assay, reducing power assay, and nitric oxide assay reveals the inhibition concentrations of 667.75 μg/ml, 734.25 μg/ml, 791.58 μg/ml, and 899.67 μg/ml (Priya et al., 2021).

18.3.4 ANTI-INFLAMMATORY ACTIVITY

The *H. arifolia* aqueous extracts prevented heat-induced hemolysis of RBCs and demonstrated stabilization as a further mechanism of their anti-inflammatory action; this effect may also prevent neutrophil lysosomal content from being released at the site of inflammation (Antonysamy et al., 2017).

18.3.5 ANTI-CANCER ACTIVITY

The phytochemical properties of *H. arifolia* have been evaluated using mass spectroscopy and gas chromatography, which revealed the mass spectrum of seven bioactive constituents. The biologically active compounds were found to act as chemotherapeutic agents, pharmaceuticals, dyes, nematicides, anti-fungal, anti-bacterial, and anti-cancer agents, as well as compounds that cure fibrosis, obesity, and diabetes (Gayathiri et al., 2019).

18.4 CONCLUSION

The present review concludes that by using the information presented on the phytochemical compounds and various biological properties of the extracts, further studies on *Hemionitis arifolia* can be carried out to isolate bioactive compounds that can be useful in drug preparation.

KEYWORDS

- **anti-diabetic**
- **anti-fungal**
- **antioxidant**
- **aqueous extracts**
- **chemotherapeutic agents**
- ***Hemionitis arifolia***
- **phytochemical compounds**
- **terrestrial**

REFERENCES

Ahmad, I., Mehmood, Z., & Mohammad, F. (1998). Screening of some Indian medicinal plants for their antimicrobial properties. *Journal of Ethnopharmacology, 62*(2), 183–193.

Antonysamy, J. M. A., Ramakrishnan, P., Perumal, S., & Shibila, T. (2017). Anti-inflammatory activity of selected pteridophytes from Western Ghats. *International Journal of Complementary and Alternative Medicine, 9*, 307.

Gayathiri, M., Catharin, S. S., Sujatha, S., Ramya, R. I., & Gnana, D. R. R. (2019). Analysis of bioactive compounds of *Hemionitis arifolia* (Burm.) Moore: An anti-diabetic fern using gas chromatography and mass spectroscopy. *International Research Journal of Pharmacy, 10*(9), 231–235.

https://www.exoticrainforest.com/Hemionitis%20arifolia%20pc.html (accessed on 17th November 2025).

Joy, J., & Bindu, A. R. (2017). Pharmacognostic and physico-chemical investigation of *Hemionitis arifolia* (Burm.) Moore. *International Journal of Pharmacognosy and Phytochemical Research, 9*(8), 1074–1080. https://doi.org/10.25258/phyto.v9i08.9614 (accessed on 17th November 2025).

Karmakar, J., & Mukhopadhyay, S. K. (2011). Study of antimicrobial activity and root symbionts of *Hemionitis arifolia. Physiology and Molecular Biology of Plants, 17*, 199–202.

Kumar, K. R. K., Jayaprakash, A. P., & Srinivasan, K. K. (2015). Antidiabetic evaluation of *Hemionitis arifolia* leaves by *in vitro* methods. *Manipal Journal of Pharmaceutical Sciences, 1*(1), Article 4.

Prashant, T., Bimlesh, K., Mandeep, K., Gurpreet, K., & Harleen, K. (2011). Phytochemical screening and extraction: A review. *International Pharmaceutica Sciencia, 1*(1), 98–106.

Priya, R. R., Bhadusha, N., Manivannan, V., & Gunasekaran, T. (2021). *In-vitro* antioxidant activity and preliminary phytochemical analysis of different leaf extracts of *Hemionitis arifolia. Journal of Pharmaceutical Research International, 33*(59B), 281–291.

Rakkimuthu, R., Naveenraj, B., Benitta, L., & Dhanya, K. (2018). Phytochemical screening and its antibacterial activity of the fern *Hemionitis arifolia* (Burm.) Moore. *World Journal of Pharmaceutical Research, 7*(13), 603–609.

Reddy, M. C., Rajendra, M. P., Ravivarma, D., Reddy, B. V., Bhargavi, C. H., & Naveen, Y. (2023). *Hemionitis arifolia* (Burm. f.) T. Moore: A promising medicinal plant in the fight against cancer – A review. *Biological Forum – An International Journal, 15*(10), 1052–1055.

Singh, A. P., Rawat, V. K., Behera, S. K., & Khare, P. B. (2010). Perspectives of pteridophytes biodiversity: A source of economy elevation. In *National Conference on Biodiversity, Development and Poverty Alleviation* (pp. 46–49).

CHAPTER 19

A Comprehensive Review of the Chemical Principles and Biological Properties of *Equisetum ramosissimum* Desf.

ALEKHYA VEERAMANENI,* THIYAGARAJAN DEEPAN, and V. N. KOTESWARARAO NERELLA

Department of Pharmacy, GIET School of Pharmacy, Chaitanya Knowledge City, Rajahmundry, Andhra Pradesh, India

*Corresponding authors

ABSTRACT

Equisetum ramosissimum Desf. of family Equisetaceae, is commonly called as branched horse tail. The plant is extensively distributed throughout South Asia, Africa, Australia, and the United States. The plant's underground creeping rhizome allows it to thrive specifically along sandy riverbanks in evergreen forests. Traditionally, the plant's decoction made from rhizomes is used to treat infertility, and paste made from branches is used to treat bone fractures. Aerial parts are used to treat several diseases and are widely used by tribes and rural people. In folklore medicine, *Equisetum ramosissimum* is used for curing tuberculosis, bladder and kidney disorders, catarrh, menstruation, gastric hemorrhages, CNS issues, diuretics, anticancer treatments, cytotoxicity, hepatoprotection, fractures, and so forth. The plant is reported to contain phytoconstituents like flavonoids, alkaloids, phenolic compounds, triterpenoids, saponins, and phytosterols.

Bioactives and Pharmacology of Pteridophytes. T. Pullaiah (Ed.)

DOI: 10.1201/9781779642738-19

19.1 INTRODUCTION

Equisetum ramosissimum Desf. of the Equisetaceae family is a perennial fern commonly known by vernacular names such as African horsetail, Bewarasiegras, and Ishobalehashi. Moist soils are best for the growth of these plants. Usually, the stems are vertical, and branching is horizontal in nature. Aerial stems can grow up to one meter and are hollow, standing upright above the ground. The branches become thinner as they grow. The name "branched horsetail" is derived from the whorls at each stem node. Above every node along the aerial stem, a whorled sheath of scale-like leaves forms. These leaves have one vein, are narrowly lanceolate, and frequently have a glossy black color (Roux, 2003). The plant is extensively distributed throughout Asia, India, China, and South Africa, mostly in terrestrial regions beside streams or perennial rivers (Roux, 2003).

The plant is utilized in South Africa to treat a variety of illnesses, including infertility, broken or misplaced bones, and has diuretic, hemostatic, antiviral, and fungal characteristics. The sap is used to treat toothaches and wounds. According to a survey conducted among Malayali tribal people in the Kolkhills, *E. ramosissimum* is used to treat urinary-related diseases, like UTIs and kidney problems. Likewise, the plant has other traditional values, such as treating inflammation, rheumatism, fever, and blood pressure (Altundag et al., 2011; Sureshkumar et al., 2018).

19.2 BIOACTIVES

Equisetum ramosissimum is known to contain a number of significant phytochemical constituents that are medicinally active. Early phytochemical analyses of various extracts using solvents such as petroleum ether, ethyl acetate, chloroform, methanol, and aqueous solutions revealed the presence of flavonoids, alkaloids, phenolic compounds, steroids, triterpenoids, and tannins (Sureshkumar et al., 2021). 24 compounds were identified by Sureshkumar et al. (2021) after using GC-MS to analyze the phytochemicals of *E. ramosissimum* stem extracts. These compounds included palmitic acid, hexacosane, n-tridecane, methyl linoleate, methyl palmitate, linoleic acid, octacosane, and neophytadiene.

The following phytoconstituents have been identified in the whole plant extract of *E. ramosissimum*: loliolide, cycloart-24, epoxy-b-ionone-3-O-b-D-glucopyranoside's 5a and 6a, cycloart-22(23)-ene-3b-ol, ergost-6,22, friedelinol, apigenin, genkwanin, 5a, 8a-triol, -ene-3b-ol-diene-3b, adenine, b-sitosterol, and b-daucosterol; apigenin-5-O-b-D-glucopyranoside; luteolin; quercetin-3-O-b-D-glucopyranoside; kaempferol-3-b-D-glucopyranoside; kaempferol-3-O-b-D-glucose-7-O-b-D-glucopyranoside (Wang and Jia, 2005).

E. ramosissimum aerial parts were hydro distilled and analyzed through GC-MS, and 37 compounds were detected, showing the presence of high contents of monoterpenoids and sesquiterpenoids. The two major compounds reported are α-bisabolol oxide A and cumin aldehyde. Other important components reported are carvacrol, thymol, carvenone, linalool, cumin aldehyde, methyl eugenol, γ-curcumene, α-humulene, hexadecenoic acid, and so forth. (Alebous et al., 2016).

Aerial parts of *E. ramosissimum* ethanolic and methanolic extracts were studied for UHPLC analysis, and the results demonstrated the presence of phytoconstituents in methanolic extracts, including 7-Glu chrysoeriol, succinic acid, 7-methoxy-4'-hydroxyflavone, myristic acid, linoelaidic acid, isoferulic acid, iso-orientin, and rutin (Abdullah et al., 2024).

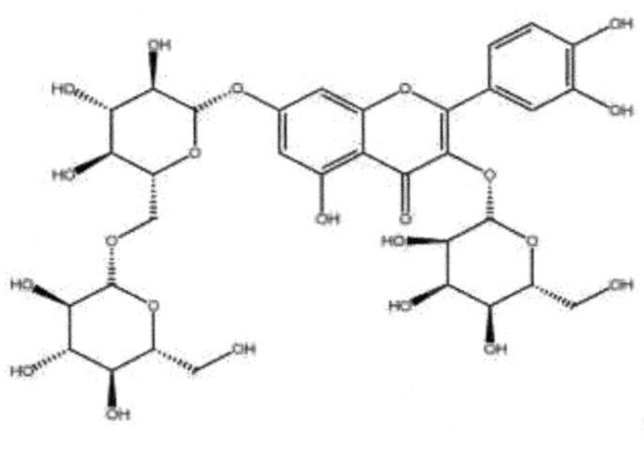

3-Glu-7-Rha Quercetin

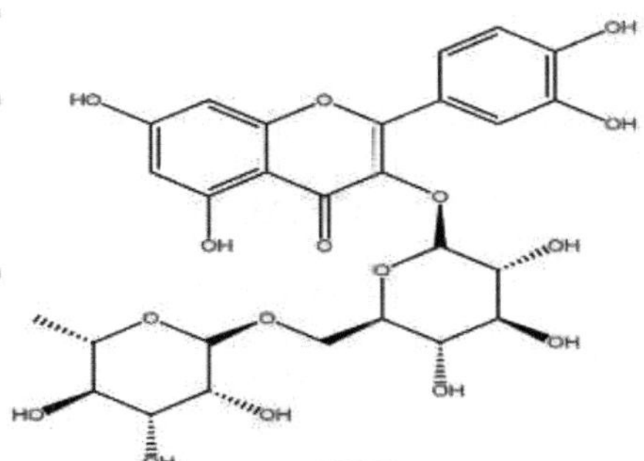

Rutin

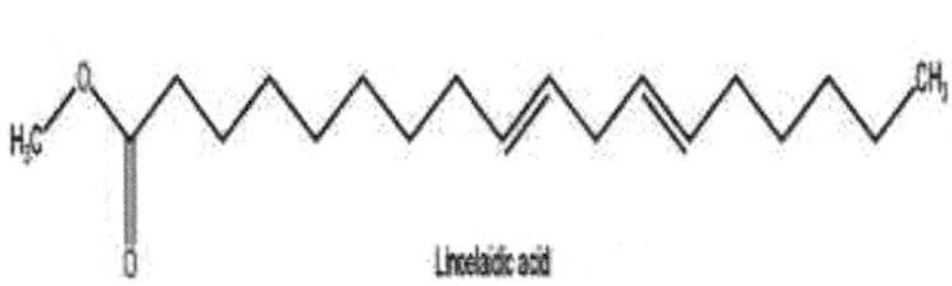

Linoelaidic acid

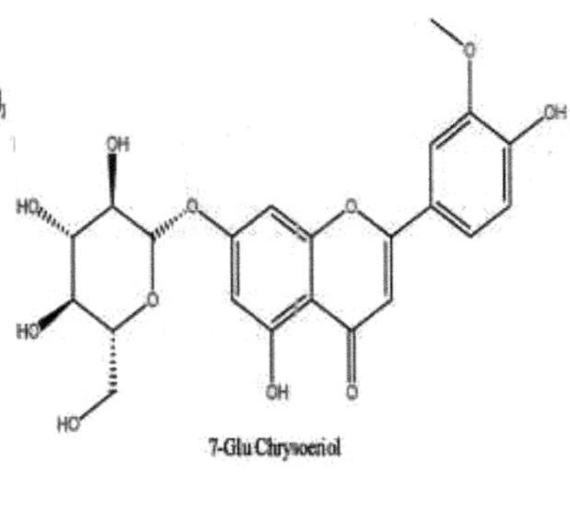

7-Glu Chrysoeriol

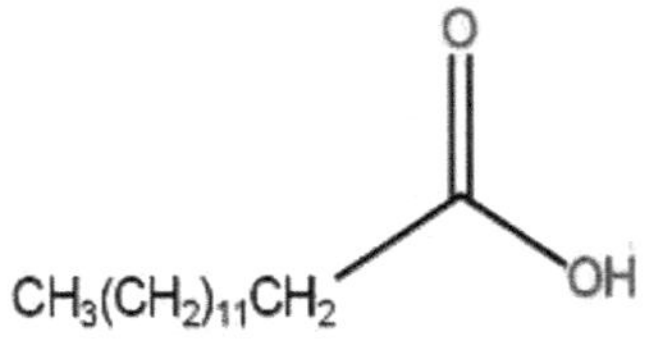

Myristic acid

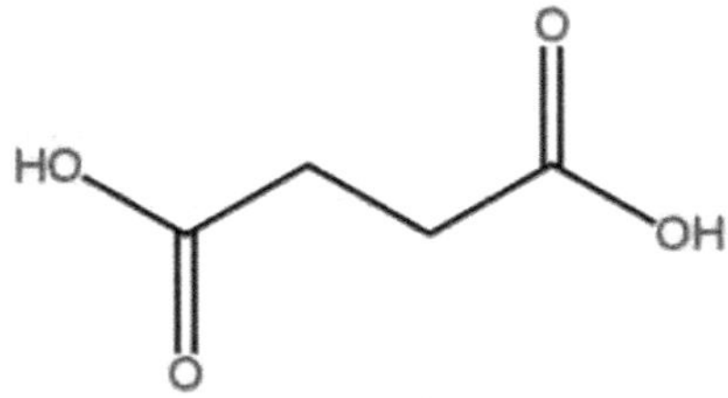

Succinic acid

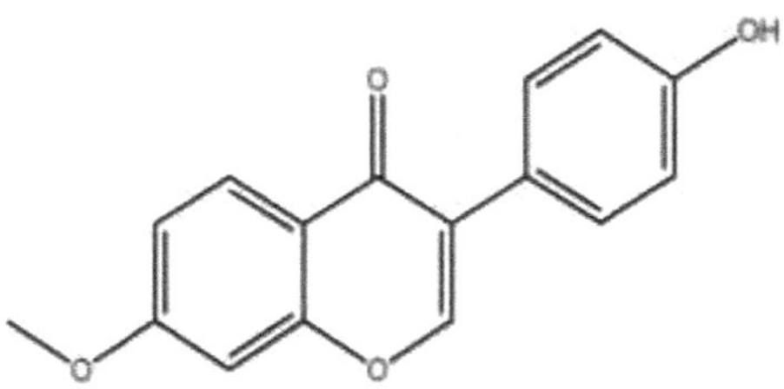

7-Methoxy-4'-hydroxyflavone

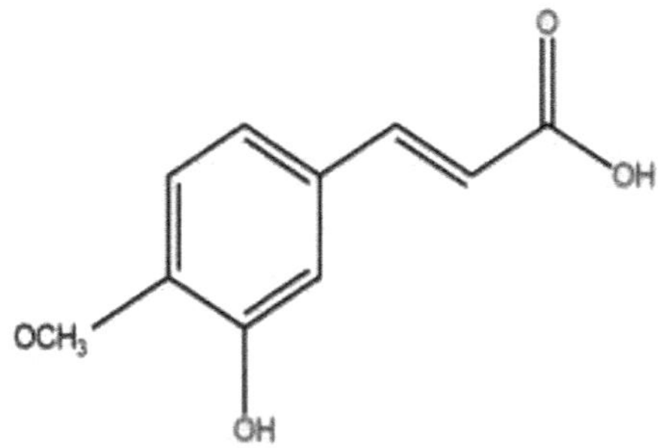

3-Hydroxy-4-methoxycinnamic acid
(Iso ferulic acid)

19.3 BIOLOGICAL ACTIVITIES

19.3.1 ANTIOXIDANT ACTIVITY

The antioxidant activity of *E. ramosissimum* aerial parts phosphate buffer extracts was assessed by measuring the activities of the antioxidant enzymes SOD (superoxide dismutase), GSH-Px (glutathione peroxidase), GPx (guaiacol peroxidase), and CAT (catalase). The FRAP assay was used to estimate the total antioxidant ability in plasma. These results indicated that catalase activity showed high levels of antioxidant activity and low levels of FRAP units (Stajner et al., 2009). In a different study, the antioxidant activity of a whole plant methanolic extract was investigated using diphenyl picrylhydrazyl (DPPH) and 2,2'-azinobis-3-ethylbenzothiazoline-6-sulfonic acid (ABTS) radical scavenging assays. The outcomes showed that antioxidants were abundant in *E. ramosissimum* (Paulsamy et al., 2013).

19.3.2 ANTI-INFLAMMATORY AND ANALGESIC ACTIVITY

The effects of methanolic and butanolic extracts of *E. ramosissimum* on inflammation, oxidative stress, and pain were investigated. Acetic acid in mice and the hot plate test in rat models were used to analyze the extracts' activities. Rat paw and ear edema were assessed for anti-inflammatory properties; radical scavenging activity in 2,2-diphenyl-1-picrylhydrazyl radical (DPPH), decreasing power; and assays for the bleaching of β-carotene/linoleic acid for antioxidant properties. The significant analgesic and anti-inflammatory effects were demonstrated by the results (Sissi et al., 2023).

19.3.3 ANTICANCER ACTIVITY

Calderon-Montano et al. (2011) revealed that the consumption of medicinal plants, particularly *Equisetum* species, offers a beneficial effect in lowering the chance of contracting long-term illnesses like cancer. This is attributed to the presence of essential phytochemical constituents like kaempferol and kaempferol glycosides in these plants, indicating their potential health benefits. Research on kaempferol has shown a wide range of pharmacological activities, including antioxidant, anti-inflammatory, antimicrobial, anticancer, neuroprotective, antidiabetic, cardioprotective, and antiallergic properties. It also exhibits molecular anticancer activity by preventing the growth and angiogenesis of malignant cells (Chen and Chen, 2013).

19.3.4 NEPHROPROTECTIVE EFFECTS

The effects of the ethanolic extract on kidney function parameters, such as uric acid, creatinine, urea, and fasting blood glucose in rats with diabetic nephropathy induced by streptozotocin, as well as investigations conducted on the liver's histopathological changes. Ethanol extract from *E. ramosissimum* was used in diabetic rats given STZ fasting blood glucose, urea, and creatinine levels. Extracting ethanol from *E. ramosissimum* has demonstrated potential in reducing some of the detrimental effects of diabetes on renal function (Abdullah et al., 2024).

19.3.5 URINARY TRACT INFECTIONS

The effects of extracts of *Equisetum ramosissimum* (aqueous, methanol, ethyl acetate, chloroform, and petroleum ether) against specific pathogenic bacteria, such as Methicillin-resistant *Staphylococcus aureus* (B23), *Bacillus subtilis* (MTCC441), Gram-negative strains such as *Salmonella typhi* (MTCC3224), *Vibrio cholerae* (MTTC3904), *Escherichia coli* (MTCC40), *Staphylococcus aureus* (MTTC2558), *Pseudomonas aeruginosa* (MTCC1748), *Klebsiella oxytoca* (B847), and *Citrobacter freundii* (B15) were investigated for urinary-related disorders. The outcomes demonstrated that uropathogenic bacterial growth was effectively inhibited by petroleum ether, methanolic, and aqueous extracts (Sureshkumar et al., 2021).

19.3.6 ANTI-HYPERLIPIDEMIC EFFECT

Equisetum ramosissimum ethanolic leaf extract is tested for antihyperlipidemic effects on male Wistar rats with high fat-induced hyperlipidemia, with or without atorvastatin. This study supports that the use of this combination helps in improving the lipid profile and parameters like triglycerides and total cholesterol. These findings are supported by previous studies, as *E. ramosissimum* extract possesses antioxidant activity due to its marked phenolic and flavonoid contents (Al-Bayati et al., 2023).

KEYWORDS

- **alkaloids**
- **anticancer**
- **cumin aldehyde**
- **diuretics**
- ***Equisetum ramosissimum***
- **flavonoids**
- **hepatoprotective**
- **hyperlipidemia**
- **phenolic compounds**
- **streptozotocin**

REFERENCES

Abdullah, R. K., Issa, R. A., Abu-Samak, M., Mohammad, B. A., Abbas, M. A., & Awwad, S. H. (2024). Nephroprotective effects of *Equisetum ramosissimum* L. extract in streptozotocin-induced diabetic rats. *Pharmacia, 71*, 1–11.

Al-Bayati, M., Issa, R., Abu-Samak, M., Alnsour, L., & Awwad, S. (2023). Phytochemical analysis and evaluation of anti-hyperlipidaemic effect for ethanolic leaf extract of *Equisetum ramosissimum* L.: *In vivo* study on rats' models. *Pharmacia, 70*(3), 557–568.

Alebous, H., Huaib, M., Hudeb, A., Sober, S., Gray, R., & Johnson, M. D. (2016). Chemical composition of essential oil from *Equisetum ramosissimum. European Journal of Medicinal Plants, 13*(2), 1–5.

Altundag, E., & Ozturk, B. (2011). Ethnomedicinal studies on the plant resources of East Anatolia, Turkey. *Procedia – Social and Behavioral Sciences, 19*, 756–777.

Calderon-Montano, J. M., Burgos-Morón, E., Pérez-Guerrero, C., & López-Lázaro, M. (2011). A review on the dietary flavonoid kaempferol. *Mini Reviews in Medicinal Chemistry, 11*(4), 298–344.

Chen, A. Y., & Chen, Y. C. (2013). A review of the dietary flavonoid, kaempferol on human health and cancer chemoprevention. *Food Chemistry, 138*(4), 2099–2107.

Paulsamy, S., Moorthy, D., Nandakumar, K., & Saradha, M. (2013). Evaluation of *in vitro* antioxidant potential of methanolic extracts of the ferns, *Actiniopteris radiata* (Sw.) Link and *Equisetum ramosissimum* Desf. *International Journal of Pharmaceutical Sciences and Research, 2*(3), 451–455.

Roux, J. P. (2003). *Swaziland Ferns and Fern Allies*. Southern African Botanical Diversity Network Report No. 19. SABONET, Pretoria.

Sissi, S., Dra, L. A., Ouhaddou, S., Ahmed, O., Larshini, M., & Markouk, M. (2023). *In vitro* antioxidant potential and *in vivo* analgesic and anti-inflammatory activities of Moroccan *Equisetum ramosissimum* source. *Natural Products Journal, 13*(3), 48–59.

Stajner, D., Popovic, B. M., Canadanovic-Brunet, J., & Anackov, G. (2009). Exploring *Equisetum arvense* L., *Equisetum ramosissimum* L. and *Equisetum telmateia* L. as sources of natural antioxidants. *Phytotherapy Research, 23*(4), 546–550.

Sureshkumar, J., Amalraj, S., & Murugan, R. (2021). Chemical profiling and antioxidant activity of *Equisetum ramosissimum* Desf. stem extract, a potential traditional medicinal plant for urinary tract infections. *Future Journal of Pharmaceutical Sciences, 7*, 192.

Sureshkumar, J., Silambarasan, R., Bharati, K. A., Krupa, J., Amalraj, S., & Ayyanar, M. (2018). A review on ethnomedicinally important pteridophytes of India. *Journal of Ethnopharmacology, 219*, 269–287.

Wang, X., & Jia, Z. (2005). Chemical constituents of *Equisetum ramosissimum. Acta Botanica Boreali-Occidentalia Sinica, 25*(12), 2524–2528.

CHAPTER 20

Phytochemistry and Bioactive Potential of *Glaphyropteridopsis erubescens* (Wall. ex Hook.) Ching

MITTA RAGHAVENDRA,[1,*] V. V. RAJESHAM,[2] SIBBALA SUBRAMANYAM,[1] SURYA PRABHA MATANGI,[1] and SHARABU RAVICHANDRA[3]

[1]*Department of Pharmaceutical Sciences, School of Biotechnology and Pharmaceutical Sciences, Vignan's Foundation for Science, Technology and Research, Guntur, Andhra Pradesh, India*

[2]*Department of Pharmacology, CMR College of Pharmacy, Hyderabad, Telangana, India*

[3]*Department of Pharmaceutical Chemistry, Chebrolu Hanumaiah Institute of Pharmaceutical Sciences, Guntur, Andhra Pradesh, India*

*Corresponding author

ABSTRACT

Glaphyropteridopsis erubescens (Wall. ex Hook) Ching is a well-known plant belonging to the family Thelypteridaceae. It is often used as a traditional herbal medicine in Asia. *Glaphyropteridopsis erubescens* has the efficacy of removing toxic substances from the body, improving digestion, arresting bleeding, inhibiting pain, suppressing cough, lowering blood pressure, reducing blood sugar, and so on. Several sulfated heteropolysaccharides have been isolated from *Glaphyropteridopsis erubescens*. They have some pharmacological activities, such as antioxidant activity

Bioactives and Pharmacology of Pteridophytes. T. Pullaiah (Ed.)

DOI: 10.1201/9781779642738-20

and antitumor activity. In addition, saponins and flavonoids, including erubosides, are isolated from *Glaphyropteridopsis erubescens*, and there are pharmacology reports. However, the study of the *Glaphyropteridopsis erubescens* plant is limited. This review has summarized several kinds of active substances in *Glaphyropteridopsis erubescens* and their pharmacological activities.

20.1 INTRODUCTION

Since prehistoric times, all human civilizations have utilized plants as a natural source of food, fuel, shelter, and medicine. Plants were the source of prophylaxis and treatment for a variety of illnesses and disorders prior to the development of iatrochemistry in the 16th century (Hardy, 2021). Plant-based innovative medication therapy is increasingly relying on Pteridophytes (aka, ferns and fern allies) due to their well-known medicinal and therapeutic properties. Numerous species within this plant division are often disregarded, despite evidence suggesting they possess a wide range of potent secondary metabolites with antimicrobial properties (Baskaran et al., 2018).

Pteridophytes have a variety of secondary metabolites, which make them useful for treating a wide range of illnesses when taken in either direct or extracted form. Ferns are used as vital food and medicine to prevent illness and maintain the health of humans and animals. They also play a significant role in folklore treatment (Bandyopadhyay and Dey, 2022). *Glaphyropteridopsis erubescens* (Wall. ex Hook.) Ching, also known as *Thelypteris erubescens* (Wall. ex Hook.) Ching, a type of Angelica, is a perennial plant that belongs to the Thelypteridaceae family and is mainly scattered in Nepal, Pakistan, India, Bhutan, Burma, the Philippines, Japan, Vietnam, and China (Xing et al., 1999).

Glaphyropteridopsis erubescens mainly grows in forests at 2200–3500 m altitude and is typically found in shady, moist, and steep environments near water. The plants of this species grow to heights exceeding 2–3 meters, supported by robust, woody rhizomes that are smooth and devoid of hairs (glabrous). The fronds are clustered together, characterized by sturdy stipes that can reach thicknesses of 1–2 meters. These stipes are ribbed, smooth, and typically have a straw-like coloration with occasional reddish hues. The laminae of this fern species are large, measuring between 100 to 200 cm in length and 25 to 50 cm in width. They are either pinnatifid or subbipinnate in structure, characterized by acuminate tips and pinnatifid divisions at their apices. The pinnae are arranged opposite each other along the rachis and are sessile. The 50 pairs of segments are arranged pectinately, spreading out in a linear-lanceolate manner, and are slightly falcate in shape. The sori are rather large, placed nearer the mid-vein than the margin, and are rarely produced towards the apex of the segments. The venation is peculiar, and the disposition of the sori depends mainly on the peculiarities of the venation.

The root and rhizomes of *Glaphyropteridopsis erubescens* have been widely used as a traditional Chinese medicine that has been effective for the treatment of syphilis, neuropathic pain, and other related diseases. Therefore, *Glaphyropteridopsis erubescens* is an herb with high medicinal value and commercial prospects. Phytochemical research has revealed that flavonoids are the dominant phytochemical constituents of *Glaphyropteridopsis erubescens*, which also contains other types of compounds, such as ferulic acid and glycosides. In addition, these bioactive compounds have the activities of anti-inflammation, anti-oxidation, antivirus, and neuronal protection effects on nervous system injury repair at the molecular and cellular levels.

According to Nwosu (2002), the *Glaphyropteridopsis erubescens* fern has remarkable therapeutic potential since, in Nigeria, scorpion bites can be treated with root powder, leaf paste can be used externally for rheumatism, and leaf decoction can be used to cure dyspepsia.

20.2 BIOACTIVES

The major bioactive compounds present in the *Glaphyropteridopsis erubescens* are flavan-4-ol glucosides, 3-deoxyanthocyanins, and flavones with modified B-ring, flavan-4-ol glucosides, euberins A-C, afzelin, abacopterin I, kaempferol, kaempferol 3-O-α-L-rhamnoside, quercetin, astragalin, quercetin 3-O-β-D-glucopyranoside, quercitrin, flavokawain B, vicianin, vanillic acid, p-hydroxybenzaldehyde, β-sitosterol arachidate, β-sitosterol, Daucosterol, 24-methylenecycloartanyl stearate, dodecanoic acid 2, 3-dihydroxypropyl ester (Jiang et al., 2013; Kashyap et al., 2017).

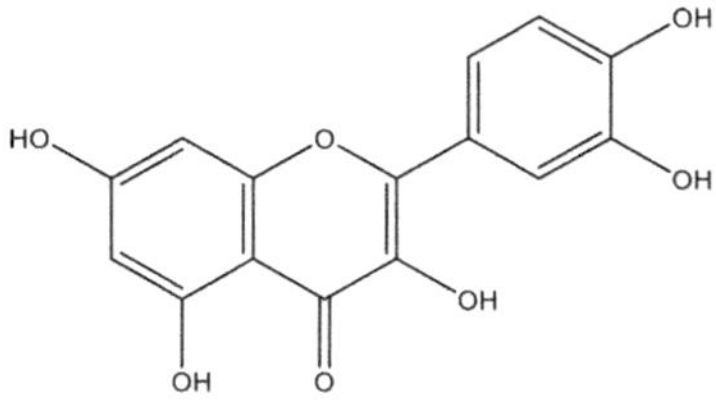

Quercetin

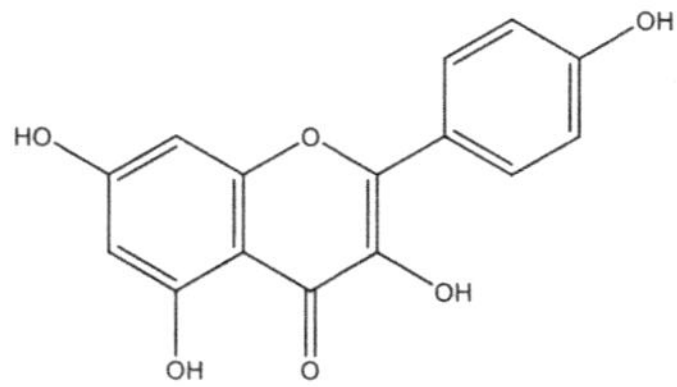

Kaempferol

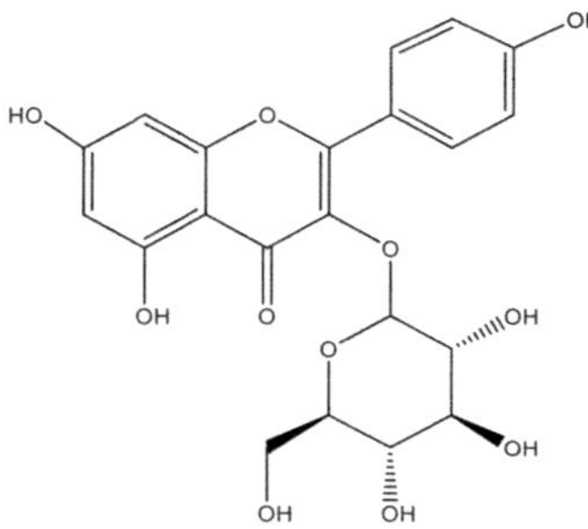

Astragalin

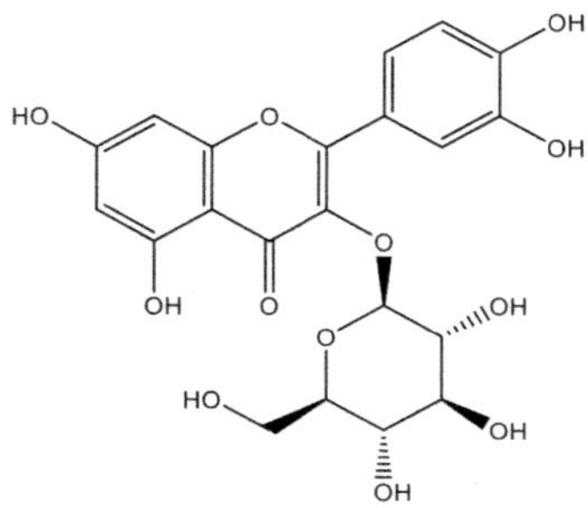

Quercetin 3-O-b-D-glucopyranoside

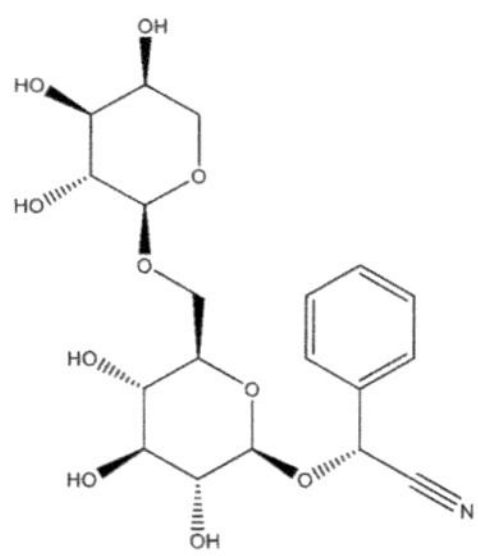

Vicianin

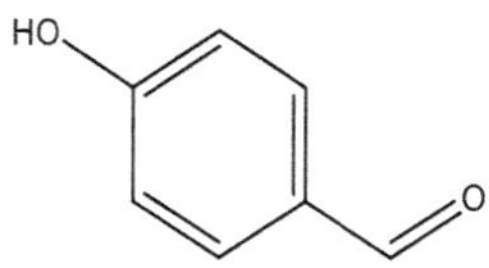

p-hydroxybenzaldehyde

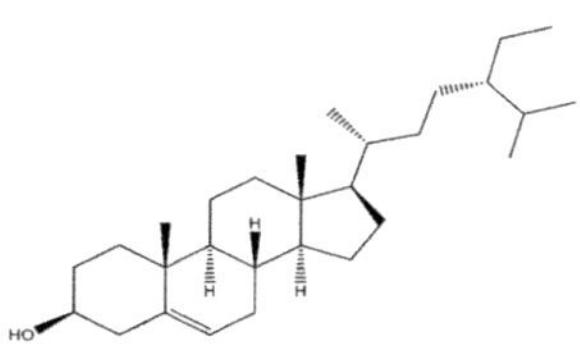

Sitosterol

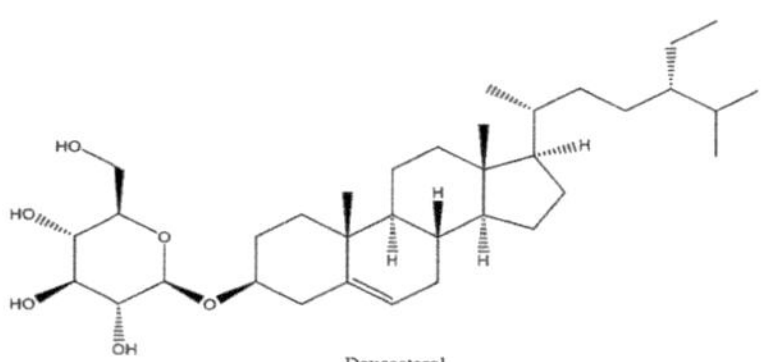

Daucosterol

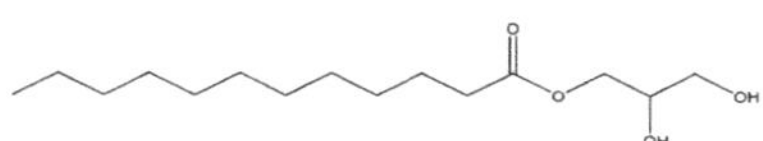

Dodecanoic acid 2, 3- dihydroxypropyl ester

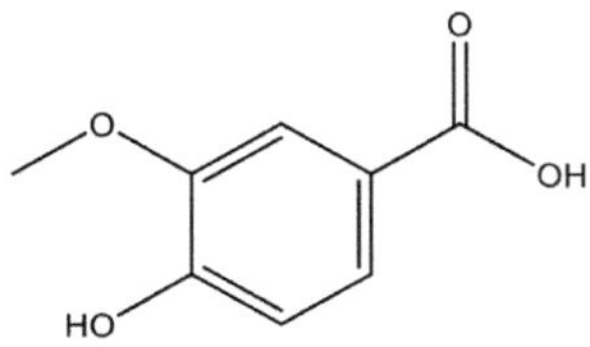

Vanillic acid

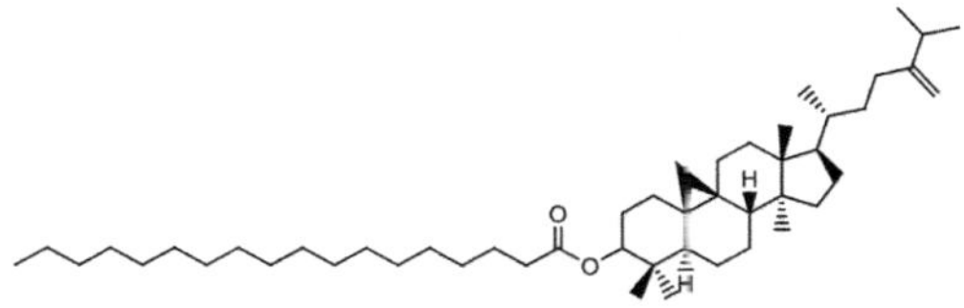

24-methylenecycloartanyl stearate

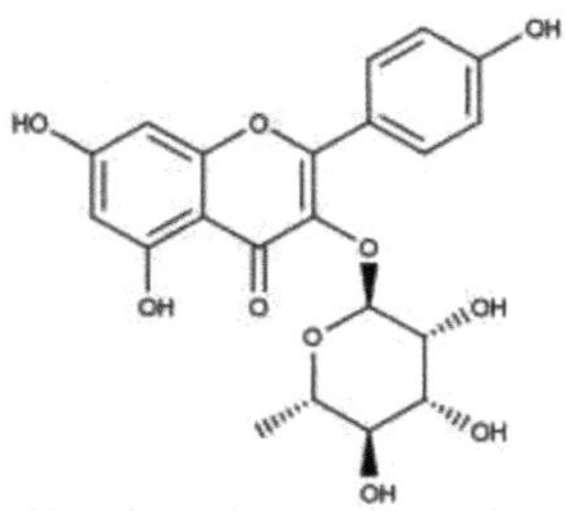

Kaempferol 3-O-a-L-rhamnoside

20.3 PHARMACOLOGY

20.3.1 ANTI-GONORRHEA ACTIVITY

Patra and Das (2021) said Pteridophytes have been more significant in plant-based innovative medicine therapy due to their recognized medicinal and therapeutic qualities. Despite being often disregarded, a number of species within this plant division have been found to possess a diverse range of possible secondary metabolites that have protective activity against various diseases. *Glaphyropteridopsis erubescens* was one of the neglected plants in the Pteridophytes category. The decoction of *Glaphyropteridopsis erubescens* young fronds was traditionally used in the treatment of gonorrhea, particularly leucorrhea. However, there was no scientific evidence to support the treatment of gonorrhea. Gaur and Bhatt (1994) reported that the people of the Deoprayag area in the Garhwal Himalayas, India, have traditionally used *Glaphyropteridopsis erubescens* rhizome powder combined with rice water as a treatment for gonorrhea, particularly leucorrhea.

20.3.2 ANTIMICROBIAL ACTIVITY

Arulanandam and Britto (2010) reported the preliminary phytochemical screening and antimicrobial activity of various extracts of *Glaphyropteridopsis erubescens* fronds with spores. The active phytochemical ingredients found in *Glaphyropteridopsis erubescens* extracts prepared in hexane, ethyl acetate, and methanol include tannins, reducing sugars, phenolic compounds, coumarins, steroids, and saponins. The antibacterial activity of selected extracts of *Glaphyropteridopsis erubescens* was tested against *Xanthomonas oryzae, Salmonella typhi-B, Enterococcus faecalis, Candida albicans, Staphylococcus aureus, Bacillus subtilis, Erwinia amylovora, Proteus vulgaris, Escherichia coli*, and *Klebsiella pneumoniae*. Among the pathogens tested, the growth of the gram-positive bacterium *Staphylococcus aureus* was inhibited in the methanol extract with a 20 mm inhibition zone, followed by the ethyl acetate extract with an 18 mm inhibition zone, whereas the hexane extract showed no activity against the same pathogen. The ethyl acetate extract showed significant activity against *Enterococcus faecalis* with a 14 mm inhibition zone, but the hexane and methanol extracts of *Glaphyropteridopsis erubescens* did not have any activity against the same pathogen.

20.3.3 ANTICANCER ACTIVITY

Kashyap et al. (2017) said that kaempferol derivatives, including kaempferol-3-O-α-L-rhamnoside, which are implicated in multiple signaling pathways, can be naturally obtained from *Glaphyropteridopsis erubescens*. Numerous signaling pathways that are triggered in a wide range of unfavorable clinico-pathological situations may be considerably modulated by kaempferol.

KEYWORDS

- **antidote**
- **anti-gonorrhea activity**
- **anti-inflammation**
- **bioactive compounds**
- ***Glaphyropteridopsis erubescens***
- **phytochemical screening**
- **pteridophytes**
- **sulfated heteropolysaccharides**

REFERENCES

Arulanandam, J. P., & John De Britto, A. (2010). Phytochemical screening and antimicrobial property of *Glaphyropteridopsis erubescens* (Hook.) Ching, Western Ghats, South India. *Pharmacologyonline, 2*, 91–96.

Bandyopadhyay, A., & Dey, A. (2022). Medicinal pteridophytes: Ethnopharmacological, phytochemical, and clinical attributes. *Beni-Suef University Journal of Basic and Applied Sciences, 11*(1), 113.

Baskaran, X. R., Geo Vigila, A. V., Zhang, S. Z., Feng, S. X., & Liao, W. B. (2018). A review of the use of pteridophytes for treating human ailments. *Journal of Zhejiang University Science B, 19*(2), 85–119.

Das, D., & Patra, B. (2021). Traditional pteridophytic herbal medicines and reproductive health disorders in women – A review. *The Journal of Phytopharmacology, 10*(6), 490–495.

Gaur, R. D., & Bhatt, B. P. (1994). Folk utilization of some pteridophytes of Deoprayag area in Garhwal Himalaya: India. *Economic Botany, 48*, 146–151.

Hardy, K. (2021). Paleomedicine and the evolutionary context of medicinal plant use. *Revista Brasileira de Farmacognosia, 31*(1), 1–5.

Jiang, J., Tian, L., Wang, L., Liu, Y., & Chen, Y. (2013). Phenolic compounds from the fern *Glaphyropteridopsis erubescens* (Hook.) Ching. *Biochemical Systematics and Ecology, 50*, 136–138.

Kashyap, D., Sharma, A., Tuli, H. S., Sak, K., Punia, S., & Mukherjee, T. K. (2017). Kaempferol – A dietary anticancer molecule with multiple mechanisms of action: Recent trends and advancements. *Journal of Functional Foods, 30*, 203–219.

Nwosu, M. O. (2002). Ethnobotanical studies on some pteridophytes of Southern Nigeria. *Economic Botany, 56*, 255–259.

Xing, G. X., Lin, Y. X., Qiu, P. X., & Yao, G. H. (1999). *Flora Reipublicae Popularis Sinicae* (Vol. 4(1), p. 127). Science Press, Beijing.

CHAPTER 21

Nephrolepis exaltata (L.) Schott: A Multipurpose Ornamental Pteridophyte

PRASHANT KARADAKATTI,[1] H. N. KARTHIK,[2] SIDDAPPA B. KAKKALAMELI,[1] DIVAKAR K. MESTA,[3] and VINAYAK UPADHYA[2,*]

[1]*Department of Studies in Botany, Davangere University, Shivagangotri, Davangere, Karnataka, India*

[2]*Department of Forest Products and Utilization, College of Forestry, University of Agricultural Sciences, Dharwad, Sirsi, Karnataka, India*

[3]*Department of Botany, Carmel College of Arts, Science and Commerce for Women, Nuvem, Goa, India*

*Corresponding author

ABSTRACT

Nephrolepis exaltata is a pteridophyte commonly known as Sword fern or Boston fern. It belongs to the family Nephrolepidaceae. This rhizomatic plant is traditionally used to treat various disorders by traditional practitioners. Scientific studies indicate that the plant consists of several potential chemical compounds, viz., dehydrovomifoliol, quercitrin, kaempferol, phytol, α-ionone, β-ionone, α-sitosterol, and so forth. The presence of bioactive compounds in various parts of the plant is responsible for biological activity such as antimicrobial activity against hazardous or pathogenic microorganisms, anticancer activity, and cytotoxic activity. The species is proven to be a good agent in the preparation of herbal face masks, which reduce the effects of air pollution and protect respiratory function. This makes *N. exaltata* a potential plant with essential components in pharmacological and industrial fields.

Bioactives and Pharmacology of Pteridophytes. T. Pullaiah (Ed.)

DOI: 10.1201/9781779642738-21

21.1 INTRODUCTION

Pteridophytes are considered vascular cryptogams existing on this earth from the Silurian period of the late Paleozoic era of GTS (Kumar, 2024). Because of their diverse habit and habitat, they have attracted human attention and found their use in the day-to-day life of human beings. A number of pteridophytic species are economically used directly or indirectly for a variety of purposes, mainly as ornamentals, medicine, biofertilizers, food, and so forth. (Bajracharya and Bajracharya, 2022; Kholia and Balkrishna, 2022; Kumar, 2024). Both codified and non-codified systems of traditional medicine use pteridophytes as medicine from ancient times. Literature indicates the use of ferns in Ayurveda as medicine or as an ingredient in medicinal formulations for the treatment of human illness (Goswami, 2009; Pathak et al., 2017). The genus *Nephrolepis* comprises about 30 species known for their ornamental appeal and medicinal properties. *Nephrolepis exaltata* (L.) Schott (family Nephrolepidaceae) is one among the genus *Nephrolepis*. *Nephrolepis exaltata* is a perennial herbaceous fern native to the South American continent. It is commonly known as Sword fern or wild Boston fern and usually grows as epiphytic or terrestrial in forested to open habitats (Bárrios and Copeland, 2021). This fern plant is known for its luxuriant lush green sword-shaped leaves with curled leaflets. Fronds are sterile, short, erect, and 1 – 1.7 mm wide; indusia are reniform to horseshoe-shaped (Anonymous, 2024). Because of its impressive habit and robustness, *N. exaltata* is cultivated as an indoor and outdoor ornamental plant. Scientific studies were carried out to assess its suitability as an ornamental plant (Der Marderosian et al., 1976). A study by Kobayashi et al. (2007) indicates that *N. exaltata* plants can be cultivated as an indoor plant because they remove chemical vapor at the rating of 9 out of a scale of 10 and also release adequate quantities of moisture into the air. Thus, it helps in cleaning the air in indoor areas.

21.2 ETHNOMEDICINAL USES

Apart from its ornamental habit, *N. exaltata* is also known for its ethnomedicinal use. Different plant parts, viz., rhizome, tuber, and leaves of *N. exaltata*, are reported to be used in the treatment of human disorders. It is used for treating cough and nose blockage, sinus problems, toothache, kidney and liver problems, jaundice, female fertility and menstrual disorders, and female sterility in the Indian subcontinent. It is referred to treat skin disorders (Cambie and Ash, 1994; Singh and Singh, 2012). It is also used as an insect repellent species in the Maharashtra state of India (Masal, 2010). The tribal community of Madhya Pradesh state uses the rhizome as a birth aid in parturition (Singh, 2005; Singh and Singh, 2013). Ho et al. (2010) reported the use of *N. exaltata* for umbilical cicatrization in newborns. Ebeed et al. (2024) referred to the use of the plant for fever, cough, and other conditions.

21.3 BIOACTIVE COMPONENTS

Medicinal properties are directly related to the phytochemicals present in a plant species (Al-Khayri et al., 2022). Ferns have been proven medicines for ages and are used in many traditional systems of medicine. *Nephrolepis exaltata* is also a good medicinal plant species because it contains diverse biologically active phytochemicals. The leaves and stem of *N. exaltata* are said to be light green and subtle in color with an agreeable odor in dry form and are found to be a good source of energy with 359.766 Kcal/100 g and 313.266 Kcal/100 g, respectively (Sharma et al., 2020). Research studies indicate the presence of phytochemicals belonging to a variety of groups, viz., alkaloids, glycosides, saponins, phytosterols, phenols, tannins, flavonoids, carbohydrates, and proteins (Sharma et al., 2021). Sharma et al. (2021) reported that extract percent yield was higher for methanol and ethyl acetate for the stem and leaf parts of *N. exaltata*, respectively. Bassey et al. (2020) also reported 5.177% of extractive value and the presence of the same functional groups for *N. exaltata*. Micromeritic studies on bulk and tapped densities, Hausner's ratio, compressibility index (Carr's index), flow rate, and the angle of repose were also conducted by Bassey et al. (2020). An ash content of 8.8 g/100 g was reported from leaf extracts by Suzuki et al. (2020).

Dave et al. (2023) studied RP-HPTLC fingerprinting of secondary metabolites from *N. exaltata* plant extracts prepared in a methanol: ethyl acetate: n-hexane (1:1:1 v/v/v) solvent system through the cold extraction method. They aimed to develop a fingerprint for the plant, which will help in identification and authentication. They reported terpenoid and flavonoid bands (between Rf 0.004 to 0.910) with a total of 15 flavonoid spots.

El-Tantawy et al. (2015a) studied the chemical constituents present in the above-ground parts of essential oil extracted from *N. exaltata*. The study found a dominance of oxygenated compounds and identified 23 compounds belonging to aldehydes, alcohols, ketones, and phenol functional groups. The following were the 23 various chemical components identified using GS-MS: 2,4-Hexadien-1-ol, Nonanal, 2-Nonenal, β-ionone, Thymol, Nonanoic acid, 2,3-Dimethyl-undec-1-en-3-ol, 1-Octen-3-ol, Benzyl alcohol, p-Mentha-1,5,8-triene, Dodecane, β-cyclocitral, 2,6,10-Trimethyldodecane, Tetradecane, Hexadecane, Heptadecane, 3-Methylheptadecane, 2-Methylheptadecane, Octadecane, Nonadecane, Methyl palmitate, Heneicosane, and Docosane (Figure 21.1).

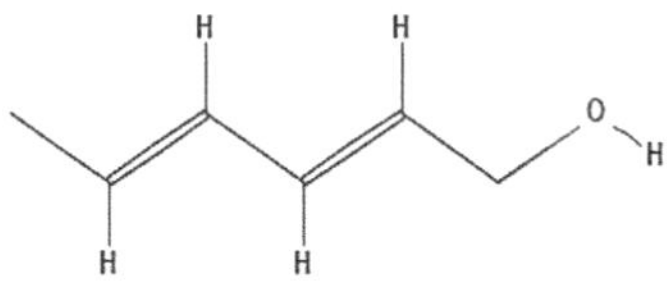

2,4-Hexadien-1-OL (1)

Nonanal

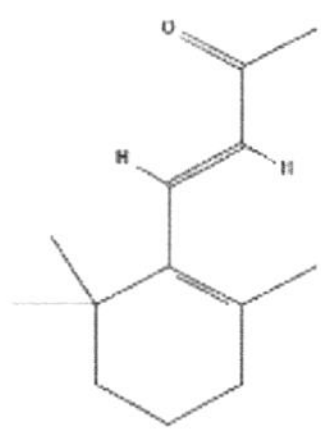

β-Ionone

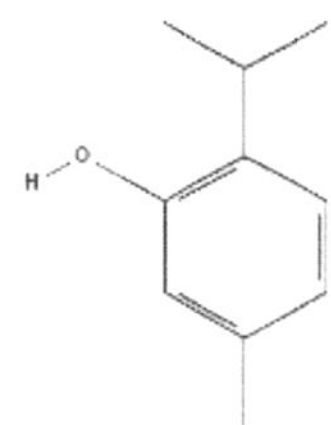

Thymol

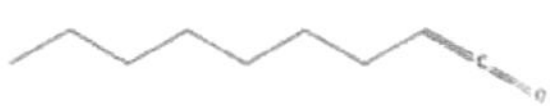

1-Nonenal

Nonanoic acid

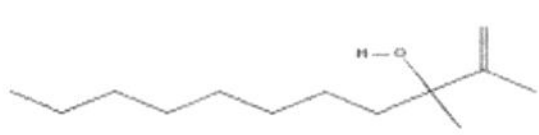

2, 3-Dimethyl-undec-1en-3-ol

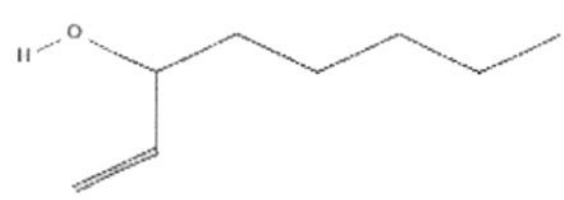

1-Octen-3-OL

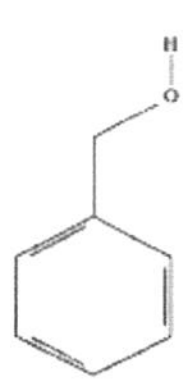

Benzyl Alcohol

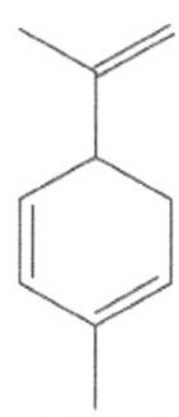

p-Mentha-1,5,8-triene

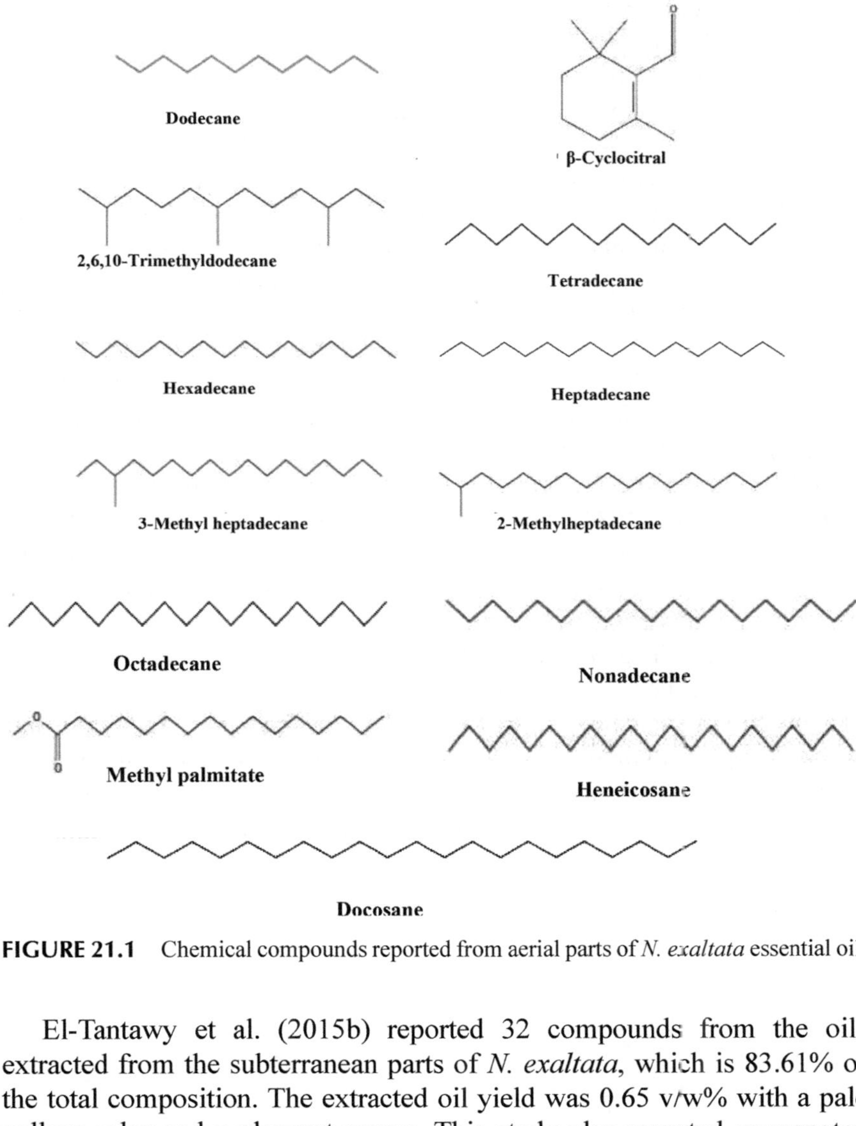

FIGURE 21.1 Chemical compounds reported from aerial parts of *N. exaltata* essential oil.

El-Tantawy et al. (2015b) reported 32 compounds from the oils extracted from the subterranean parts of *N. exaltata*, which is 83.61% of the total composition. The extracted oil yield was 0.65 v/w% with a pale yellow color and a pleasant aroma. This study also reported oxygenated compounds (33.85%) in line with the study by El-Tantawy et al. (2015a), and linalool, thymol, methyl palmitate, α-cadinol, geraniol, and eugenol were the major constituents. A total of 14 compounds were commonly reported in both studies. However, El-Tantawy et al. (2015b) reported p-cymene, linalool, durene, pentyl cyclohexadiene, 1,2,3,4-tetrahydronaphthalene, 2,6-dimethyl undecane, geraniol, 1,2,3,4-tetrahydro-5-methyl naphthalene, eugenol, 3-methyl tridecane, 3-methyl tetradecane, pentadecane, methyl 7,9-tridecadienyl ether, 3-methyl pentadecane, α-cadinol, 2,6,11,15-tetramethyl hexadecane, and dibutyl phthalate additionally.

The above oxygenated compounds have a key role in various fields. The chemical compound allylic alcohol, 2,4-hexadien-1-ol, was used as a flavoring agent and also as a fragrance cosmetic agent (Zhong et al., 2009), and β-ionone was reported for its potential role in perfume manufacturing (Paparella et al., 2021).

Oyawaluja et al. (2024) isolated and identified 21 compounds from the ethanol extracts of dried leaves of *N. exaltata*. The ethanol extract showed different classes of derivatives, such as furan, phenol, megastigmens, iridoid glycosides, flavonoids, and lignin glycosides. The authors used 1D and 2D NMR spectroscopic methods for the elucidation of isolated compounds (Oyawaluja et al., 2024). The study reported the presence of phloretic acid from *N. exaltata* for the first time. The other reported compounds were dehydrovomifoliol, quercitrin, kaempferol, afzelin, quercetin, luteolin, 3-O-methylquercetin, and demethoxymatteucinol (Figure 21.2) for the first time from the genus *Nephrolepis*. The chapter also mentions the first report of the following compounds: methoxygaetneroside, (2S)-5,7,3,'5'-tetrahydroxyflavanone, epigallocatechin, lanicepside A, pinoresinol 4'-O-β-D-glucopyranoside, and glochidiobioside from the family, along with commonly observable compounds, viz., p-hydroxybenzoic acid, p-coumaric acid, and gallic acid (Oyawaluja et al., 2024).

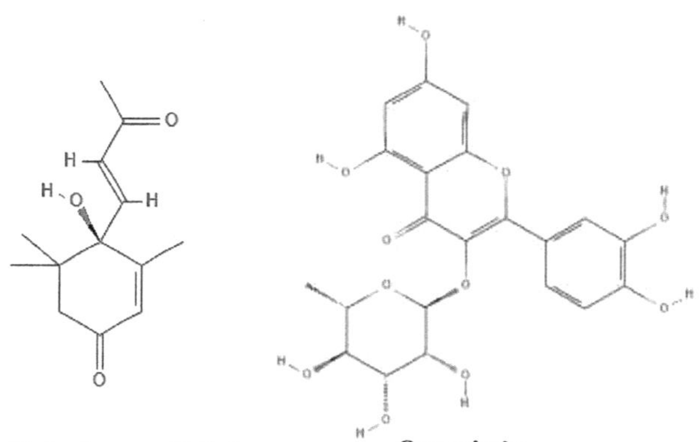

Dehydrovomifoliol **Quercitrin**

FIGURE 21.2 (Continued)

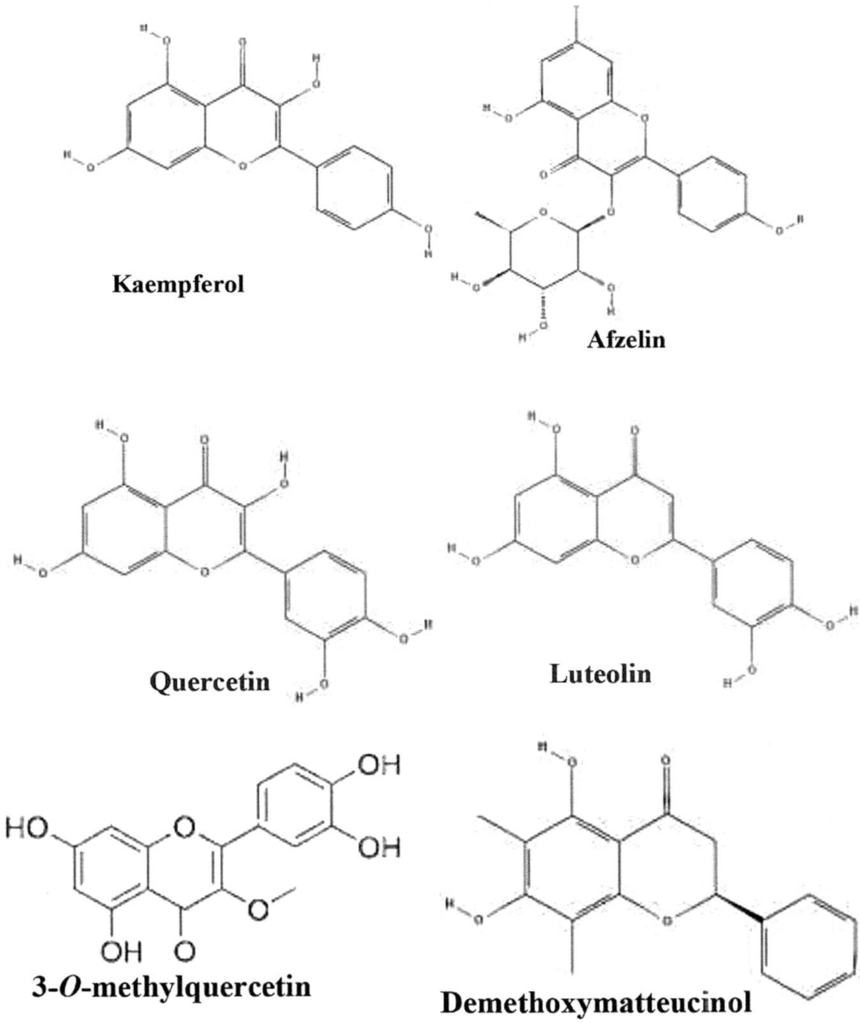

FIGURE 21.2 Reported compounds from *N. exaltata* ethanol leaf extract.

Sharma et al. (2021) studied the chemical constituents from leaf and stem extracts using the GC-MS method and reported 13 and 25 components, respectively (Figure 21.3). The major constituents they reported are phytol, neophytadiene, 3,7,11,15-tetramethyl-2-hexadecen-1-ol, glycerol 1-palmitate, nonacosane, heptacosane, squalene, hexadecanoic acid methyl ester, 2-hexadecene, astaxanthin, vitamin E, 3-O-methyl-d-glucose, and a few steroidal compounds, viz., dihydrobrassicasterol-H_2O, clionasterol, campesterol, stigmasterol, and α-sitosterol, and so forth. The mentioned compounds are reported for antioxidant, antiviral, antibacterial, antihypertensive, vasodilator, anti-wrinkle, immunosuppression activities, and other related medicinal properties (Sharma et al., 2021).

Neophytadiene

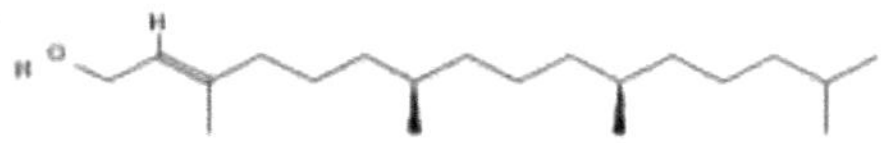

Phytol

Nonacosane

3,7,11,15-Tetramethyl-2-hexad

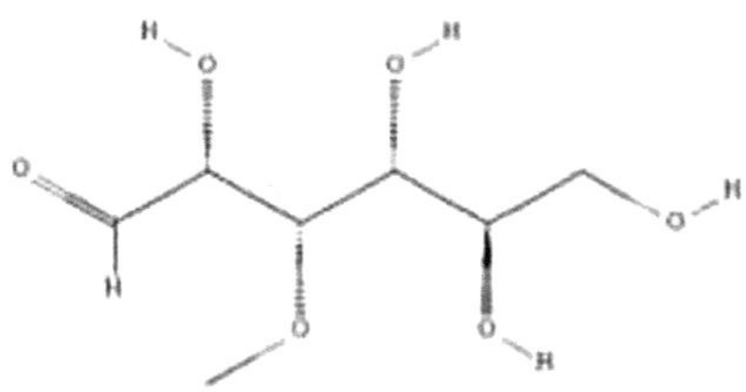

3-O-Methyl-d-glucose

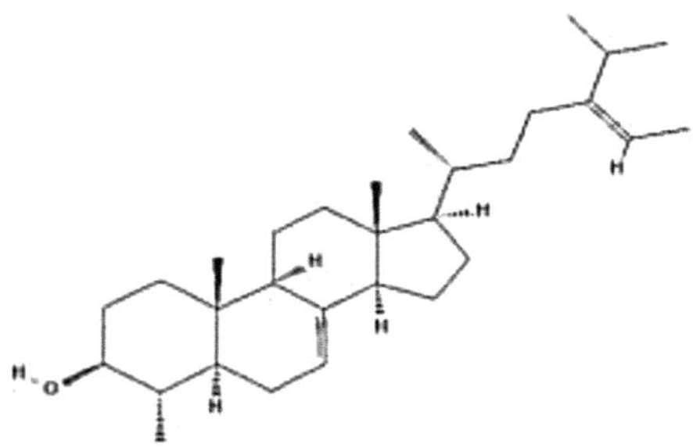

á -Sitosterol

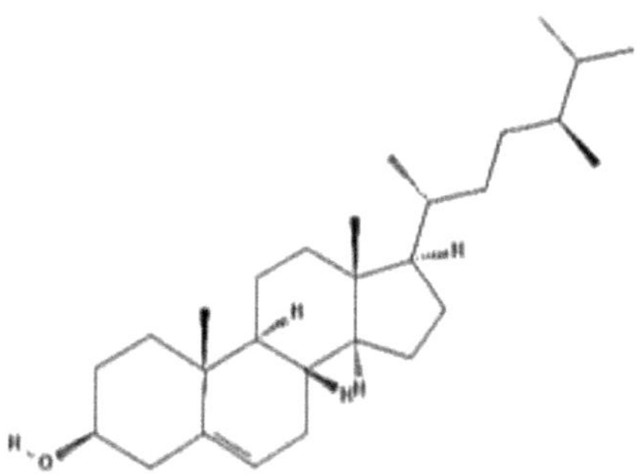

Dihydrobrassicasterol-H_2O

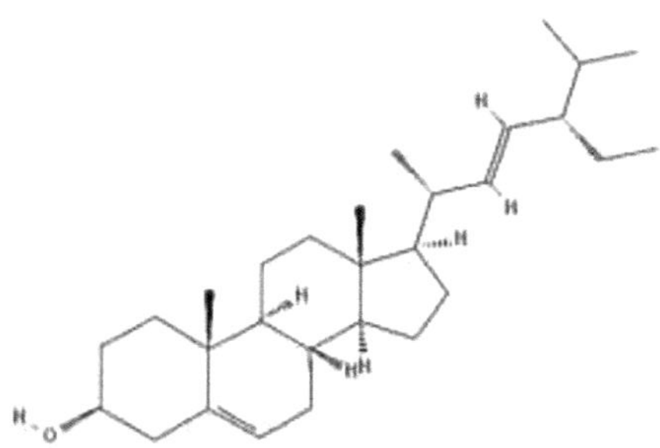

Stigmasterol

FIGURE 21.3 Compounds from *N. exaltata* leaf and stem extracts.

Sanchez et al. (2018) reported the presence of terpenoids in the aqueous extract of the *N. exaltata* plant, but the authors did not identify the compounds.

21.4 NANOPARTICLE SYNTHESIS AND BIOACTIVITIES

Green synthesis of nanoparticles is the production of nanoparticles without the use of hazardous chemicals or toxic solvents, but from natural sources, such as plants and microorganisms. This process is considered a sustainable and eco-friendly technique and has gained attention in recent years. Scientific studies indicated the absorption capability of different metals by this species. Hence, the phytoremediation property led the researchers to think about the ability of nanoparticle production using this fern (Samuel et al., 2022).

A study by Nadeem et al. (2022) indicated the non-toxic synthesis of 16 nm diameter-sized spherical crystalline iron oxide nanoparticles (FeO NPs) using the plant extract of *N. exaltata* as a capping and reducing agent. The elemental composition of the produced nanoparticles was identified by Energy Dispersive X-ray spectrophotometer. The authors reported a high efficiency of the antiplasmodial property of FeO nanoparticles. Furthermore, they believe that synthesized nanoparticles can be effectively utilized in pharmaceutical and medical applications as they lack toxic effects. The *in vitro* cytotoxicity was assessed in this study using human peripheral blood mononuclear cells.

Synthesis of spherical-shaped 60 to 80 nm particle-sized zinc oxide nanoparticles (ZnO-NPs) using $ZnCl_2 \cdot 2H_2O$ salt precursor and an aqueous extract of *N. exaltata* was described by Aboul-Soud et al. (2023). The authors evaluated antiplatelet activity by inhibiting the platelet aggregation induced by platelet activation factor and arachidonic acid and found it effective. The synthesized ZnO-NPs were reported to have good cytotoxic activity, which was evaluated in the human lung cancer cell line (A549) under *in vitro* conditions (Aboul-Soud et al., 2023).

Silver nanoparticles were synthesized using the leaflet extract of *N. exaltata* and were tested against human bacterial pathogens, viz., *Bacillus subtilis* NCIM 2063, *Corynebacterium diphtheriae, Escherichia coli, Klebsiella pneumoniae* NCIM 2719, *Proteus morgani* NCIM 2719, *Pseudomonas testosteroni* NCIM 5098, and *Xanthomonas axonopodis* pv. *punicae*. The study results indicated the antibacterial activity of silver nanoparticles against human bacterial pathogens (Bhor et al., 2014).

21.5 PHARMACOLOGICAL ACTIVITY

The presence of diverse phytoconstituents is responsible for the medicinal properties in a given plant species. *Nephrolepis exaltata* consists of diverse phytoconstituents and is also mentioned for a number of medicinal uses. The various pharmacological activities assessed through standard methods using crude extracts or isolated compounds are discussed below.

21.5.1 ANTI-MICROBIAL ACTIVITY

El-Tantawy et al. (2015a) studied antibacterial and antifungal activity against bacterial and fungal strains using *N. exaltata* volatile oil extracted from aerial parts. The authors used three (*Enterococcus faecalis, Staphylococcus aureus*, and *Staphylococcus pneumoniae*) gram negative bacterial strains, six gram positive bacterial strains (*Proteus vulgaris*, *Salmonella typhimurium*, *Klebsiella pneumoniae*, *Pseudomonas aeruginosa*, *Shigella flexneri* and *Escherichia coli*) and three fungi (*Microsporum gypseum*, *Tricopyton metagrophytes* and *Tricophyton rubrum*) for the assessment. Hydro-distilled volatile oils were screened using the agar diffusion and minimum inhibitory concentration (MIC) method to check the antimicrobial activities. Gentamicin, ampicillin, and amphotericin B were used as standards for comparison for gram-negative bacteria, gram-positive bacteria, and fungi, respectively, during the experiment. The MIC values ranged from 0.98 to 1.95 μg/mL for gram-positive bacteria and 0.24 to 7.81 μg/mL for gram-negative bacteria. The MIC values showed that all the tested microorganisms were sensitive to the volatile oil except for *Pseudomonas aeruginosa*, where the MIC is 125 μg/mL. Less inhibition activity was reported on gram-positive bacteria compared to ampicillin. The MIC values ranged between 0.24 to 1.95 μg/mL against the fungal strains. El-Tantawy et al. (2015b) in another study evaluated antibacterial and antifungal activity against the same bacterial and fungal strains with the same standards using hydro-distilled volatile oil of *N. exaltata* subterranean parts. The study results indicated inhibition activity against only *Staphylococcus aureus* (MIC – 7.81 μg/mL), *Klebsiella pneumoniae* (MIC – 0.98 μg/mL), and *Escherichia coli* (MIC – 3.90 μg/mL) bacterial strains. The subterranean parts of the volatile oil did not inhibit other bacterial strains effectively. Among the three tested dermatophytes, the oil showed activity with an MIC value of 3.9 μg/mL against *Trichophyton mentagrophytes*. The authors discussed that the antimicrobial activity shown by the oil is due to several oxygenated and other phytoconstituents present.

21.5.2 ANTI-NUTRIENT ACTIVITY

Bassey et al. (2020) studied the anti-nutrient composition for the *N. exaltata* plant collected from Nigeria. They reported the presence of oxalates (oxalic acid – 265.62 mg/100g), hydrocyanide (13.66 mg/100 g), phytates (phytic acid – 10.10±0.05 mg/100 g), and tannin (92.37±0.06 mg/100 g) antinutrients. However, the concentration of oxalic acid was higher than the permissible limit.

21.5.3 CHOLINESTERASE INHIBITION ASSAY

Suzuki et al. (2020) assessed cholinesterase inhibition (U/L) using hydroalcoholic leaf extracts of *N. exaltata* at concentrations of 1 mg/mL and 3 mg/mL, and the activity was compared with the positive control neostigmine (1 mg/mL) and a negative control Biocontrol N. The percentage of enzyme inhibition indicated that the *N. exaltata* leaf extracts have no inhibitory effect on cholinesterase enzyme activity. An *ex vivo* experiment using mouse phrenic nerve-diaphragm preparations was conducted to analyze the pharmacological facilitatory effects of the anticholinesterase agent present in *N. exaltata* extracts. The study reported that all the concentrations tested showed comparable results with the Tyrode control, indicating that the plant extract is not toxic to biological preparations (Suzuki et al., 2020).

21.5.4 PRELIMINARY TOXICITY ASSAY

Suzuki et al. (2020) studied preliminary toxicity through the *Salmonella*/microsome assay using the TA100 strain against five varying concentrations from 625 μg/plate to 5000 μg/plate in comparison with positive and negative controls. The study described the absence of toxicity in *N. exaltata* leaf extract in all tested concentrations. A brine shrimp lethality assay was performed on *Artemia franciscana* using the aqueous and ethanolic extracts with concentrations ranging from 625 mg/L to 100000 mg/L. The study reported partial or full lethality at concentrations higher than 5000 mg/L for both extracts. Hence, it was concluded that *N. exaltata* has a limited cytotoxic effect (Popovici et al., 2018).

21.5.5 GENO TOXICITY ASSAY

The genotoxicity of *N. exaltata* extracts at concentrations of 2, 1, and 0.5 g/L was tested on scraped *Allium cepa* bulbs, and the results were compared with the positive control (paracetamol, 750 mg/L) and negative control (potable tap water). The study indicated that the extracts did not show genotoxicity in the *Allium cepa* assay (Suzuki et al., 2020).

21.5.6 ANTIOXIDANT ACTIVITY

Srivastava et al. (2005) conducted a superoxide assay using the extracts of fronds, roots, and rhizomes of *N. exaltata* plants treated with various concentrations of arsenic. The study showed significant antioxidant activity with the extracts of fronds and roots. The authors also conducted catalase activity tests and found no significant increase in catalase activity in the extracts of frond or root tissues of *N. exaltata*, while higher activity was found in rhizome tissues that were exposed to 300 μM arsenate.

21.5.7 CARDIOTOXICITY TEST

A study was carried out to test the effect of *N. exaltata* leaf aqueous extracts on the heartbeats of cockroaches (*Nauphoeta cinerea*) by Sanchez et al. (2018). A concentration of 0.2% pure leaf extracts and diluted extracts with water at the ratios of 1:1, 1:2, 1:4, and 1:8 was tested along with control groups of saline and acetylcholine chloride on the immobilized semi-isolated heart of *N. cinerea*. The results showed an irreversible obstruction of heart beating in pure aqueous and diluted 1:2 and 1:1 leaf extracts. However, 1:4 and 1:8 extracts showed effects similar to the saline control. The study concluded that cardiotoxicity is dosage-dependent in *N. exaltata*.

21.5.8 ANTITUMOR ACTIVITY

The antitumor activity was determined by MTT assay based on cell viability against the essential oils distilled from the aerial parts of *N.*

exaltata using human cancer cell lines of colon (HCT-116), breast (MCF7), and lung (A-549). The viability percentages noted were 31.47%, 19.64%, and 16.75% for HCT-116, MCF7, and A-549, respectively. The authors concluded the presence of antitumor activity in the essential oil (El-Tantawy et al., 2015a).

Similarly, the essential oil extracted from subterranean parts was also tested for antitumor activity against the cell lines mentioned in the earlier paragraph. The human cancer cell viability percentages noted were 63.59%, 17.86%, and 13.97% for colon (HCT-116), breast (MCF7), and lung (A-549), respectively. The anticancer activity was noted against breast and lung cancer cell lines (El-Tantawy et al., 2015b).

21.5.9 CYTOTOXIC ACTIVITY

The cytotoxic activity of the aerial and subterranean parts oil was evaluated using a viability assay on human cancer cell lines. The 50% inhibitory concentration (IC_{50}) values for aerial parts were 89.4 µg/mL, 44 µg/mL, and 43.5 µg/mL; for subterranean parts, >200 µg/mL, 43.2 µg/mL, and 24.3 µg/mL for human cancer cell lines of colon (HCT-116), breast (MCF7), and lung (A-549), respectively (El-Tantawy et al., 2015a; El-Tantawy et al., 2015b).

21.5.10 INSECTICIDAL ACTIVITY

A study was carried out by Sanchez et al. (2018) on cockroaches (*Nauphoeta cinerea*) using *N. exaltata* leaf aqueous extracts (0.2% and diluted extracts 1:1, 1:2 with water). The article reported permanent obstruction of heartbeats, indicating its potential use as an insecticide against cockroaches.

Ebeed et al. (2024) evaluated the repellent potential and larvicidal activity of *N. exaltata* crude leaf extracts in ethanol, ethyl acetate, and n-hexane against the mosquito species *Culex pipiens* adults and larvae. The study reported the highest larval mortality with the hexane crude extract. The alkaline phosphatase activity and glutathione-S-transferase activity were also assessed in larvae treated with the hexane crude extract, and the results exhibited a decline in protein content and a reduction in glutathione-S-transferase activity. Finally, molecular docking and dynamic simulation techniques indicated the extract's content affinity to interfere with the enzyme detoxification functions.

21.5.11 *HERBAL MASKS FROM N. EXALTATA AND ITS EFFECTS*

The use of face masks on a daily basis plays a crucial role in human health and hygiene. Face masks protect individuals from air pollution and allergens, prevent and lower the risk of germ spread, and also protect immunocompromised persons (Lee et al., 2020). There are a variety of face masks available in the market depending on individual needs at different costs, and herbal face masks are one among them (Das et al., 2021). In the preparation of herbal masks, a number of plant-origin herbal preparations that are mentioned and proven to have biological activity were used (Manikandan and Mani, 2023; Wang et al., 2017).

Nephrolepis exaltata is considered one of the indoor ornamental plants that have the ability to absorb a high amount of air pollutants, such as formaldehyde, xylene, trichloroethylene, and carbon monoxide (Prasetyo et al., 2017; Wolverton et al., 1989). Hence, this plant was studied as one of the options in the preparation of herbal face masks. There are several preclinical and clinical studies that assessed the efficacy of herbal masks prepared from *N. exaltata*. The masks prepared using *N. exaltata* herbal laminate alone or in combination with other plant extracts, such as *Hibiscus rosa-sinensis*, were tested for their ability to serve as efficient face masks for protection (Nurhayadi et al., 2023; Prasetyo et al., 2017; Widiono et al., 2019).

A preclinical study of the *N. exaltata* herbal mask proved to prevent blood glucose elevation in Nanosilica-inhaled, streptozotocin-induced Wistar rats (Widiono et al., 2019). The face masks laminated with *N. exaltata* spray absorb volatile organic compounds (Nurhayadi et al., 2023), reduce reactive oxygen species significantly (Prasetyo et al., 2020), increase nasal IgA levels and pulmonary function (Prasetyo et al., 2019), and are beneficial for individuals working in the textile industry to avoid air pollution. It has also been demonstrated to protect sinuses, marked by reduced levels of IgA and IL-6, and improve lung function in motorcycle taxi drivers (Prasetyo et al., 2020).

The above significant findings led to the development of a brand patent for an herbal mask extract, Acchadana® *N. exaltata* – *Hibiscus rosa-sinensis*, by the Diponegoro University Applied Sciences Laboratory. These herbal masks were tested to protect the respiratory tract from exposure to pollutants, viruses, pathogens, and free reactive compounds (Ariani et al., 2023).

21.6 CONCLUSION

Nephrolepis exaltata is a popular indoor ornamental pteridophyte. It is used in the treatment of several human ailments by traditional systems of medicine. Phytochemical investigations indicate the presence of a diverse group of chemical constituents responsible for a variety of biological activities. The species is used in the preparation of herbal face masks that inhibit the number of air pollutants and help in the protection of human health. This makes *N. exaltata* a potential plant with essential components in pharmacological and industrial fields.

KEYWORDS

- **anticancer**
- **bioactive compound**
- **herbal face mask**
- **nephrolepidaceae**
- ***Nephrolepis exaltata***
- **pteridophyte**

REFERENCES

Aboul-Soud, M. A., Siddique, R., Fozia, F., Ullah, A., Rashid, Y., Ahmad, I., Zaghloul, N. S., Al-Rejaie, S. S., & Mohany, M. (2023). Antiplatelet, cytotoxic activities and characterization of green-synthesized zinc oxide nanoparticles using aqueous extract of *Nephrolepis exaltata. Environmental Science and Pollution Research, 30*(29), 73870–73880.

Al-Khayri, J. M., Upadhya, V., Pai, S. R., Naik, P. M., Al-Mssallem, M. Q., & Alessa, F. M. (2022). Comparative quantification of the phenolic compounds, piperine content, and total polyphenols along with the antioxidant activities in the *Piper trichostachyon* and *P. nigrum. Molecules, 27*(18), 5965.

Anonymous. (2024). *Flora of North America* [Internet]. http://www.efloras.org/florataxon.aspx?flora_id=1&taxon_id=200004792 (accessed on 17th November 2025).

Ariani, R., Prasetyo, A., Puspita, R., Kholis, F. N., Sadhana, U., & Istiadi, H. (2023). Which are the best in protecting the upper respiratory tract, KN95® or Acchadana®-herbal masks on IL-6 perspective? *Bali Medical Journal, 12*(1), 1049–1053.

Bajracharya, G. B., & Bajracharya, B. (2022). A comprehensive review on Nepalese wild vegetable food ferns. *Heliyon, 8*(11), e11687.

Bárrios, S., & Copeland, A. (2021). *Nephrolepis exaltata*. The IUCN Red List of Threatened Species [Internet]. https://dx.doi.org/10.2305/IUCN.UK.20213.RLTS.T192003740A192138518.en (accessed on 17th November 2025).

Bassey, M. E., Johnny, I. I., Umoh, O. T., & Douglas, F. T. (2020). Phytomedicinal potentials of species of *Nephrolepis* (Schott.). *World Journal of Pharmaceutical Research, 9*(4), 1400–1410.

Bhor, G., Maskare, S., Hinge, S., Singh, L., & Nalwade, A. (2014). Synthesis of silver nanoparticles by using leaflet extract of *Nephrolepis exaltata* L. and evaluation of antibacterial activity against human and plant pathogenic bacteria. *Asian Journal of Pharmaceutical Technology and Innovation, 2*(7), 23–31.

Cambie, R. C., & Ash, J. (1994). *Fijian Medicinal Plants*. CSIRO Publishing.

Das, S., Sarkar, S., Das, A., Das, S., Chakraborty, P., & Sarkar, J. (2021). A comprehensive review of various categories of face masks resistant to COVID-19. *Clinical Epidemiology and Global Health, 12*, 100835.

Dave, R. S., Sharma, D. K., & Shah, K. R. (2023). RP-HPTLC fingerprinting of secondary metabolites from *Nephrolepis exaltata* and *Cycas revoluta*. *Journal of the Indian Chemical Society, 100*(4), 100941.

Der Marderosian, A. H., Giller, F. B., & Roia, F. C. Jr. (1976). Phytochemical and toxicological screening of household plants potentially toxic to humans. *Journal of Toxicology and Environmental Health, 1*, 939–953.

Ebeed, H., Baz, M., Habib, E., Prabhu, S., & Ceasar, S. A. (2024). Integrated metabolomic analysis and molecular docking: Unveiling the potential of *Nephrolepis exaltata* (L.) Schott phytocompounds for mosquito control via glutathione-S-transferase targeting. *International Journal of Biological Macromolecules, 5*, 133072.

El-Tantawy, M. E., Afifi, M. S., & Shams, M. M. (2015b). Chemical composition, antimicrobial and cytotoxic activities of volatile constituents from the subterranean organs of *Nephrolepis cordifolia* (L.) C. Presl and *Nephrolepis exaltata* (L.) family Nephrolepidaceae grown in Egypt. *Canadian Journal of Pure and Applied Sciences, 9*, 3365–3370.

El-Tantawy, M. E., Shams, M. M., & Afifi, M. S. (2015a). Chemical composition and biological evaluation of the volatile constituents from the aerial parts of *Nephrolepis exaltata* (L.) and *Nephrolepis cordifolia* (L.) C. Presl grown in Egypt. *Natural Product Research, 30*(10), 1197–1201.

Goswami, H. K. (2009). Non-angiospermic plants are also ancient medicinal plants: Conserve and explore them. *Bionature, 29*(2), 95–107.

Ho, R., Teai, T., Bianchini, J., Lafont, R., & Raharivelomanana, P. (2010). Ferns: From traditional uses to pharmaceutical development, chemical identification of active principles. In H. Fernandez, A. Kumar, & M. A. Revilla (Eds.), *Working with ferns: Issues and applications* (pp. 321–346). Springer Science, Berlin Heidelberg/New York.

Kholia, B. S., & Balkrishna, A. (2022). Pteridophytes used by peoples of Indian Himalayan region and Northern India: An overview. In J. Marimuthu, H. Fernández, A. Kumar, & S. Thangaiah (Eds.), *Ferns – Biotechnology, propagation, medicinal uses and environmental regulation* (pp. 379–412). Springer Nature Singapore.

Kobayashi, K. D., Kaufman, A. J., Griffis, J., & McConnell, J. (2007). Using houseplants to clean indoor air. *Ornamentals and Flowers, 39*, 1–7.

Kumar, R. (2024). A review on diversity, distribution and morphology of selected Indian ornamental ferns and fern allies (Pteridophytes). *Authorea Preprints*, 1–9.

Lee, L. Y., Lam, E. P., Chan, C. K., Chan, S. Y., Chiu, M. K., Chong, W. H., Chu, K. W., Hon, M. S., Kwan, L. K., Tsang, K. L., & Tsoi, S. L. (2020). Practice and technique of using face mask amongst adults in the community: A cross-sectional descriptive study. *BMC Public Health, 20*, 1–1.

Manikandan, K., & Mani, M. (2023). Concoction of herbal extract and application on fabric to develop facemask—a potential element against respiratory infections. *International Journal of Pharmaceutical Research, 15*(3), 17. https://doi.org/10.31838/ijpr/2023.15.03.003 (accessed on 17th November 2025).

Masal, V. P. (2010). *Ethnomedicinal and phytochemical studies of some pteridophytes of Ratnagiri district from Maharashtra* [Dissertation]. Shivaji University, Kolhapur, India.

Nadeem, F., Fozia, F., Aslam, M., Ahmad, I., Ahmad, S., Ullah, R., Almutairi, M. H., Aleya, L., & Abdel-Daim, M. M. (2022). Characterization, anti-plasmodial and cytotoxic activities of green synthesized iron oxide nanoparticles using *Nephrolepis exaltata* aqueous extract. *Molecules, 27*(15), 4931.

Nurhayadi, T., Prasetyo, A., & Miranti, I. P. (2023). *Nephrolepis exaltata* herbal-mask better than regular cloth-mask in decreasing nasal mucociliary transport-time of textile industry workers. *Jurnal Biomedika dan Kesehatan, 6*(1), 51–61.

Oyawaluja, A. A., Oyawaluja, B. O., Aydogan, F., Ali, Z., Khan, I. A., Zhao, J., Odukoya, O. A., Samuel, T. A., & Coker, H. A. (2024). Chemical constituents from the leaves of *Nephrolepis exaltata. Biochemical Systematics and Ecology, 1*(112), 104772.

Paparella, A., Shaltiel-Harpaza, L., & Ibdah, M. (2021). β-Ionone: Its occurrence and biological function and metabolic engineering. *Plants, 10*(4), 754.

Pathak, A., Gaurav, N., Singh, A., Kumar, A., & Singh, A. P. (2017). A review on ethno-medicinally important species of pteridophytes. *International Journal of Current Science Research, 3*(3), 1186–1201.

Popovici, P. C., Ancuceanu, R., Olaru, O., Stoicescu, C., & Dinu, M. (2018). Toxicity assessment of *Nephrolepis exaltata* (L.) Schott. *Acta Medica Marisiensis, 64*, 23.

Prasetyo, A., Adi Rahardja, A., Azzahro, D. T., Miranti, I. P., Saraswati, I., & Kholis, F. N. (2019). *Nephrolepis exaltata* herbal mask increases nasal IgA levels and pulmonary function in textile factory workers. *Advances in Preventive Medicine, 2019*(1), 5687135.

Prasetyo, A., Ariani, R., Antari, A. L., Sadhana, U., Miranti, I. P., & Kristina, T. N. (2020). *Nephrolepis exaltata–Hibiscus rosa-sinensis* combination herbal mask decreases levels of reactive oxygen species in nasal wash. *Universa Medicina, 39*(3), 184–191.

Prasetyo, A., Sabila, A. A., Dairiyanti, A., Aufa, M. H., Antari, A. L., Sadhana, U., Miranti, I. P., & Kristina, T. N. (2020). Combination of *Nephrolepis exaltata–Hibiscus rosa-sinensis* herbal mask protects the sinonasal immune system and increases lung function of motorcycle taxi drivers. *Journal of Physiological Anthropology, 39*(3), 1–18.

Prasetyo, A., Sulistyo, H., Sadhana, U., Miranti, I. P., Amarwati, S., & Istiadi, H. (2017). The blocking effects of *Nephrolepis exaltata* on the sinonasal pathogenesis. In *Proceedings of the 10th Asia Pacific IAP Congress*, Bali, Indonesia (pp. 24–27).

Samuel, M. S., Ravikumar, M., John, J. A., Selvarajan, E., Patel, H., Chander, P. S., Soundarya, J., Vuppala, S., Balaji, R., & Chandrasekar, N. (2022). A review on green synthesis of nanoparticles and their diverse biomedical and environmental applications. *Catalysts, 12*(5), 459.

Sanchez, D. M. E., de Souza, J., Oshima-Franco, Y., & Rostelato-Ferreira, S. (2018). Pharmacological effects of *Nephrolepis exaltata* L. (fern) aqueous extract on an insect-based model (*Nauphoeta cinerea*). *Journal of Plant Science and Phytopathology, 2*(1), 31–36.

Sharma, D. K., Dave, R. S., & Shah, K. R. (2020). Proximate analysis and evaluation of energy value from leaves and stem of sword fern: *Nephrolepis exaltata. International Research Journal of Pure and Applied Chemistry, 21*(17), 10–17.

Sharma, D. K., Dave, R. S., & Shah, K. R. (2021). Phytochemical screening and characterization of volatile compounds by gas chromatography-mass spectrometry from *Nephrolepis exaltata. Asian Journal of Pharmaceutical and Clinical Research, 14*, 93–98.

Singh, S. (2005). *Pteridophytic flora of Central India* [Dissertation]. Dr. Hari Singh Gour Vishwavidyalaya, Sagar, India.

Singh, S., & Singh, R. (2012). Ethnomedicinal use of pteridophytes in reproductive health of tribal women of Pachmarhi Biosphere Reserve, Madhya Pradesh, India. *International Journal of Pharmaceutical Sciences and Research, 3*(12), 4780–4790.

Singh, S., & Singh, R. (2013). Utilization of pteridophytes of Achanakmar-Amarkantak Biosphere Reserve, Central India in women's health and beauty care practices. *International Research Journal of Pharmacy, 4*(1), 235–240.

Srivastava, M., Ma, L. Q., Singh, N., & Singh, S. (2005). Antioxidant responses of hyper-accumulator and sensitive fern species to arsenic. *Journal of Experimental Botany, 56*(415), 1335–1342.

Suzuki, E. M., Corrêa, D. F., Oliveira, I. C., Miura, R. Y., Yoshida, E. H., Silva, M. G., Rostelato-Ferreira, S., & Oshima-Franco, Y. (2020). The *in vitro*, *ex vivo*, and *in vivo* experimental findings of *Nephrolepis exaltata. African Journal of Plant Science, 14*(8), 315–324.

Wang, Y. F., Kang, F., You, S. J., Tsai, C. H., & Lin, G. L. (2017). Preparation and characteristic of antibacterial facemasks with Chinese herbal microcapsules. *Aerosol and Air Quality Research, 17*(8), 2119–2128.

Widiono, E. S., Setyadi, A., Awal, P. U. S., & Subagio, A. (2019). Protective effects of *Nephrolepis exaltata* mask against blood glucose level (a study of Wistar rats with chronic exposure to nanosilica). *Indian Journal of Science and Technology, 12*(36), 1–5.

Wolverton, B. C., Johnson, A., & Bounds, K. (1989). *Interior landscape plants for indoor air pollution abatement* [Internet]. https://ntrs.nasa.gov/api/citations/19930073077/downloads/19930073077.pdf (accessed on 17th November 2025).

Zhong, L., Liu, Z., Yan, R., Johnson, S., Zhao, Y., Fang, X., & Cao, D. (2009). Aldo–keto reductase family 1 B10 protein detoxifies dietary and lipid-derived alpha, beta-unsaturated carbonyls at physiological levels. *Biochemical and Biophysical Research Communications, 387*(2), 245–250.

CHAPTER 22

Lycopodium complanatum L. (Lycopodiaceae): A Comprehensive Review of Its Secondary Metabolites and Pharmacological Properties

R. KRIPASREE,[1] BINCE MANI,[2] and SINJUMOL THOMAS[1,*]

[1]*Department of Botany, Carmel College (Autonomous), Mala, Thrissur, Kerala, India*

[2]*Department of Botany, St. Thomas College Palai (Autonomous), Kottayam, Kerala, India*

*Corresponding authors

ABSTRACT

Lycopodium complanatum is an important species of Pteridophyte commonly used in traditional medicinal systems. They are good sources of different bioactive compounds and have significant pharmacological properties. Serratane terpenoids and lycopodium alkaloids are the two major classes of bioactive compounds present in *L. complanatum*. It exhibits a wide range of biological activities such as cytotoxicity, anti-inflammatory, anti-tumor, antiviral, and other antimicrobial activities. Recent advancements in scientific investigations revealed it as a valuable source of diverse secondary metabolites of pharmacological importance.

22.1 INTRODUCTION

From the beginning of human history, approximately 80% of the world's population maintains their health and medical care from traditional medicines. Pteridophytes, being the first known vascular plant species, garnered

Bioactives and Pharmacology of Pteridophytes. T. Pullaiah (Ed.)

DOI: 10.1201/9781779642738-22

attention and began to be investigated and utilized for their beneficial characteristics about two millennia ago. This early interest marked the beginning of human exploration into the uses of Pteridophytes and paved the way for a deeper understanding of plant biology and the ecological significance of these early vascular plants (Patil, 2020).

It has been observed that the family Lycopodiaceae (club mosses), which possess lycopodium alkaloids, have a prominent influence on diverse biological effects on various physiological processes. The discovery of various compounds from Lycopodiaceae serves as regulators of different facets of the immune system, which improve the medical system (Gísladóttir, 2022). An extremely valuable herbal treatment based on lycopods was prevalent in several countries throughout the part of primitive cultures. The spores of lycopods were used to apply directly to open wounds based on the conventional therapeutic way of treating injuries and lesions. The Blackfoot tribe and Iroquois explored *Lycopodium complanatum* in different ways; the former used it to treat lung diseases, and the latter used it to trigger pregnancy (Moerman, 2009).

Centuries ago, club mosses were used as valuable herbal medicine in the traditional system of medicine all around the world in their ethnic part of life. Club mosses produce various types of secondary metabolites called lycopodium alkaloids, which are incredibly effective in medicinal systems due to their bioactive properties. A total of 508 secondary metabolites have been identified so far from the 46 species of the genus *Lycopodium*. The two major classes of bioactive compounds present among the secondary metabolites are serratane terpenoids and lycopodium alkaloids. There are various applications of club mosses relevant to our scientific medicinal knowledge. The benefits of club moss spores from *Lycopodium clavatum* L. are used directly to heal wounds and rashes and are a very popular traditional practice among the natives of North America and Europe (Wang et al., 2021).

The first discovered lycopodium alkaloid is lycopodine, which was isolated from the plant *Lycopodium complanatum*. This most scientifically significant discovery was made by the scientist Bodeker during the early studies dating back to 1881. The isolation of lycopodine marks the beginning of research on lycopodium alkaloids. Since then, further research has uncovered many other lycopodium alkaloids with diverse biological properties. This established the significance of lycopodine and considered it a pivotal element in the study of natural product chemistry (Ma and Gang, 2004).

Lycopodium complanatum L. [Syn: *Diphasiastrum complanatum* (L.) Holub, *Diphasium complanatum* (L.) Rothm., *Lepidotis complanata* (L.) C. Presl.] is commonly known as ground cedar. It is a subshrub primarily seen in temperate biomes. The native range of the plant belongs to the subarctic and temperate northern hemispheres (POWO, 2024). Morphologically, the size of the plant ranges from small to medium-sized with spathulate or ovate, membranous or thickly membranous leaves. Aerial shoots are creeping. Its lateral branches are sub-erect and dichotomously branched with well-differentiated, flattened, and highly branched branchlets. At the terminal region of the peduncle, pale yellow strobili are present. Sporophylls are broadly ovate and imbricate, having a membranous margin with sporangia enclosed in it (eFlora, 2013).

In this elaborate model, we thoroughly explore the intricate biochemical features of *L. complanatum*, elucidating their respective alkaloids, flavonoids, terpenoids, and other bioactive compounds, and discuss their implications for human health and therapeutic interventions. Through this comprehensive examination, we aim to provide insights into the remarkable biochemical diversity of *L. complanatum*, paving the way for further exploration and utilization of this plant in the pursuit of novel pharmaceutical research and development.

22.2 BIOACTIVES

A class of naturally occurring compounds known as lycopodium alkaloids has significant interest from synthetic, biological, and biogenetic perspectives due to their distinctive ring structures. After looking through club moss for structurally intriguing compounds, complanadine A, a novel dimeric alkaloid with a skeleton like lycodine, was isolated from *L. complanatum* (Kobayashi et al., 2000). The analysis of the content of the lycopodium alkaloid fraction of *L. complanatum* by capillary gas chromatography-mass spectrometry (GC-MS) revealed that lycopodine was the major alkaloid (60.8%) found in *L. complanatum* (Orhan et al., 2009).

Triterpene compounds such as serratenediol and tohogenol were isolated by Inubushi et al. (1967) from *L. complanatum*. The whole plants collected from New Guinea were extracted with methanol, and subsequent fractionation through column chromatography and thin-layer chromatography aided the isolation of serratenediol and tohogenol. The obtained compounds were in the form of diacetates.

Kobayashi et al. (2001) identified two intermediates of lycopodium alkaloids, lyconadin A ($C_{16}H_{20}N_2O$) and 11-hydroxylycodine ($C_{16}H_{23}N_2O$), isolated from *L. complanatum*. They obtained 0.0003% of lyconadin A and 0.0003% of 11-hydroxylycodine, as a colorless solid, together with 0.003% of complanadine A and 0.0005% lycodine. Lyconadin A and 11-hydroxylycodine showed distinct cytotoxic activity against murine lymphoma L1210 cells and human epidermoid carcinoma KB cells.

Morita et al. (2005) isolated and elucidated the structures of a novel molecule of lycopodium alkaloids, a dimer of $C_{16}N_2$ type, complanadine B from the methanol extract of *L. complanatum*. The extract yielded 0.0002% of complanadine B along with known compounds such as complanadine A and lycopodine. Ishiuchi et al. (2006a,b; 2007; 2009a,b; 2011a,b) discovered different lycopodium alkaloids from *L. complanatum*. The compounds such as $C1_6N$-type alkaloid lycopladine A (Ishiuchi et al., 2006a) and lycopladines B–D and lyconadin B (Ishiuchi et al., 2006b) were identified independently and tested for their cytotoxic activity. Two dimeric alkaloids, complanadines C and D, were isolated, and their structure and stereochemistry were elucidated by using spectroscopic data (Ishiuchi et al., 2007). $C_{16}N_2$-type alkaloids such as lycopladines F and G with an additional C4N unit (Ishiuchi et al., 2009a) and lycopladine H, with a fused-tetracyclic skeleton (Ishiuchi et al., 2009b), were characterized from the methanol extract of *L. complanatum*, and the structure and relative stereochemistry of all the compounds were elucidated based on the analysis of spectral data. The lycopodium alkaloids, lyconadin C and F, having a $C_{16}N_2$-type structure with a unique fused-tetracyclic ring system and cycloheptene ring fused to decahydroquinoline and pyridone rings, were isolated and elucidated. The structure and relative stereochemistry of both compounds were depicted based on spectroscopic data analysis (Ishiuchi et al., 2011a). In the same year, they isolated and characterized three new lycopodium alkaloids (lyconadins D, E, and complanadine E) from the aqueous methanol extract of *L. complanatum*. This study reported the occurrence of fastigiatine-type alkaloids (lyconadins D and E) and an unsymmetrical dimeric alkaloid forming a piperidine ring at C-1–C-5 (complanadine E) from *L. complanatum* (Ishiuchi et al., 2011b).

Kubota et al. (2007) isolated the $C_{16}N_1$-type alkaloid, lycopladine E, from *L. complanatum*, and the structure was elucidated based on spectroscopy and chemical correlation. Lycopladine E showed neurotrophic factor biosynthesis in 1321N1 human astrocytoma cells and was found to enhance the mRNA expressions for nerve growth factor (NGF).

Five novel polyhydroxyserratane triterpenoids were isolated and characterized along with nine known molecules from the aqueous methanol (1:9) extract of *D. complanatum*. Polyhydroxy derivatives such as serratane-3α,14α,15α,20b,21β,24,29-heptol, 3α,20β,21b-trihydroxyserrat-14-en-24-oic acid, 3β,20β,21β-trihydroxyserrat-14-en-24-oic acid, 3α,20β,21β-trihydroxy-16-oxoserrat-14-en-24-oic acid, and 16-oxolyclanitin-29-yl E-40-hydroxyl-30-methoxycinnamate were obtained as novel compounds, and their structures were elucidated based on the analysis of spectroscopic and literature data. Nine known compounds, including lycoclavanol, serratenediol, epi-serratenediol, lycoclaninol, wightianol B, lycernuic acid A, 16-oxolyclanitin-29-yl p-coumarate, lycoclavanin, and a-onocerin, were also isolated (Yan et al., 2008).

Wu et al. (2010) isolated and characterized diphaladine A as a colorless oil from the ethanol (95%) extract of *D. complanatum*. By considering the characteristic NMR data, along with the knowledge about the structural types of lycopodium alkaloids, they ascribed diphaladine A to be a lycopodine-type lycopodium alkaloid. They isolated obscurumine A (2), L20, des-N-methyl-β-obscurine, des-N-methyl-α-obscurine, lycoposerramine-K, phlegmariurine B, and lycoflexion for the first time from this species. A bioactive C_{15}N-type lycopodium alkaloid, lycospidine A, together with the known 12-deoxyhuperzine O, was characterized from the methanol extract of the whole plant by Cheng et al. (2013) as optically active colorless needle crystals. Lycospidine A was the first characterized lycopodium alkaloid that contains an unusual five-membered A ring.

Isolation, characterization, and analysis of the cytotoxic effects of four serratene-type triterpenes were conducted by Yan et al. (2009). They identified compounds such as complanatumol A, complanatumol B, complanatumol C, and complanatumol D. Among these triterpenes, complanatumol D was the only compound that showed notable cytotoxic activity against human leukemia K562/S and the doxorubicin-resistant K562/R cell lines. All other triterpenes were inactive. Tang et al. (2015) isolated the lycopodine-type lycopodium alkaloid, dehydroisofawcettiine N-oxide, and 11 known analogues: dehydroisofawcettiine, lycopodine, lycopodine N-oxide, 6α-hydroxylycopodine, clavolonine, flabelliformine, 6-epi-8β-acetoxylycoclavine, acetylfawcettiine, lycoclavine, 12-deoxyhuperzine O, and lycodine from the methanol (95%)

extract of the whole plant of *L. complanatum*. By spectroscopic 2D NMR techniques, they established the structural configuration of the molecules. They deduced the absolute configurations of dehydroisofawcettiine and its new N-oxide derivative by chemical transformation in conjunction with cotton effects in their electronic circular dichroism (ECD) spectra.

Two rare lyconadin-type lycopodium alkaloids, lyconadins G and H, along with four known alkaloids, lyconadin A and lyconadins C–E, were isolated by Cheng et al. (2016). Based on spectroscopic analysis of the structure and X-ray crystallographic examination, the absolute configuration of lyconadins G was established. This was the first X-ray diffraction experiment that determined the exact structure of a lyconadin-type lycopodium alkaloid. Furthermore, these results could offer additional insights into the biosynthesis of lyconadins.

10 different serratene-type triterpenoids were isolated by Nguyen et al. (2017) from *L. complanatum*. Among these, the two compounds, 21β,24-dihydroxyserrat-14-en-3α-yl acetate and 16-oxo-21β,24-dihydroxyserrat-14-en-3α-yl acetate, are novel, while the others are known compounds such as 3α,21β-dihydroxyserrat-14-en-16-one, 21α-hydroxyserrat-14-en-3β-yl acetate, serrat-14-en-3β,21α-diol, 16-oxo-21β-hydroxyserrat-14-en-3α-yl acetate, 21β-hydroxyserrat-14-en-3α-yl acetate, 21β-hydroxyserrat-14-en-3β-yl acetate, lycoclavanol, and serrat-14-en-3β,21β-diol. The compounds 21β,24-dihydroxyserrat-14-en-3α-yl acetate and 16-oxo-21β-hydroxyserrat-14-en-3α-yl acetate show a beneficial therapeutic effect in treating Alzheimer's disease. A $C_{16}N$-type lycopodium alkaloid called lycoplanine A, with a 6/9/5 tricyclic ring skeleton, was discovered by Zhang et al. (2017). It is an unusual ring system having a unique 1-oxa-6-azaspiro nonane moiety and an unprecedented 3-azabicyclo dodecane unit. The structural configuration of this compound was elucidated by analyzing the data from NMR spectroscopy and single-crystal X-ray diffraction.

Weng et al. (2018) isolated five abietane diterpenoids, complanatins A-E, along with two known abietane diterpenoids, xanthoperol and sugiol. The cytotoxic effects of five compounds, complanatins A-D and sugiol, in three human lung cancer cell lines (MSTO-211H, NCI-H2052, and NCI-H226) were also evaluated, and it was found that complanatin C and D exhibited cytotoxic activities against the three cell lines. In addition, a biogenetic pathway for all the compounds was proposed.

Three novel lycopodium alkaloids, lycoplanines B, C, and D, were isolated by Zhang et al. (2018a) from the ethanol extract of the whole plant of *L. complanatum*. Lycoplanine B is a $C_{17}N$ lycopodium alkaloid, with an unusual formyl group. Known compounds such as des-N-methyl-a-obscurine, lycoplanine A, lycopodine N-oxide, and lycopodine were also isolated and characterized from the same extract. The structure of the compounds was elucidated by comprehensive 1D and 2D NMR studies and spectral data comparison with known counterparts. Additionally, they also proposed a potential biosynthetic pathway for lycoplanine B.

L. complanatum var. *glaucum* was tested for its bioactives and discovered lycogladines A-H. It includes five new fawcettimine-type lycopodium alkaloids (lycogladines A-E) and three compounds obtained from natural sources for the first time (lycogladines F-H). Along with these molecules, they also found four known analogs such as fawcettimine, fewcettidine, lycopoclavamine B, and lycopladine B. The structures of these compounds were elucidated based on single-crystal X-ray diffraction and spectroscopic analysis data by HRESIMS, 1D, and 2D NMR methods. Lycogladines A has a new ring system by the linkage of C-13-O-C-2, which is a very rare structural arrangement in lycopodium alkaloids (Zhang et al., 2018b). Phytochemical investigation of *L. complanatum* conducted by Dong et al. (2019) was able to isolate and characterize six new polyhydroxy serratene triterpenoids, serrat A, serrat B, serrat C, serrat D, serrat E, and serrat F, along with a known analogue, 16-oxolyclanitin. They elucidated the structures and configurations by HRESIMS, NMR, and comparisons with other known spectroscopic data. They also tested the cytotoxic activities of all the compounds against HepG2 cells, MCF-7 cells, and a series of human lung cancer cell lines, and all the compounds showed cytotoxic activity to varying extents. In another study, eight new serratane triterpenoids, lycomplanatums A-H, along with five known analogues, were isolated from the ethyl acetate (EA) fraction of the ethanol (95%) extract of whole plants of *L. complanatum*. Additionally, five more known phytoconstituents were also characterized from the EA fraction, such as lycernuic A, 3α,21β,29-trihydroxyserrat-14-en-16-one, 3α,21β,24-trihydroxyserrat-14-en-16-one, 3β,21β,24-trihydroxyserrat-14-en-16-one, and 3β,21β-dihydroxyserrat-14-en-16-one. By extensive spectroscopic methods, including 1D/2D NMR, HRESIMS, and DFT GIAO 13C NMR calculation, structures were elucidated (Zhou et al., 2020). Hirasawa et al. (2021) isolated a new lycopodium alkaloid, complanadine F, a seco-complanadine A type, from ethyl acetate fractions of the methanol extract of the whole plant. The planar structure and relative configuration of the molecule were depicted based on spectroscopic data. They also reported the moderate inhibitory effect of this compound against acetylcholinesterase. Zhang et al. (2023) isolated and characterized a novel abietane diterpenoid, complanatin F, together with three known compounds, complanatins A, B, and C, from the ethyl acetate (EtOAc) extract of *L. complanatum*. Complanatin F was obtained as a white amorphous powder with a structure elucidated as 1β, 11-epoxyabieta-12-hydroxy-8, 11, 13-triene-7-one. By the analysis of NMR spectroscopy and HRESIMS data, and by comparing with previous spectral data, they confirmed its structure. Inhibitory activities of all the compounds against A549, HepG2, and MCF-7 tumor cell lines were also examined. Many of the lycopodium alkaloids are present in almost all other genera of the Lycopodiaceae family. However, *Diphasiastrum complanatum* is exceptional in producing distinct compounds such as dimers complanadine A–E and additionally, some novel unclassified structures like lycospidine, lycopladines A and F, and lyconadines A, B, C, and F (Halldorsdottir et al., 2015).

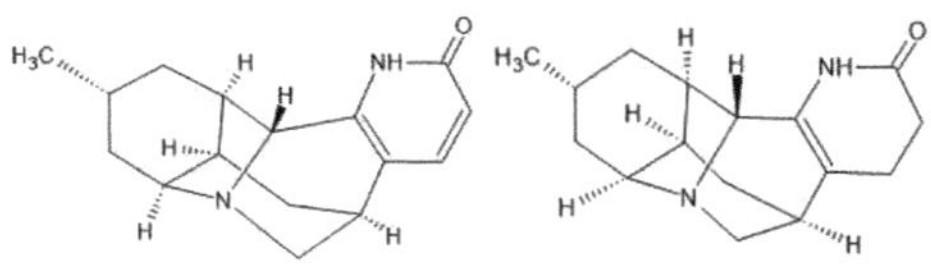

Lyconadin A and B

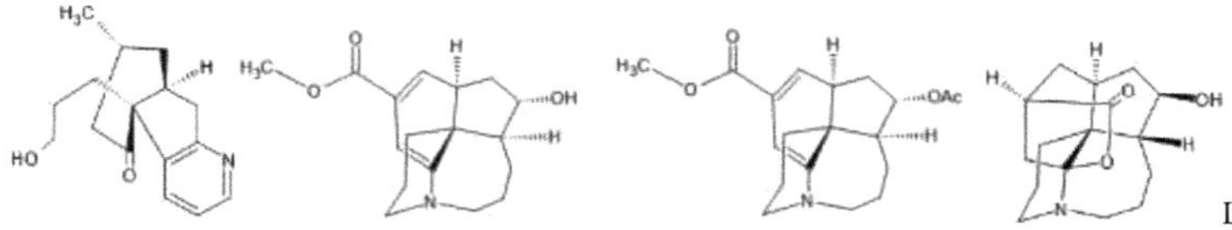

Lycopladines

A-D

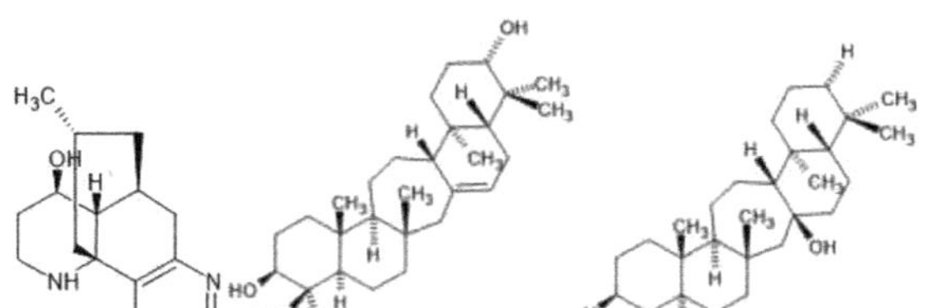

11-hydroxy lycodine Serratenediol Tohogenol

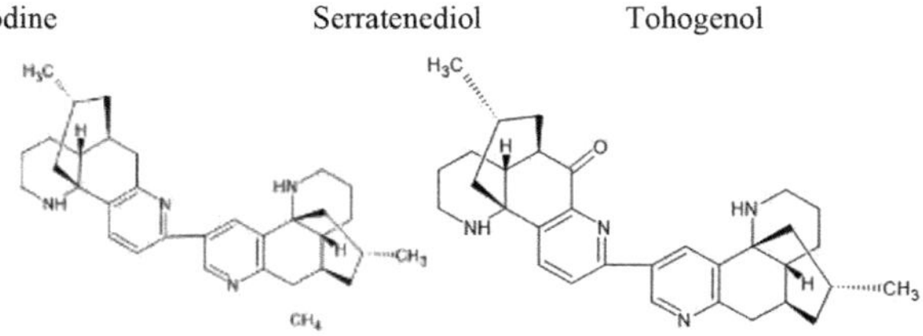

Complanadine A and B

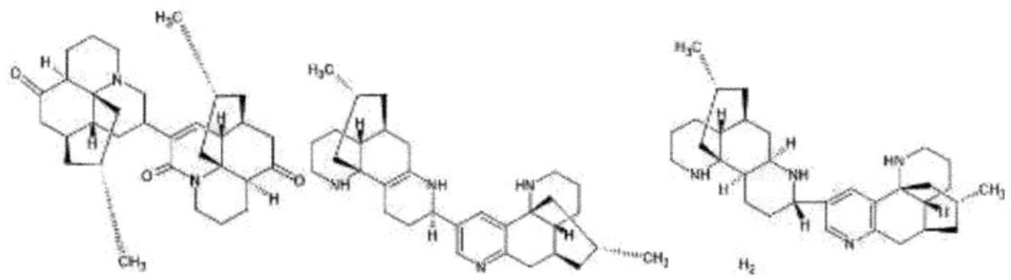

Complanadine C- E

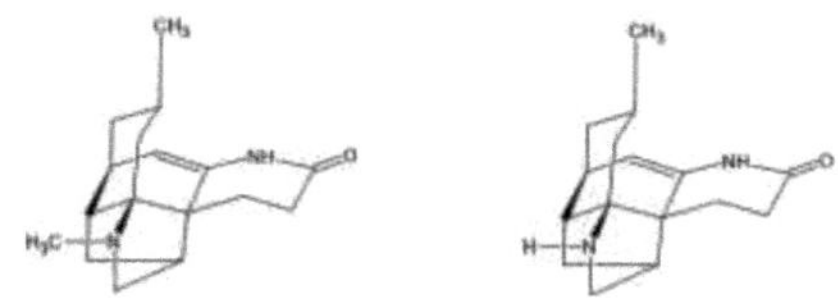

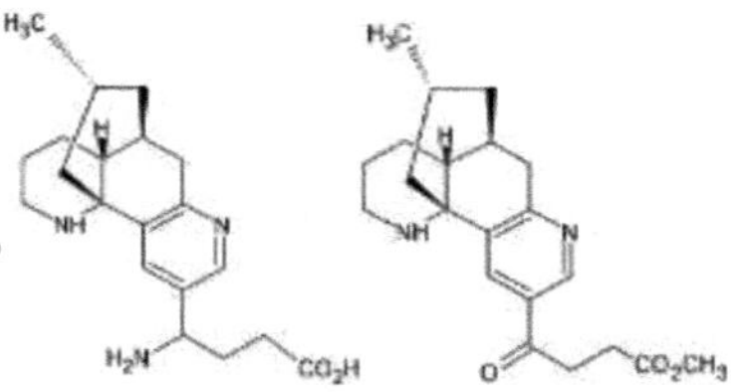

Lyconadins D and E Lycopladine F and G

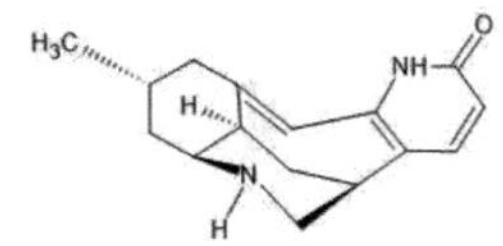

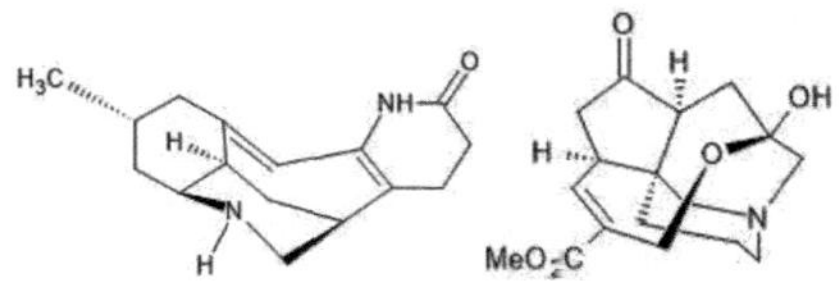

Lyconadins G and H Lycogladine A

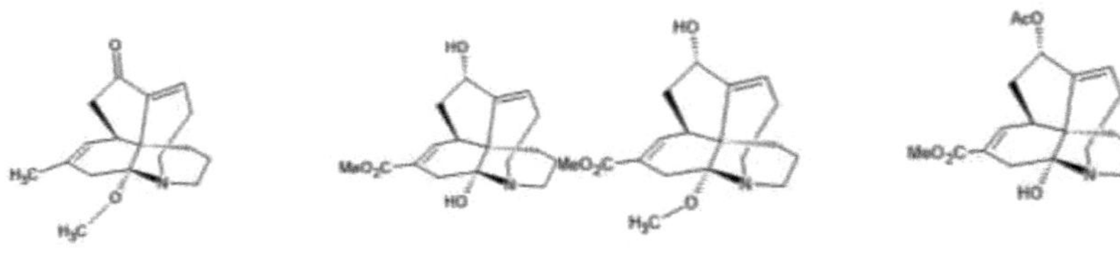

Lycogladines B- E

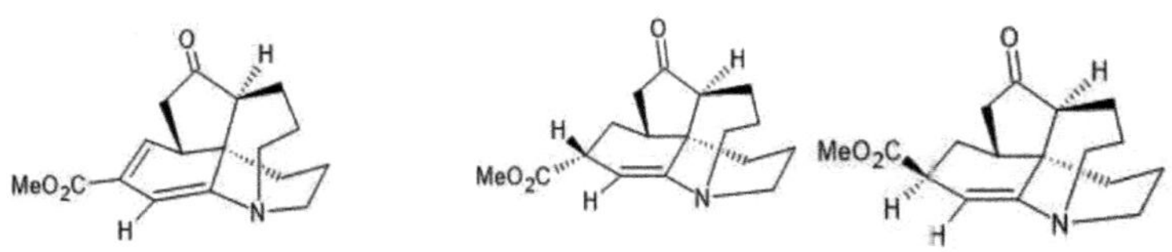

Lycogladines F-H

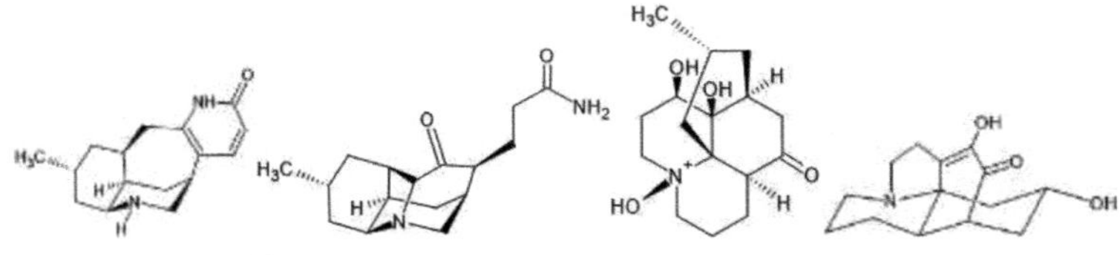

Lyconadin C Lyconadin F Diphaladine A Lycospidine A

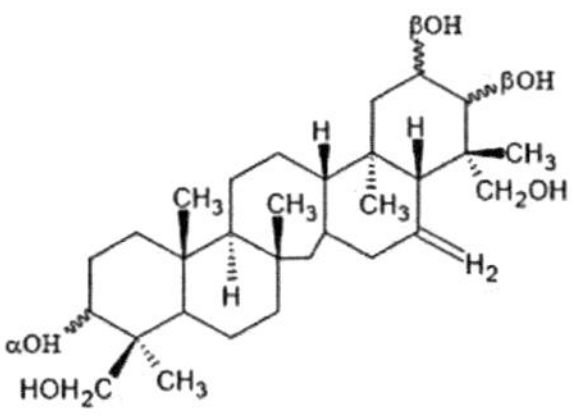

serratane-3α,14α,15α,20β,21b,24,29-heptol

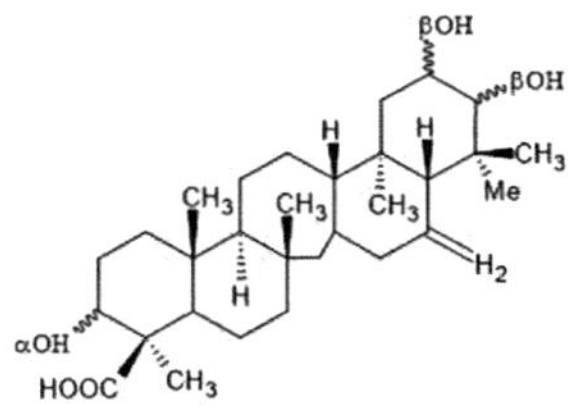

3α,20β,21β-tri hydroxyserrat-14-en-24-oic acid,

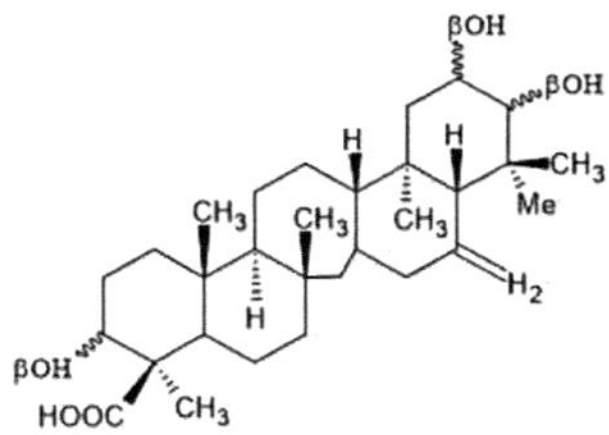

3β,20β,21β-trihydroxyserrat-14-en-24-oic acid

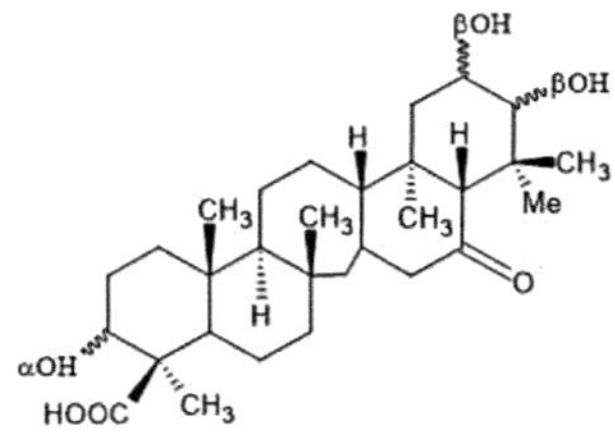

3α,20β,21β-trihydroxy-16-oxoserrat-14-en-24-oic acid.

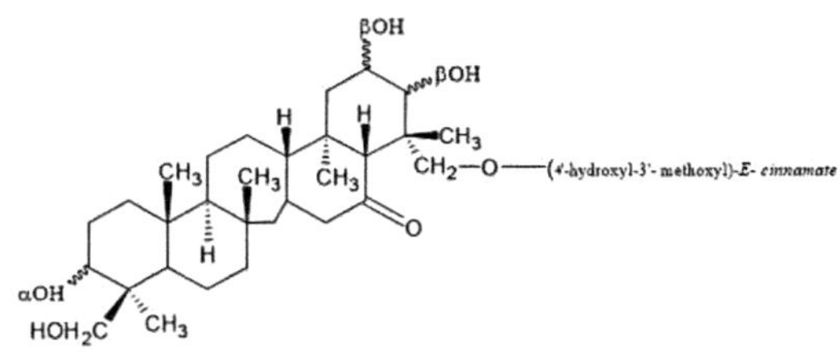

16-oxolyclanitin-29-yl E-40-hydroxyl-30-methoxycinnamate

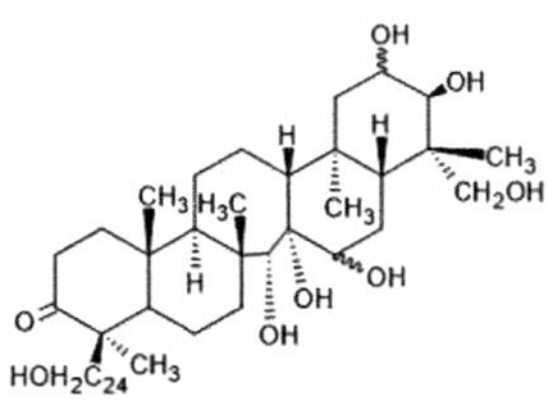

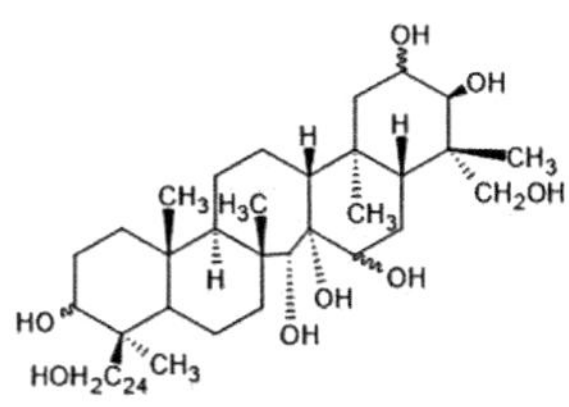

Complanatumols A and B

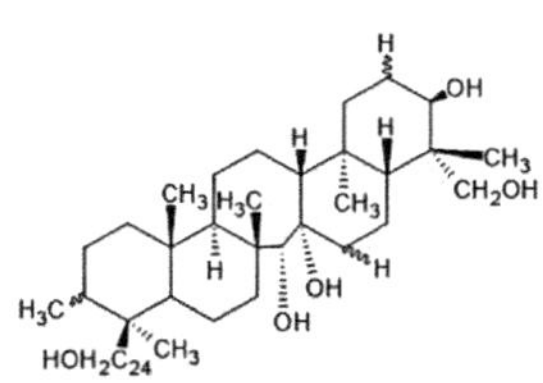

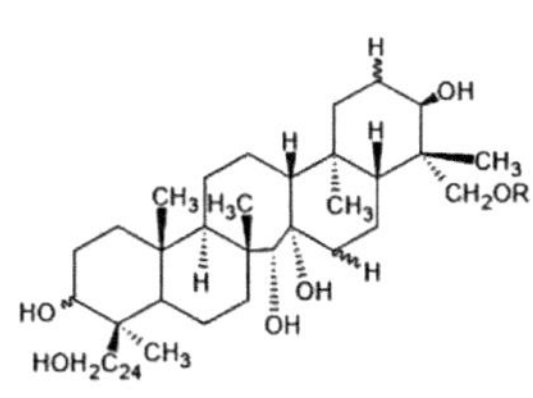

Complanatumols C and D

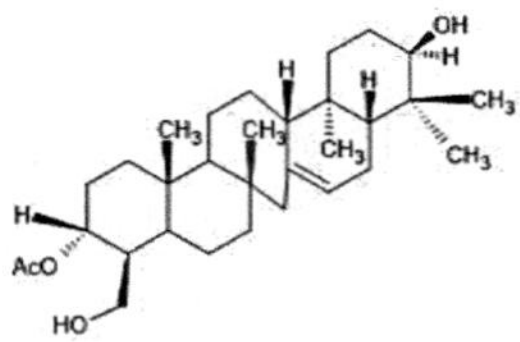

21β,24-dihydroxyserrat-14-en-3α-yl acetate

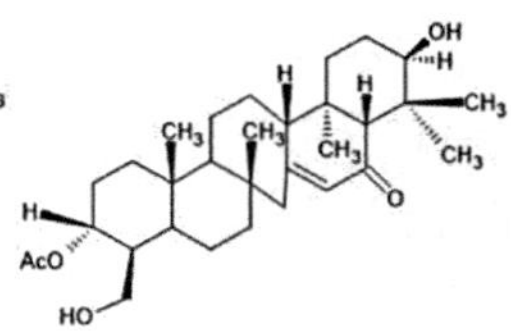

16-oxo-21β,24-dihydroxyserrat-14-en-3α-yl acetate

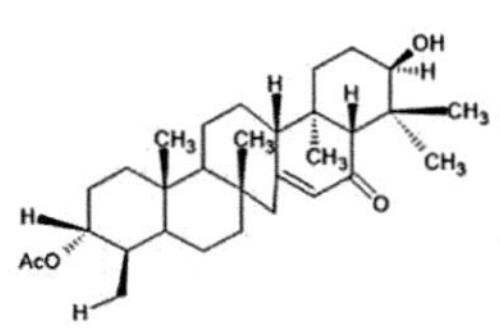

16-oxo-21β-hydroxyserrat-14-en-3α-yl acetate

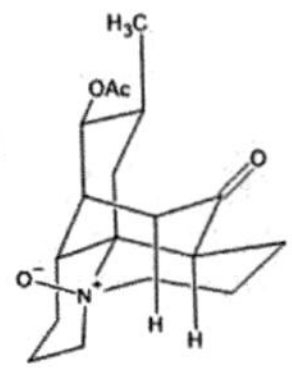

Dehydroisofawcettiine N-oxide

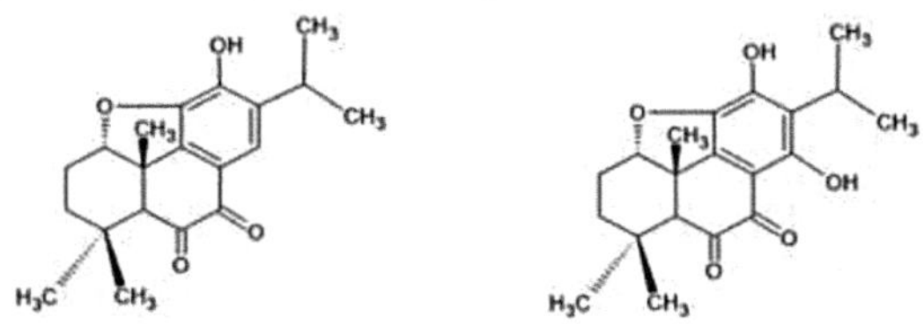

Complanatins A and B

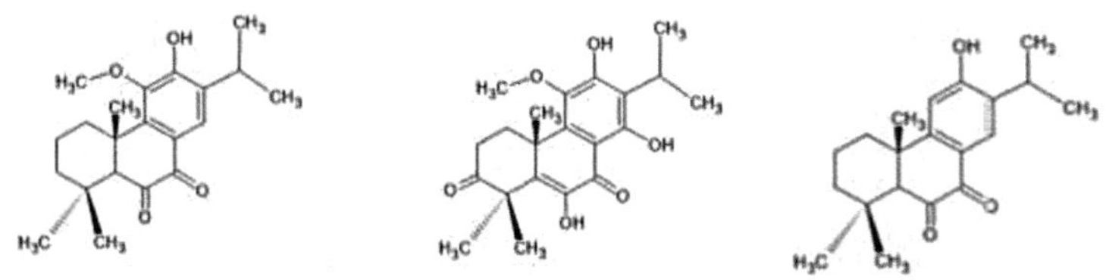

Complanatins C-E

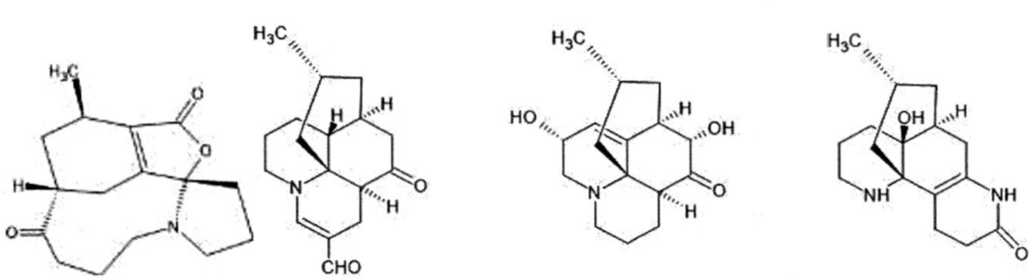

Lycoplanines A- D

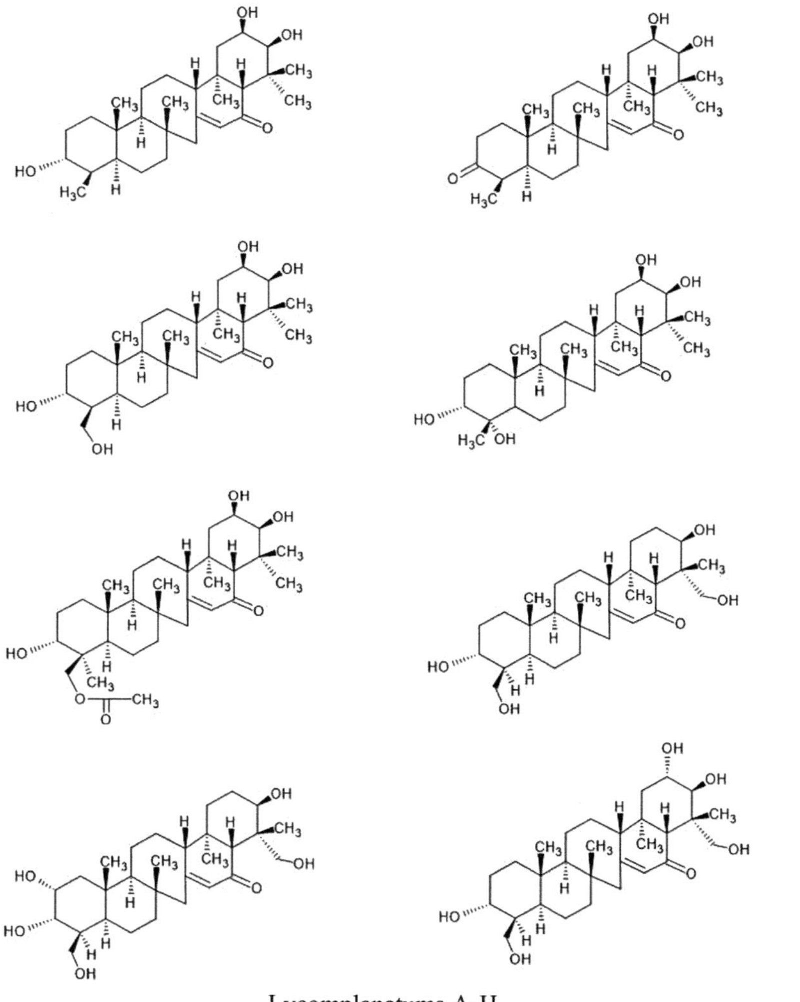

Lycomplanatums A-H

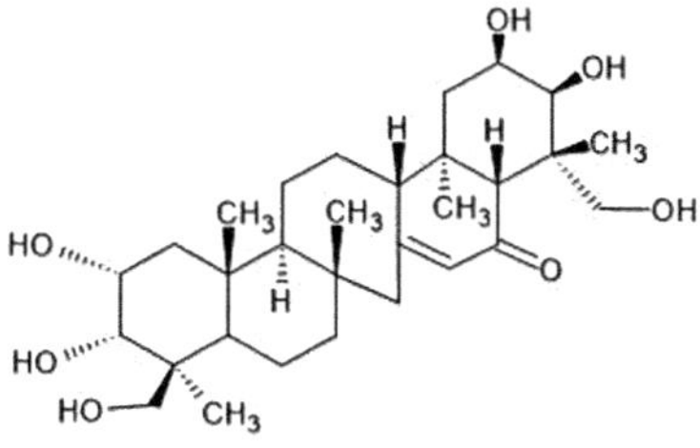

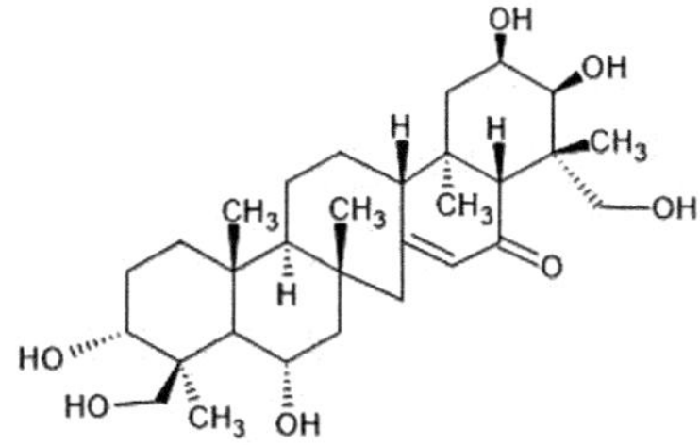

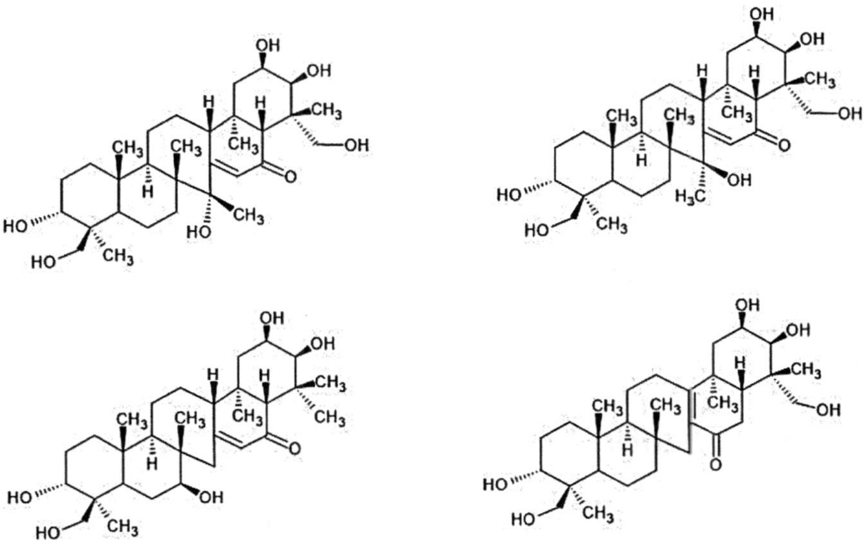

Serrat A-F

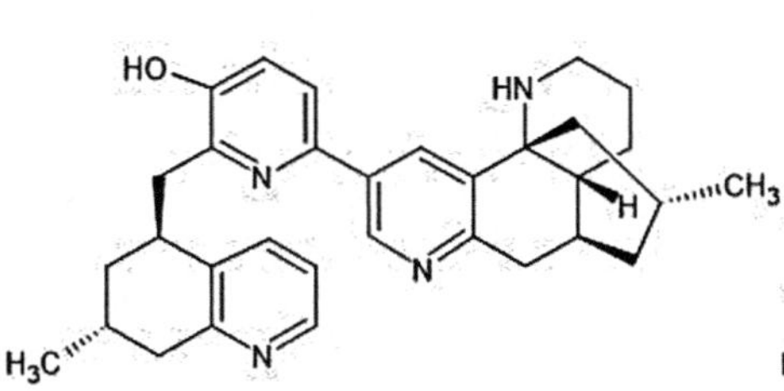

Complanadine F

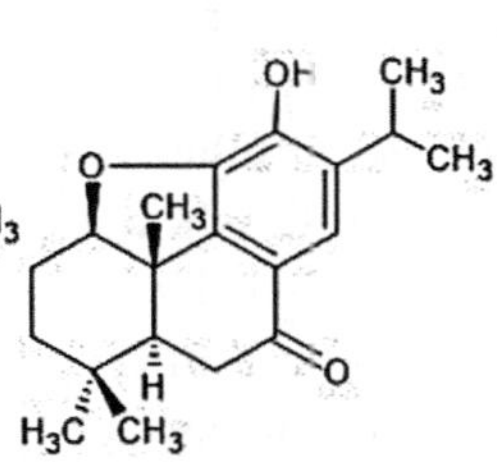

Complanatin F

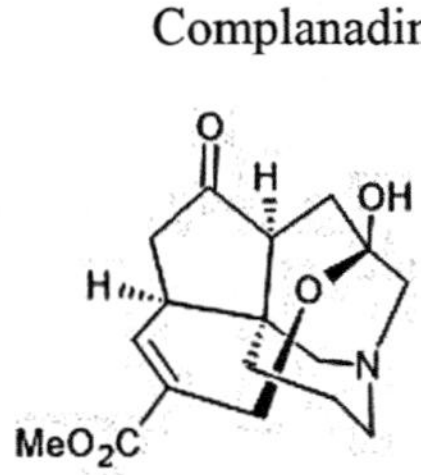

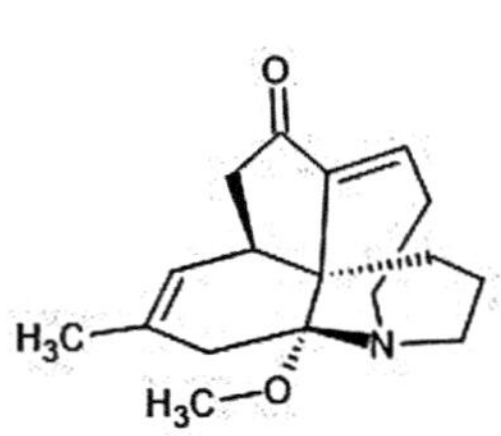

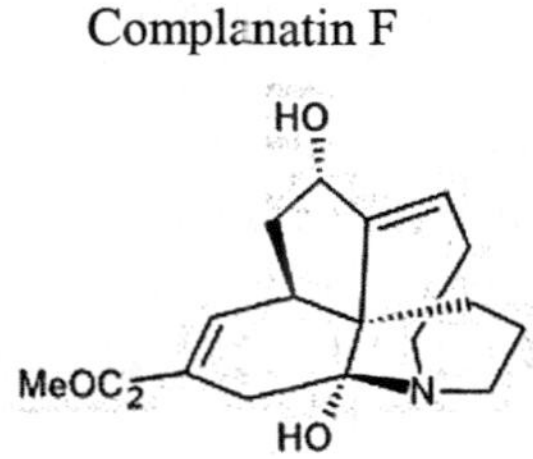

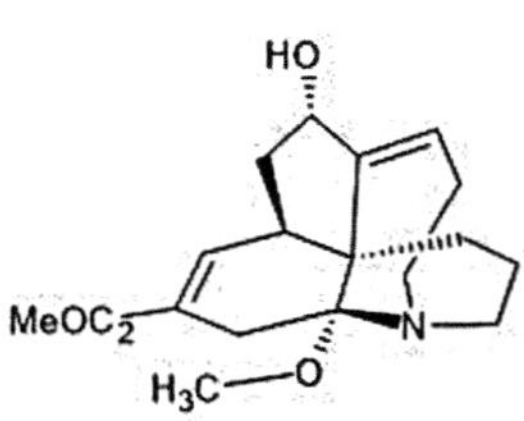

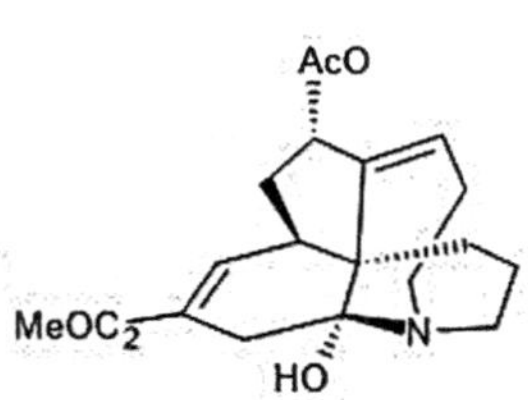

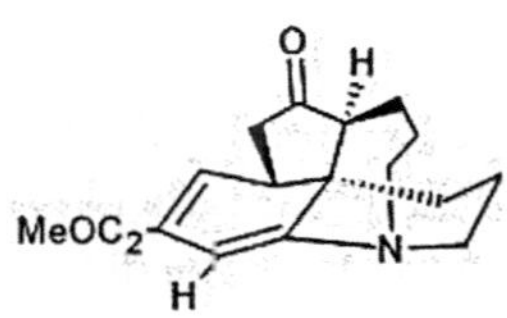

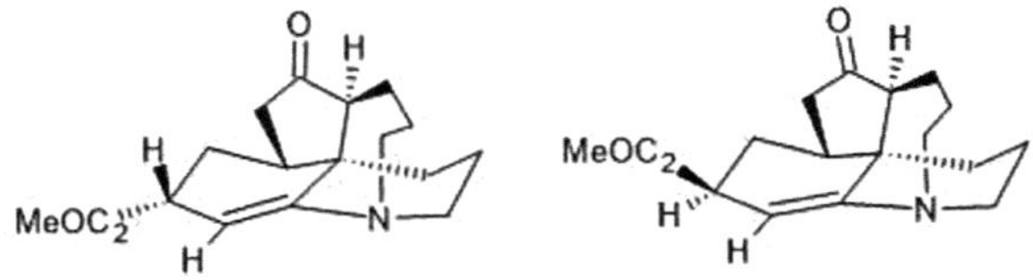

Lycogladines A-H

22.3 PHARMACOLOGY

The pharmacological importance of *L. complanatum* has been elucidated through various studies (Table 22.1) as outlined below.

22.3.1 ANTICHOLINESTERASE ACTIVITY

Orhan et al. (2009) studied the anticholinesterase activity of *L. complanatum* ssp. *chamaecyparissus* (A. Br.) Doll. Acetylcholinesterase (AChE) and butyrylcholinesterase (BChE) inhibitory activities of *L. complanatum* were evaluated by modifying the spectrophotometric method developed by Ellman et al. (1961). The extracts were prepared with petroleum ether (LCC-PE), chloroform (LCC-CHL), and methanol (MET), as well as the alkaloid fraction, and were screened against both AChE and butyrylcholinesterase (BChE) at concentrations of 0.2 and 1 mg/mL using a microplate reader assay. LCC-PE showed the highest activity of all the extracts examined against AChE, with 76.5% inhibition at 1 mg/mL concentration; the other extracts were shown to be ineffective at 0.2 and 1 mg/mL. Interestingly, LCC-PE also appeared to be quite potent against BChE at the two tested concentrations (51.3% and 69.6%). The remaining extracts showed no inhibition on AChE or BChE, even though the LCC-CHL extract was demonstrated to moderately inhibit BChE at 1 mg/mL with 48.7% inhibition (Orhan et al., 2009).

22.3.2 ANTIOXIDANT ACTIVITY

The evaluation of the antioxidant activity of the LCC extracts was measured based on the radical scavenging effect of 1,1-diphenyl-2-picrylhydrazyl (DPPH) free radical at 0.2 mg/mL using a microplate reader assay. Butylated hydroxyanisole (BHA) was taken as a reference drug. It showed only a small range of radical-scavenging activity between 22.7% and 38.6%, while BHA displayed 92.7% of scavenging activity against DPPH (Orhan et al., 2009).

22.3.3 ANTIBACTERIAL ACTIVITY

The antibacterial activity of the plant extracts (LCC) was screened against standard strains and their clinical isolates such as *Escherichia coli* (ATCC 35218), *Pseudomonas aeruginosa* (ATCC 10145), *Proteus mirabilis* (ATCC 7002), *Acinetobacter baumannii* (RSKK 02026), *Klebsiella pneumoniae* (RSKK 574), and *Staphylococcus aureus* (ATCC 25923). The extract was also compared to ampicillin, ofloxacin, and levofloxacin as the reference drugs by the microdilution method. The extracts were more effective against the ATCC strains than their isolated counterparts and had the same minimum inhibitory concentrations (MIC) against all the bacteria tested. However, they were relatively active against the ATCC strain of *S. aureus*, which had the lowest MIC value of 4 mg/mL (Orhan et al., 2009).

22.3.4 ANTIFUNGAL ACTIVITY

Antifungal activities of the plant extracts (LCC) were screened against two yeast-like fungi, *Candida albicans* (ATCC 10231) and *C. parapsilosis* (ATCC 22019) by the microdilution method. These fungi were screened by an array of antifungal tests, taking ketoconazole and fluconazole as standards. LCC extracts showed significant antifungal activity against both *C. albicans* and *C. parapsilosis*, with a MIC value of 16 mg/mL (Orhan et al., 2009).

22.3.5 ANTIVIRAL ACTIVITY

Antiviral analysis of LCC extracts was tested using both DNA and RNA viruses. Herpes simplex virus (HSV) and Parainfluenza-3 virus (PI-3) were selected for DNA and RNA viruses, respectively. Madin-Darby Bovine Kidney (MDBK) and Vero cell lines were used for the assessment of the antiviral activity tests, and the cells were cultured in Eagle's minimal essential medium (EMEM) containing 10% fetal calf serum (FCS), 100 mg/mL streptomycin, and 100 IU/mL penicillin. Acyclovir for HSV and oseltamivir for PI-3 were used as reference drugs. Both strains were evaluated based on the Frey and Liess method as TCID50. A cell culture microscope (x400) was used to determine the cells, and it was compared with treated and untreated control cultures and with acyclovir and oseltamivir as the control agents. Further, the maximum cytopathogenic effect (CPE) concentration was analyzed. They reported that the extracts were completely inactive against PI-3, and only the petroleum ether extract showed significant activity against HSV (Orhan et al., 2009).

22.3.6 ANTIPROTOZOAL ACTIVITY

Trypanosoma brucei rhodesiense, T. cruzi, Leishmania donovani, and *Plasmodium falciparum* are major disease-causing infectious parasite in various parts of the world. These protozoans cause sleeping sickness, Chagas disease, leishmaniasis, and malarial disease, respectively. The antiprotozoal activity of the plant extracts of *L. complanatum* subsp. *chamaecyparissus* (LCC) was tested against *Trypanosoma brucei rhodesiense, T. cruzi, Leishmania donovani,* and *Plasmodium falciparum.* The extracts of petroleum ether (PE), chloroform ($CHCl_3$), methanol (MeOH), and alkaloid (ALK) fractions were taken for the evaluation. The growth of *T. brucei rhodesiense* was inhibited by all the extracts, with IC50 values of 9.3 to 47.0 μg/ml. The LCC-PE extract showed the greatest antileishmanial activity, with an IC50 value of 4.5 μg/ml. The most potent activity against *P. falciparum* was exhibited by LCC-ALK, with an IC50 value of 2.7 μg/ml, and LCC-PE, with an IC50 value of 2.8 μg/ml extracts. Thus, the results reveal that the non-polar and semi-polar extracts, that is, LCC-PE, LCC-$CHCl_3$, and LCC-ALK, are generally more active than the highly polar MeOH extract, which showed no activity (Orhan et al., 2013). The effective treatment of these diseases can be improved by this finding through the synthesis of novel, easily accessible, and secure drugs.

22.3.7 CYTOTOXIC ACTIVITIES

Cytotoxic activities of 13 serratane triterpenoids, such as lycomplanatum (A-H), 3α,21β,29-trihydroxyserrat-14-en-16-one, 3β,21β-dihydroxyserrat-14-en-16-one, 3α,21β,24-trihydroxyserrat-14-en-16-one, 3β,21β,24-trihydroxyserrat-14-en-16-one, and lycernuic A, in six different concentrations against seven human cancer cell lines were tested for their effects on the proliferation of SKVO-3, DU-145, NCI-1975, BEL-7402,

MKN-45, PC-3, and MCF-7 cells by MTT assay. It showed moderate antiproliferative effects towards all cancer cell lines, especially by lycomplanatum B and lycernuic A against MCF-7 cells, with IC50 values of 13.8–44.7 μM. So, the result interprets that the growth of cancer cells can be inhibited by these compounds at a certain level (Zhou et al., 2020).

Following the identification of the novel abietane-type diterpenoid, complanatin F, along with three known compounds, complanatin A, B, and C, Zhang et al. (2023) tested the anti-tumor activity of all these compounds against A549, HepG2, and MCF-7 tumor cell lines. They evaluated the cell lines by seeding them in a 96-well plate after 24 h of pre-incubation. The antiproliferative property of the compounds was examined by comparing them with staurosporine (STS), taken as the positive control. The overall result showed that only complanatin C has inhibitory action. Its inhibition was observed against A549, with an IC50 value of 29.44 μM, and HepG2, with 43.70 μM. This finding marks a relevant step in developing a medicinally important system for the treatment of cancer.

Along with the identification of two new alkaloids, lyconadin A and 11-hydroxylycodine, Kobayashi et al. (2001) tested the cytotoxicity of these new alkaloids. They exhibited remarkable cytotoxicity in the test against murine lymphoma L1210 cells, with IC50 values of 0.46 and 5.5 μg/mL, respectively, for lyconadin A and 11-hydroxylycodine. They also showed cytotoxic properties against human epidermoid carcinoma KB cells, with IC50 values of 1.7 and >10 μg/mL, respectively. This discovery unveiled the cytotoxic potential of these alkaloids for the further advancement of cancer-curing strategies.

Polyhydroxy serratene triterpenoids, such as serrat A, serrat B, serrat C, serrat D, serrat E, serrat F, and 16-oxolyclanitin, isolated from the extract of *L. complanatum*, were tested for their cytotoxic activities against HepG2 cells, MCF-7 cells, and a series of human lung cancer cell lines, such as A549, Calu-6, NCI-H441, NCI-H226, and NCI-H1975, by CCK-8 assay. The results showed that serrat A, serrat C, and serrat E have proliferation inhibition effects on human lung cancer cell lines to varying extents. Serrat B, serrat E, and serrat G have moderate inhibitory activities in HepG2 cells. Furthermore, in MCF-7 cells, every compound exhibits specific cytotoxic activities with IC50 values < 30 μM, whereas compounds such as serrat C and serrat D had IC50 values < 20 μM. The results indicate that every compound of polyhydroxy serratene triterpenoids has a remarkable capacity to combat cancer development (Dong et al., 2019).

22.3.8 ANTI-INFLAMMATORY ACTIVITIES

The inhibition of nitric oxide (NO) production is potentially beneficial for reducing inflammation. By using the Griess assay, the anti-inflammatory properties of 10 molecules from *L. complanatum* [lycomplanatum (A-E, H), 3α,21β,29-trihydroxyserrat-14-en-16-one, 3β,21β-dihydroxyserrat-14-en-16-one, 3β,21β,24-trihydroxyserrat-14-en-16-one, and lycernuic A] were tested (25, 50, and 100 μM concentrations) for the inhibition of intracellular NO that was produced after treating RAW 264.7 cells with LPS (lipopolysaccharide), along with the cytotoxic effects of these compounds on LPS-treated RAW 264.7 cells. At higher concentrations (≥50 μM), lycomplanatum B, E, 3β,21β-dihydroxyserrat-14-en-16-one, and lycernuic A showed cytotoxic effects on RAW 264.7 cells. All the tested phytoconstituents were found to be effective in inhibiting the LPS-induced NO production in a concentration-dependent manner. Furthermore, the inhibitory effects of lycomplanatum B at 50 μM and lycernuic A at 25 μM on NO production were noticeable compared with other molecules (Zhou et al., 2020). It specifically lays out the pharmacodynamic principle behind the anti-inflammatory properties of *L. complanatum.*

22.3.9 BETA-SECRETASE 1 (BACE1) INHIBITION ACTIVITY

Serratene-type triterpenoids such as 21β,24-dihydroxyserrat-14-en-3α-yl acetate, 16-oxo-21β,24-dihydroxyserrat-14-en-3α-yl acetate, 3α,21β-dihydroxyserrat-14-en-16-one, 21α-hydroxyserrat-14-en-3β-yl acetate, serrat-14-en-3β,21α-diol, 16-oxo-21β-hydroxyserrat-14-en-3α-yl acetate, 21β-hydroxyserrat-14-en-3α-yl acetate, 21β-hydroxyserrat-14-en-3β-yl acetate, lycoclavanol, and serrat-14-en-3β,21β-diol isolated from *L. complanatum* were tested for BACE1 inhibition activity. The β-amyloid precursor protein (APP) will not cleave into β peptides when BACE1 is inhibited by these alkaloids, thereby reducing the risk of Alzheimer's disease. The two alkaloids, 21β,24-dihydroxyserrat-14-en-3α-yl acetate and 16-oxo-21β-hydroxyserrat-14-en-3α-yl acetate, have IC50 values of 2.79 ± 0.28 and 2.49 ± 0.12 μM, respectively, showing that they have significant BACE1 inhibition activity and therapeutic potential to regulate Alzheimer's disease (Nguyen et al., 2017).

22.3.10 T-TYPE CALCIUM CHANNEL (TTCC) INHIBITOR ACTIVITY

In modern drug development systems, T-type calcium channels (Cav3.1–3.3) have great interest in treating some of the major therapeutic problems related to neuropathic pain, absence epilepsy, insomnia, and Parkinson's disease. Therefore, a C16N-type lycopodium alkaloid, lycoplanine A, isolated from *L. complanatum*, was evaluated for its TTCC inhibitor activity. The ethanol extract of the compound showed a distinct inhibitory action on Cav3.1 TTCC, reaching a peak current with an IC50 value of 6.06 μM. The Cav3.1 inhibitory effect was a marked discovery in the medicinal system for neuropathic treatment, including arthritis (Zhang et al., 2017) (Table 22.1).

TABLE 22.1 Major Phytochemicals and Their Bioactivities of *L. complanatum*

Sl. No.	Phytochemical	Bioactivity	Reference
1.	Complanadine A	Induction of neurotrophic factor secretion, anti-hyperalgesic, neurotrophic factor biosynthesis in 1321N1 human astrocytoma cells	Johnson and Siegel (2014); Kobayashi et al. (2000); Yuan et al. (2010)
2.	Lyconadin A	Cytotoxic activity against murine lymphoma L1210 and human epidermoid carcinoma KB cells	Kobayashi et al. (2001) Loertscher et al. (2013)
3.	11-hydroxy lycodine	Cytotoxic activity against murine lymphoma L1210 cells and human epidermoid carcinoma KB cells	Kobayashi et al. (2001)
4.	Lycopladine A	weak cytotoxicity against murine lymphoma L1210 cells	Ishiuchi et al. (2006a)
5.	Complanadine C	antimicrobial activity against *Cryptococcus neoformans* and *Aspergillus niger*	Ishiuchi et al. (2007)
6.	Complanadine D	Antimicrobial activity against *Cryptococcus neoformans* and *Aspergillus niger*, cytotoxicity against murine leukemia L1210 cells, neurotrophic factor biosynthesis in 1321N1 human astrocytoma cells	Ishiuchi et al. (2007); Ishiuchi et al. (2011b)
7.	Lycomplanatums A-H	Moderate antiproliferative effects in all cancer cell lines and Anti-inflammatory activities	Zhou et al. (2020)
8.	Lycopladine E	Neurotrophic factor biosynthesis in 1321N1 human astrocytoma cells and enhances the mRNA expressions for NGF	Kubota et al. (2007)
9.	Complanadine E	Neurotrophic factor biosynthesis in 1321N1 human astrocytoma cells	Ishiuchi et al. (2011)b

TABLE 22.1 *(Continued)*

Sl. No.	Phytochemical	Bioactivity	Reference
10.	Complanatumol D	Cytotoxic activity D against human leukemia K562/S and the doxorubicin-resistant K562/R cell lines	Yan et al. (2009)
11.	Lyconadin C	Anti-neurodegenerative agents	Yang et al. (2014)
12.	Complanatin C and D	Cytotoxic activities against human lung cancer cell lines (MSTO-211H, NCI-H2052 and NCIH226)	Weng et al. (2018)
13.	Serrat A, Serrat C, and Serrat E	Moderate cytotoxicities against MCF-7 cells and a series of lung cancer cell lines.	Dong et al. (2019)
14.	Serrat B, Serrat E, and Serrat G	Cytotoxicities against MCF-7 cells and inhibited HepG2 cells moderately	Dong et al. (2019)
15.	Complanadine F	Moderate cholinesterase (from bovine erythrocyte) inhibitory activity	Hirasawa et al. (2021)
16.	1β 11-epoxyabieta-12-hydroxy-8, 11, 13-triene-7-one	Inhibitory activities against A549, HepG2 and MCF-7 tumor cell lines	Zhang et al. (2023)
17.	21β,24-dihydroxyserrat-14-en-3α-yl acetate and 16-oxo-21β-hydroxyserrat-14-en-3α-yl acetate	To treat Alzheimer's disease by Beta-secretase 1(BACE1) inhibition activity	Nguyen et al. (2017)
18.	Lycoplanine A,	Cav3.1 T-type calcium channel (TTCC) inhibitor with an IC_{50} value of 6.06 μM	Zhang et al (2017)

KEYWORDS

- **ampicillin**
- **bioactives**
- **club moss**
- **cytopathogenic effect**
- **leishmaniasis**
- **lycopodine**
- **lycopodium alkaloids**
- **xanthoperol**

REFERENCES

Cheng, J. T., Liu, F., Li, X. N., Wu, X. D., Dong, L. B., Peng, L. Y., Huang, S. X., He, J., & Zhao, Q. S. (2013). Lycospidine A, a new type of lycopodium alkaloid from *Lycopodium complanatum*. *Organic Letters, 15*(10), 2438–2441. https://doi.org/10.1021/ol400907v.

Cheng, J. T., Zhang, Z. J., Li, X. N., Peng, L. Y., Luo, H. R., Wu, X. D., & Zhao, Q. S. (2016). Lyconadins G and H, two rare lyconadin-type lycopodium alkaloids from *Lycopodium complanatum*. *Natural Products and Bioprospecting, 6*(6), 279–284. https://doi.org/10.1007/s13659-016-0111-9.

Dong, Q., Zou, Z., Jia, X., Yu, X., Li, J., Zhou, W., Sun, H., Wu, W., Tan, G., & Xu, K. (2019). Cytotoxic polyhydroxy serratene triterpenoids from *Lycopodium complanatum*. *Bioorganic Chemistry, 87*, 373–379. https://doi.org/10.1016/j.bioorg.2019.03.011.

eFlora. (2024). *Lycopodium complanatum* in Flora of China @ efloras.org. http://www.efloras.org/florataxon.aspx?flora_id=2&taxon_id=242330823.

Ellman, G. L., Courtney, K. D., Andres, V., & Featherstone, R. M. (1961). A new and rapid colorimetric determination of acetylcholinesterase activity. *Biochemical Pharmacology, 7*, 88–95. https://doi.org/10.1016/0006-2952(61)90145-9.

Gísladóttir, G. (2022). *Lycopodium alkaloids and immunomodulation—Comparative alkaloid fingerprinting and immunomodulating activity of extracts from club mosses*. Doctoral dissertation, University of Iceland, Reykjavík, Iceland].

Halldorsdottir, E., Kowal, N., & Olafsdottir, E. (2015). The genus *Diphasiastrum* and its lycopodium alkaloids. *Planta Medica, 81*(12/13), 995–1002. https://doi.org/10.1055/s-0035-1546182.

Hirasawa, Y., Kakimoto, M. A., Zaima, K., Uchiyama, N., Goda, Y., & Morita, H. (2021). Complanadine F, a novel dimeric alkaloid from *Lycopodium complanatum*. *Journal of Natural Medicines, 75*(2), 403–407. https://doi.org/10.1007/s11418-020-01476-9.

Inubushi, Y., Sano, T., & Price, J. (1967). Triterpene constituents of *Lycopodium complanatum* L. from New Guinea. *Australian Journal of Chemistry, 20*(2), 387–387. https://doi.org/10.1071/ch9670387.

Ishiuchi, K., Kubota, T., & Morita, H., & Kobayashi, J. (2006a). Lycopladine A, a new C16N alkaloid from *Lycopodium complanatum*. *Tetrahedron Letters, 47*(19), 3287–3289. https://doi.org/10.1016/j.tetlet.2006.03.027.

Ishiuchi, K., Kubota, T., Hayashi, S., Shibata, T., & Kobayashi, J. (2009a). Lycopladines F and G, new C16N2-type alkaloids with an additional C4N unit from *Lycopodium complanatum*. *Tetrahedron Letters, 50*(29), 4221–4224. https://doi.org/10.1016/j.tetlet.2009.04.139.

Ishiuchi, K., Kubota, T., Hayashi, S., Shibata, T., & Kobayashi, J. (2009b). Lycopladine H, a novel alkaloid with fused-tetracyclic skeleton from *Lycopodium complanatum*. *Tetrahedron Letters, 50*(47), 6534–6536. https://doi.org/10.1016/j.tetlet.2009.09.035.

Ishiuchi, K., Kubota, T., Hoshino, T., Obara, Y., Nakahata, N., & Kobayashi, J. (2006b). Lycopladines B–D and lyconadin B, new alkaloids from *Lycopodium complanatum*. *Bioorganic & Medicinal Chemistry, 14*(17), 5995–6000. https://doi.org/10.1016/j.bmc.2006.05.028.

Ishiuchi, K., Kubota, T., Ishiyama, H., Hayashi, S., & Sato, T., & Kobayashi, J. (2011a). Lyconadins C and F, new lycopodium alkaloids from *Lycopodium complanatum*. *Tetrahedron Letters, 52*(2), 289–292. https://doi.org/10.1016/j.tetlet.2010.11.024.

Ishiuchi, K., Kubota, T., Ishiyama, H., Hayashi, S., Sato, T., Mori, K., Obara, Y., Nakahata, N., & Kobayashi, J. (2011b). Lyconadins D and E, and complanadine E, new lycopodium alkaloids from *Lycopodium complanatum*. *Bioorganic & Medicinal Chemistry, 19*(2), 749–753. https://doi.org/10.1016/j.bmc.2010.12.025.

Ishiuchi, K., Kubota, T., Mikami, Y., Obara, Y., Nakahata, N., & Kobayashi, J. (2007). Complanadines C and D, new dimeric alkaloids from *Lycopodium complanatum*. *Bioorganic & Medicinal Chemistry, 15*(1), 413–417. https://doi.org/10.1016/j.bmc.2006.09.043.

Johnson, T., & Siegel, D. (2014). Complanadine A, a selective agonist for the Mas-related G protein-coupled receptor X2. *Bioorganic & Medicinal Chemistry Letters, 24*(15), 3512–3515. https://doi.org/10.1016/j.bmcl.2014.05.060.

Kobayashi, J., Hirasawa, Y., Yoshida, N., & Morita, H. (2000). Complanadine A, a new dimeric alkaloid from *Lycopodium complanatum*. *Tetrahedron Letters, 41*(47), 9069–9073. https://doi.org/10.1016/s0040-4039(00)01630-0.

Kobayashi, J., Hirasawa, Y., Yoshida, N., & Morita, H. (2001). Lyconadin A, a novel alkaloid from *Lycopodium complanatum*. *Journal of Organic Chemistry, 66*(17), 5901–5904. https://doi.org/10.1021/jo0103874.

Kubota, T., Yahata, H., Ishiuchi, K., Obara, Y., Nakahata, N., & Kobayashi, J. (2007). Lycopladine E, a new C16N1-type alkaloid from *Lycopodium complanatum*. *Heterocycles, 74*(1), 843–843. https://doi.org/10.3987/com-07-s(w)76.

Loertscher, B. M., Zhang, Y., & Castle, S. L. (2013). Exploration of an epoxidation–ring-opening strategy for the synthesis of lyconadin A and discovery of an unexpected Payne rearrangement. *Beilstein Journal of Organic Chemistry, 9*(1), 1179–1184. https://doi.org/10.3762/bjoc.9.132.

Ma, X., & Gang, D. R. (2004). The lycopodium alkaloids. *Natural Product Reports, 21*(6), 752–772. https://doi.org/10.1039/B409720N (accessed on 17th November 2025).

Moerman, D. (2009). *Native American Medicinal Plants: An Ethnobotanical Dictionary* (p. 799). Timber Press (OR).

Morita, H., Ishiuchi, K., Haganuma, A., Hoshino, T., Obara, Y., Nakahata, N., & Kobayashi, J. (2005). Complanadine B, obscurumines A and B, new alkaloids from two species of *Lycopodium*. *Tetrahedron, 61*(8), 1955–1960. https://doi.org/10.1016/j.tet.2005.01.011.

Nguyen, V. T., Zhao, B. T., Seong, S. H., Kim, J. A., Woo, M. H., Choi, J. S., & Min, B. S. (2017). Inhibitory effects of serratene-type triterpenoids from *Lycopodium complanatum* on cholinesterases and β-secretase 1. *Chemico-Biological Interactions, 274*, 150–157. https://doi.org/10.1016/j.cbi.2017.07.00.

Orhan, İ. E., Özçelık, B., Aslan, S., Kartal, M., Karaoğlu, T., Şener, B., Terzioğlu, S., & Choudhary, M. I. (2009). *In vitro* biological activity screening of *Lycopodium complanatum* L. ssp. *chamaecyparissus* (A. Br.) Döll. *Natural Product Research, 23*(6), 514–526. https://doi.org/10.1080/14786410802163743.

Orhan, I. E., Sener, B., Kaiser, M., Brun, R., & Tasdemir, D. (2013). Antiprotozoal activity and cytotoxicity of *Lycopodium clavatum* and *Lycopodium complanatum* subsp. *chamaecyparissus* extracts. *Turkish Journal of Biochemistry, 38*(4), 403–408. https://doi.org/10.5505/tjb.2013.07379.

Patil, S. P. (2020). Recently isolated lycodine-type lycopodium alkaloids and their total synthesis: A review. *Future Journal of Pharmaceutical Sciences, 6*(1). https://doi.org/10.1186/s43094-020-00108-z.

POWO. (2024). *Lycopodium complanatum* L. *Plants of the World Online*. https://powo.science.kew.org/taxon/urn:lsid:ipni.org:names:147262-2.

Tang, Y., Xiong, J., & Hu, J.-F. (2015). Lycopodium alkaloids from *Diphasiastrum complanatum*. *Natural Product Communications, 10*(12), 1934578X1501001. https://doi.org/10.1177/1934578x1501001219.

Wang, B., Guan, C., & Fu, Q. (2021). The traditional uses, secondary metabolites, and pharmacology of *Lycopodium* species. *Phytochemistry Reviews, 21*, 1–79. https://doi.org/10.1007/s11101-021-09746-4.

Weng, Y., Yu, X., Li, J., Dong, Q., Li, F., Cheng, F., Zhang, Y., Yao, C., Zou, Z., Zhou, W., Tan, G., & Xu, K. (2018). Abietane diterpenoids from *Lycopodium complanatum*. *Fitoterapia, 128*, 135–141. https://doi.org/10.1016/j.fitote.2018.05.014.

Wu, X. D., He, J., Xu, G., Peng, L. Y., Song, L. D., & Zhao, Q. S. (2010). Diphaladine A, a new lycopodium alkaloid from *Diphasiastrum complanatum* (Lycopodiaceae). *Acta Botanica Yunnanica, 31*(1), 93–96. https://doi.org/10.3724/SP.J.1143.2009.08197.

Yan, J., Sun, L., Li, W., Zhou, L., Li, Z., Zhang, X., Yang, L., & Qiu, M. (2009). Cytotoxic serratane triterpenes from *Diphasiastrum complanatum* with a hydroxy group at C-27. *Planta Medica, 76*(4), 353–357. https://doi.org/10.1055/s-0029-1186168.

Yan, J., Yi, P., Chen, B., Lu, L., Li, Z., Zhang, X., Zhou, L., & Qiu, M. (2008). Polyhydroxyserratane triterpenoids from *Diphasiastrum complanatum*. *Phytochemistry, 69*(2), 506–510. https://doi.org/10.1016/j.phytochem.2007.07.024.

Yang, Y., Haskins, C. W., Zhang, W., Low, P. L., & Dai, M. (2014). Divergent total syntheses of lyconadins A and C. *Angewandte Chemie International Edition, 53*(15), 3922–3925. https://doi.org/10.1002/anie.201400416.

Yuan, C., Chang, C. T., Axelrod, A., & Siegel, D. (2010). Synthesis of (+)-complanadine A, an inducer of neurotrophic factor excretion. *Journal of the American Chemical Society, 132*(17), 5924, 5925. https://doi.org/10.1021/ja101956x.

Zhang, Y. G., Cheng, F., Tan, J. B., Wu, J. P., Weng, Y. M., Zhou, W. H., Wan, Z. Y., Guo, K., Zhang, Q., Lu, Y. T., He, X. A., Zhu, G. Z., Liu, Y. H., & Xu, K. P. (2023). A new abietane diterpenoid from *Lycopodium complanatum*. *Journal of Asian Natural Products Research, 25*(12), 1217–1222. https://doi.org/10.1080/10286020.2023.2209567.

Zhang, Z. J., Nian, Y., Zhu, Q. F., Li, X. N., Su, J., Wu, X. D., Yang, J., & Zhao, Q. S. (2017). Lycoplanine A, a C16N lycopodium alkaloid with a 6/9/5 tricyclic skeleton from *Lycopodium complanatum*. *Organic Letters, 19*(17), 4668–4671. https://doi.org/10.1021/acs.orglett.7b02293.

Zhang, Z. J., Qi, Y.-Y., Wu, X.-D., Su, J., & Zhao, Q.-S. (2018b). Lycogladines A–H, fawcettimine-type lycopodium alkaloids from *Lycopodium complanatum* var. *glaucum* Ching. *Tetrahedron, 74*(14), 1692–1697. https://doi.org/10.1016/j.tet.2018.02.034.

Zhang, Z. J., Zhu, Q. F., Su, J., Wu, X. D., & Zhao, Q. S. (2018a). Lycoplanines B–D, three lycopodium alkaloids from *Lycopodium complanatum*. *Natural Products and Bioprospecting, 8*(3), 177–182. https://doi.org/10.1007/s13659-018-0161-2.

Zhou, W., Kang, F., Huang, L., Li, J., Wang, W., Xiao, L., Wen, Q., Yu, X., Xu, Y., Zou, Z., Zhou, H., Zang, H., Chen, S., & Xu, K. (2020). Serratane triterpenoids from *Lycopodium complanatum* and their anti-cancer and anti-inflammatory activities. *Bioorganic Chemistry, 101*, 103959–103959. https://doi.org/10.1016/j.bioorg.2020.103959.

CHAPTER 23

Phytochemical and Pharmacological Review of the Medicinal Fern *Pyrrosia* Mirbel (Polypodiaceae)

SACHIN M. PATIL[1] and BHARATKUMAR JADHAV[2]

[1]*Department of Botany, Dr. Ghali College, Gadhinglaj (affiliated to Shivaji University, Kolhapur), Maharashtra, India*

[2]*Department of Botany, Balasaheb Desai College, Patan, Maharashtra, India*

*Corresponding authors

ABSTRACT

Pyrrosia is an epiphytic fern that grows wildly in tropical and temperate regions and has been used as traditional medicine by aboriginal people to cure various health problems such as colds, coughs, fevers, dysentery, breast cancer, soothing gallstones, urinary diseases, and skin rashes, or used as pain relievers and diuretics. The present inclusive review on the phytochemistry and pharmacology of the genus *Pyrrosia* enhances our understanding and emphasizes the importance of future investigation and a foundation for future research in this genus.

23.1 INTRODUCTION

The genus *Pyrrosia* Mirbel belongs to the family Polypodiaceae and has more than 100 species distributed in tropical and subtropical regions of Africa, Asia, and Oceania, extending north to the Himalayas, central China, and Japan in Asia. The species are often found attached to aged,

Bioactives and Pharmacology of Pteridophytes. T. Pullaiah (Ed.)

DOI: 10.1201/9781779642738-23

large trees or on moist rocky substratum (Hovenkamp, 1986). The genus is epiphytic or epilithic, characterized by having a long creeping rhizome, with simple, monomorphic to dimorphic fronds, and stellate hairs on the lamina. Red-colored sori are present either on half or the entire ventral surface of the lamina. Several reports have shown that the frond or whole plant decoction of different *Pyrrosia* species has been used as traditional medicine for the treatment of coughs, fevers, acute pyelonephritis, chronic bronchitis, bronchial asthma, dysentery, skin diseases, urinary calculi, and bloody urine (Chen et al., 2015; Cheng et al., 2014; Cui et al., 2016; George et al., 2017; Patil et al., 2013, 2020; Seno et al., 2022; Yang et al., 2003). Therefore, pteridologists around the world have screened the plants of *Pyrrosia* for phytochemical constituents and different activities like antioxidant, antibacterial, antifungal, and anti-inflammatory.

23.2 PHYTOCHEMICAL CONSTITUENTS

Chemical constituents of *P. gralla* from China were studied by Zheng et al. (1998). Phytochemical constituents using the ethanol fraction from the whole plant extract of *Pyrrosia calvata* were carried out by Chen et al. (2015), who isolated 17 compounds, of which one is a novel flavanone glycoside. The phytoconstituents include 3′,5′,5,7-tetrahydroxy-6-C-β-D-glucopyranosyl-flavanone, (R/S)-eriodictyol-8-C-β-D-glucopyranoside, quercetin-3-O-α-D-rhamnosyl (1′→3′)-β-D-glucopyranoside, hemipholin, 4β-carboxymethyl-(-)-epicatechin methyl ester, kaempferol, quercetin, mangiferin, chlorogenic acid, 1,5-O-dicaffeoylquinic acid, 3,5-O-dicaffeoylquinic acid, 3-O-caffeoylquinic acid methyl ester, 1-O-caffeoyl glycoside, 4-O-β-D-glucopyranosyl-caffeic acid, 3′-O-methylepicatechin-7-O-β-D-glucopyranoside, hop-22(29)-en-30-ol, and diploptene.

Pyrrosia davidii (Baker) Ching was evaluated for phytochemical constituents, and about 20 compounds were isolated from the fern. These compounds included three steroids, one pigment, 11 flavonoids, two terpenoids, one phenolic acid, and two fatty acids (Wang et al., 2023). *P. lingua* (Thunb.) Farwell is a Japanese folk medicine used as a diuretic. The hexane fraction of fresh rhizomes of *P. lingua* yielded six phytoconstituents, of which five are new dammarane triterpenoids, identified as octanordammarane, (18S)-18-hydroxydammar-21-ene, (18S)-Pyrrosialactone, (18S)-Pyrrosialactol, and 3-deoxyocotillol, along with the known dammara-18(28),21-diene (Yamashita et al., 1998). The diuretic potential compounds were identified from the 95% ethanol fraction of *P. petiolosa* (Christ) Ching. Among these was a new benzanilide, that is, petiolide A, and others were phenolic derivatives barbatic acid and kaempferol (Lang et al., 2021).

A phytoconstituent, *Pyrrosia*e *folium*, was isolated along with other phytoconstituents like chlorogenic acid, mangiferin, isomangiferin, rutin, and kaempferol from the aerial parts of *Pyrrosia sheareri* (Bak.) Ching, *Pyrrosia lingua* (Thunb) Farwell, and *Pyrrosia petiolosa* (Christ) Ching. The *Pyrrosia*e *folium* has been used in the Chinese Pharmacopoeia and in clinics to treat acute pyelonephritis, chronic bronchitis, and bronchial asthma (Cui et al., 2016). The whole plant extract of *P. sheareri* (Baker) Ching was used to study the phytoconstituents and isolated six compounds, of which five are known compounds (mangiferin, kaempferol, quercetin, chlorogenic acid, diploptene, etc.) and a new xanthone glycoside, 3,5,7,8-tetramethoxyxanthone-1-O-β-D-glucopyranoside (Cui et al., 2016).

The chemical components of essential oil from *Pyrrosia tonkinensis* were evaluated by Xin et al. (2016). They reported that 28 compounds, representing 88.1% of the total essential oil, were identified. The major volatile components were trans-2-hexenal (22.1%), followed by nonanal (12.8%), limonene (9.6%), phytol (8.4%), 1-hexanol (3.8%), 2-furancarboxaldehyde (3.5%), and heptanal (3.1%). The whole plant extract of *Pyrrosia sheareri* was used to study the phytochemical constituents by He et al. (2022). They isolated six compounds, that is, 3,5,7,8-tetramethoxyxanthone-1-O-β-D-glucopyranoside, mangiferin, kaempferol, quercetin, chlorogenic acid, and diploptene, of which 3,5,7,8-tetramethoxyxanthone-1-O-β-D-glucopyranoside is a new xanthone glycoside.

23.3 PHARMACOLOGY

23.3.1 ANTIOXIDANT PROPERTIES

Antioxidant properties like DPPH, FRAP, and ABTS of different species of *Pyrrosia* were carried out by different researchers from around the world. DPPH scavenging activity from different fractions (ethyl acetate, methanol, and hexane) of the epiphytic fern, *Pyrrosia heterophylla*, was studied by George et al. (2017). The ethyl acetate (PHE) and methanolic (PHM) fractions of *P. heterophylla* exhibited significant scavenging of DPPH radicals. The results showed a concentration-dependent reaction and were statistically significant ($p < 0.001$). PHE exhibited the highest antioxidant capacity in the DPPH radical scavenging assay with 86.63 ± 0.85% inhibition at the maximum tested concentration. PHM showed a maximum inhibition of 66.36 ± 1.05%, whereas the *P. heterophylla* hexane fraction was found to have low inhibitory activity with a maximum inhibition of 7.56 ± 0.40%.

Antioxidant activity of *P. lanceolata* was measured using the DPPH method, and as IC50 and AAI values are shown, the ethanolic extract of *P. lanceolata* (1.4 ± 0.0 μg/mL) has moderate antioxidant activity (Komala et al., 2015). The antioxidant activity of *P. lanceolata* from different fractions (n-hexane, dichloromethane, ethyl acetate, aqueous, and ascorbic acid) was measured with a free radical scavenging assay using DPPH as the radical. The n-hexane extract showed no activity with IC50 values greater than 500 μg/mL. However, the ethyl acetate extract demonstrated a high level of antioxidant activity with an IC50 value of 12.08 ± 0.27 μg/mL (Indriani et al., 2023).

The ethanol extract of *P. petiolosa* showed antioxidant activity as determined by free radical scavenging. The IC50 value (the concentration required to inhibit radical formation by 50%) of *P. petiolosa* extract in the DPPH assay is 140 μg/mL (Hsu, 2008). Chen et al. (2023) studied the DPPH scavenging activity of *P. petiolosa* collected from Guizhou (GZ) and Guangxi (GX) provinces of China. About 13 flavonoid compounds were identified as potential biomarkers for distinguishing between the two major regions, Guizhou (GZ) and Guangxi (GX). The activity in the samples from different geographic locations showed significant variation, suggesting that variation in the flavonoid profile might significantly affect DPPH scavenging activity. Akhmadjon et al. (2020) studied the DPPH radical scavenging activity from different fractions of the whole plant of *P. lingua*. DPPH radical scavenging activity in water and ethanol fractions showed higher activity than in other fractions. Cao et al. (2015) studied the DPPH radical scavenging activity of different species of *Pyrrosia*. They reported that *P. nummulariifolia* and *P. petiolosa* showed higher DPPH radical scavenging activity than *P. muda* and *P. lingua.*

The FRAP antioxidant assay of *P. heterophylla* was studied by George et al. (2017). The antioxidant compounds are capable of donating an electron to reduce Fe^{3+} to Fe^{2+}, and the amount of Fe_{2+} complex formed can be determined by measuring the absorbance of the blue-colored ferric ferrocyanide complex at 700 nm. Higher absorbance of the extracts indicates greater activity and vice versa. The highest reducing power was exhibited by *P. heterophylla* ethanolic, whereas low reducing power was observed in the n-hexane fraction.

Chen et al. (2023) studied the FRAP antioxidant activity of *P. petiolosa* collected from different locations in Guizhou (GL, ZN, ZY, AL, PD) and Guangxi (GC, GY, QX, QZ) provinces of China. FRAP values of the Guizhou samples were significantly higher than those of the Guangxi samples. The activity in the samples from the two provinces showed significant variation, suggesting that variation in the flavonoid profile might significantly affect FRAP activity. Samples from location ZN of Guizhou province had the most potent FRAP activity (328.12 ± 10.75 mg TE/g dw), indicating that material from the ZN location is probably the best for medicinal use.

The radicals of ABTS are generated in the aqueous phase when potassium persulfate is present. This assay measures the relative ability of an antioxidant to scavenge the ABTS radicals thus formed. ABTS radical scavenging activity of *P. heterophylla* was studied by George et al. (2017). The result showed that the highest scavenging was exhibited in the ethanolic fraction of *P. heterophylla* (89.48 ± 2.08%), followed by the methanolic fraction (53.25 ± 0.86%), and very low activity with minimum inhibition (9.04 ± 0.48%) was observed in the n-hexane fraction.

Chen et al. (2023) studied the ABTS scavenging activity of *P. petiolosa* collected from different locations in Guizhou (GL, ZN, ZY, AL, PD) and Guangxi (GC, GY, QX, QZ) provinces of China. ABTS activity in the samples from the two provinces showed significant variation. The Guizhou samples were significantly higher than those of the Guangxi samples. Samples from location ZN of Guizhou province had the most potent ABTS value, indicating that material from the ZN location is probably the best for medicinal use.

ABTS radical scavenging activity of different ferns (including *Pyrrosia* species) was carried out by Cao et al. (2015). All the fractions from 15 ferns showed a remarkable variation in their ABTS radical scavenging potential, and the value of IC50 ranged from 6.4 to 31.0 mg/L. The fractions from *P. nummulariifolia* and *P. petiolosa* exhibited very high ABTS radical scavenging potential, whereas *P. lingua* did not reach up to 50%, which means they had very low ABTS radical scavenging activity. The extracts from 15 ferns with higher total flavonoid content showed higher ABTS radical scavenging potential.

23.3.2 ANTIBACTERIAL PROPERTY

Microorganisms developing resistance to traditional antibiotics is the major challenge in the treatment of diseases. Therefore, nowadays, scientists are searching for prominent drugs from plants. Traditionally, humans used crude extracts of various plant species to cure diseases. Pteridophytes are also used as traditional medicine in the Indian and Chinese traditional medicine systems. George et al. (2017) studied the antimicrobial properties of the epiphytic fern, *P. heterophylla* (L.) M. G. Price, from different solvent fractions. The tested microorganisms used for the study are human pathogenic species that cause different diseases. The results revealed that all the solvent fractions inhibited the growth of the tested microorganisms. Among all tested solvent fractions, the hexane extract exhibited the maximum inhibition against *E. coli* (18.67±1.15 mm), *S. aureus* (11.67±0.58 mm), and *A. niger* (19.33±1.15 mm). However, the ethyl acetate extract possessed higher inhibition against *S. mutans* (17±1.73 mm), the methanol extract was the most effective against *K. pneumoniae* (24.67±0.58 mm), and *P. aeruginosa* (21.00±0 mm). The MIC values of all the extracts against the tested bacteria were found to be below 40 µg/ml; hence, all these extracts of *P. heterophylla* are proved to exhibit good antibacterial activity. The antibacterial activity of the different fractions (aqueous, n-butanol, ethanol, ethyl acetate, and petroleum ether) of a traditional Chinese medicinal fern, *Pyrrosia petiolosa,* was carried out by Cheng et al. (2014). The plant parts were commonly used as traditional Chinese medicine for the treatment of acute pyelonephritis, chronic bronchitis, and bronchial asthma. The ethanolic extract showed good antibacterial activity with MIC values ranging from 1.25 to 10.00 mg/mL, and most of the MBC values were equal to or twice as high as the corresponding MIC values. Also, results revealed that petroleum ether and n-butanol fractions inhibited all the tested microorganisms with the maximum inhibition zone (15.25±0.35 mm). The ethyl acetate fraction exhibited good antibacterial activity except against *Pseudomonas aeruginosa* ATCC 27853. Song et al. (2016) studied the antibacterial activity of the *P. petiolosa* ethyl acetate fraction against *Staphylococcus aureus* to decrease the hla and sea virulence genes. The results of the antibacterial assay revealed that the *P. petiolosa* ethyl acetate fraction had strong inhibitory activity against *S. aureus*, with MIC and MBC of 7.8 and 15.6 mg/mL, respectively. Also, they observed that the transcriptional levels of hla and sea virulence genes were reduced to 14.33% and 46.39%

at the MIC compared to the control. The essential oils were extracted from six *Pyrrosia* species (*P. dividii, P. gralla, P. porosa, P. lingua, P. subfurfuracea, P. tonkinensis*) and tested against gram-positive and gram-negative bacteria (Fan et al., 2020). The results showed a broad spectrum of antibacterial activity at the concentration tested, but there were significant differences among *Pyrrosia* species. However, the essential oils extracted from fern species like *P. lingua, P. porosa,* and *P. subfurfuracea* showed the highest antibacterial activity against gram-positive *Staphylococcus aureus* (ATCC 25923) with MICs of 2.5 μL/mL, 5.0 μL/mL, and 5.0 μL/mL, respectively. Also, they showed the highest antibacterial activity against the gram-negative bacteria *Shigella flexneri* (12022) with an MIC of 5.0 μL/mL, respectively. Fan et al. (2020) extracted essential oils from *P. petiolosa* samples collected at different localities and tested them against gram-positive and gram-negative bacteria. The results demonstrated that the antibacterial activity of *P. petiolosa* varied in different producing areas. The highest antimicrobial activity was observed in the *P. petiolosa* sample collected from Sichuan and Henan. Also, it was observed that there was significantly higher antibacterial activity against *Shigella flexneri* and *Staphylococcus aureus* (MICs of 2.5 μL/mL).

23.3.3 ANTIFUNGAL PROPERTY

Many ferns were used to cure fungal diseases in traditional medicine systems like Ayurveda, Unani, and Chinese medicine systems. Recently, different types of alkaloids, flavonoids, and phenolic compounds were isolated from ferns, which have antifungal properties. The whole plant extracts of *P. heterophylla* were used to study the antifungal activity, which was determined by the agar well diffusion method. Two different species of fungi, such as *Candida albicans* and *Aspergillus niger*, were swabbed. From the results of the antimicrobial assay, ethyl acetate and hexane extracts possessed higher inhibition against *C. albicans* (23.67±0.58 mm) and *A. niger* (19.33±1.15 mm) compared to other extracts (George et al., 2017). The antifungal analysis of essential oil extracted from *P. petiolosa* collected from different localities was studied by Fan et al. (2020). Samples collected from Henan localities recorded higher antimicrobial activity against *Candida albicans* with a MIC value of 2.5 μL/mL. Also, Fan et al. (2020) extracted essential oils from six *Pyrrosia* species (*P. dividii, P. gralla, P. porosa, P. lingua, P. subfurfuracea, P. tonkinensis*) and tested against *Candida albicans*. *P. subfurfuracea* showed antifungal activity against *Candida albicans* (ATCC 90028) with a MIC of 5 μL/mL.

ACKNOWLEDGMENT

The author, Dr. Sachin Patil, is very much thankful to the Principal, Dr. Ghali College, Gadhinglaj, and Dr. Bharatkumar Jadhav is thankful to Balasaheb Desai College, Patan, for providing the facilities.

KEYWORDS

- **antimicrobial assay**
- **antioxidant**
- **bronchial asthma**
- **fern**
- **pharmacology**
- **phytochemistry**
- **terpenoids**

REFERENCES

Akhmadjon, S., Hong, S. H., Lee, E. H., Park, H. J., & Cho, Y. J. (2020). Biological activities of extracts from tongue fern (*Pyrrosia lingua*). *Journal of Applied Biological Chemistry, 63*(3), 181–188. https://doi.org/10.3839/jabc.2020.025.

Cao, J., Zheng, Y., Xia, X., Wang, Q., & Xiao, J. (2015). Total flavonoid contents, antioxidant potential and acetylcholinesterase inhibition activity of the extracts from 15 ferns in China. *Industrial Crops and Products, 70*(B), 135–140.

Chen, J., Ning, S., Lu, X., Xiang, W., Zhou, X., Bu, Y., Li, L., & Huang, R. (2023). Variation in flavonoid and antioxidant activities of *Pyrrosia petiolosa* (Christ) Ching from different geographic origins. *Frontiers in Plant Science, 14*. https://doi.org/10.3389/fpls.2023.1173489 (accessed on 17th November 2025).

Chen, Y. J., Xie, G. Y., Xu, G. K., Dai, Y. Q., Shi, L., & Qin, M. J. (2015). Chemical constituents of *Pyrrosia calvata*. *Natural Product Communications, 10*(7), 1191–1193.

Cheng, D., Zhang, Y., Gao, D., & Zhang, H. (2014). Antibacterial and anti-inflammatory activities of extract and fractions from *Pyrrosia petiolosa* (Christ et Bar.) Ching. *Journal of Ethnopharmacology, 155*(2), 1300–1305.

Cui, L., Zhang, Y., Shao, W., & Gao, D. (2016). Analysis of the HPLC fingerprint and QAMS from *Pyrrosia* species. *Industrial Crops and Products, 85*, 29–37.

Fan, Y., Feng, H., Liu, L., Zhang, Y., Xin, X., & Gao, D. (2020). Chemical components and antibacterial activity of the essential oil of six *Pyrrosia* species. *Chemistry and Biodiversity, 17*(10), e2000526. https://doi.org/10.1002/cbdv.202000526.

George, M., Joseph, L., & Josekumar, V. S. (2017). Antioxidant, anti-lipid peroxidative and antimicrobial properties of the epiphytic fern, *Pyrrosia heterophylla* (L.) M. G. Price. *International Journal of Pharmacy and Pharmaceutical Sciences, 9*(4), 87–93.

He, K., Fan, L. L., Wu, T. T., & Du, J. (2019). A new xanthone glycoside from *Pyrrosia sheareri. Natural Product Research, 33*(20), 2982–2987.

Hovenkamp, P. H. (1986). *A monograph of the fern genus Pyrrosia (Polypodiaceae). Leiden Botanical Series* (p. 280).

Hsu, C. Y. (2008). Antioxidant activity of *Pyrrosia petiolosa. Fitoterapia, 79*(1), 64–66. https://doi.org/10.1016/j.microc.2020.105687.

Indriani, I. R., Teruna, H. Y., & Hendra, R. (2023). Antioxidant and toxicological activities of *Pyrrosia lanceolata* (L.) Farw. extracts. *Pharmacy Education, 23*(2), 174–178. https://doi.org/10.46542/pe.2023.232.174178.

Komala, I., Yardi, A., Betha, O. S., Muliati, F., & Ni'mah, M. (2015). Antioxidant and anti-inflammatory activity of the Indonesian ferns, *Nephrolepis falcata* and *Pyrrosia lanceolata. International Journal of Pharmacy and Pharmaceutical Sciences, 7*(12), 162–165.

Lang, T. Q., Zhang, Y., Chen, F., Luo, G. Y., & Yang, W. D. (2021). Characterization of chemical components with diuretic potential from *Pyrrosia petiolosa. Journal of Asian Natural Products Research, 23*(8), 764–771.

Patil, S. M., Masal, V. P., & Dongare, M. M. (2013). In search of ethnomedicinal pteridophytes from the Western Ghats of Maharashtra (India). *Indian Fern Journal, 30*, 69–77.

Patil, S. M., Patil, M. N., & Patil, S. S. (2020). Diversity and ethnomedicinal pteridophytes of Kolhapur district, Maharashtra. *Indian Fern Journal, 37*(1, 2), 226–238.

Seno, S. H., Sembiring, P. D. B., Kurniatin, A., & Nurcholis, W. (2022). Total phenolic, total flavonoid contents, and radical scavenging activity of different concentrations of ethanol extract of *Pyrrosia piloselloides* (L.) M. G. Price. *Tropical Journal of Natural Product Research, 6*(4), 484–487.

Song, L., Cao, M., Chen, C., Qi, P., Li, N., Wu, D., Peng, J., Wang, X., Zhang, M., Hu, J., & Zhao, J. (2017). Antibacterial activity of *Pyrrosia petiolosa* ethyl acetate extract against *Staphylococcus aureus* by decreasing *Hla* and *sea* virulence genes. *Natural Product Research, 31*(11), 1347–1350.

Wang, Q., Pei, X., Dai, X., Wang, Q., Huang, G., & Cao, J. (2023). Chemical constituents of *Pyrrosia davidii* (Baker) Ching. *Biochemical Systematics and Ecology, 108*, 104646.

Xin, X. W., Liu, Q. S., Zhang, Y. Y., & Gao, D. M. (2016). Chemical composition and antibacterial activity of the essential oil from *Pyrrosia tonkinensis* (Giesenhagen) Ching. *Natural Product Research, 30*(7), 853–856.

Yamashita, H., Masuda, K., Kobayashi, T., Ageta, H., & Shiojima, K. (1998). Dammarane triterpenoids from rhizomes of *Pyrrosia lingua. Phytochemistry, 48*(8), 2461–2466.

Yang, C., Shi, J. G., Mo, S. Y., & Yang, Y. C. (2003). Chemical constituents of *Pyrrosia petiolosa. Journal of Asian Natural Products Research, 5*, 143–150.

Zheng, X., Xu, Y., & Xu, J. (1998). Chemical studies on *Pyrrosia gralla* (Gies.) Ching. *Zhongguo Zhong Yao Za Zhi, 23*(2), 98, 99, 128, 129.

CHAPTER 24

Bioactive Insights from *Adiantum lunulatum* Burm. f.: A Fern with Potential Therapeutic Applications

YOGESHREE PATIL, SHREEDHAR OTARI, and SAVALIRAM G. GHANE*

Department of Botany, Plant Physiology Laboratory, Shivaji University, Kolhapur, Maharashtra, India

*Corresponding author

ABSTRACT

Adiantum lunulatum (Family: Adiantaceae) is an important medicinal fern with worldwide occurrence. It is known as the Walking Maidenhair Fern. Traditionally, the entire plant has been used to treat asthma, infections, and blood disorders. It contains a variety of important phytochemicals, including tannins, flavonoids, steroids, saponins, anthocyanins, alkaloids, phenols, terpenoids, anthraquinones, and glycosides, which contribute to its therapeutic properties. Extracts obtained from its leaves and roots have shown multiple biological activities, such as antimicrobial, anti-diabetic, anti-ulcer, and antioxidant. This chapter details the traditional uses and phytochemicals of different parts of the plant, as well as emphasizing their biological activities.

24.1 INTRODUCTION

Adiantum lunulatum Burm.f. (Syn: *Adiantum philippense* L.) belongs to the family Adiantaceae. It is commonly known as Walking Maidenhair Fern in English, Kali Jhant in Hindi, Hansavati in South India, and Hansraj

Bioactives and Pharmacology of Pteridophytes. T. Pullaiah (Ed.)

DOI: 10.1201/9781779642738-24

in Mumbai (Brahmachari and Chatterjee, 2002; Wagay and Kaloo, 2019). In India, this plant is often used ornamentally and is commonly called Hanspadi (Kakadia et al., 2020). It is widespread in tropical regions worldwide (POWO, 2024). *Adiantum lunulatum* is a leafy fern with 10–15 cm long, wiry, tufted, chestnut-brown stipes. Fronds are often elongated and measure 15–30 x 7.5 cm. It has a simple pinnate lamina that is 12–30 cm long, with a rachis that has a large terminal pinna, sometimes expanded with an apical vegetative bud. Pinnules are subdivided, semi-orbicular, shallowly lobed with 4–8 lobes, and the lower margin of each lobe is slightly concave. Petioles are 6–13 mm long, having a herbaceous texture. Sori are distributed continuously along the edges of the lobe, which is a 2 mm broad crescent-shaped cup (Singh et al., 2013).

24.2 TRADITIONAL USES

Traditionally, the entire *Adiantum lunulatum* plant is used to treat a variety of ailments, including fever, asthma, cough, pimples, bone fractures, cachexia, atrophy, emaciation, and watery diarrhea. Tribal peoples believe that *A. lunulatum* can cure various diseases such as microbial infections, blood-related diseases, leprosy, and centipede bites. In Ayurveda, it is utilized as an anti-dysenteric, febrifugal, soothing agent in erysipelas, and to alleviate burning sensations in the body (Moorthy et al., 2014).

24.3 PHYTOCHEMICALS FROM A. LUNULATUM

Several researchers have reported that *Adiantum lunulatum* is a rich source of various phytoconstituents, including alkaloids, glycosides, tannins, flavonoids, steroids, phenols, and saponins (Table 24.1) (Bhasimon et al., 2023; Jenat and Suresh, 2018; Mengane, 2016; Mithraja et al., 2012; Moorthy et al., 2014; Paul, 2018; Prakash et al., 2022). According to Sravani and Sunitha (2023), the methanol extract of *A. lunulatum* demonstrated the highest phenolic (0.45 ± 0.006) and flavonoid (0.26 ± 0.002) contents at a concentration of 500 μg/ml. In contrast, the hexane extract showed the lowest phenolic (0.29 ± 0.010) and flavonoid (0.17 ± 0.003) contents at the same concentration. The presence of phenolics and flavonoids may contribute to the extensive medicinal properties, making it a valuable resource in both traditional and modern herbal medicines.

TABLE 24.1 Phytochemicals Reported from *A. lunulatum*

Plant Part	Phytochemicals	Extract	References
Leaf	Alkaloids, glycosides, tannins, flavonoids, steroids, phenols	Hydroalcoholic	Prakash et al. (2022)
Whole plant	Phenols, tannins, terpenoids, saponins, glycosides, proteins	Aqueous, ethanol, methanol, chloroform	Jenat and Suresh (2018)
	Carbohydrates, phenols, saponins, tannins, xanthoproteins	Aqueous, ethanol, petroleum ether	Moorthy et al. (2014)
	Alkaloids, phenols, terpenoids, glycosides, carbohydrates, steroids	Methanol	Bhasimon et al. (2023)
	Tannin, flavonoids, steroids, saponin, anthocyanin, emodins, phlobatannins, alkaloids, phenols, terpenoids, anthraquinones, glycosides	Methanol, n- hexane, water	Mengane (2016)
	Phenols, saponins, tannins, xanthoproteins, carboxylic acid, carbohydrates	Acetone, benzene, chloroform, aqueous, ethanol, petroleum ether	Mithraja et al. (2012)
	Alkaloids, cardiac glycosides, coumarin, diterpenes, emodins, flavonoids, saponins, steroids, tannins, triterpenoids	Methanol, acetone, benzene, hexane	Paul (2018)

24.4 MAJOR BIOACTIVE COMPOUNDS FROM *A. LUNULATUM*

Adiantum lunulatum, a terrestrial fern known for its diverse medicinal properties due to a wide range of bioactive compounds (Table 24.2). Research has identified several key phytoconstituents in this plant, including alkaloids, phenols, terpenoids, glycosides, carbohydrates, and steroids (Bhasimon et al., 2023). These compounds contribute to the plant's extensive pharmacological potential. Among the notable compounds found in *Adiantum lunulatum* were hopane triterpenoids, such as 6α-acetoxy-16β,22-dihydroxy-3-ketoisohopane and mollugogenol A, which have been identified in its petrol extract (Brahmachari and Chatterjee, 2002). Furthermore, the HPTLC analysis of the methanol extract has detected 0.051% quercetin (Pathania et al., 2012). Additionally, Mukherjee et al. (2003) isolated a terpene compound, 3β-acetoxy-6α-hydroxy-hop-15,17(21)-diene from the petrol extract. The petroleum ether extract of *Adiantum lunulatum* root revealed the presence of various essential amino acids, including alanine, glycine, valine, proline, leucine, and tyrosine (Wagay and Kaloo,

2019). These amino acids are vital for numerous physiological processes and contribute to the plant's nutritional and therapeutic values. Moreover, the hexane extract of the whole plant has been found to contain several triterpenoid compounds. These include fern-9(11)-ene, fern-9(11)-en-25-oic acid, fern-9(11)-en-28-ol, filicenol-B, adiantone, and fern-9(11)-en-6α-ol. Notably, researchers have also identified a new triterpenoid, 22,29ξ-epoxy-30-norhopane-13β-ol in the hexane extract (Reddy et al., 2001). These triterpenoids are significant due to their potential therapeutic effects, including anti-inflammatory, antimicrobial, and anticancer activities. The extensive range of phytochemicals found in *Adiantum lunulatum* underscores its potential as a valuable resource in both traditional and modern medicine. These bioactive compounds not only contribute to the medicinal properties but also highlight its potential for developing new therapeutic agents for various health conditions (Figure 24.1).

TABLE 24.2 Major Bioactive Compounds Reported from *A. lunulatum*

Plant Part	Extract	Compounds	References
Whole plant	Petrol	Mollugogenol A (3β,6α,16β,22tetrahydroxyisohopane), 6α-acetoxy-16β,22-dihydroxy-3-ketoisohopane	Brahmachari and Chatterjee (2002)
Whole plant	Petrol	3β-acetoxy-6α-hydroxy-hop-15,17(21)-diene	Mukherjee et al. (2003)
Whole plant	Hexane	30-Norhopan-22-one, (21α) (adiantone), 22,29ξ-epoxy-30-norhopane-13β-ol, fern-9(11)-en-28-ol, fern-9(11)-en-25-oic acid, fern-9(11)-en-6α-ol, 25-hydroxyfilic-3-ene (filicenol B), fern-9(11)-ene (fernene, davallene)	Reddy et al. (2001)
Roots	Petroleum ether	Amino acids like alanine, glycine, valine, proline, leucine and tyrosine	Wagay and Kaloo (2019)

24.5 BIOLOGICAL ACTIVITIES

24.5.1 ANTIBACTERIAL ACTIVITY

The antibacterial activity of *Adiantum lunulatum* was evaluated using the agar well diffusion method against four bacterial strains: *Staphylococcus aureus*, *Pseudomonas aeruginosa*, *Proteus vulgaris*, and *Escherichia coli*. Extracts were prepared in several solvents (methanol, ethanol, chloroform, and water) and further tested for their antibacterial properties. The zone

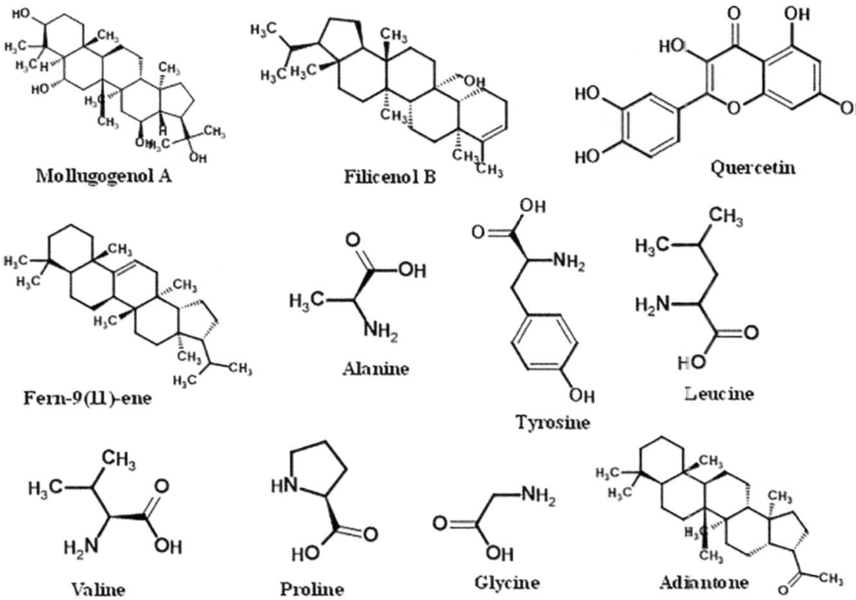

FIGURE 24.1 Major bioactive compounds reported from *A. lunulatum*.

of inhibition (ZOI) was measured and compared to the standard chloramphenicol. The ethanolic extract of *A. lunulatum* showed the highest ZOI (13±11 mm) against *Staphylococcus aureus*. Similarly, the chloroform extract also exhibited the highest ZOI against *Pseudomonas aeruginosa* (14±12 mm) and *Escherichia coli* (11±10 mm). In the case of *Proteus vulgaris*, both ethanol and methanol extracts demonstrated the highest ZOI (13±11 and 13±7 mm, respectively) (Jenat and Suresh, 2018). Additionally, the *in vitro* antimicrobial activity of *Adiantum lunulatum* was assessed using the disc diffusion method against 14 different microorganisms, including 2 fungi. The petroleum ether extract exhibited stronger inhibitory activity compared to the aqueous and ethanol extracts. Specifically, it showed moderate zones of inhibition against the tested organisms, viz. *S. flexneri* (13.20 mm), *S. typhimurium* (12.40 mm), *C. albicans* (11.20 mm), and *K. pneumoniae* (10.60 mm) (Moorthy et al., 2014). All the compounds, including fern-9(11)-ene, fern-9(11)-en-25-oic acid, and adiantone, were evaluated for their antibacterial activity against both gram-positive bacteria (*Bacillus sphaericus, Bacillus subtilis,* and *Staphylococcus aureus*) and gram-negative bacteria (*Escherichia coli, Pseudomonas aeruginosa*, and

Salmonella typhi) using the disc susceptibility test. These compounds exhibited significant activity against gram-negative bacteria, with notable effectiveness against all tested strains except *Escherichia coli.* The antibacterial efficacy of these compounds was comparable to that of the positive control kanamycin (Reddy et al., 2001). The antibacterial activity of *A. lunulatum* extracts has been investigated using the disc diffusion method. According to Thomas (2013), the methanol extract of the roots demonstrated the highest effectiveness against *Pseudomonas aeruginosa* and *Staphylococcus aureus*. However, the petroleum ether and aqueous extracts failed to exhibit any antibacterial activity. Additionally, Parihar et al. (2010) found that both aqueous and alcoholic extracts of *A. lunulatum* leaves showed significant zones of inhibition against *Escherichia coli, Salmonella arizonae, Salmonella typhi*, and *Staphylococcus aureus*. However, these extracts were ineffective against *Agrobacterium tumefaciens.* This study indicated that *A. lunulatum* possesses significant antibacterial activity against both gram-positive and gram-negative bacteria.

24.5.2 ANTIOXIDANT ACTIVITY

Semwal et al. (2013) investigated the antioxidant activity of the methanol extract of *A. lunulatum* leaves using *in vitro* methods, including DPPH and superoxide scavenging assays. They reported that *A. lunulatum* showed superior antioxidant activity compared to other plants, with IC50 values of 0.4 ± 0.23 and 0.52 ± 0.18 mg/ml, respectively. Furthermore, Sravani and Sunitha (2023) evaluated the antioxidant activity by ABTS, DPPH, and hydroxyl radical scavenging assays. These activities were assessed for the methanol, chloroform, and hexane extracts of aerial parts of *A. lunulatum*. For the ABTS radical scavenging assay, the chloroform and methanol extracts exhibited responses in terms of IC50 values of 163.09 and 62.60 μg/ml compared to the standard quercetin (3.1795 μg/ml). Similarly, the DPPH and hydroxyl radical scavenging activities of the methanol extract showed remarkable responses (IC50 61.63 and 54.19 μg/ml, respectively). The hexane extract showed comparatively poor antioxidant activity. This study highlighted that the methanol extract of *A. lunulatum* has great antioxidant potential.

24.5.3 ANTI-DIABETIC ACTIVITY

The *in vitro* antidiabetic potential of the methanolic extract from the aerial parts of *A. lunulatum* was evaluated through the α-amylase and α-glucosidase inhibitory activity, followed by a glucose uptake experiment using the L-6 cell line (Sravani and Katta, 2024). The study found that the extract exhibited significant inhibition of α-amylase (71.23 ± 0.97%) and α-glucosidase (61.07%) enzymes at a concentration of 800 μg/ml. In the case of the glucose uptake experiment, the extract showed maximum utilization of glucose (45.24%) in the L-6 cell line. These findings collectively indicated that the methanol extract of *A. lunulatum* possesses remarkable anti-diabetic activity.

24.5.4 ANTIULCER ACTIVITY

The results revealed that *A. lunulatum* showed significant ulcer healing activity in different ulcer-inducing rat models. At a 500 mg/kg dose, the ethanolic extract of *A. lunulatum* (EEAL) showed a reduction in the ulcer index (43.55, 24.49, and 48.50%) in ethanol-induced, cold stress restraint, and aspirin-induced ulcer models, respectively. Also, EEAL showed protection of the gastric mucosal layer in the pylorus-ligated induced ulcer model and proved its antiulcer potential (Deshpande and Balekar, 2019).

24.5.5 ANTIFERTILITY ACTIVITY

The effect of the alcohol extract and decoction of the whole plant on he reproductive organs (testis, epididymis, vas deferens) and accessory sex organs (seminal vesicle and prostate gland) of male albino rats with different dosages and durations was observed, and the antifertility activity was studied (Bhatia et al., 2010). It was found that administration of 500 mg/kg b.w. extract for 30 days stopped the process of spermatogenesis at the secondary spermatocyte stage. At the same dosage, it showed degenerative changes in seminiferous tubules and in the cauda epididymis after 60 days of drug administration. The lumen of the seminiferous tubules was filled with cellular debris and edematous fluid. The thickness of the lamina propria was reduced, followed by the disappearance of spermatozoa and stereocilia. The luminal epithelium showed nuclear pyknosis, particularly in the region of the folds. Creeping of mucosal folds and less secretion resulted in a reduction of lumen size. The muscle layer and lamina propria were disturbed, and the mucosal crypts vanished. Also, it showed significant alteration in prostate gland histology. The alveoli were deteriorated, the lumen had less secretion, and the mucosal layer was found to be damaged. The changes were restored when treatment was withdrawn. The above results concluded that *A. lunulatum* could be used as an anti-androgenic agent (Bhatia et al., 2010).

24.5.6 HEPATOPROTECTIVE ACTIVITY

The effect of ethanol extract of *A. lunulatum* on serum enzymes like serum glutamic-pyruvic transaminase (SGPT), serum glutamic-oxaloacetic transaminase (SGOT), alkaline phosphatase (ALP), bilirubin, and total protein level in Simvastatin-intoxicated rats was evaluated. It was observed that SGPT (120.34±10.19 U/l), SGOT (134.82±5.15 U/l), ALP (186.14±5.69 U/l), and bilirubin (1.48±0.06 mg/l) enzyme levels increased significantly, while the total protein level (3.09±0.22 mg/l) decreased after 30 days of administration of simvastatin (20 mg/kg). The ethanol extract reversed the enzyme and protein levels. The extract (400 mg/kg) highlighted a significant decrease in superoxide dismutase (SOD, 5.87±0.005) and glutathione (GSH, 16.87±0.03) in the simvastatin-intoxicated group when compared with the normal group. Histological changes like steatosis and perivenular fibrosis were prevented by the administration of ethanol extract (200 and 400 mg/kg). This study illustrated that *A. lunulatum* has potent hepatoprotective activity (Swathi and Harindran, 2018). Similarly, the hepatoprotective effect of ethanol extract of the plant on serum enzymes against ethanol-induced hepatotoxicity in rats was investigated. The results revealed that treatment with the extract at 250 and 500 mg/kg showed a significant reduction in the level of serum enzymes like SGOT, SGPT, ALP, total protein, and total bilirubin as compared to the disease control group. Furthermore, the standard and plant extract at 250 and 500 mg/kg dosage showed increased levels of SOD (67.30 ± 4.52 and 75.23 ± 4.22), catalase (CAT) (234.83 ± 11.59 and 240.80 ± 10.65), and glutathione peroxidase (GPx) (135.66 ± 10.20 and 145.9 ± 12.22) as compared to the diseased control group. Also, it showed remarkable improvement in cellular necrosis, fibrosis, inflammation, and vacuolization, which evidenced the hepatoprotective activity (Kakadia et al., 2020).

24.5.7 ANTI-HYPERLIPIDEMIC ACTIVITY

The effect of the leaf extract in triton-induced hyperlipidemic rats was studied. At 300 mg/kg p.o., the total cholesterol (TC) and triglycerides

(TG) levels (208.56±8.91 and 131.51±5.43 mg/dl, respectively) declined significantly. Fecal cholesterol (8.50±0.13 mg/g) and fecal bile acid (3.97±0.10 mg/g) levels of fecal matter increased at 300 mg/kg treatment. As the fecal bile acid level significantly increased, it might have assisted in cholesterol excretion (Prakash et al., 2022).

24.6 CONCLUSION

Adiantum lunulatum is abundant in various phytochemicals, including phenols, tannins, saponins, terpenoids, anthraquinones, and glycosides. These compounds have been isolated from the fronds and roots of the plant. Traditional medicine has long recognized the importance of *A. lunulatum*, and modern pharmacological studies further validated its medicinal potential. The diverse phytoconstituents found in *A. lunulatum* exhibit a wide range of biological activities, including antioxidant, antimicrobial, antidiabetic, antihyperlipidemic, antifertility, and hepatoprotective effects. These phytochemicals are instrumental in developing significant therapeutics and hold promise for creating innovative treatments for various diseases.

KEYWORDS

- **anthraquinones,**
- **cachexia**
- **chloroform**
- **glutathione**
- **gram-negative bacteria**
- **herbaceous texture**
- **methanol extract**

REFERENCES

Bhasimon, P., Priyalatha, B., & Vimala, K. S. (2023). The pharmacognostical and phytochemical analysis of *Adiantum lunulatum* Burm. *Journal of Pharmacognosy and Phytochemistry, 12*(5), 221–228.

Bhatia, D. K., Sharma, A. K., Pathania, P. C., & Khanduri, N. C. (2010). Antifertility effects of crude different of *Adiantum lunulatum* Burm. on reproductive organs of male albino rats. *Biological Forum, 2*(2), 88–93.

Brahmachari, G., & Chatterjee, D. (2002). Triterpenes from *Adiantum lunulatum. Fitoterapia, 73*(5), 363–368.

Deshpande, M. N., & Balekar, N. (2019). Gastroprotective and ulcer healing activity of ethanolic extract of *Adiantum lunulatum. Indian Drugs, 56*(3), 68–73.

Jenat, P. J., & Suresh, S. N. (2018). Phytochemical analysis and antimicrobial activity of fern *Adiantum lunulatum* Burm. f. *World Journal of Pharmaceutical Research, 7*(3), 640–648.

Kakadia, N. P., Amin, M. A., & Deshpande, S. S. (2020). Hepatoprotective and antioxidant effect of *Adiantum lunulatum* Burm. f. leaf in alcohol-induced rat model. *Journal of Complementary and Integrative Medicine, 17*(3), 20180063.

Mengane, S. K. (2016). Phytochemical analysis of *Adiantum lunulatum. International Journal of Current Microbiology and Applied Sciences, 5*(11), 351–356.

Mithraja, M. J., Marimuthu, J., Mahesh, M., Paul, Z. M., & Jeeva, S. (2012). Inter–specific variation studies on the phytoconstituents of *Christella* and *Adiantum* using phytochemical methods. *Asian Pacific Journal of Tropical Biomedicine, 2*(1), S40–S45.

Moorthy, K., Vinodhini, R., Senthilkumar, B., Punitha, T., Senbagam, D., & Thajuddin, N. (2014). Phytochemical analysis, *in vitro* anti-microbial activity of Indian medicinal ferns *Adiantum lunulatum* and *Hemionitis arifolia. Journal of Pharmaceutical Research, 8*(11), 1671–1676.

Mukherjee, K. S., Mondal, S., Sohel, S. M. A., Mukherjee, P., Chatterjee, D., & Brahmachari, G. (2003). A new triterpene from *Adiantum lunulatum* Burm. *Indian Journal of Chemistry, 42B*, 2665–2667.

Parihar, P., Parihar, L., & Bohra, A. (2010). *In vitro* antibacterial activity of fronds (leaves) of some important pteridophytes. *Journal of Microbiology and Antimicrobials, 2*(2), 19–22.

Pathania, S., Kumar, P., Singh, S., Khatoon, S., Rawat, A. K. S., Punetha, N., Jensen, D. J., Lauren, D. R., & Somvanshi, R. (2012). Detection of ptaquiloside and quercetin in certain Indian ferns. *Current Science, 102*(12), 1683–1691.

Paul, J. P. (2018). Screening of preliminary phytochemicals of *Adiantum lunulatum* Burm. – An important pteridophyte. *World Journal of Pharmaceutical Research, 7*(6), 732–737.

POWO. (2024). *Plants of the World Online*. Facilitated by the Royal Botanic Gardens, Kew. http://www.plantsoftheworldonline.org/ (accessed on 17th November 2025).

Prakash, S., Ghatuary, S. K., Prasad, S., & Parmar, M. (2022). Extraction, phytochemical investigation and study of antihyperlipidemic potential of leaves extract of *Adiantum lunulatum. Asian Journal of Pharmaceutical Education and Research, 11*(3), 135–146.

Reddy, V. N., Ravikanth, V., Rao, T. P., Diwan, P. V., & Venkateswarlu, Y. (2001). A new triterpenoid from the fern *Adiantum lunulatum* and evaluation of antibacterial activity. *Phytochemistry, 56*(2), 173–175.

Semwal, A., Mamta, S. F., Upreti, K., Bhatt, S. P., & Kumud, U. (2013). Evaluation of antioxidant activity of some pteridophytes. *International Journal of Herbal Medicine, 1*, 2–5.

Singh, S., Khatoon, S., Singh, H., Behera, S. K., Khare, P. B., & Rawat, A. K. S. (2013). A report on pharmacognostical evaluation of four *Adiantum* species, Pteridophyta, for their authentication and quality control. *Revista Brasileira de Farmacognosia, 23*(2), 207–216.

Sravani, T., & Katta, S. (2024). Efficacy of quercetin and kaempferol from methanolic extracts of *Adiantum lunulatum* Burm. f. and *Polystichum acrostichoides* as antidiabetic activity. *Asian Journal of Chemistry, 36*(3), 749–754.

Sravani, T., & Sunitha, K. (2023). Assessing the effect of solvent extraction on the antioxidant activity of *Polystichum acrostichoides* and *Adiantum lunulatum* aerial parts: A comparative study. *Biomedicine, 43*(5), 1409–1417.

Swathi, P. P., & Harindran, J. (2018). Evaluation of hepatoprotective activity of ethanolic extract of *Adiantum lunulatum* Burm. f. *World Journal of Pharmaceutical Research, 7*(17), 1198–1207.

Thomas, T. (2013). Antibacterial activity of *Adiantum lunulatum* Burm. f. towards bacteria implicated in cutaneous infections. *Journal of Biological Scientific Opinion, 1*(4), 334–336.

Wagay, S. A., & Kaloo, M. A. (2019). A phyto-chemical study of the roots of *Adiantum lunulatum* Burn. *International Journal of Advanced Research in Chemical Science, 6*(6), 1, 2.

CHAPTER 25

Phytochemistry and Bioactive Potential of *Phymatosorus scolopendria* (Burm. f.) Pic. Serm

MITTA RAGHAVENDRA,[1,*] V. V. RAJESHAM,[2] PULALA RAGHUVEER YADAV,[3] SIBBALA SUBRAMANYAM,[1] SURYA PRABHA MATANGI,[1] and SHARABU RAVICHANDRA[4]

[1]*Department of Pharmaceutical Sciences, School of Biotechnology and Pharmaceutical Sciences, Vignan's Foundation for Science, Technology and Research, Guntur, Andhra Pradesh, India*

[2]*Department of Pharmacology, CMR College of Pharmacy, Hyderabad, Telangana, India*

[3]*Department of Biotechnology, Indian Institute of Technology Hyderabad, Sangareddy, Telangana, India*

[4]*Department of Pharmaceutical Chemistry, Chebrolu Hanumaiah Institute of Pharmaceutical Sciences, Guntur, Andhra Pradesh, India*

*Corresponding author

ABSTRACT

Phymatosorus scolopendria (Burm. f.) Pic.Serm., commonly known as the monarch fern or wart fern, is a perennial fern native to tropical regions, particularly in the Pacific and Indian Ocean islands, Southeast Asia, and parts of Africa. It is characterized by its glossy, leathery fronds and creeping rhizomes. This fern is often found in humid, shaded environments such as forests and along streams. The fronds resemble centipedes, which is reflected in its species name, "scolopendria." Due to its attractive foliage,

Bioactives and Pharmacology of Pteridophytes. T. Pullaiah (Ed.)

DOI: 10.1201/9781779642738-25

P. scolopendria is a popular ornamental plant frequently used in gardens and as ground cover in shaded areas. *P. scolopendria* contains flavonoids, tannins, and alkaloids as secondary metabolites. These bioactives are believed to provide anti-inflammatory, antimicrobial, and antioxidant effects. In traditional medicine, the fern and its parts are used to treat various ailments, such as fever, cough, skin infections, and wounds. The rhizomes and fronds are commonly employed in herbal remedies, highlighting the plant's significance in the traditional pharmacopeias of many cultures. Beyond its medicinal uses, the fern holds cultural importance in various Pacific Island communities, where it is used in ceremonies and decorations.

25.1 INTRODUCTION

The second-largest class of vascular cryptogams, pteridophytes reproduce by spores rather than seeds. As decorative and therapeutic plants, ferns have long been a significant aspect of Polynesians' daily lives (Mannan et al., 2008). *Phymatosorus scolopendria*, known as the monarch fern, wart fern, or serpent fern, is classified within the family Polypodiaceae. It is called by different names: *Microsorum scolopendria* (Burm. f.) Copel, *Polypodium phymatodes* L., *Pleopeltis phymatodes* (L.) T. Moore. *Phymatosorus scolopendria* is native to many tropical and subtropical regions, including the Pacific and Indian Ocean islands, Southeast Asia, and parts of Africa. It thrives in humid, shaded environments such as forests, along stream banks, and in coastal areas. The fern is adaptable, growing terrestrially on the forest floor and epiphytically on trees, allowing it to inhabit diverse ecosystems. Its broad distribution indicates its resilience and ability to colonize various habitats. It is notable for its glossy, dark green fronds that are leathery and pinnate, resembling centipedes, which is reflected in its species name, "scolopendria." The fronds can grow up to one meter in length and often have wavy margins. The plant features creeping rhizomes covered with brown scales, which help it spread across the ground or cling to trees (WHO, 1998).

This musk fern is a visually striking fern with glossy, dark green, pinnate fronds. The fronds have a leathery texture and often feature wavy margins. The nature and intensity of various biotic and abiotic stress factors negatively influence plant growth, development, and productivity. The brackens are 15–60 cm long and spaced 2–6 cm apart, occasionally concerted on short brachyblasts. The stipe is 5–28 cm long, pale, sleek, and hair is not present except for a few scales at the base. The lamina is profoundly pinnatifid with 1–4 pairs of convexities and a similar terminal lobe, though it is rarely straightforward. It is whitish green, shriveled, rugged, and shaven except for a few scattered scales on the midrib and costae. The sideward lobes are nearly linear, 4–18 cm in length and 1–3.5 cm in width, relenting to an acute or dullish apex and typically gradually narrowing towards the base. The selvage is entire or undulate, and the terminal segment is often outstretching. Sori, up to 4 mm in diameter, are rounded or slightly elliptical, arranged in one (sporadically two) row on either side of the littorals, with additional sori situated in the wings between the lobes. They are slightly recessed into the lamina, creating truncated projections on the upper surface. Coupled with its attractive, glossy fronds, these features make the plant popular in landscaping and ornamental gardening, enhancing its overall aesthetic appeal. It is commonly used as ground cover in shaded garden areas, in hanging baskets, and as an epiphyte on trees or trellises. Its ability to thrive in low-light conditions makes it an ideal plant for indoor decoration and shaded outdoor spaces. The fern's robust nature and minimal maintenance requirements further enhance its appeal in residential and commercial landscapes (Bostock and Spokes, 1998; Claussen, 2005).

25.2 BIOACTIVES

The major bioactive compounds present in *Phymatosorus scolopendria* are Makisterones A and C, Ecdysone, Inokosterone, Amarasterone A, 25-epimer of amarasterone A, 20-hydroxyecdysone, Poststerone, 24,28-diepi-cyasterone, 20-deoxymakisterone A, and 25-deoxy-ecdysone 22-glucoside (Snogan et al., 2007), Dodecanoic Acid, 1,2,3-Propane-triyl ester, 4-nitrophenyl laurate, Myristic acid vinyl ester, Hexadecanoic acid, 4-nitrophenyl ester (Sujatha et al., 2020), 20-hydroxyecdysone, Makisterone A, Inokosterone, Ecdysone, 2-Deoxy-20-hydroxyecdysone, 2-Deoxyecdysone, and Makisterone C (Ho et al., 2007).

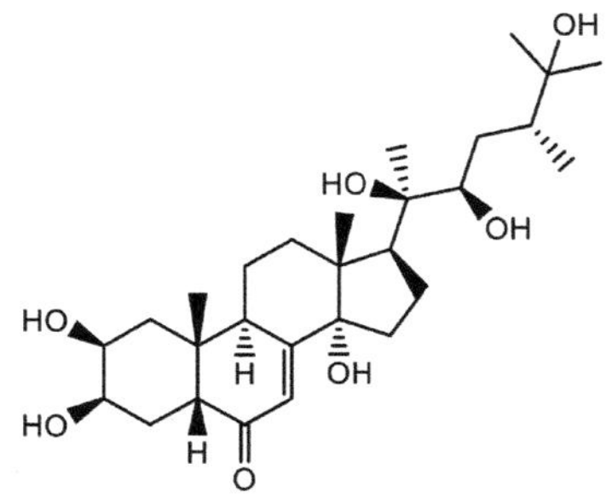

Makisterone A

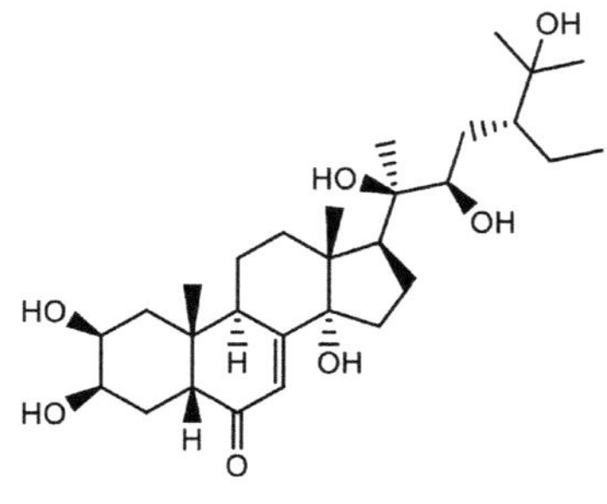

Makisterone C

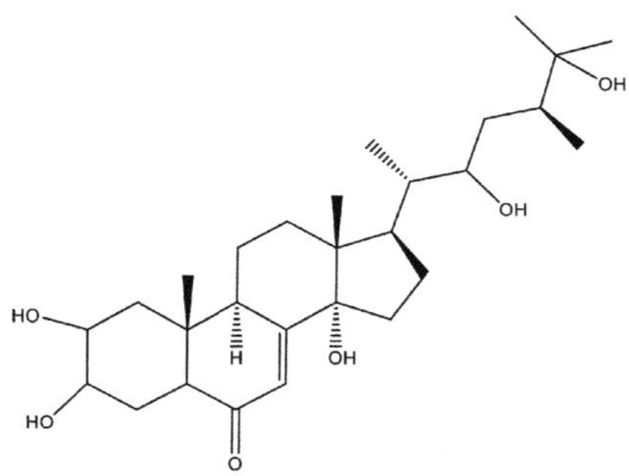

20-Deoxy makisterone

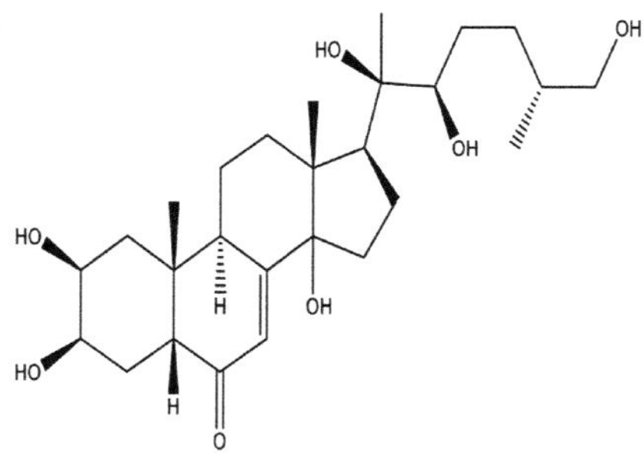

Inokosterone

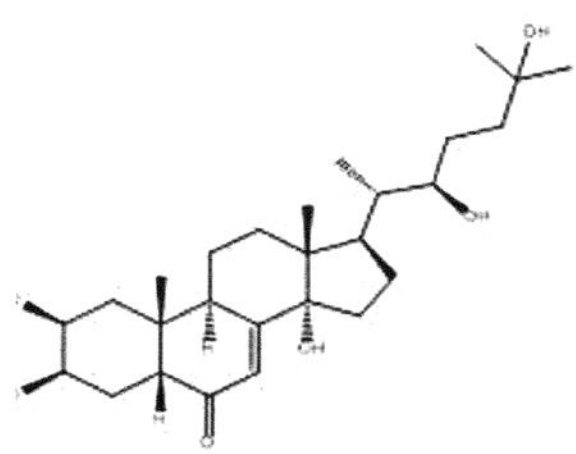

Ecdysone

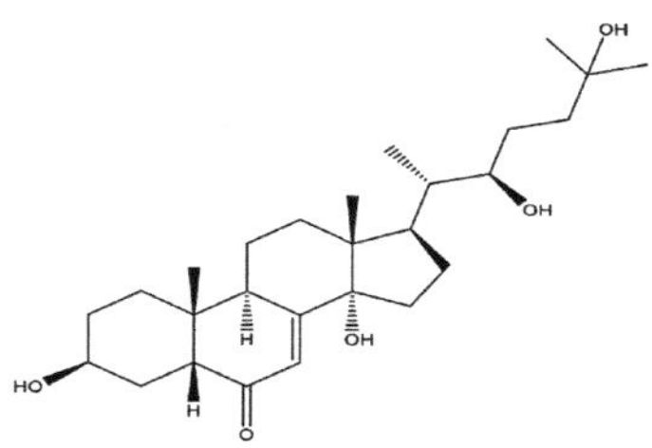

2-deoxy ecdysterone

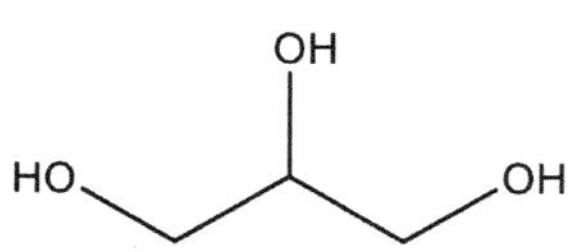

Propane-1,2,3-triol

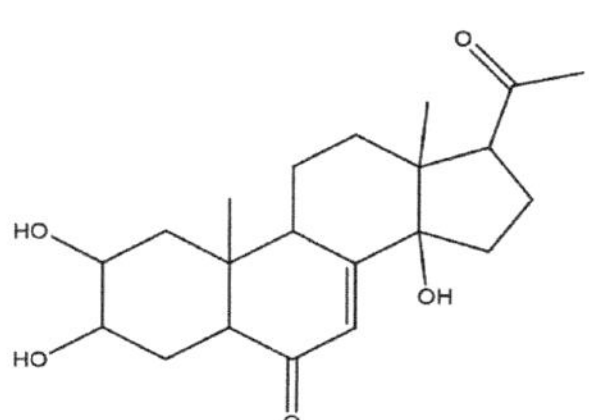

Poststerone

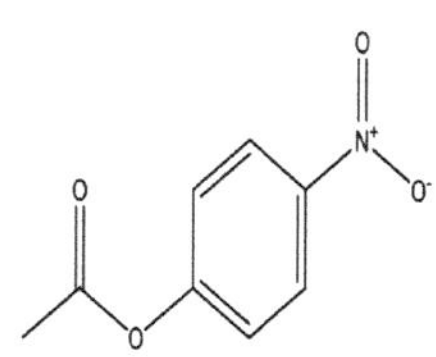

4-Nitrophenyl acetate

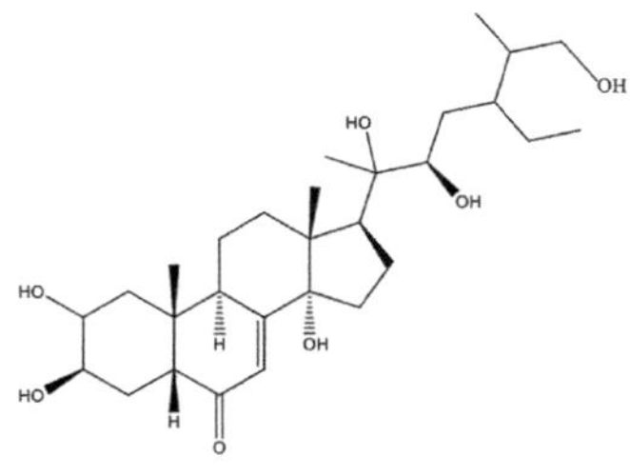

Amarasterone-A

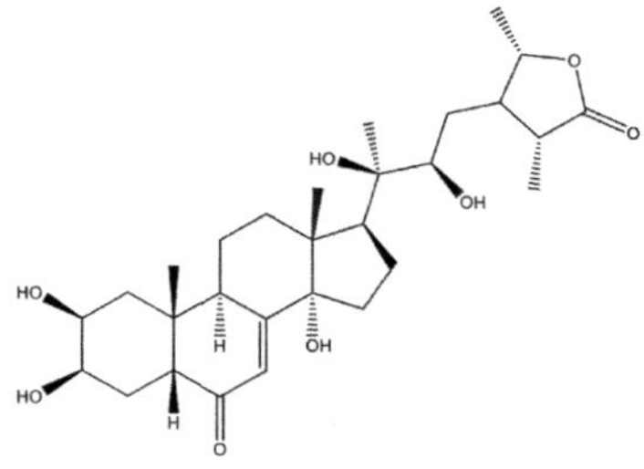

24,25-Diepi-Cyasterone

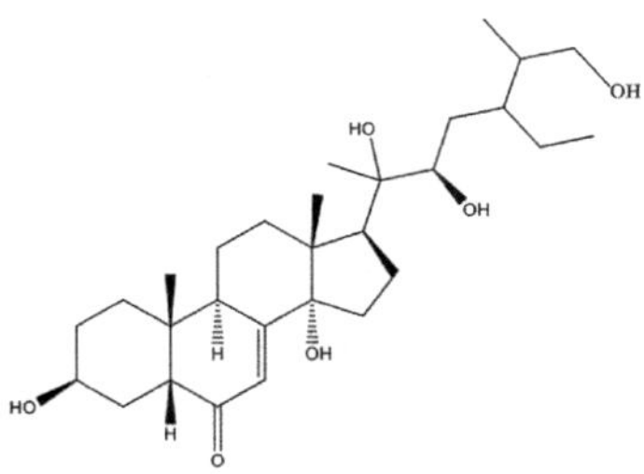

2-Deoxy-20 hydroxy ecdysterone

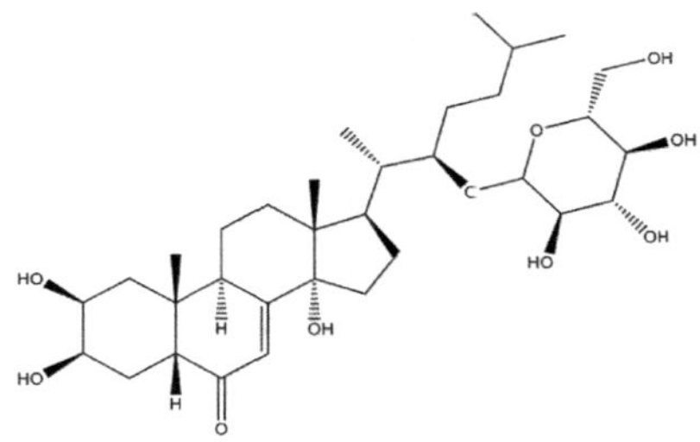

25-Deoxy ecdysone-22-glucoside

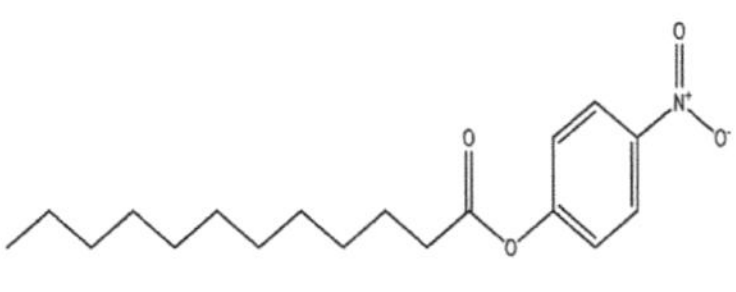

4-nitrophenyl laurate

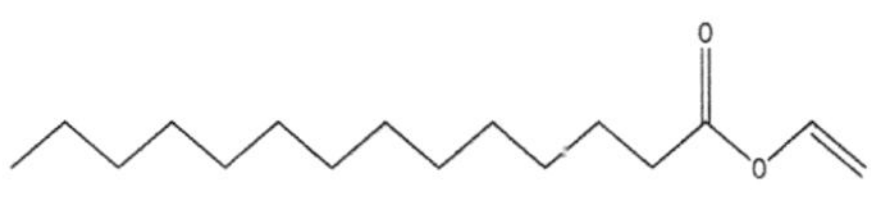

Myristic acid vinyl ester

Hexadecanoic acid

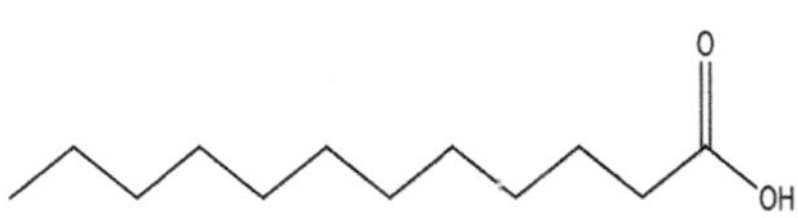

Dodecanoic Acid,

25.3 PHARMACOLOGY

Balada et al. (2022) investigated the chemistry, antioxidant, antimicrobial, and anti-inflammatory properties of *Phymatosorus scolopendria* (Syn.: *Microsorum scolopendria*) rhizome and leaf ethyl acetate extracts.

25.3.1 ANTIOXIDANT ACTIVITY

Using RP-HPLC coupled with mass spectroscopy, protocatechuic acid 4-O-glucoside, p-coumaroyl feruloyl tartaric acid, tartaric acid, kaempferol 3-O-glucuronide, isoxanthohumol, cirsimaritin, kaempferide, daidzein, daidzin, resveratrol, catechol, pyrogallol, and 1,4-naphthoquinone were identified as secondary metabolites. The total polyphenol content of rhizome and leaf ethyl acetate extracts was determined to be 57.13 ± 0.81 and 14.77 ± 0.52 mg gallic acid per gram of sample. The flavonoid content was calculated as 29.01 ± 0.65 and 10.58 ± 0.56 mg kaempferol per gram of sample, respectively. The antioxidant potential was confirmed through *in vitro* DPPH radical scavenging and ORAC-FL assays. The rhizome and leaf extracts' IC50 values in the DPPH inhibition experiment were found to be 12.05 and 20.34 μg/mL, respectively. The ORAC-FL assay revealed that the rhizome extract had potential radical scavenging capacity with an ORAC value of 1.63, while the leaf extract's ORAC value was 1.54. In comparison to the leaf extract, the *Phymatosorus scolopendria* rhizome extract exhibits the highest level of free radical scavenging activity (Balada et al., 2022).

25.3.2 ANTIMICROBIAL ACTIVITY

The antibacterial efficacy of rhizome and leaf ethyl acetate extracts of *Microsorum scolopendria* against *Staphylococcus aureus* and *S. epidermidis* was tested using the biofilm formation inhibition and disruption assays. Both extracts were capable of interfering with the production of bacterial biofilms, although they did not have any inhibitory effects on microbial growth. Following a 48-hour incubation period, bacteria exposed to extracts at concentrations varying between 0 and 512 μg/mL exhibited a dose-dependent inhibition of biofilm formation. However, the rhizome extract's activity against *Staphylococcus epidermidis* was more noticeable at a concentration of 128 μg/mL than the leaf extract, which might be due to the presence of isoxanthohumol, resveratrol, and kaempferol, which induced downregulation of rsbU and spa gene expression, responsible for adhesion, protection, and communication of bacteria (Balada et al., 2022). In the biofilm disruption assay, the leaf extract was effective in disrupting biofilm formation inhibition rather than inhibiting their formation. *Staphylococcus aureus* viability was unaffected by either extract, but bacterial biofilm development and disruption were somewhat impacted (Balada et al., 2022).

25.3.3 ANTI-INFLAMMATORY ACTIVITY

The COX-1 enzyme, which is constitutively produced, is a maintenance protein that keeps cells operating physiologically, whereas an inflammatory environment or tissue injury can induce the expression of the enzyme COX-2. The results of COX inhibition were supported by the DPPH radical scavenging and ORAC assays, as both extracts have the ability to act on the generation and functionality of radical molecules. The rhizome extract had a selectivity index of 9.96 and an IC50 of 3.14 for COX-2. Additionally, the ORAC and DPPH tests showed the highest levels of radical-scavenging activity. The leaf extract showed a selectivity index of 6.84 and an IC50 of 3.52 µg/mL for the COX-2 enzyme. The results indicated that both extracts selectively inhibit the COX-2 and COX-1 enzymes, but the rhizome extract has a higher selectivity index than the leaf extract (Balada et al., 2022).

25.3.4 CYTOTOXICITY EFFECT -MTS ASSAY

In the first method, the cytotoxic effect of rhizome and leaf extracts of *Microsorum scolopendria* was screened on the HDFa cell line using the MTS assay. The study clearly revealed that both extracts had no toxic effects on HDFa cell lines. In the second method, HDFa cells were infected with *Staphylococcus aureus* and *Staphylococcus epidermidis*. Damage to the cell membrane was indicated by the release of the LDH enzyme. At six hours after *Staphylococcus aureus* incubation, LDH release rose by about 60% in control group cells, indicating cell death. Nevertheless, this effect was significantly lessened when cells were treated with bacteria and extracts at the same time. In tests when both *Staphylococcus epidermidis* and extracts were in close proximity to cells, there was a decrease in the release of LDH (Balada et al., 2022).

25.3.5 ANTIDIARRHEAL ACTIVITY

Ragone (1997) said that *Phymatosorus scolopendria* has been traditionally used in various cultures for its medicinal properties, including the treatment of chronic diarrhea. In traditional medicine practices, especially within Polynesian and Southeast Asian communities, *Phymatosorus scolopendria* is prepared in decoctions or infusions to alleviate symptoms of persistent diarrhea, which may be attributed to its astringent and anti-inflammatory properties.

25.3.6 INSECT REPELLENT ACTIVITY

Mbong et al. (2023) said that *Phymatosorus scolopendria* has been investigated for its potential insect repellent properties. This fern, native to tropical regions, has been traditionally used in various cultures for its medicinal properties. Recent studies have identified the presence of bioactive compounds in *Phymatosorus scolopendria* that exhibit significant insect repellent activity. These compounds, including essential oils and various secondary metabolites, have been shown to effectively repel mosquitoes and other insects, offering a natural alternative to synthetic repellents.

25.3.7 ANTIBACTERIAL ACTIVITY

Frankova et al. (2021) investigated the ethanolic extract of *Phymatosorus scolopendria's in vitro* antibacterial activity against pathogens linked to burns, wounds, and sores. They also assessed the impact on human fibroblast survival, migration, and proliferation using MTT tests and wound healing scratches. For *Staphylococcus aureus*, the extract exhibited a minimum inhibitory concentration (MIC) of ≥ 4 μg/mL. The extract did not exhibit any toxicity towards the standard lines of NHDF-Ad ($IC_{80} >$ 128 μg/mL).

25.3.8 ANTICANCER ACTIVITY

Srinivasan et al. (2020) reported that *Phymatosorus scolopendria* has explored bioactive compounds such as dodecanoic acid, myristic acid vinyl ester, 1,2,3-propanetriyl ester, 4-nitrophenyl laurate, and hexadecanoic acid, 4-nitrophenyl ester, using GC-MS. Molecular docking analyses showed strong binding affinities to key lung cancer protein targets (2ITO protein), suggesting potential inhibitory effects on cancer cell proliferation. Additionally, these bioactive compounds were evaluated for drug-likeliness based on Lipinski's Rule of Five, demonstrating favorable pharmacokinetic properties. These findings indicated that *Phymatosorus scolopendria* holds promise as a source of natural compounds for developing new anticancer therapies.

KEYWORDS

- **antidiarrheal activity**
- **bacterial biofilms**
- **biofilm formation inhibition**
- **cancer cell proliferation**
- **cytotoxic effect of rhizome**
- **insect repellent activity**
- **pteridophytes**

REFERENCES

Balada, C., Diaz, V., Castro, M., Echeverria-Bugueno, M., Marchant, M. J., & Guzman, L. (2022). Chemistry and bioactivity of *Microsorium scolopendria* (Polypodiaceae): Antioxidant effects on an epithelial damage model. *Molecules, 27*(17), 2–20.

Bostock, P. D., & Spokes, T. M. (1998). Polypodiaceae. In P. M. McCarthy (Ed.), *Flora of Australia* (Vol. 48, pp. 468–495). ABRS: Canberra / CSIRO Publishing: Collingwood. (*Microsorum*, pp. 479–486).

Claussen, J. (2005). *Native plants of Christmas Island. Flora of Australia Supplementary Series* No. 22. Australian Biological Resources Study: Canberra / Christmas Island Natural History Association.

Frankova, A., Vistejnova, L., Merinas-Amo, T., Leheckova, Z., Doskocil, I., Soon, J. W., Kudera, T., Laupua, F., Alonso-Moraga, A., & Kokoska, L. (2021). *In vitro* antibacterial activity of extracts from Samoan medicinal plants and their effect on proliferation and migration of human fibroblasts. *Journal of Ethnopharmacology, 264*, 113220.

Ho, R., Teai, T., Loquet, D., Bianchini, J. P., Girault, J. P., Lafont, R., & Raharivelomanana, P. (2007). Phytoecdysteroids in the genus *Microsorum* (Polypodiaceae) of French Polynesia. *Natural Product Communications, 2*(8), 803–806.

Mannan, M. M., Maridass, M., & Victor, B. (2008). A review on the potential uses of ferns. *Ethnobotanical Leaflets, 12*, 281–285.

Mbong, E. O., Ivon, E. A., Idio, E., Utuk, K. E., Okon, J. E., & Anwana, E. D. (2023). Correlating habitat dynamism with foliar anatomical modulations: A study with *Phymatosorus scolopendria* (Burm. f.) Ching. *World Journal of Applied Science and Technology, 15*(1), 61–68.

Ragone, D. (1997). Ethnobotany of *Phymatosorus scolopendria*: Uses in Polynesia and Southeast Asia. *Economic Botany, 51*(4), 402–405.

Snogan, E., Vahirua-Lechat, I., Ho, R., Bertho, G., Girault, J. P., Ortiga, S., Maria, A., & Lafont, R. (2007). Ecdysteroids from the medicinal fern *Microsorum scolopendria* (Burm. f.). *Phytochemical Analysis, 18*, 441–450.

Srinivasan, S., Sivaraman, C. S., Issac, R. R., Mahalingam, G., & Roke, G. D. (2020). Docking against lung cancer protein and drug-likeliness of bioactive compounds from *Phymatosorus scolopendria* (Burm. f.) Pic. Serm. *International Journal of Pharmaceutical Sciences and Drug Research, 12*(5), 437–441.

Sujatha, S., Sara, S. C., Gayathiri, M., Roselin, I. R., & Ruby, R. G. D. (2020). Analysis of bioactive compounds present in methanolic extract of *Phymatosorus scolopendria* (Burm. f.) Pic. Serm. through gas chromatography and mass spectroscopy. *International Journal of Pharmaceutical Sciences and Research, 11*(7), 3294–3299.

World Health Organization. (1998). *Medicinal Plants in the South Pacific* (p. 141). WHO Regional Publications, Manila.

CHAPTER 26

Bioactive Compounds and Pharmacology of the Genus *Athyrium* Roth (Athyriaceae): A Review

SURAJ P. DEVKAR,[1] SACHIN B. PATIL,[2,*] SAGAR A. DESHMUKH,[1,2] and SACHIN M. PATIL[3]

[1]*Department of Botany, The New College, Kolhapur (affiliated to Shivaji University, Kolhapur), Maharashtra, India*

[2]*Department of Botany, Anandibai Raorane Arts, Commerce and Science College, Vaibhavwadi, Sindhudurg (affiliated to the University of Mumbai), Maharashtra, India*

[3]*Department of Botany, Dr. Ghali College, Gadhinglaj (affiliated to Shivaji University, Kolhapur), Maharashtra, India*

*Corresponding author

ABSTRACT

A lady fern genus, *Athyrium*, belongs to the family Athyriaceae and has a rich history of traditional medicinal use. Traditionally, the plant is used to treat various diseases like coughs, fever, bronchitis, pneumonia, urinary ailments, and skin disorders. These therapeutic properties are attributed to their diverse phytochemical composition, which includes flavonoids, phenolics, terpenes, terpenoids, alkaloids, and amino acids.

26.1 INTRODUCTION

Genus *Athyrium* Roth. is a cosmopolitan fern genus comprising 202 species distributed in the wild flora of the globe, including North and South

Bioactives and Pharmacology of Pteridophytes. T. Pullaiah (Ed.)

DOI: 10.1201/9781779642738-26

America, Europe, Asia, Africa, and some islands like Hawaii, Kuril Islands, Ogasawara-shoto, and Prince Edward Island (POWO, 2024). Genus *Athyrium* is feathery-like with finely divided yellow to green foliage about 75 cm long and 25 cm wide and grows as sciophytes in damp soil. The distinguishing feature of the genus is horseshoe-shaped spore-producing clusters covered by a protective structure called indusium. Young leaves and rhizomes are edible and used for cooking in many regions of India (Patil et al., 2020). Some species have ornamental value, while some are used as traditional medicine to treat various ailments. Scientific investigations have revealed its antioxidant, antimalarial, antimicrobial, anticancer, anti-inflammatory, anti-apoptotic, and cytotoxic properties (Britannica, 2024; Salehi et al., 2019). The key bioactive compounds found in *Athyrium* include kaempferol, quercetin, chlorogenic acid, caffeic acid, ecdysteroids, phenolic glycosides, procyanidin, and n-caffeoyl-phenylalanine.

26.2 ETHNOBOTANY

Since ancient times, the genus *Athyrium* has been traditionally utilized for treating various illnesses. A decoction prepared from the leaves of *A. atkinsonii* Bedd., combined with Acacia gum powder, is used to treat respiratory disorders such as asthma, bronchitis, emphysema, pneumonia, urinary ailments, and various skin disorders. Additionally, the herbal tonic is used to alleviate body weakness. The fronds and rhizome of *A. hohenackerianum* T. Moore are used as a decoction against rheumatic pain as well as an anthelmintic (Patil et al., 2020). The people of South India used the fronds and roots of *A. falcatum* Bedd. in their traditional medicine practices, whereas the young fronds were consumed to treat internal ailments such as cancer, and the roots were consumed as an anthelmintic. Similarly, the whole plant of *A. falcatum* has been used as an anthelmintic by the local people in Madhya Pradesh (Singh and Upadhyay, 2014). The roots of *A. lanceum* T. Moore were used to cure breast pain during childbirth and increase milk flow. Also, the paste of dried powder is effective against sores. Moreover, it has been used as a remedy for burns and scalds, against intestinal fever, and as an anti-inflammatory herb for the treatment of ascariasis in Malaysia (Chai et al., 2013). *A. multidentatum* (Döll) Ching is a well-known medicinal species and nutritious potherb that has been used in Chinese traditional medicine for its tranquilizer, antihypertensive, and diuretic properties (Liu et al., 2013). *A. parasnathense*

(C.B. Clarke) Ching ex Mehra & Bir is used to cure respiratory disorders such as asthma, bronchitis, emphysema, and pneumonia. *A. pectinatum* (Wall.) Presl., young leaves are used as vegetables, and the rhizome has a strong anthelmintic property (Parihar et al., 2006). The extracts from *A. schimperi* Moug. ex Fee subsp. *biserrulatum* (Christ) Fraser Jenk. leaves have been reported to effectively treat ulcers and intestinal disorders (Irfan et al., 2021). *A. sinense* and *A. monomachii* are commonly found ferns in the Amur region of Russia, where they are consumed as food by the local population. In some regions of China, the young fronds of these ferns are also used as a food source (Kreshchenok et al., 2021). *A. acutipinnulum* is a native plant from South Korea, traditionally used to treat epilepsy, gonorrhea, and nerve disorders, and also in some regions, it is consumed as food (Park et al., 2023).

26.3 BIOACTIVE COMPOUNDS

The various bioactive compounds belonging to different groups (like flavonoids, alkaloids, phenolics, saponins, and tannins) are derived from different *Athyrium* species. About 37 compounds are found in various species, and structures are drawn here (Figures 26.1–26.3). Flavonoids were identified as the predominant secondary metabolites in pteridophytes. Umikalsom et al. (1994) utilized two-dimensional paper chromatography to analyze the ethanolic extract of boiled and dried pinnae from *Athyrium filix-femina* (L.) Roth. Their findings revealed the presence of flavonoids and phenolic glycosides, specifically identifying procyanidin, kaempferol 3-O-rutinoside, kaempferol 3-O-glucoside, quercetin 3-O-rutinoside, and quercetin 3-O-glucoside in the extract. Further, *A. filix-femina* fronds extracted in chloromethane and methanol by utilizing column chromatography (Sephadex LH-20), vacuum liquid chromatography (VLC), and High-Performance Liquid Chromatography (HPLC) with ethyl acetate fractions reported caffeic acid amides including n-caffeoyl-phenylalanine, n-caffeoyl-tryptophan, 5-O-caffeoyl-shikimic acid, 3,4-di-O-caffeoyl-quinic acid, and chlorogenic acid (Adam, 1995). Similarly, volatile organic compounds investigated by GC-MS in diethyl ether solvent of *A. filix-femina* exhibited the presence of isoprenoids with 2-phenylethanal and 1-octen-3-ol (Fons et al., 2010). Ecdysteroids, recognized as insect molting hormones with significant biological activity, have also been identified in *Athyrium* species, with several ecdysteroids isolated including the novel 24-epi-pterosterone phytoecdysone, along with ecdysone, ecdysterone, pterosterone, and ponasterone-A, C from the aqueous root extract of *A. yokoscense* (Franch. & Sav.) Christ. (Ohta et al., 1996). Further, flavonoid glycosides isolated in aqueous methanol extraction of *A. solenopteris* (Kunze) T. Moore fronds showed the presence of kaempferol 3–O–glucoside, kaempferol 3–O–diglucoside, quercetin 3–O–glucoside, quercetin 3–O–diglucoside, and quercetin 3–O–diglucoside–3'–O xyloside (Samy et al., 2004). The HPLC-MS study on *A. multidentatum* (Doll.) Ching identified chlorogenic acid and its isomer, eriodictyol o-hexoside, and n-caffeoyl-phenylalanine, along with one unknown compound, from the methanolic extract (Qi et al., 2017). While HPLC analysis on *A. multidentatum* ethanolic extract revealed the presence of previously known bioactive compounds including flavonoids like myricetin, quercetin, kaempferol, chlorogenic acid, ferulic acid, aspidinol B, and caffeic acid, which further contribute to the anti-inflammatory activity (Han et al., 2018). Jing et al. (2019) isolated polysaccharides from the ethanolic extract of *A. multidentatum* rhizome, which were primarily composed of arabinose, fucose, galactose, glucuronic acid, glucose, mannose, and rhamnose, contributing to varying biological activities. Eicosapentaenoic acid, an important polyunsaturated fatty acid (PUFA), was isolated from the mature fronds of *A. sinense* using methanol and chloroform (Nekrasov et al., 2019). Further, Liu et al. (2016) isolated striatisporolide A, a butenolide derivative from the rhizome of *A. multidentatum*. Further study revealed that *A. multidentatum* showed the highest content of total amino acids and total essential amino acids when comparing other ferns under study, that is, *Matteuccia struthiopteris, Osmunda cinnamomea,* and *Pteridium aquilinum* (Wang et al., 2019). The GC-MS analysis on the ethanolic extract of *A. hohenackerianum* showed the presence of a mixture of fatty esters, various terpenes, fatty acyls, alkaloids, fatty acids, and phytosterols. They include butanoic acid, 2,4,6-trimethyloctane, trans-Linalool oxide, hexadecane, 1,10-Decanediol, loliolide, n-hexadecanoic acid, n-Nonadecanol-1, heneicosanoic acid, methyl ester, 10-Nonadecanone, Phytol, octacosanol, propanoic acid, 2-(benzoylamino)-3-phenyl-m, Methyl (Z)-5,11,14,17-eicosatetraenoate, 4,8,12,16-Tetramethylheptadecan-4-olide, γ-sitosterol, hexadecanoic acid, 2-hydroxy-1-(hydroxymethyl), octadecanoic acid, 2,3-dihydroxypropyl ester, and androsta-1,4-diene-3,17-dione (Elasbali et al., 2022). The methanolic extract and ethanolic fraction of *A. acutipinnulum* revealed the presence of several compounds contributing to antioxidant activities, including methyl 2-hydroxy-3-phenylpropanoate, protocatechualdehyde, caffeic acid, trans-p-coumaric acid, 5-O-caffeoyl shikimic acid, 5-O-caffeoyl quinic acid, quercetin 3-O-β-glucoside, naringenin-7-O-β-glucoside, ponasterone A, ecdysone, and esculetin. Additionally, (-)-4-E-caffeoyl-L-threonic acid and sutchenoside A were identified for the first time in the *Athyrium* species (Park et al., 2023) (Figure 26.1)

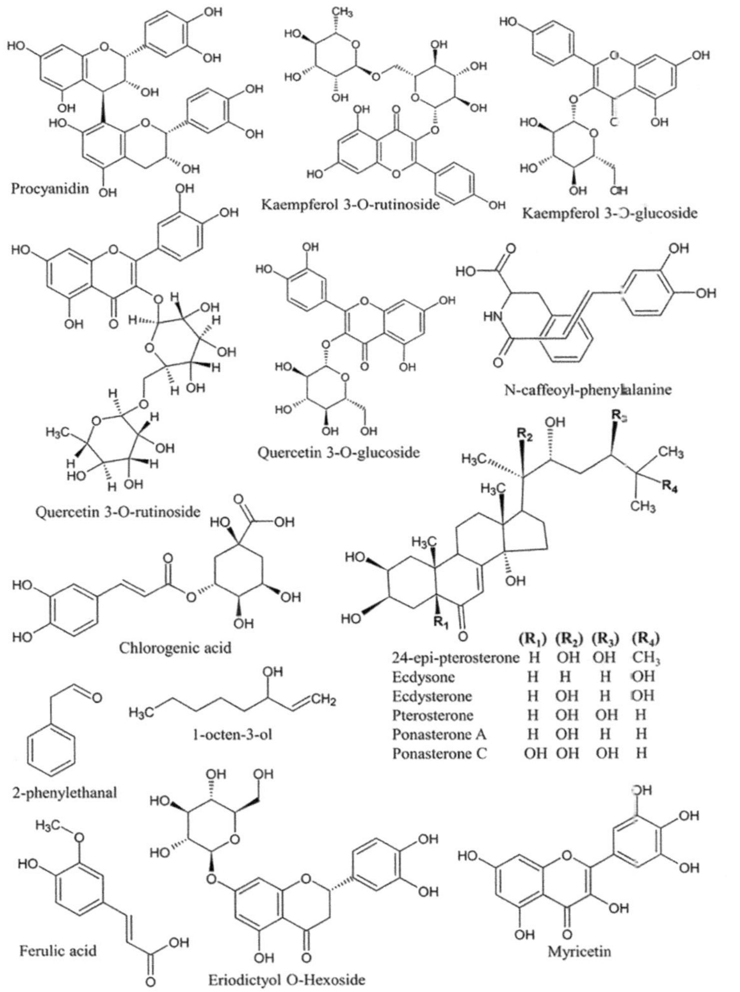

FIGURE 26.1 Isolated phytochemicals from Genus *Athyrium*.

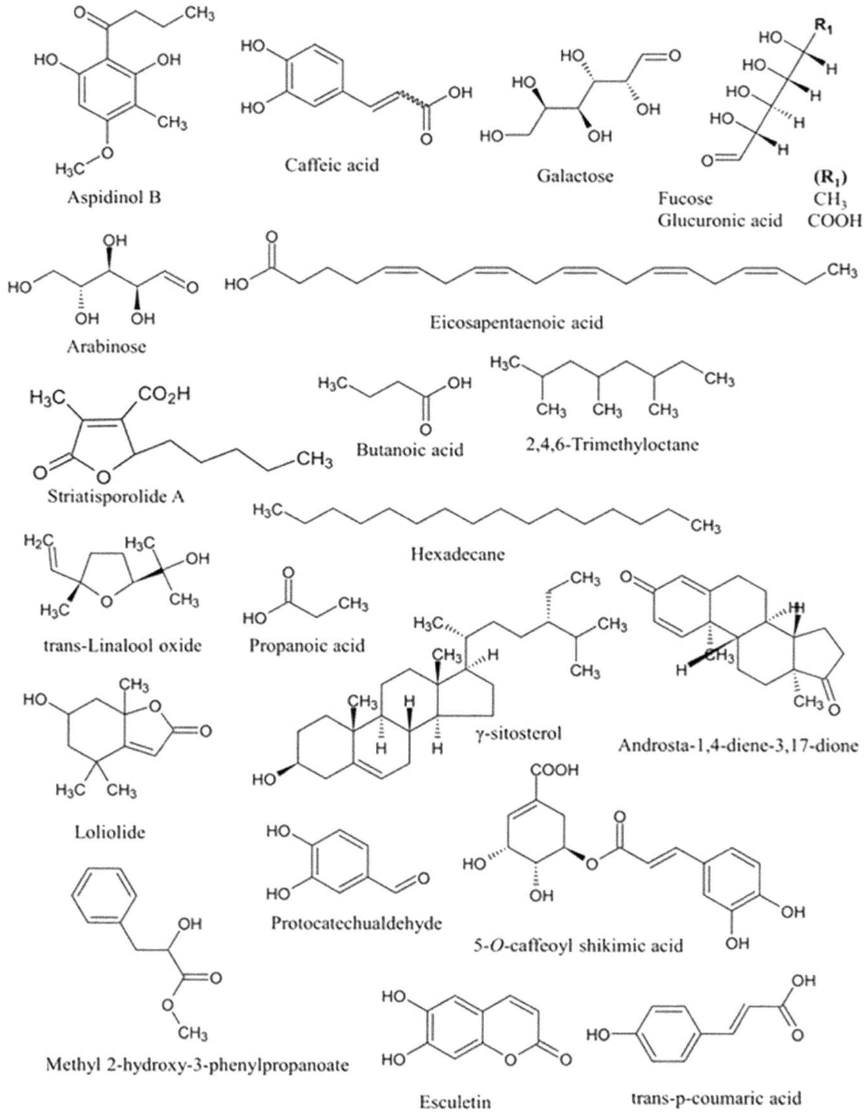

FIGURE 26.2 Isolated phytochemicals from Genus *Athyrium*.

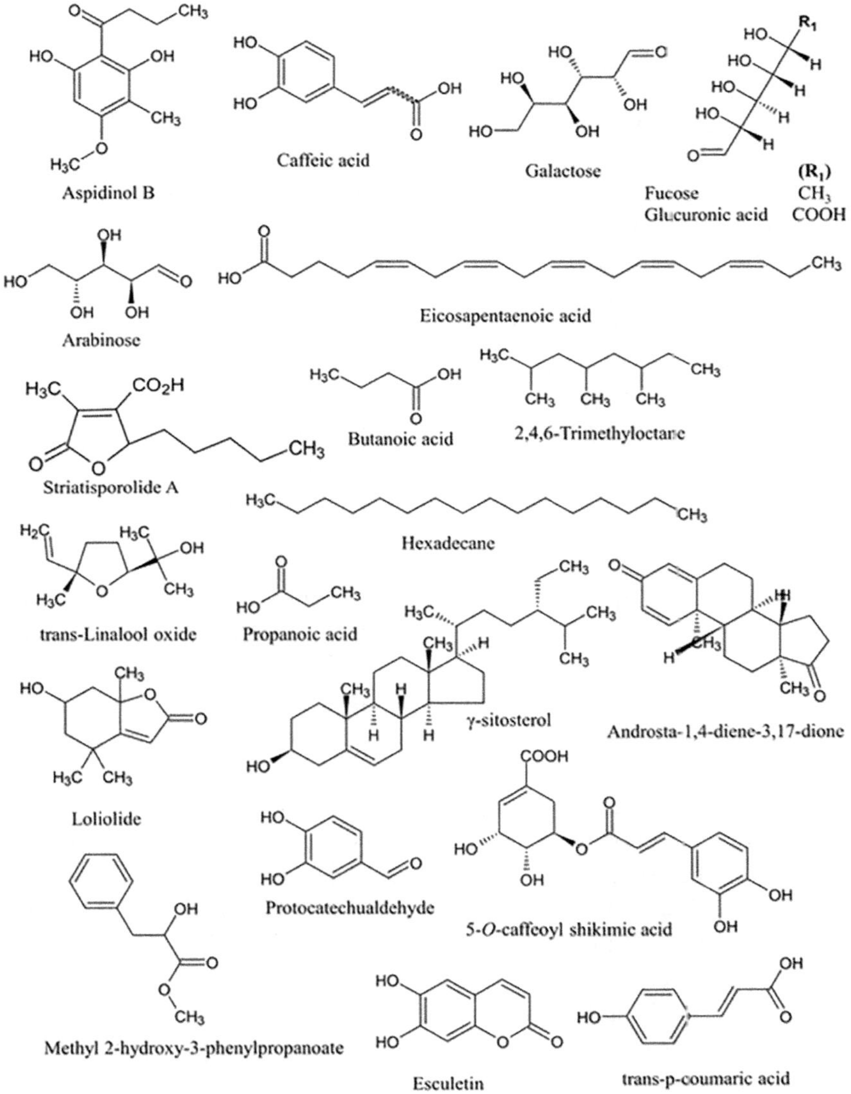

FIGURE 26.3 Isolated phytochemicals from Genus *Athyrium*.

26.4 BIOACTIVITY

26.4.1 ANTIOXIDANT POTENTIAL

The antioxidant potential from the whole plant of *A. esculenta* Copel was evaluated using total polyphenolic content (TPC), ferric reducing antioxidant power (FRAP), and 2,2'-azino-bis (3-ethylbenzothiazoline-6-sulfonic acid) (ABTS) radical scavenging assays. Results showed TPC as 129.2 ± 13.4 μg GAE/mg, FRAP as 0.55 ± 0.002 mM, and TEAC of 1.12 ± 0.007 mM, indicating its high polyphenol content and antioxidant capacity (Razab and Aziz, 2010). The methanolic extract of fresh leaves of *A. filix-femina* was evaluated for its antioxidant activity using various assays. The results showed an oxygen radical absorbance capacity (ORAC) of 186.53 μmol TE/g FW and TPC of 915.53 ± 25.16 GAE/100 g (Soare et al., 2012). Valizadeh et al. (2016) evaluated DPPH and ABTS radical scavenging activity from methanolic leaf and rhizome extract, showing DPPH activity (34.17 ± 1.12 and 12.12 ± 0.57 μg/ml, respectively) by comparing butylated hydroxy toluene (BHT = 9.96 ± 0.45 μg/ml) and the strongest ABTS activity (18.78 ± 1.02 and 20.85 ± 1.39 μmol Trolox/g, respectively). Polysaccharide fractions isolated and fractionated in aqueous ethanol from the rhizome of *A. multidentatum* (Doll.) Ching reported concentration-dependent superoxide radical scavenging activity (4.0 to 20.0 μg/mL). Similarly, for hydroxyl radical scavenging activity at 10.0–50.0 μg/mL, the fractions also demonstrated the strongest antioxidant activity, with EC50 = 16.81 and 7.97 μg/mL, respectively. Additionally, regarding the chelating effect on Fe^{2+} ions, which are the most powerful pro-oxidants, the highest chelating powers of polysaccharide fractions and EDTA were 40.48%, 35.71%, and 47.62% at concentrations of 20.0, 12.0, and 16.0 mg/mL, respectively (Liu et al., 2013). Furthermore, polysaccharides were isolated from the rhizome of *A. multidentatum* (CPA) and degraded using hydrogen peroxide and ascorbic acid, resulting in four derivatives: CPA-1, CPA-2, CPA-3, and CPA-4. CPA-1 exhibited the strongest DPPH radical scavenging activity with a 63.35% inhibition rate at 30 μg/mL, followed by CPA-2, CPA-3, and CPA-4. For hydroxyl radical scavenging, tested using Fenton reactions, CPA-4 demonstrated the highest activity with a 9.95% scavenging effect at 120 and 200 μg/mL, followed by CPA-3, CPA-2, and CPA-1 (Sheng and Sun, 2014). Similarly, Jing et al. (2019) isolated polysaccharides that demonstrated protective effects against D-galactose-induced oxidative stress and aging in male Kunming mice through activation of the Nrf2/HO-1 and PI3K/Akt pathways, highlighting their potential as a natural remedy for oxidative stress-related conditions. Striatisporolide A (SA), isolated from the rhizomes of *A. multidentatum*, inhibits the intracellular overproduction of reactive oxygen species (ROS) in human umbilical vein endothelial cells (HUVECs) at a 100 μM dose (Liu et al., 2016).

The promising cytoprotective activity was shown by isolated heteropolysaccharide fractions (PS-1, PS-2, PS-3, PS-4, and PS-5) from the methanolic extract of *A. multidentatum* rhizome, using anion-exchange column chromatography. The fractions demonstrated antioxidant activity and activated specific cellular pathways (PI3K/Akt/FOXO3a and Nrf2/HO-1). Testing their ability to protect cells from damage resulted in increased cell viability and activities of antioxidant enzymes (superoxide dismutase and catalase). The increased indicators of reduced cell apoptosis, such as nitric oxide content, mitochondrial membrane potential levels, and the Bcl-2/Bax ratio (Jing et al., 2020).

26.4.2 ANTIMICROBIAL ACTIVITY

Soare et al. (2012) investigated the antibacterial activity of methanolic leaf extracts of *A. filix-femina* using the disc diffusion method. The study demonstrated moderate zones of inhibition against gram-negative pathogenic bacteria, including *Escherichia coli* with an inhibition zone of 8.00 ± 1.41 mm, *Pseudomonas aeruginosa* with 8.50 ± 0.70 mm, and *Salmonella abony* with 8.50 ± 0.70 mm. For gram-positive bacteria, *Staphylococcus aureus* and *Enterococcus faecalis* exhibited an inhibition zone of 7.00 mm each. Regarding saprophytic microorganisms, *A. filix-femina* showed an inhibition zone of 8.00 mm against *Brevibacterium flavum*, 9.00 mm against *Bacillus cereus*, 10.00 mm against *Saccharomyces cerevisiae*, and 8.00 mm against *Aspergillus niger*. The minimum inhibitory concentration (MIC) of 60 μg/mL revealed the highest inhibition against *Salmonella abony*. In the case of saprophytes, *A. filix-femina* was most effective against *Saccharomyces cerevisiae* and *Bacillus cereus*, showing moderate activity against *Brevibacterium flavum* and *Aspergillus niger* with an MIC of 600 μg/mL, and exhibited no activity against *Sarcina* sp.

Striatisporolide A (SA), isolated from the methanolic extract of *A. multidentatum* rhizomes, was evaluated for its *in vitro* antibacterial activity against *Escherichia coli* (*E. coli*). The antibacterial activity was assessed using 2,3,5-triphenyltetrazolium chloride (TTC) and Kirby-Bauer disc agar diffusion assays. The highest growth inhibition of *E. coli* observed was 32.74 ± 0.058% at a 200 μM concentration of SA and 31.24 ± 0.065% at a 400 μM concentration (Sheng et al., 2019).

26.4.3 ANTICANCER ACTIVITY

A. multidentatum was evaluated for anticancer activity in HepG2 (human liver cancer cell line) and normal liver cells (HL7702) using MTT and colony formation assay; promising results were shown as dose and time dependent manner that is, 50, 100, and 200 μg/mL viability of HepG2 decreases as 84.5, 68.3, 59.6% in 24 hours and 67.09, 42.92, 25.72% for 48 hours with IC50 220.32 ± 10.32 μg/mL for 24 h and 113.51 ± 7.45 μg/mL for 48 h. In case of HL7702 no significant toxic effects were observed which showing IC50 332.25 ± 15.17 μg/mL and 303.98 ± 20.68 μg/mL for 24 h and 48 h. In case of colony formation *A. multidentatum* extract showed potent and dose-dependent inhibition of colony formation in HepG2 cells (Qi et al., 2017). The ethanolic extract of *A. hohenackerianum* (AHEE) has been studied as a potential anticancer agent in breast, lung, and colon cancer cells. Notably, AHEE exhibited significant activity against HCT-116 colon cancer cells, with an IC50 value of 123.90 μg/mL. In comparison, the standard anticancer drug Cisplatin has an IC50 value of 25.95 μg/mL these findings suggest that AHEE holds promise as a potential basis for developing new anticancer drugs (Elasbali et al., 2022).

26.4.4 ANTI-INFLAMMATORY ACTIVITY

In an *in vivo* study, *A. multidentatum* 95% ethanolic whole plant extract was investigated for its anti-inflammatory effects on female mice with acute lung injury (ALI). The results showed a significant reduction in bronchoalveolar lavage fluid (BALF) cell count and cytokine levels, specifically TNF-α, IL-1β, and IL-6, in a dose-dependent manner, especially at 10 mg/kg. These effects were compared with dexamethasone as an anti-inflammatory drug (Han et al., 2018).

26.4.5 CYTOTOXICITY

A. yokoscense whole plant extract (AYE) at a concentration of 100 μg/mL showed a protective effect against BPA-induced cytotoxicity in A549 human lung cells. AYE also reduced intracellular ROS and inhibited the expression of inflammatory markers like cyclooxygenase-2 (COX-2), extracellular signal-regulated kinase 1/2 (Erk1/2), and nuclear factor kappa-light-chain-enhancer of activated B cells (NF-κB). This suggests that AYE mitigates oxidative stress and inflammation by modulating the pathways in lung epithelial cells (Lee et al., 2023).

26.4.6 ANTI-APOPTOSIS

Striatisporolide A (SA), isolated from the rhizomes of *A. multidentatum*, exhibited potent anti-apoptotic activity in human umbilical vein endothelial cells (HUVECs). At a 100 μM dose, SA significantly reduced apoptosis, with rates of 2.17 ± 0.20% for SA alone and 3.1 ± 0.34% for SA combined with H_2O_2 (Liu et al., 2016).

26.4.7 TOXICITY

A toxicity study was conducted using *A. filix-femina* leaf extract against brine shrimp larvae, demonstrating dose-dependent activity (LC50 = 6.1 μg/mL) at 30 μg/mL, with podophyllotoxin serving as the positive control (Valizadeh et al., 2016).

KEYWORDS

- **antimicrobial**
- **antioxidant**
- **cyclooxygenase-2**
- **flavonoids**
- **insect molting hormones**
- **kappa-light-chain-enhancer**
- **lady fern**

REFERENCES

Adam, K. P. (1995). Caffeic acid derivatives in fronds of the lady fern (*Athyrium filix-femina*). *Phytochemistry*, *40*, 1577, 1578. http://dx.doi.org/10.1002/chin.199613223.

Britannica, T. Editors of Encyclopaedia. (2024). Lady fern. *Encyclopaedia Britannica*. https://www.britannica.com/plant/lady-fern (accessed on 17th November 2025).

Chai, T. T., Elamparuthi, S., Yong, A. L., Quah, Y. H. C., & Wong, F. C. (2013). Antibacterial, anti-glucosidase, and antioxidant activities of selected highland ferns of Malaysia. *Botanical Studies*, *54*(1), 55.

Elasbali, A. M., Al-Soud, W. A., Al-Oanzi, Z. H., Qanash, H., Alharbi, B., Binsaleh, N. K., Alreshidi, M., Patel, M., & Adnan, M. (2022). Cytotoxic activity, cell cycle inhibition, and apoptosis-inducing potential of *Athyrium hohenackerianum* (lady fern) with its phytochemical profiling. *Evidence-Based Complementary and Alternative Medicine*, *2022*, 2055773. https://doi.org/10.1155/2022/2055773.

Fons, F., Froissard, D., Bessière, J. M., Buatois, B., & Rapior, S. (2010). Biodiversity of volatile organic compounds from five French ferns. *Natural Product Communications*, *5*, 1655–1658. http://dx.doi.org/10.1177/1934578X1000501028.

Han, X. Z., Ma, R., Chen, Q., Jin, X., Jin, Y. Z., An, R. B., Piao, X. M., Lian, M. L., Quan, L. H., & Jiang, J. (2018). Anti-inflammatory action of *Athyrium multidentatum* extract suppresses the LPS-induced TLR4 signaling pathway. *Journal of Ethnopharmacology*, *217*, 220–227. http://dx.doi.org/10.1016/j.jep.2018.02.031.

Irfan, M., Jan, G., Murad, W., Jan, F. G., Rauf, A., Alsayari, A., Almarhoonf, Z. M., & Mabkhot, Y. N. (2021). Ethnomedicinal and traditional uses of the ferns of Khyber Pakhtunkhwa, Pakistan. *Brazilian Journal of Biology*, *84*, e250256. https://doi.org/10.1590/1519-6984.250256.

Jing, L., Jiang, J. R., Liu, D. M., Sheng, J. W., Zhang, W. F., Li, Z. J., & Wei, L. Y. (2019). Structural characterization and antioxidant activity of polysaccharides from *Athyrium multidentatum* (Doll.) Ching in D-galactose-induced aging mice via PI3K/AKT pathway. *Molecules*, *24*, 3364. http://dx.doi.org/10.3390/molecules24183364.

Jing, L., Sheng, J., Jiang, J., Wang, Y., Shen, X., Liu, D., Zhang, W., & Mao, S. (2020). Chemical characteristics and cytoprotective activities of polysaccharide fractions from *Athyrium multidentatum* (Doll.) Ching. *International Journal of Biological Macromolecules*, *158*, 1227–1238. https://doi.org/10.1016/j.ijbiomac.2020.05.053.

Kreshchenok, I., Lesik, E., Toushkin, A., & Toushkina, A. (2021). Edible ferns of the Amur region and their rational use. *IOP Conference Series: Earth and Environmental Science*, *937*, 022138. http://dx.doi.org/10.1088/1755-1315/937/2/022138.

Lee, J. K., Choi, W. S., Song, J. Y., Kwon, O. S., Lee, Y. J., Lee, J. S., Lee, S., Choi, S. R., Lee, C. H., & Lee, J. Y. (2023). Anti-inflammatory effects of *Athyrium yokoscense* extract via inhibition of the Erk1/2 and NF-κB pathways in bisphenol A-stimulated A549 cells. *Toxicological Research*, *39*, 135–146.

Liu, D. M., Sheng, J. W., Wang, S. H., Zhang, W. F., Zhang, W., & Zhang, D. J. (2016). Cytoproliferative and cytoprotective effects of Striatisporolide A isolated from rhizomes of *Athyrium multidentatum* (Doell.) Ching on human umbilical vein endothelial cells. *Molecules*, *21*, 1280. http://dx.doi.org/10.3390/molecules21101280.

Liu, D., Sheng, J., Li, Z., Qi, H., Sun, Y., Duan, Y., & Zhang, W. (2013). Antioxidant activity of polysaccharide fractions extracted from *Athyrium multidentatum* (Doll.)

Ching. *International Journal of Biological Macromolecules*, *56*, 1–5. https://doi.org/10.1016/j.ijbiomac.2013.01.023.

Nekrasov, E. V., Svetashev, V. I., Khrapko, O. V., & Vyssotski, M. V. (2019). Variability of fatty acid profiles in ferns: Relation to fern taxonomy and seasonal development. *Phytochemistry*, *162*, 47–55. https://doi.org/10.1016/j.phytochem.2019.02.015.

Ohta, S., Guo, J. R., Hiraga, Y., & Suga, T. (1996). 24-Epi-pterosterone: A novel phytoecdysone from the roots of *Athyrium yokoscense*. *Phytochemistry*, *41*, 745–747. http://dx.doi.org/10.1016/0031-9422(95)00688-5.

Parihar, P., Parihar, L., & Bohra, A. (2006). Antibacterial activity of *Athyrium pectinatum* (Wall.) Presl. *Natural Product Radiance*, *5*(4), 262–265.

Park, H. J., Ryu, S. H., Yeon, S. W., Turk, A., Lee, S., Lee, H. H., Hwang, B. Y., & Lee, M. K. (2023). Antioxidant constituents of *Athyrium acutipinnulum*. *Korean Journal of Pharmacognosy*, *54*, 53–60. https://doi.org/10.22889/KJP.2023.54.2.53.

Patil, S. M., Patil, M. N., & Patil, S. S. (2020). Diversity and ethnomedicinal pteridophytes of Kolhapur district, Maharashtra. *Indian Fern Journal*, *37*(1–2), 226–238.

POWO. (2024). *Plants of the World Online*. Facilitated by the Royal Botanic Gardens, Kew. https://powo.science.kew.org/.

Qi, G., Liu, Z., Fan, R., Yin, Z., Mi, Y., Ren, B., & Liu, X. (2017). *Athyrium multidentatum* (Doll.) Ching extract induces apoptosis via mitochondrial dysfunction and oxidative stress in HepG2 cells. *Scientific Reports*, *7*, 2275. http://dx.doi.org/10.1038/s41598-017-02573-8.

Razab, R., & Aziz, A. A. (2010). Antioxidants from tropical herbs. *Natural Product Communications*, *5*, 441–445. http://dx.doi.org/10.1177/1934578X1000500321.

Salehi, B., Ezzat, S. M., Fokou, P. V. T., Albayrak, S., Vlaisavljevic, S., Sharifi-Rad, M., & Sharifi-Rad, J. (2019). *Athyrium* plants—Review on phytopharmacy properties. *Journal of Traditional and Complementary Medicine*, *9*, 201–205. http://dx.doi.org/10.1016/j.jtcme.2018.09.001.

Samy, R. A., Manickam, V. S., Wallace, J. W., & Subarayan, S. B. G. (2004). Flavonoid glycosides of *Athyrium solenopteris* (Kunze) T. Moore of Palni Hills of South India. *Journal of the Indian Chemical Society*, *81*(1), 77–78.

Sheng, J. W., Liu, D. M., Jing, L., Xia, G. X., Zhang, W. F., Jiang, J. R., & Tang, J. B. (2019). Striatisporolide A, a butenolide metabolite from *Athyrium multidentatum* (Doll.) Ching, as a potential antibacterial agent. *Molecular Medicine Reports*, *20*, 198–204. https://doi.org/10.3892/mmr.2019.10244.

Sheng, J., & Sun, Y. (2014). Antioxidant properties of different molecular weight polysaccharides from *Athyrium multidentatum* (Doll.) Ching. *Carbohydrate Polymers*, *108*, 41–45. https://doi.org/10.1016/j.carbpol.2014.03.011.

Singh, B. P., & Upadhyay, R. (2014). Medicinal pteridophytes of Madhya Pradesh. *Journal of Medicinal Plants Studies*, *2*(4), 65–68.

Soare, L. C., Ferdes, M., Stefanov, S., Denkova, Z., Nicolova, R., Denev, P., Bejan, C., & Paunescu, A. (2012). Antioxidant activity, polyphenols content and antimicrobial activity of several native pteridophytes of Romania. *Notulae Botanicae Horti Agrobotanici Cluj-Napoca*, *40*, 53–57. http://dx.doi.org/10.15835/nbha401668.

Umikalsom, Y., Grayer-Barkmeijer, R. J., & Harborne, J. B. (1994). A comparison of the flavonoids in Athyriaceae and Aspleniaceae. *Biochemical Systematics and Ecology*, *22*, 587–594. https://doi.org/10.1016/0305-1978(94)90071-X.

Valizadeh, H., Sonboli, A., Kordi, F. M., Dehghan, H., & Bahadori, M. B. (2016). Cytotoxicity, antioxidant activity and phenolic content of eight fern species from north of Iran. *Pharmaceutical Sciences*, *21*, 18–24. http://dx.doi.org/10.15171/PS.2015.12.

Wang, Y., Gao, S., He, X., Li, Y., Li, P., Zhang, Y., & Chen, W. (2019). Growth, secondary metabolites and enzyme activity responses of two edible fern species to drought stress and rehydration in Northeast China. *Agronomy*, *9*, 137. http://dx.doi.org/10.3390/agronomy9030137.

CHAPTER 27

Pharmacological and Bioactive Profile of *Lepisorus* Roth (Polypodiaceae): A Comprehensive Review

SACHIN B. PATIL,[1] SURAJ P. DEVKAR,[2,*] SACHIN M. PATIL,[3,*] and SAGAR A. DESHMUKH[1,2]

[1]*Department of Botany, Anandibai Raorane Arts, Commerce and Science College, Vaibhavwadi, Sindhudurg (affiliated to the University of Mumbai), Maharashtra, India*

[2]*Department of Botany, The New College, Kolhapur (affiliated to Shivaji University, Kolhapur), Maharashtra, India*

[3]*Department of Botany, Dr. Ghali College, Gadhinglaj (affiliated to Shivaji University, Kolhapur), Maharashtra, India*

*Corresponding authors

ABSTRACT

The genus *Lepisorus* (Polypodiaceae), comprising approximately 90 species, is distributed across Eastern and Southern Asia, the Pacific islands, and Madagascar. It is distinguished by its creeping rhizomes, simple leaves, and sori with scale-like paraphyses. Ethnobotanical uses of *Lepisorus* species include traditional medicines employed for their diuretic and hemostatic properties in Korean medicine, as well as for treating fractures and kidney disorders. Pharmacologically, *Lepisorus* species exhibit significant bioactivities. Compounds from various *Lepisorus* species show strong antioxidant and anti-cancer properties, including DPPH radical scavenging and inhibition of liver cancer cell proliferation. The present comprehensive review on phytochemistry and pharmacology continues to

Bioactives and Pharmacology of Pteridophytes. T. Pullaiah (Ed.)

DOI: 10.1201/9781779642738-27

enhance our understanding of *Lepisorus*, underscoring the need for further exploration and providing a foundation for future research.

27.1 INTRODUCTION

The genus *Lepisorus* was initially described by J. Smith and subsequently put forward as a section of *Drynaria* (Smith 1846). Ching (1933) elevated *Lepisorus* to generic status, selecting the previously recognized species *Polypodium lineare* Thunb. as *L. thunbergianus* as the type specimen. At the outset, *Lepisorus* is divided into two sections based on sorus shape: section *Pseudovittaria*, characterized by linear sori found in species like *L. sinensis* and *L. vittarioides*, and section *Eulepisorus*, encompassing the rest of the genus with discrete sori. A revised infrageneric classification of *Lepisorus* was put forth by Yu and Lin (1997). They introduced a total of six sections primarily distinguished by characteristics of rhizome scales with soral paraphyses. Lin (2000) later simplified this classification into two sections: section *Lepisorus* and section *Hymenophyton*, categorized by whether the leaves are evergreen or deciduous. Zhang and Qi (2009) revised the genus *Lepisorus* (J. Sm.) Ching sect. *Lepisorus* from China. *Lepisorus* is distributed across Southern and Eastern Asia, with extensions into the Pacific islands and tropical regions of Madagascar, including Hawai'i Island. This expansive geographic range underscores the genus's adaptability to diverse tropical and subtropical environments (Wang et al., 2012; Zink, 1993;). It comprises approximately 90 species grouped into 18 sections, making it a genus within the Polypodiaceae family that contains a high diversity of species (PPGI, 2016; Zhao et al., 2020a). These ferns are easily distinguished by their unique morphological features, such as creeping rhizomes adorned with lattice-like scales, simple leaves, and sori surrounded by scale-like paraphyses (Ching, 1933; Qi et al., 2013; Zhao et al., 2020b). Identification of *Lepisorus* species is based on distinctive traits such as the shape of the leaves, location of sori, and structure of rhizome scales and paraphyses (Fujiwara et al., 2018).

27.2 ETHNOBOTANY

Lepisorus bicolor (Takedal) Ching is known as *dhule uniu* in Nepal, and local people take frond or pinnae juice orally twice a day (for 3–4 days) to cure rheumatic pains (Ojha and Deakota, 2015). Karma et al. (2017)

noted that the leaves of *Lepisorus contortus* (Christ) Ching were traditionally used in Sowa Rigpa medicine to treat bone fractures, burns, wounds, kidney disorders, trauma, and injuries in Bhutan. *L. mehrae* is used to cure fever, cough, and body aches in Nepal (Pradhan et al., 2020; Subba and Thapa, 2018). Additionally, the boiled juice of pinnae was used to treat lung disease, cough and cold, sinusitis, headache, wounds, and body pain (Pandey and Pradhan 2020). The whole plant is used for gastrointestinal disorders (Uprety et al., 2010). *L. longifolius* (Blume) Holttum is a traditional medicinal fern used to treat cancer in Indonesia (Silalahi et al., 2015). *L. loriformis* is commonly known as 'Surtipaat' and is found on tree trunks or in rock crevices in forests. Traditionally, the aqueous extract of the root of this plant is used against urinary disorders. *L. thunbergianus* is a folk medicine of Korea used as a diuretic or hemostatic and is used to treat human breast, liver, and colon cancer (Choi et al., 1999). *L. ussuriensis* (Regel et Maack.) Ching is employed in Korean traditional medicine for its diuretic, hemostatic, and antitussive properties (Kim et al., 1996).

27.3 CHEMICAL CONSTITUENTS

The genus *Lepisorus* contains a variety of photo-active constituents, including diglycosides, flavonoids, and other bioactive compounds. These compounds are enlisted below following their structure. Phytochemical compounds in *L. contortus* are β-(4-hydroxyphenyl) ethyl-4-O-E-caffeoyl-O-β-D-apiofuranosyl-(1-2)]-β-D-glucopyranoside; β-(3,4-dihydroxyphenyl) ethyl-6-O-E-caffeoyl-O-[β-D-apiofuranosyl-(1-2)]-β-D-glucopyranoside; β-(3,4-dihydroxyphenyl) ethyl-4-O-E-caffeoyl-O-[β-D-apiofuranosyl-(1-2)]-β-D-glucopyranoside; β-(3,4-dihydroxyphenyl) ethyl-3-O-E-caffeoyl-O-[β-D-apiofuranosyl-(1-2)]-β-D-glucopyranoside; β-(4-hydroxyphenyl) ethyl-3-O-E-caffeoyl-O-β-D-apiofuranosyl-(1-2)]-β-D-glucopyranoside; Calceolarioside B; β-(3,4-dihydroxyphenyl) β-D-glucopyranoside; Arauridine; Threo-1-(4-hydroxy-3-methoxyphenyl)-2-{4-[-(E)-3-hydroxy-1-propenyl]-2-methoxyphenoxy}-1,3-propanediol;40,5,7-trihydroxy-3-methoxyflavone-7-O-R-L-arabinofuranosyl-(1-6)-β-D-glucopyranoside; 30,40,5,7-tetrahydroxy-3-methoxyflavone-7-O-R-L-rhamnopyranosyl-(1-6)-β-D-glucopyranoside;40,5,7-trihydroxy-3-methoxyflavone-7-O-rutinoside; Transilin; 40,5,7-trihydroxy-3-methoxyflavone-7-O-β-D-glucopyranoside; Quercetin-3-O-β-D-glucoside; Syringic acid; Vanillic acid; Phloretic acid; Diplopterol; β-sitosterol (Yang et al., 2011). Phytochemical compounds in

L. mehrae are Protocatechuic acid; Gallic acid; 7,8,3′,4′-Tetrahydroxyflavanone; Rhamnocitrin; Quercetin; Hyperoside; Quercetin-7-glucoside; Rutin; Rhamnetin; Aromadendrin; Camphor; Hexanoic acid (Pandey and Pradhan, 2020). Phytochemical compounds in *L. thunbergianus* are Caffeic acid; Chlorogenic acid; Epicatechin 7-O-beta-D-glucoside (Lee, 1998). Isovitexin; Orientin; Iso-orientin; Chlorogenic acid (Yang et al., 2015b). Phytochemical compounds in *L. ussuriensis* are Quercetin 3-methyl ether 7-O-L-arabinofuranosyl glucopyranoside; Quercetin 3-methyl ether glucoside; Vitexin; Orientin; Eriodyctiol glucoside; 2a,3b-(22R)-trihydroxycholestan-6-one-22-O-β-D-glucopyranosyl-(1-2)-α-L-arabinopyranoside; alpha-ecdysone; Ecdysterone; 7-O-methylnaringenin-(4′→O→6″)-scutellarein; Epicatechin; Epigallocatechin; Catechin; Naringenin; Dihydroquercetin; Quercetin; Kaempferol; Diosmetin; Luteolin; Baicalein; 7,8-dihydroxyflavone (Kim et al., 1996, 1999; Luo et al., 2015). The chemical structures of the compounds reported in the studied *Lepisorus* species are illustrated in Figures 27.1–27.4.

27.4 PHARMACOLOGY

27.4.1 ANTI-OXIDANT ACTIVITY

Yang et al. (2015a) isolated orientin and isoorientin from *L. thunbergianus* (Kaulf.) Ching, revealing DPPH radical scavenging activities with IC50 values of 15–17 μM and reducing reactive oxygen species (ROS) production by 50% at 100 μM. The findings suggested that chlorogenic acid notably inhibited intracellular ROS and nitric oxide (NO) production, which showed strong DPPH scavenging activity and reduced gene expression of iNOS, COX-2, IFN-β, and TNF-α. Isoorientin at 100 μM achieved 50% inhibition of human liver cancer cells (Huh7 and HepG2), suggesting that isolated compounds may help to prevent liver cancer. Pandey and Pradhan (2020) analyzed total flavonoid content (TFC) and total phenolic content (TPC). They noted that TFC values ranged from 71.94 ± 2.73 to 135.57 ± 1.18 mg QE/g DW across different solvent extracts of *L. mehrae*. The aqueous extract of *L. mehrae* exhibited the highest antioxidant activity, with 76.34 ± 0.41% scavenging against DPPH and 88.63 ± 0.21% scavenging against ABTS at a concentration of 100 μg/mL. However, this extract showed the least inhibition of α-glucosidase, with an IC50 value of 0.527 ± 0.004 mg/mL (46.09 ± 0.51% inhibition at 0.50 mg/mL).

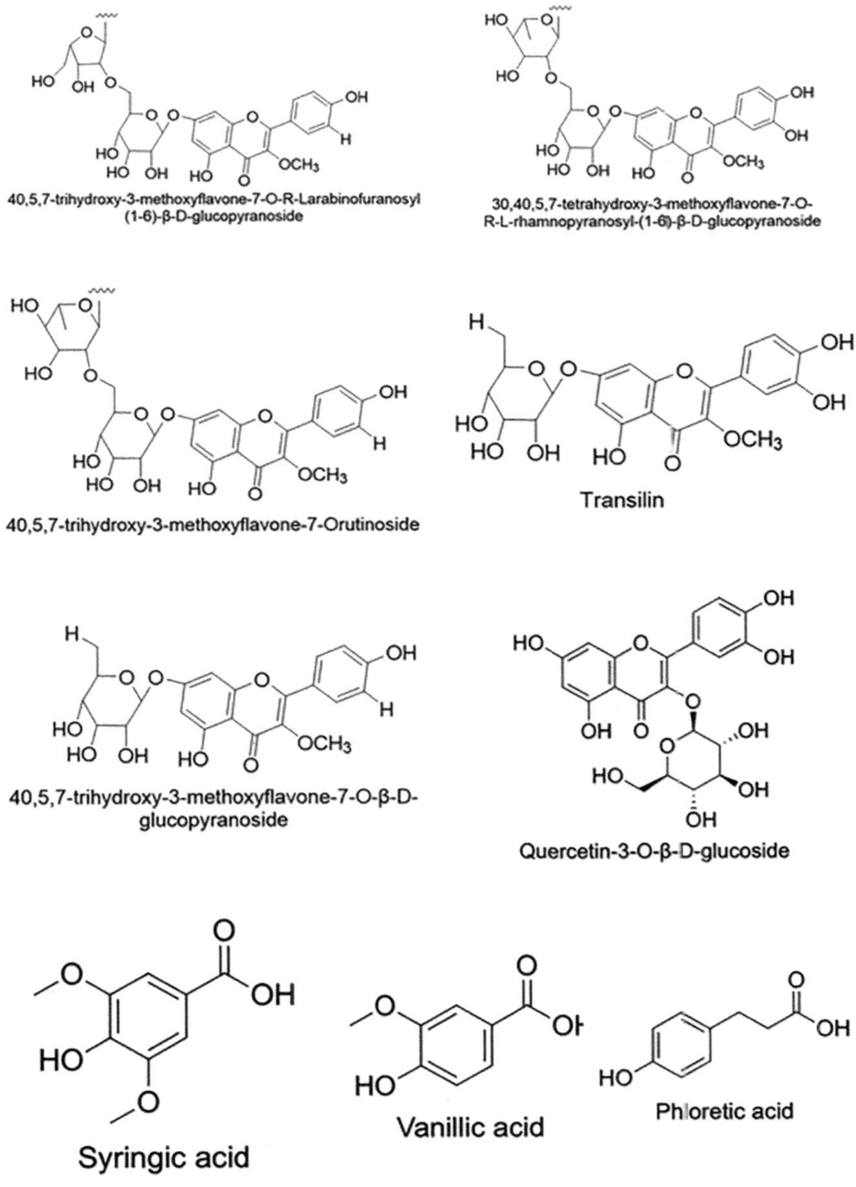

FIGURE 27.1 Chemical structure of compounds are reported in *Lepisorus* species.

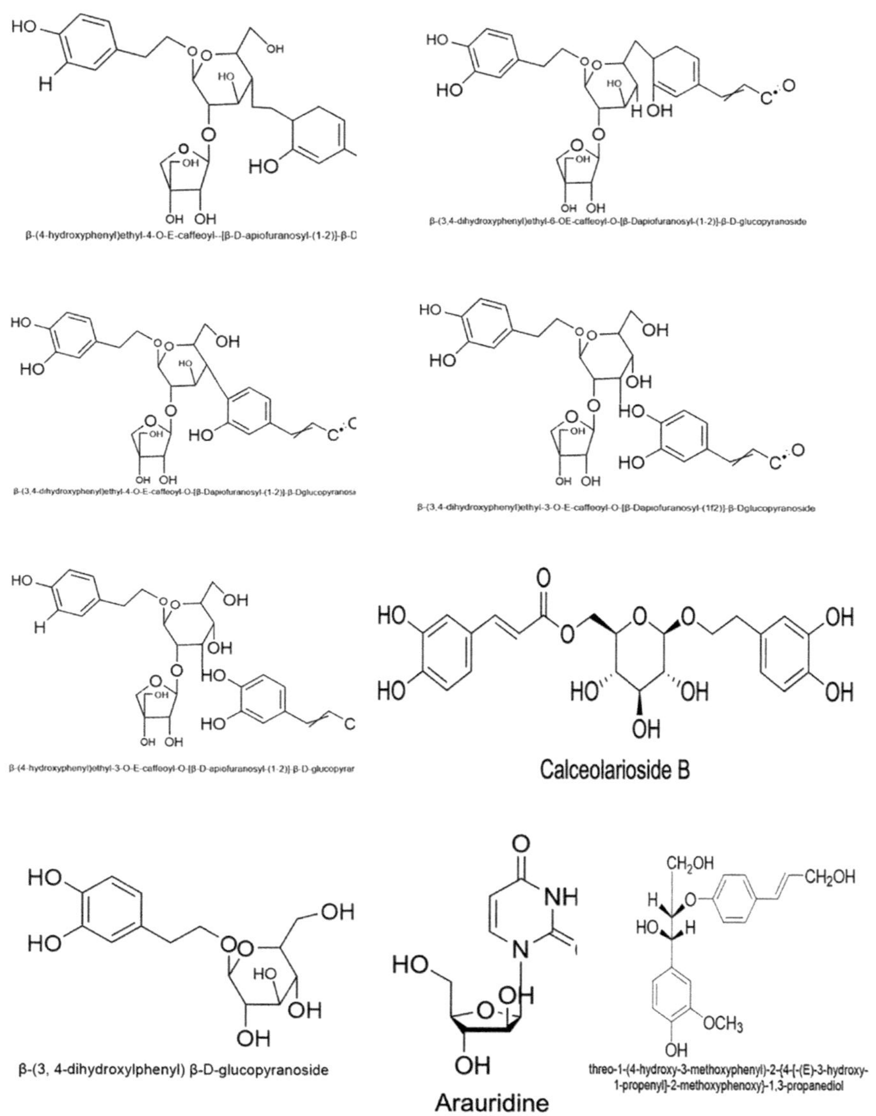

FIGURE 27.2 Chemical structure of compounds reported in *Lepisorus* species.

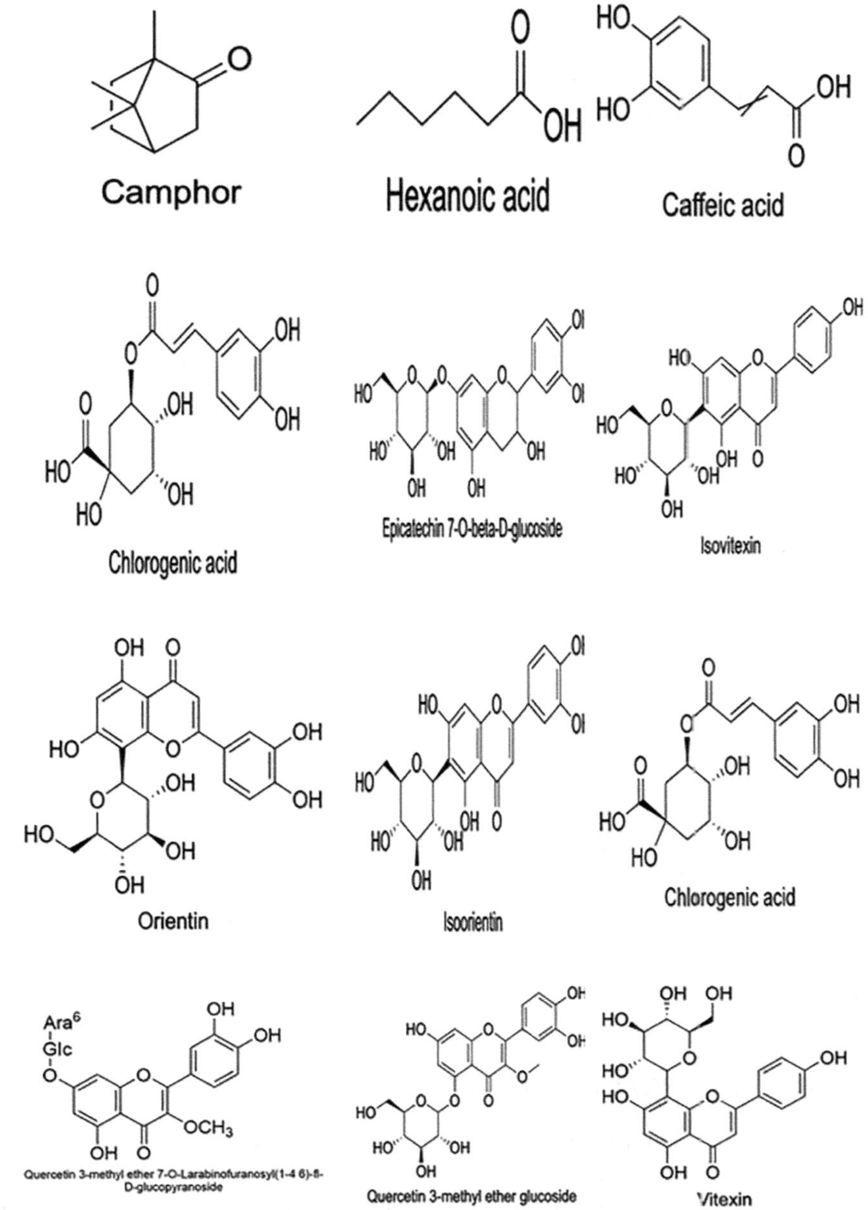

FIGURE 27.3 Chemical structure of compounds reported in *Lepisorus* species.

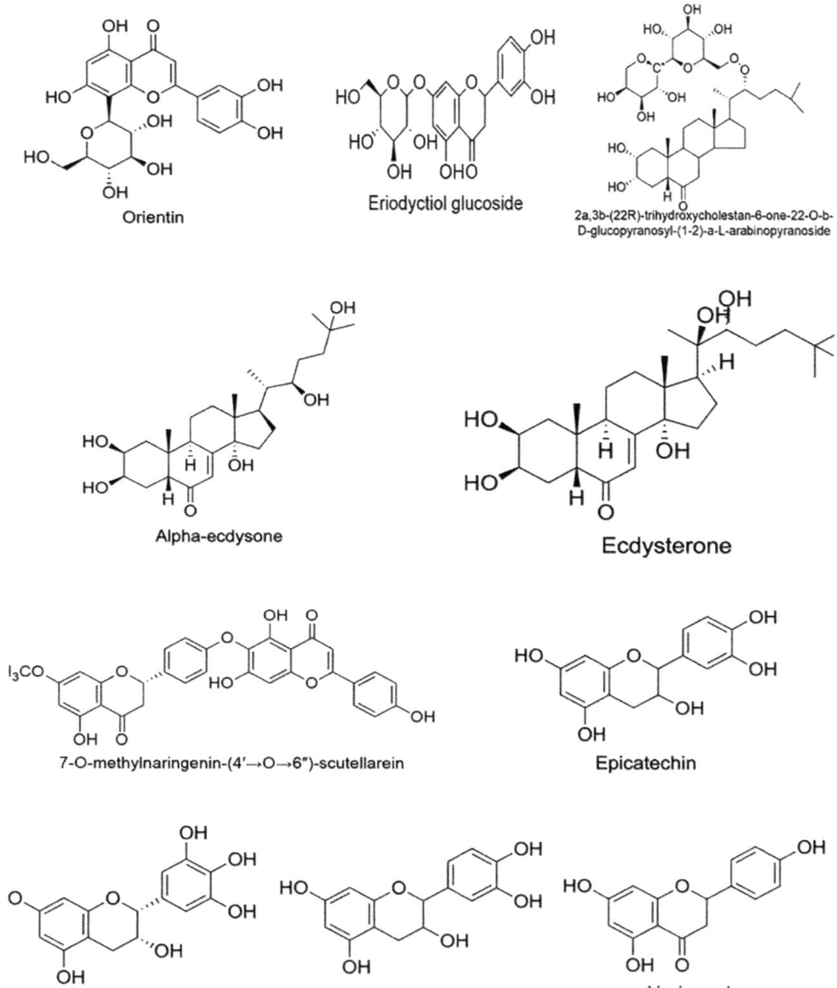

FIGURE 27.4 Chemical structure of compounds reported in *Lepisorus* species.

27.4.2 ANTI-INFLAMMATORY ACTIVITY

Dried whole plants of *L. ussuriensis* were powdered and subjected to extraction with 70% ethanol at room temperature, followed by filtration, combination, and evaporation of the solvent. The extract was suspended in water, resulting in the isolation of chemical compounds like 7-O-methylnaringenin-scutellarein, epicatechin, epigallocatechin, catechin, naringenin, dihydroquercetin, quercetin, kaempferol, diosmetin, luteolin, baicalein, and 7,8-dihydroxyflavone. The anti-inflammatory activities of these compounds were assessed by measuring NF-κB-dependent luciferase activity using a dual luciferase reporter assay system. HeLa cells were seeded in 96-well plates for 24 hours and then treated with the compounds. Quercetin, diosmetin, and luteolin demonstrated significant inhibition of TNF-α-induced NF-κB reporter gene expression at concentrations of 30 μM and 100 μM (Luo et al., 2015).

27.4.3 ANTI-CANCER ACTIVITY

L. contortus was collected, air-dried and powdered before being extracted with 95% ethanol and evaluated its potential for cancer chemoprevention, various assays were performed, including NF-κB luciferase, aromatase, quinone reductase 2 (QR2), cyclooxygenase-1 (COX-1), and cyclooxygenase-2 (COX-2). These assays were assessed to inhibit tumor necrosis factor-alpha (TNF-α)-induced NF-κB activity, nitric oxide (NO) production, and the activities of aromatase, QR2, COX-1, and COX-2. They found Quercetin-3-O-β-D-glucoside emerged as a notable active compound, demonstrating significant inhibition of QR2 with an IC50 value of 3.84 μM. This finding confirmed that kaempferol and quercetin glycosides were effective inhibitors of QR2. Moreover, quercetin-3-O-β-D-glucoside showed substantial inhibition of NF-κB activity, with an IC50 value of 33.6 μM. Several other compounds, including β-(4-hydroxyphenyl)ethyl-4-O-E-caffeoyl-O-[β-D-apiofuranosyl-(1-2)]-β-D-glucopyranoside, β-(3,4-dihydroxyphenyl)ethyl-6-O-E-caffeoyl-O-[β-D-apiofuranosyl-(1-2)]-β-D-glucopyranoside, β-(3,4-dihydroxyphenyl)ethyl-3-O-E-caffeoyl-O-[β-D-apiofuranosyl-(1-2)]-β-D-glucopyranoside, and calceolarioside B displayed significant aromatase activity, with IC50 values of 30.7, 32.3, 26.8, and 35.3 μM, respectively. These findings underscore the promising potential of *L. contortus* as a source of bioactive compounds for cancer chemoprevention, targeting multiple molecular pathways. In *in vitro* studies, Hepa1c1c7 and MCF-7 cells were cultured in specific media with supplements and incubated with sample solutions at 37°C with 5% CO_2 for 72 hours. Aqueous DMSO was the control. After fixation, staining, and solubilizing the dye, optical density was measured at 515 nm. Paclitaxel (23.4 μM) inhibited MCF-7 by 64% and Hepa1c1c7 by 57%, while vinblastine (49.4 μM) inhibited MCF-7 by 74% and Hepa1c1c7 by 90%. Yang et al. (2015a) investigated the n-hexane fraction for cytotoxic activity, revealing a marked suppression of proliferation in colon cancer, liver, and human breast cancer cells. Notably, this study provides the first evidence that an extract of *L. thunbergianus* has been found to induce apoptosis in SW620 colon cancer cells, suggesting its potential as a therapeutic agent for colon cancer.

27.4.4 ANTI-TYROSINASE ACTIVITY

Kim et al. (2020) revealed the anti-tyrosinase activity and found that 5-caffeoylquinic acid inhibited α-MSH-induced melanogenesis. This was confirmed by melanin content and tyrosinase activity assays. 5-caffeoylquinic acid inhibited melanogenesis by chelating copper from the copper-tyrosinase complex. *L. thunbergianus* extract also demonstrated strong radical scavenging and anti-melanogenesis effects without significant cytotoxicity. The n-BuOH fraction, rich in 5-caffeoylquinic acid, showed potential as a natural skin-whitening agent.

27.5 BIOINDICATORS

Kono and Tomiyusu (2009) observed that fronds of the fern *L. thunbergianus* are effective for estimating atmospheric mercury levels. The mercury concentration in mature fronds closely correlates with ambient air mercury levels over several months. Using these fronds, atmospheric mercury levels of 3–8 ng/m³ can be estimated, with a 95% confidence interval indicating an estimated concentration of 5 ng/m³.

KEYWORDS

- **antioxidant**
- **atmospheric mercury levels**
- **colon cancer cells**
- **pharmacology**
- **phytochemical**
- **syringic acid**
- **taxonomy**

REFERENCES

Ching, R. C. (1933). The studies of Chinese ferns IX. *Bulletin of the Fan Memorial Institute of Biology, Botany*, *4*, 47–106.

Choi, Y. H., Kim, J. W., & Choi, Y. H. (1999). A steroidal glycoside from *Lepisorus ussuriensis*. *Phytochemistry*, *51*, 453–456. https://doi.org/10.1016/S0031-9422(98)00734-4.

Fujiwara, T., Serizawa, S., & Watano, Y. (2018). Phylogenetic analysis reveals the origins of tetraploid and hexaploid species in the Japanese *Lepisorus thunbergianus* (Polypodiaceae) complex. *Journal of Plant Research*, *131*(6), 945–959. https://doi.org/10.1007/s10265-018-1057-1.

Karma, Y., Shakuli, K., Pema, Y., & Phurpa, W. (2017). Taxonomical identification of Himalayan edible medicinal plants in Bhutan and the phenolic contents and antioxidant activity of selected plants. *Journal of Biological Active Products from Nature*, *7*(2), 89–106. https://doi.org/10.1080/22311866.2017.1366664.

Kim, H., Kim, J., Kim, J., Jung, S., & Lee, K. (2020). Characterization of caffeoylquinic acids from *Lepisorus thunbergianus* and their melanogenesis inhibitory activity. *ACS Omega*, *5*(48), 30946–30955. https://doi.org/10.1021/acsomega.0c03783.

Kim, J., Choi, Y., & Choi, Y. (1999). A steroidal glycoside from *Lepisorus ussuriensis*. *Phytochemistry*, *51*, 453–456. https://doi.org/10.1016/S0031-9422(98)00734-4.

Kim, J., Choi, Y., Lim, Y., & Yeo, H. (1996). A flavonoid diglycoside from *Lepisorus ussuriensis*. *Phytochemistry*, *43*(5), 1111–1113. https://doi.org/10.1016/S0031-9422(96)00390-6.

Kono, Y., & Tomiyasu, T. (2009). Biomonitoring of atmospheric mercury levels with the epiphytic fern *Lepisorus thunbergianus* (Polypodiaceae). *Chemosphere*, *77*(10), 1387–1392. https://doi.org/10.1016/j.chemosphere.2009.09.042.

Lee, M. (1998). Phenolic compounds from *Lepisorus thunbergianus*. *Korean Journal of Pharmacognosy*, *29*(2), 142–145.

Lin, Y. (2000). Lepisorioideae. In Y. X. Lin, S. G. Lu, X. C. Zhang, & L. Shi (Eds.), *Flora of China* (pp. 32–115). Science Press.

Luo, J., Zhou, W., Cao, S., Jin, M., Zhang, C., Jin, X., & Li, G. (2015). A new biflavonoid from the whole herb of *Lepisorus ussuriensis*. *Natural Product Research*, *30*(13), 1470–1476. https://doi.org/10.1080/14786419.2015.1057726.

Ojha, R., & Deakota, H. P. (2015). Edible and medicinal pteridophytes of Nepal: A review. *Ethnobotany Research and Applications*, *22*, 1–16.

Pandey, B., & Pradhan, S. (2020). Chemical composition, *in vitro* antioxidant, and enzymes inhibitory potential of three medicinally important plants from Nepal (*Lepisorus mehrae*, *Pleurospermum benthamii*, and *Roscoea auriculata*). *Advances in Traditional Medicine*, *20*, 1–16. https://doi.org/10.1007/s13596-020-00481-7 (accessed on 17th November 2025).

PPGI. (2016). A community-derived classification for extant lycophytes and ferns. *Journal of Systematics and Evolution*, *54*(6), 563–603. https://doi.org/10.1111/jse.12229.

Pradhan, S., Chaudhary, R., Sigdel, S., & Pandey, B. (2020). Ethno-botanical knowledge of Khandadevi and Gokulganga rural municipality of Ramechhap District of Nepal. *Ethnobotany Research and Applications*, *20*, 1–32.

Qi, X., Zhang, X., Lin, Y., Gilbert, M., & Hovenkamp, P. (2013). *Lepisorus* (J. Smith) Ching. In *Flora of China (Pteridophytes)* (Vol. 2, pp. 808–824). Science Press; Missouri Botanical Garden Press.

Silalahi, M., Nisyawati, Walujo, E. B., Supriatna, J., & Mangunwardoyo, W. (2015). The local knowledge of medicinal plant traders and diversity of medicinal plants in the Kabanjahe traditional market, North Sumatra, Indonesia. *Journal of Ethnopharmacology, 175*, 432–443. https://doi.org/10.1016/j.jep.2015.09.006.

Smith, J. (1846). An enumeration of ferns cultivated at Kew, with characters and observations on some of the genera and species. *Botanical Magazine (Comp.), 72*, 13.

Subba, B., & Thapa, S. (2018). Analysis of phytoconstituents and biological activities on the selected medicinal plants of Dolakha and Sindhupalchowk district of Nepal. *Journal of the Institute of Science and Technology, 22*(2), 140–147.

Uprety, Y., Asselin, H., Boon, E., Yadav, S., & Shrestha, K. (2010). Indigenous use and bio-efficacy of Rasuwa District, Central Nepal. *Journal of Ethnobiology and Ethnomedicine, 6*(3), 1–10. https://doi.org/10.1186/1746-4269-6-3.

Wang, L., Schneider, H., Zhang, X., & Xiang, Q. (2012). The rise of the Himalaya enforced the diversification of SE Asian ferns by altering the monsoon regimes. *BMC Plant Biology, 12*(1), 210. https://doi.org/10.1186/1471-2229-12-210.

Yang, J., Kondratyuk, T., Jermihov, K., Marler, L., Qiu, X., Choi, Y., Cao, H., Yu, R., Sturdy, M., Huang, R., Liu, Y., Wang, L., Mesecar, A., van Breemen, R., Pezzuto, J., Fong, H., Chen, Y., & Zhang, H. (2011). Bioactive compounds from the fern *Lepisorus contortus*. *Journal of Natural Products, 74*, 129–136. https://doi.org/10.1021/np100598b.

Yang, J., Kwon, Y., & Kim, M. (2015b). Isolation and characterization of bioactive compounds from *Lepisorus thunbergianus* (Kaulf.). *Arabian Journal of Chemistry, 8*(6), 713–719. https://doi.org/10.1016/j.arabjc.2011.07.027.

Yang, J., Kwon, Y., Lim, J., Yu, C., & Kim, M. (2015a). Antioxidant and anticancer properties of the extracts from *Lepisorus thunbergianus* (Kaulf.) Ching. *Korean Journal of Medicinal Crop Science, 23*(4), 324–333.

Yu, S., & Lin, Y. (1997). A study on systematics of the genus *Lepisorus* (Polypodiaceae). *Acta Phytotaxonomica Sinica, 35*, 341–347.

Zhang, X., & Qi, X. (2009). Taxonomic revision of *Lepisorus* (J. Sm.) Ching sect. *Lepisorus* (Polypodiaceae) from China. *Journal of Systematics and Evolution, 47*(6), 581–598. https://doi.org/10.1111/j.1759-6831.2009.00055.x.

Zhao, C., Wei, R., Xiang, Q., & Zhang, X. (2020b). The origin of allotetraploid *Lepisorus inaequibasis* (Polypodiaceae) and paternal bias in its morphology and abiotic niche. *Taxon, 69*(9), 1–13. https://doi.org/10.1002/tax.12345.

Zhao, C., Wei, R., Zhang, X., & Xiang, Q. (2020a). Backbone phylogeny of *Lepisorus* (Polypodiaceae) and a novel infrageneric classification based on the total evidence from plastid and morphological data. *Cladistics, 36*(3), 235–258. https://doi.org/10.1111/cla.12367.

Zink, M. (1993). *Systematics of the fern genus Lepisorus (J. Smith) Ching (Polypodiaceae–Lepisoreae)* (Doctoral dissertation). University of Zürich, Switzerland.

CHAPTER 28

Nephrolepis cordifolia (L.) C. Presl (Nephrolepidaceae): A Bioactive-Rich Sword Fern

SURAJ B. PATEL and SAVALIRAM G. GHANE*

Department of Botany, Plant Physiology Laboratory, Shivaji University, Kolhapur, Maharashtra, India

*Corresponding author

ABSTRACT

Nephrolepis cordifolia (L.) C. Presl belongs to the family Nephrolepidaceae, commonly called sword fern, native to tropical and subtropical regions of Asia. It is a rhizomatous fern, and the rhizome is covered with brownish scales. The plant's tuber is used as a wild snack, and its extract is used for stomach upset and urinary problems. Young leaves are commonly used as vegetables and are traditionally used to treat fever, cough, and skin diseases. Whole plant extract has shown promising results as a contraceptive and is also used in the treatment of liver and skin-related diseases. Bioactive compounds like cardiac glycosides, phenolic acids, flavonoids, steroids, tannins, and terpenoids have been isolated from both the underground and aerial parts of this plant. *N. cordifolia* possesses several potent biological activities, including anthelmintic, antibacterial, antifungal, allelopathic, antioxidant, diuretic, anti-inflammatory, and antidiabetic properties, and so forth.

28.1 INTRODUCTION

Nephrolepis cordifolia (L.) C. Presl is conventionally popular by various names such as tuberous sword fern, ladder fern, erect sword fern, lemon

Bioactives and Pharmacology of Pteridophytes. T. Pullaiah (Ed.)

DOI: 10.1201/9781779642738-28

butter fern, fish bone fern, and narrow sword fern. It belongs to the family Nephrolepidaceae (Chettri et al., 2020). In Tamil, it is well known as moothirakilangu; in the hilly region of Sikkim, it is known as pani amala, while in Malaysia, it is called Sarawak (Chettri et al., 2020). Its homotypic synonyms are *Aspidium cordifolium, Nephrodium cordifolium*, and *Polypodium cordifolium* (POWO, 2024). This plant is native to tropical and subtropical regions throughout the world, including Asia, Africa, Australia, and America (Bajracharya and Bajracharya, 2022). It is a rhizomatous fern, with a rhizome covered with brownish scales; it has egg-shaped fleshy tubers that grow from a network of creeping stalks. Stolons are slender and wiry, having some scales. Plants grow in clumps. Fronds are upright or hanging downward, with pinnae arranged in two rows and toothed. Sori are present on the lower side of the older leaves near the leaf margin, having a persistent pale green indusium (Pal, 2021).

Pteridophyta is one of the major divisions used in Ayurvedic systems like Sushruta, Charaka, and Samhita for the treatment of various diseases. Additionally, some physicians also use them in the Unani system of medicine (Uddin et al., 1998). Some native doctors in China use pteridophytes in the traditional Chinese system of medicine. In Nepal, raw or roasted tubers of *N. cordifolia* are eaten as a wild snack. Tuber extract is used in the treatment of stomach upset and urinary problems. Reports also show that tubers are used in the treatment of chest congestion, rheumatism, and anorexia (Adebiyi, 2016). In the local market of India, the consumption of *N. cordifolia* tuber extract during the menstrual period showed permanent sterility. In India, young leaves are cooked and used as a vegetable. Fresh leaf decoction is traditionally used to treat fever, cough, and skin diseases (Adebiyi, 2016). Pinnae are used in jaundice and are also found effective in the treatment of cough and wounds. Diuretic potential has been reported from the rhizome extract of *N. cordifolia* (Adebiyi, 2016). Nutrient analysis was performed by Guachan et al. (2008) and revealed that this fern is a potential commercial crop. Whole plant extract is traditionally used as a contraceptive and is also used to treat liver and skin-related diseases. Similarly, the *N. cordifolia* plant is applicable in the management of asthma, immunity, and biliousness (Adebiyi et al., 2019).

28.2 BIOACTIVES:

Chemical screening of *N. cordifolia* revealed the presence of bioactive compounds like phenolic acids, flavonoids, aromatic hydrocarbons, aldehydes, esters, phthalates, alkanes, and terpenoids (Adebiyi et al., 2019; El-Tantawy et al., 2016; Jaishee et al., 2021). El-Tantawy et al. (2016) extracted essential oil from the aerial part of *N. cordifolia* by hydro-distillation. GC-MS analysis of the oil revealed the presence of various bioactives. The highest area was occupied by nonanal (10.56%), followed by β-Ionone (8.02%), eugenol (7.17%), anethol (4.60%), hexadecane (4.11%), 2-methyl naphthalene (3.85%), durene (3.55%), azulene (3.47%), 1,7-dimethyl naphthalene (3.24%), 1,2,3,4-tetrahydro naphthalene (2.80%), pseudocumene (2.63%), methyl palmitate (2.15%), octadecane (2.04%), p-cymene (1.69%), β-cyclocitral (1.91%), nonanoic acid (0.73%), heneicosane (0.54%), docosane (0.45%), and phytol (0.36%) (El-Tantawy et al., 2016). Similarly, Jaishee et al. (2021) reported phenolic compounds from hot aqueous and methanolic extracts of *N. cordifolia*. In the investigation, a total of nine phenolic compounds (phloroglucinol, gallic acid, dihydrobenzoic acid, catechol, catechin, caffeic acid, vanillic acid, ferulic acid, and salicylic acid) were detected that showed numerous biological activities.

Further, GC-MS analysis of the hexane leaf extract of *N. cordifolia* revealed the presence of various bioactive compounds (Adebiyi et al., 2019). In this study, a total of 16 bioactives were identified, wherein the maximum area was occupied by n-hexadecanoic acid (24.42%) with a retention time (Rt) of 14.14 min. Another important compound, 1,4-benzenedicarboxylic acid (Rt 21.83 min), occupied 18.35% of the area and plays a vital role as an apoptosis inhibitor and a plasticizer. Antibacterial compounds, viz. octadecanoic acid (11.05%) and heptacosane (3.30%), were also reported at 15.96 and 20.99 min, respectively. Similarly, cis-13-octadecenoic acid (10.31%), a major bioactive compound having a vital role in medicine and surgery, has also been identified. Bis(2-ethylhexyl) phthalate (8.72%) and squalene (5.82%) have also been reported. Tetracosane (1.92%), hexadecanoic acid methyl ester (3.13%), methyl stearate (1.50%), docosyl pentafluoropropionate (0.72%), 1-nonadecene (1.29%), 2-methyl-Z,Z-3,13-octadecadienol (2.11%), and docosanoic acid (2.12%) have also been reported from the hexane extract of *N. cordifolia* leaves (Adebiyi et al., 2019) (Figure 28.1).

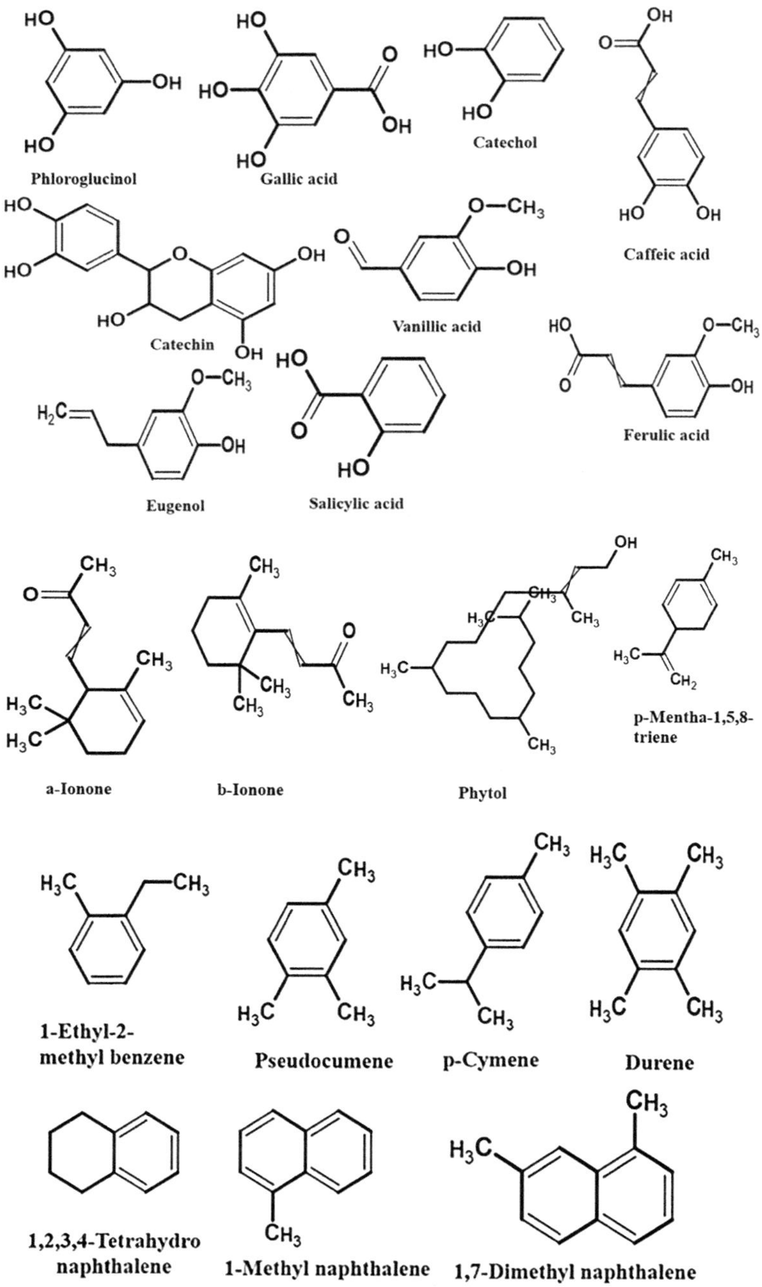
Phloroglucinol
Gallic acid
Catechol
Caffeic acid
Catechin
Vanillic acid
Ferulic acid
Eugenol
Salicylic acid
a-Ionone
b-Ionone
Phytol
p-Mentha-1,5,8-triene
1-Ethyl-2-methyl benzene
Pseudocumene
p-Cymene
Durene
1,2,3,4-Tetrahydro naphthalene
1-Methyl naphthalene
1,7-Dimethyl naphthalene

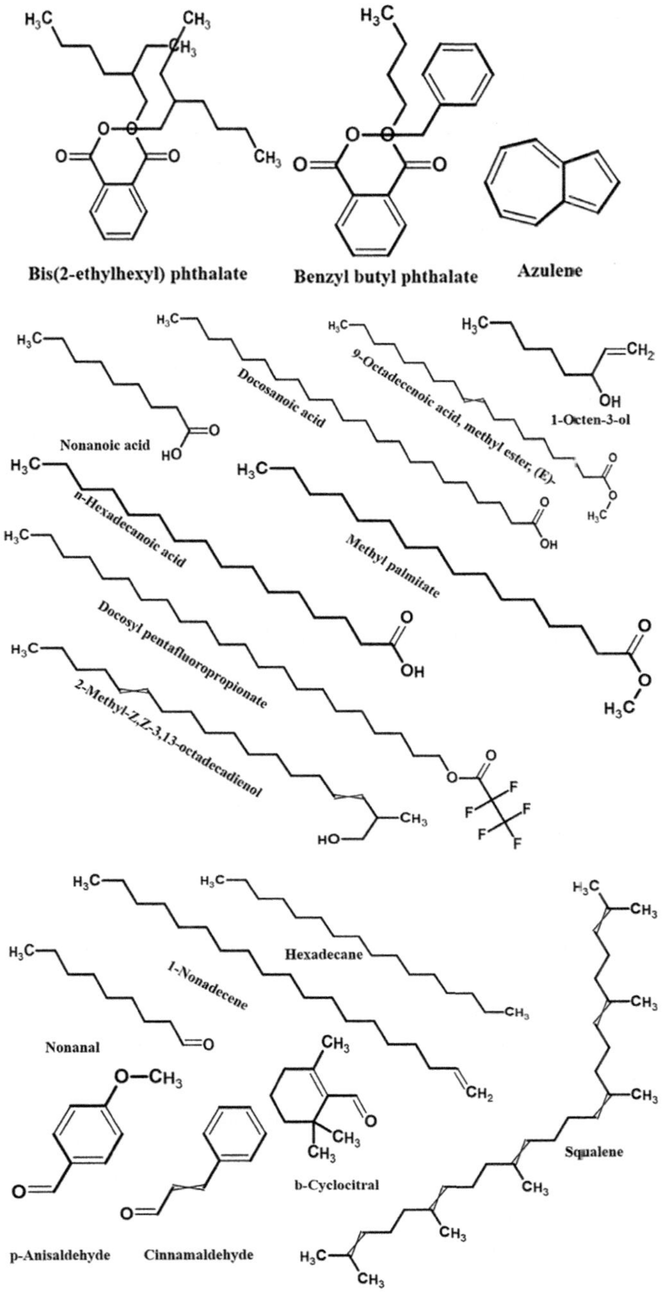

FIGURE 28.1 Bioactive compounds reported from *Nephrolepis cordifolia.*

28.3 PHARMACOLOGY

28.3.1 ANTHELMINTIC ACTIVITY

Pal (2021) investigated the anthelmintic properties of the leaves of *N. cordifolia*. In this study, the aqueous extract (4, 8, and 10 mg/mL) of *N. cordifolia* leaves was tested on earthworms (*Pheretima posthuma*), and albendazole (2–4 mg/mL) was used as a standard. The assay involved observing the paralysis and death time of the worms upon exposure to the plant extract. Different extracts and standards were taken in petri dishes, and earthworms of equal size were introduced and paralyzed. *N. cordifolia* leaf extract exhibited dose-dependent paralysis and death of the animals. Higher concentrations of the extract resulted in quicker paralysis and death. At 8 and 10 mg/mL extract concentrations, the time taken for paralysis was 80 and 50 minutes, while the time taken for death was 155 and 140 minutes, respectively. At the same time, standard albendazole at 4 mg/mL exhibited 21 and 60 minutes for paralysis and death of earthworms (Pal, 2021). The study highlights the significant anthelmintic activity exhibited by *N. cordifolia* leaves, underscoring its potential as a natural alternative for controlling parasitic infections and promoting sustainable health practices.

28.3.2 ANTIBACTERIAL ACTIVITY

Antibacterial activity was investigated by Rani et al. (2010). In the investigation, fronds of *N. cordifolia* extracted with hexane, chloroform, ethanol, and water. Based on pathogenicity, antibacterial potential of extracts was tested against nine bacterial strains viz., *Escherichia coli* (MTCC-443), *Pseudomonas aeruginosa* (MTCC-424), *Streptococcus faecalis* (MTCC 1927), *Proteus mirabilis* (MTCC-1429), *Klebsiella pneumoniae* (MTCC-109), *Bacillus cereus* (MTCC-430), *Salmonella typhimurium* (MTCC-98), *Shigella flexneri* (MTCC 1456), *Bacillus subtilis* (MTCC-121), and *Staphylococcus aureus* (MTCC-96). All bacterial strains were maintained on Mueller Hinton agar/broth and agar disc diffusion method was used to investigate the antibacterial potential. Final suspension contains 10^8 CFU/mL of each bacterium and each disc was inoculated with 10 μL of 50 mg/mL extracts. Gentamicin and erythromycin (30 μg/disc) were used as positive controls. After inoculation, plates were incubated at 37°C for 24 hours, and the visible growth inhibition zone (mm) was noted. Results showed that the aqueous extract was found to be superior to other extracting media for growth inhibition. *N. cordifolia* extract was found to be the most effective against *S. typhimurium* (20.10±0.10 mm) and *B. cereus* (20.10±0.00 mm), while the least response was noted against *P. mirabilis* (17.00±0.10 mm) and *S. faecalis* (17.10±0.10 mm). The remaining bacterial strains exhibited approximately the same inhibitory activity for the extract under investigation. The inhibitory effect of both standards was found to be superior to that of the plant extracts (Rani et al., 2010). This study highlighted the significant *in vitro* antibacterial activity of both aqueous and non-aqueous frond extracts of *N. cordifolia* and the potential of this fern as a source of natural antimicrobial agents.

28.3.3 ANTIFUNGAL ACTIVITY

The aqueous extract of *N. cordifolia* fronds was tested against three fungal species; namely, *Trichophyton rubrum* (ATCC-28188), *Microsporum gypseum* (ATCC-24102), and *Trichophyton mentagrophytes* (ATCC-9533). All the fungal strains were maintained on Saubouraud dextrose agar/broth. To study the antifungal activity, disc diffusion method was used (Rani et al., 2010). Final suspension (10^8 CFU/mL) of each fungal strain against 10 µL of 50 mg/mL extracts was used. Ketoconazole and amphotericin B (30 µg/disc) were used as positive controls. After the inoculation, plates were incubated at 27°C for 72 h, and the growth inhibition zone was noted. Results showed that the plant extract was the most effective against *T. rubrum* (11.06±0.05 mm) and the least inhibitory against *M. gypseum* (9.90±0.10 mm). The inhibitory action of both standards was found to be superior to the tested plant extract. The plant highlighted notable antifungal activity, showing potential as a natural alternative to synthetic agents (Rani et al., 2010).

28.3.4 ANTIOXIDANT ACTIVITY

DPPH radical scavenging activity of *N. cordifolia* leaves was studied by Barsal et al. (2021). The plant extract was prepared by simple maceration and dissolved separately in ethyl alcohol and water, and then used for analysis. Different concentrations of the plant extract (1, 10, and 100 µg/mL) were used along with ascorbic acid as a standard. Results showed that the ethyl acetate extract of leaves exhibited 13.3±1.01%, 21.22±1.01%, and 35.61±1.01% activity. The aqueous leaf extract (10 µg/mL) exhibited the highest activity (84.44±0.44%) compared to the standard (97.62±0.22%). Similarly, Pal et al. (2021) also investigated DPPH radical scavenging activity, for which the ethanolic leaf extract at different concentrations (5–30 µg/mL) was used. Results reported a linear relationship between the concentration of the extract and percent inhibition. Maximum DPPH radical scavenging activity (67.02%) was observed at 30 µg/mL. From these results, it is revealed that *N. cordifolia* showed reliable radical scavenging properties and could be used as a natural source of antioxidants.

28.3.5 ANTIDIABETIC ACTIVITY

Antidiabetic activity in *N. cordifolia* was assessed by alpha-amylase inhibitory and glucose diffusion inhibitory activity (Barsal et al., 2021). A simple maceration technique was used for the extraction, and extracts at different concentrations (0.25, 0.5, 1, and 2 mg/mL) were used. Voglibose was used as a standard drug. Results showed that higher activity was noted from the ethyl acetate leaf extract (50.5±0.77%). However, the water extract of leaves (73.91±0.67%) revealed the highest activity compared to the other extracting media. The glucose diffusion inhibitory assay was performed using aqueous and ethyl acetate extracts of leaves (Barsal et al., 2021). This assay was significant for assessing the ability of plant extracts to reduce glucose absorption in the gastrointestinal tract, which could be beneficial for managing postprandial blood glucose levels in diabetic patients. This activity was performed using 3,5-dinitrosalicylic acid reagent, and dextrose was used as a standard. Results were recorded in the amount of glucose diffused out of the dialysis membrane at different time intervals (30–180 min) when treated with different concentrations (20 and 40 mg) of the aqueous extract. Results showed that the leaf extract represented better glucose diffusion inhibition at lower concentrations. In the leaf extract at 20 mg/mL concentration, maximum diffusion (4.80±0.09 µL) was observed at 180 min; whereas, at 40 mg/mL, the highest diffusion (4.39±0.31 µL) was noted at 120 min. This finding suggests that *N. cordifolia* extracts could be beneficial in managing postprandial hyperglycemia, potentially aiding in the treatment of diabetes.

28.3.6 DIURETIC ACTIVITY

Rhizome juice of *N. cordifolia* was used to investigate diuretic activity on Wistar rats. Wistar rats weighing between 150–200 g were used for the experiment. In the study, parameters such as urine volume and electrolyte concentration in the urine of the treated animals were recorded. The rats were divided into different groups: a control group, a group treated with a standard diuretic drug (20 mg/kg) (furosemide), and groups treated with various doses of the rhizome juice (100 mg/kg). The control group received only water, the standard group received furosemide, and the test groups received the rhizome juice at doses of 2, 4, and 8 mL/kg body weight. Urine output was measured for 5 hours after administration. Parameters such as urine volume and electrolyte concentration (sodium, potassium, and chloride) were also analyzed. Results showed that urine volume (26.12±1.32 mL) was significantly increased after treatment (8 mL/kg body weight); in the control, urine volume was comparatively low (18.12±1.34 mL). Electrolyte content in urine was also increased with the treatment. In the control, total sodium (58.11 ± 2.42 mg/100 mL), total potassium (50.33 ± 2.35 mg/100 mL), and total chloride (81.3 ± 1.14 mg/100 mL), while after treatment, electrolytes increased significantly (97.32 ± 1.62, 82.16 ± 1.43, 100.66 ± 1.55 mg/100 mL, respectively). The diuretic effect of the rhizome juice was comparable to that of the standard diuretic drug furosemide, especially at higher doses. The study denoted that *N. cordifolia* rhizome juice acts as an effective hypernatremic, hyperchloremic, and hyperkalemic diuretic (Rajasekaran and Sivakumar, 2009).

28.3.7 ANTI-INFLAMMATORY ACTIVITY

Anti-inflammatory activity of *N. cordifolia* leaf extract was investigated by Pal et al. (2021). For this investigation, the HRBC (human red blood cell) membrane stabilization method was used. Different concentrations of the ethanolic extract (100, 250, and 500 μg/mL) and diclofenac as a reference drug were used. Results showed that the response noted was in a dose-dependent manner. Maximum inhibition (82.66%) was noted at 500 μg/mL treatment, which was comparatively higher than the inhibition by the reference drug (81.29%). Higher concentrations of the extracts demonstrated greater inhibition of protein denaturation, indicating stronger anti-inflammatory activity. The inhibition of protein denaturation by the extracts was comparable to that of standard anti-inflammatory drugs, highlighting the potential therapeutic benefits of the plant. The remaining two concentrations showed lower inhibition than the reference drug. Based on the findings, it could be effective in wound healing procedures.

KEYWORDS

- **alpha-amylase inhibitory**
- **ethyl acetate leaf extract**
- **ethyl alcohol**
- **gentamicin**
- **ladder fern**
- **nephrolepidaceae**
- **nephrolepis**
- **phytochemicals**

REFERENCES

Adebiyi, A. O. (2016). Phytochemical constituents and proximate composition of *Nephrolepis cordifolia* (L.) C. Presl grown in Nigeria. *New York Science Journal, 9*(2), 79–82.

Adebiyi, A. O., Oyeyemi, S. D., Tedela, P. O., & Ojo, V. I. (2019). GC-MS analysis of bioactive compounds from n-hexane leaf extract of a tropical fern, *Nephrolepis cordifolia* (L.) C. Presl. *East African Scholars Journal of Biotechnology and Genetics, 1*(5), 118–123.

Bajracharya, G. B., & Bajracharya, B. (2022). A comprehensive review on Nepalese wild vegetable food ferns. *Heliyon, e11687.*

Baral, R., Subedi, L., Gurung, M., Ojha, S., Shrestha, B., Chaudhary, D., & Jammarkattel, N. (2021). Phytochemical screening, free radical scavenging activity, *in-vitro* alpha-amylase enzyme and glucose diffusion inhibition activity of ethyl acetate and water extracts of selected medicinal plants of Nepal. *International Journal of Herbal Medicine, 9*(3), 18–27.

Chettri, U., Kumari, S., & Chettri, B. (2020). A review on anti-microbial and hepatoprotective properties of Himalayan wild fern *Nephrolepis cordifolia* (Pani Amla). *The Pharma Innovation Journal, 9*(9), 572–577.

El-Tantawy, M. E., Shams, M. M., & Afifi, M. S. (2016). Chemical composition and biological evaluation of the volatile constituents from the aerial parts of *Nephrolepis exaltata* (L.) and *Nephrolepis cordifolia* (L.) C. Presl grown in Egypt. *Natural Product Research, 30*(10), 1197–1201. http://dx.doi.org/10.1080/14786419.2015.1046070.

Guachan, D. P., Manandhar, D., Shrestha, S., & Suwal, K. (2008). Nutritional analyses of *Nephrolepis cordifolia* (L.) C. Presl. *Kathmandu University Journal of Science, Engineering and Technology, 1*(5), 68–72.

Jaishee, N., Lama, R., & Chakraborty, U. (2021). Evaluation of antimicrobial activity and HPLC analysis of phenolics present in *Nephrolepis cordifolia* (L.) C. (Persl). *Journal of Mycopathological Research, 59*(1), 37–45.

Pal, R. (2021). *In-vitro* evaluation of anthelmintic activity of *Nephrolepis cordifolia* leaves. *Journal of Pharmaceutical and Experimental Medicine, 1*(1), 1–6.

Pal, R., Teli, G., Sharma, B., Kumar, B., & Chawla, P. A. (2021). *In-vitro* anti-inflammatory and antioxidant activity of *Nephrolepis cordifolia* and molecular docking of its active chemical constituents. *Pharmaspire, 13*(1), 21–27.

POWO. (2024). *Plants of the World Online*. Facilitated by the Royal Botanic Gardens, Kew. http://www.plantsoftheworldonline.org/ (accessed on 17th November 2025).

Rajasekaran, A., & Sivakumar, V. (2009). Evaluation of diuretic potential of *Nephrolepis cordifolia* rhizome juice in Wistar rats. *Sains Malaysiana, 38*(1), 57–59.

Rani, D., Khare, P., & Dantu, P. (2010). *In-vitro* antibacterial and antifungal properties of aqueous and non-aqueous frond extracts of *Psilotum nudum*, *Nephrolepis biserrata* and *Nephrolepis cordifolia*. *Indian Journal of Pharmaceutical Sciences, 72*(6), 818.

Uddin, M. G., Mirza, M. M., & Pasha, M. K. (1998). The medicinal uses of pteridophytes of Bangladesh. *Bangladesh Journal of Plant Taxonomy, 5*(2), 29–41.

CHAPTER 29

Phytochemistry, Bioactive Compounds, and Pharmacological Activities of *Helminthostachys zeylanica* (L.) Hook.

PASUPULETI SIVARAMAKRISHNA,[1,2]
PULICHERLA YUGANDHAR,[3,*] and NATARU SAVITHRAMMA[1]

[1]*Department of Botany, Sri Venkateswara University, Tirupati, Andhra Pradesh, India*

[2]*Department of Botany, Government College for Men, Kadapa, Andhra Pradesh, India*

[3]*Survey of Medicinal Plants Unit, Regional Ayurveda Research Institute, Itanagar, Arunachal Pradesh, India*

*Corresponding author

ABSTRACT

Helminthostachys zeylanica (L.) Hook., a medicinal fern extensively used in traditional medicine across various Southeast Asian countries, has garnered attention for its diverse pharmacological properties. This plant is renowned for its antidiabetic, hepatoprotective, anti-inflammatory, antipyretic, anticancer, antiulcer, immunomodulatory, and anti-infective activities. The rhizome of *H. zeylanica* is particularly rich in bioactive compounds, with flavonoids and stilbenes being the predominant secondary metabolites. Among these compounds, ugonins, a type of flavonoid, have been extensively studied for their pharmacological effects.

Bioactives and Pharmacology of Pteridophytes. T. Pullaiah (Ed.)

DOI: 10.1201/9781779642738-29

29.1 INTRODUCTION

Helminthostachys zeylanica (L.) Hook. is a terrestrial sciophyte fern that is widely distributed throughout Southeast Asia. This medicinal plant has a rich history of use in traditional folk medicine across various regions, including Sri Lanka, India, South China, the Philippines, and Taiwan. It is popularly known in India as "Kamraj" (king of potency) due to its reputed ability to enhance vitality and potency. In China, the rhizome of *H. zeylanica* is referred to as "Ding-Di-U-Gon." In Ayurvedic and traditional medicinal formulations, *H. zeylanica* is valued for its properties as a potency enhancer and neurogenic tonic. The rhizomes are the most utilized part of the plant, known for their potent bioactive compounds, including flavonoids and stilbenes. These compounds contribute to the plant's various therapeutic effects, such as boosting physical and mental strength. However, the extensive use of *H. zeylanica* in traditional medicine and its exploitation by pharmaceutical companies seeking its valuable bioactive compounds has led to its status as an endangered species. The overharvesting of this fern threatens its natural populations, underscoring the need for sustainable harvesting practices and conservation efforts to ensure its availability for future generations.

29.2 ETHNOBOTANICAL USES

The ethnic groups of Maharajganj district in India utilize the young leaves (fronds) of *H. zeylanica* as a leafy vegetable. Similarly, in various ethnic communities in Northeastern India, the young leaves and fertile spike are cooked as a vegetable. This plant is traditionally used to treat sciatic neuritis due to its sedative, pain-relieving, and anti-inflammatory properties. Fronds have also been used as a euphoriant, tranquilizer, and sexual stimulant (Singh and Upadhyay, 2012). The rhizomes of *H. zeylanica* are employed for various medicinal purposes. They are used to treat pertussis and sexual dysfunction. The Kattunaikan tribe of Malappuram district, Kerala, uses the rhizome to treat snake bites (Dan and Shanavaskhan, 1991), and the aboriginals of the Nicobar Islands utilize it for treating impotence (Dagar and Dagar, 1987). It is reported to possess antiviral properties (Pandey and Bhargavan, 1980) and is traditionally used for treating jaundice (Prakash, 1998).

In Ayurvedic practices, the rhizome is boiled with cow's milk and a decoction given with black pepper (*Piper nigrum*) to treat leucorrhea over two months (Bhatt et al., 2010). Additionally, a paste made from the rhizome of *H. zeylanica* (5 g), the rhizome of *Chlorophytum tuberosum*, and the root of *Bombax ceiba* is administered for one month as a tonic for waist pain and potency enhancement (Singh et al., 1989). The rhizome also treats dysentery, colds, and early-stage pulmonary tuberculosis. Moreover, it is used alongside *Lygodium flexuosum* to treat fever, red urine, and urinary bladder pain (Sarker et al., 2012). A decoction of the plant is used to treat boils and ulcers (Peny and Metjger, 1980), and it serves as a vitality and brain tonic (Vasudeva, 1999). Rhizome powder is given for spermatorrhea and to improve memory power (Benjamin and Manickam, 2007). Traditional medicine treats tuberculosis, asthma, snake bites, diarrhea, and malaria (Yen et al., 2018). The dried leaves are smoked to treat nosebleeds. The plant is used as a mild laxative in the Moluccas (Kritikar and Basu, 1975). The Talang Mamak tribe of Indonesia uses *H. zeylanica* to treat inflammatory conditions, dysentery, cataracts, early-stage tuberculosis, syphilis, diabetes, and malaria (Ridhasya et al., 2019).

29.3 PHYTOCHEMISTRY

Phytochemical analysis of the rhizome of *H. zeylanica* has revealed the presence of various secondary metabolites, including flavonoids, stilbenes, phenolics, and glucosides (Chen et al., 2003; Huang et al., 2003). Among these, flavonoids are significant secondary metabolites found in the roots. Chen et al. (2003) isolated three new cyclized geranyl stilbenes including ugonstilbene-A, ugonstilbene-B, and ugonstilbene-C as well as a known compound, 3-hydroxyacetophenone from the ethanol extracts of *H. zeylanica*. Huang et al. (2009) isolated six known compounds and five new compounds using the n-hexane, CHC13, and EtOAc layers from the methanol extract of *H. zeylanica*. The known compounds are ugonin-J, ugonin-K, β-sitosterol, stigmasterol, hexadecanoic acid, and tetracosanoic acid. The new compounds include:

1. Acetogenin (6-hydroxy-8-pentadecyloxocane-2,7-dione)
2. Four isoprenylated flavonoids:

   i. 4″a,5″,6″,7″,8″,8″a-hexahydro-5,3′,4′-trihydroxy-5″,5″,8″a-trimethyl-4H-chromeno[2″,3″:7,6]flavone.
   ii. 4″a,5″,6″,7″,8″,8″a-hexahydro-5,3′,4′-trihydroxy-5″,5″,8″a-trimethyl-4Hchromeno[2″,3″:7,8]flavone.

iii. 2-(3,4-dihydroxyphenyl)-6-([2,2-dimethyl-6-methylenecyclohexyl]methyl)-5,7-dihydroxy-chroman-4-one.
iv. 2-(3,4-dihydroxy-2-[(2,6,6-trimethylcyclohex-2-enyl)methyl]phenyl)-3,5,7-trihydroxy-4H-chromen-4-one

Huang et al. (2009) also isolated eight new prenylated flavonoids (ugonins M-T) along with five known compounds, including ugonin-J, ugonin-K, ugonin-L, 5,4′-dihydroxy-4″,4″-dimethyl-5″-methyl-5″H-dihydrofurano[2″,3″:6,7]flavanone, and quercetin from the ethyl acetate extracts of *H. zeylanica* rhizome. Huang et al. (2017) isolated eight new flavonoids from the ethanol extract of *H. zeylanica* rhizomes. Among these, six are cyclized geranyl flavonoids (ugonin-V, ugonin-W, ugonin-X, ugonin-N, ugonin-S, ugonin-Y), and two are quercetin glucosides (quercetin-4′-O-β-D-glucopyranosyl-(1→2)-β-D-glucopyranoside and quercetin-3-O-β-D-glucopyranosyl-4′-O-β-D-glucopyranosyl-(1→2)-β-D-glucopyranoside) (Figure 29.1).

29.4 PHARMACOLOGICAL ACTIVITIES

H. zeylanica is widely recognized for its broad spectrum of pharmacological properties, which include antimicrobial, antidiabetic, anti-inflammatory, antioxidant, hepatoprotective, anti-obesity, anticancer, immunomodulatory, antiulcer, antihyperuricemic, and antimelanogenic effects. Extensive studies on the crude extracts and isolated compounds of *H. zeylanica*, utilizing both animal models and cell cultures, have revealed therapeutic efficacy comparable to standard pharmaceutical drugs. A comprehensive review of the literature on these pharmacological activities is discussed in detail below.

29.4.1 ANTICANCER ACTIVITY

Tsai et al. (2021) conducted a study to investigate the effects of *H. zeylanica* extracts (E1 and E2) on human gastric cancer cells. The study found that *H. zeylanica*-E1, at doses up to 1000 μg/mL, did not significantly impact the viability of the two gastric cancer (GC) cell lines tested (AGS and BGC823) compared to normal GES-1 cells. In contrast, *H. zeylanica*-E2 demonstrated dose and time-dependent effects on apoptosis in GC cells. This extract increased apoptosis in GC cells by modulating endogenous apoptosis-associated proteins, such as Bax, caspase 3/7, and cleaved

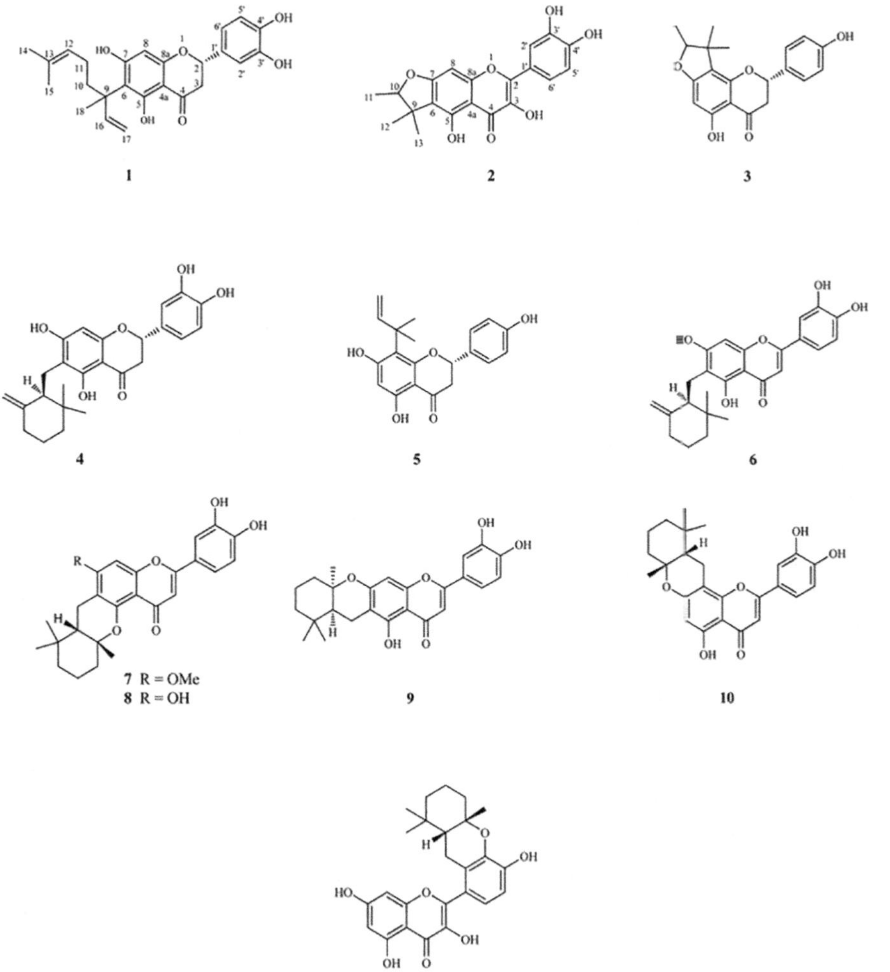

FIGURE 29.1 The compounds isolated from *H. zeylanica*: (1) Neougonin-A; (2) Neougonin-A; (3) 2-(3,4-dihydroxyphenyl)-6-((2,2-dimethyl-6-methylenecyclohexyl) methyl)-5,7-dihydroxychroman-4-one; (4) Ugonin-D; (5) Ugonin-E; (6) Ugonin-J; (7) Ugonin-L; (8) Ugonin-S; (9) Flavone; (10) Flavone; and (11) Ugonin-N (Su et al., 2016).

PARP proteins, particularly in late-stage apoptosis in BGC823 cells. *H. zeylanica*-E2 enhanced GC autophagy by regulating autophagy-associated proteins and increasing the LC3-II/LC3-I ratio. The extract also significantly reduced the accumulation of COX-2 protein in AGS cells, similar to the effect of CLC (a positive control). Furthermore, *H. zeylanica*-E2 extract exhibited significant inhibition of GC cell cycle progression in a time-dependent manner, leading to cell cycle arrest in the G0/G1 and G2/M phases in both AGS and BGC823 cells. These findings confirm the anticancer potential of *H. zeylanica*-E2 extract, suggesting its potential use as a novel adjunctive agent in treating gastric cancer. In conclusion, this study provides evidence for the anticancer properties of *H. zeylanica*-E2 extract against gastric cancer cells. The extract's ability to induce apoptosis, enhance autophagy, and inhibit cell cycle progression indicates its potential as a complementary treatment option for gastric cancer. Further research is warranted to explore the mechanisms underlying these effects and to evaluate the extract's efficacy and safety in clinical settings.

29.4.2 ANTIDIABETIC ACTIVITY

Ridhasya et al. (2019) conducted *in vitro* antidiabetic studies on various extracts of *H. zeylanica* rhizome by measuring the inhibition of α-Glucosidase, a key enzyme in glucose absorption in the gastrointestinal tract. The enzyme hydrolyzes the p-nitrophenyl-α-D-glucopyranoside (PNPG) substrate into p-nitrophenol (yellow color) and glucose. Methanol extracts exhibited significant antidiabetic properties with an IC50 of 47.86 ± 0.06 μg/mL, followed by ethanol (60.78 ± 0.02 μg/mL) and ethyl acetate (61.18 ± 0.59 μg/mL) extracts. In contrast, dichloromethane and n-hexane extracts showed the least activity. Further investigations by Ridhasya et al. (2020) focused on the antidiabetic efficiency of isolated chemical constituents by determining the inhibition of α-Glucosidase activity. Two prenylated flavonoids, ugonin-J and ugonin-K, exhibited moderate activity with IC50 values of 273.13 ± 0.402 ppm and 138.21 ± 0.263 ppm, respectively, compared to the standard drug Acarbose, an α-glucosidase inhibitor. Flavonoids such as flavonols, luteolin, myricetin, and quercetin are potent inhibitors of α-glucosidase. The study confirmed that ugonin-J and ugonin-K exert α-glucosidase inhibition similar to Acarbose.

Shah et al. (2020) examined the *in vitro* antidiabetic effect of ugonins, focusing on PTP1B inhibition, an enzyme crucial for regulating diabetes and obesity. The methanol extract of *H. zeylanica* showed significant PTP1B inhibition (80% inhibition at 30 μg/mL). Ugonin-J, ugonin-S, ugonin-L, ugonin-U, and ugonin-M effectively inhibited PTP1B, with IC50 values ranging from 0.6 to 7.3 μM. Ugonin-J was the most active compound with an IC50 of 0.6 μM, making it 26-fold more active than its parent compound luteolin (IC50–16.0 μM). All ugonins exhibited impressive α-glucosidase inhibition, with IC50 values ranging from 3.9 to 32.9 μM, with compounds ugonin-J and ugonin-S being the most active, having IC50 values of 3.9 μM and 7.4 μM, respectively. The extensive studies on the antidiabetic properties of *H. zeylanica* highlight its significant potential as a therapeutic agent. The extracts, particularly methanol, ethanol, and ethyl acetate, have demonstrated strong α-Glucosidase inhibitory activity. Isolated compounds, such as ugonin-J and ugonin-K, have shown promising results comparable to standard antidiabetic drugs like Acarbose. Moreover, the studies on PTP1B inhibition by ugonins underscore their potential in diabetes and obesity management, with ugonin-J emerging as a highly potent inhibitor.

29.4.3 ANTIHYPERURICIMIA

Fitrya and Muharni (2014) conducted a study to assess the antihyperuricemic potential of ethanol extracts derived from the rhizome of *H. zeylanica*. The extracts were administered to male Swiss mice at various concentrations (20, 40, 80, and 160 mg/kg body weight) over a period of 30 days. The researchers then measured the levels of uric acid in the blood of the experimental animals to evaluate the extract's effect on uric acid reduction. The study's results demonstrated that the ethanol extract of *H. zeylanica* exhibited significant hypouricemic activity at concentrations of 80 and 160 mg/kg body weight, with reductions in uric acid levels of 91.95% and 91.77%, respectively. These findings were compared to the effects of allopurinol, a common medication used to treat hyperuricemia, indicating that the chemical constituents of *H. zeylanica* may be more effective in managing elevated uric acid levels. In conclusion, the study provides evidence for the potent antihyperuricemic properties of *H. zeylanica* rhizome ethanol extracts. The extract's ability to significantly lower uric acid levels in mice suggests its potential as a natural alternative or complementary treatment for hyperuricemia and related conditions such as gout.

29.4.4 ANTI-INFECTION ACTIVITY

Chen et al. (2017) conducted a study to explore the effects of ugonin-U on NLRP3 inflammasome activation in human monocytic cell lines. The study used THP-1 cells (obtained from ATCC) and a variant of THP-1 cells with reduced NLRP3 activity. Ugonin-U was administered at concentrations ranging from 1 to 10 μM and was found to induce mitochondrial superoxide generation in a dose-dependent manner. The compound also activated the NLRP3 inflammasome by mobilizing calcium ions (Ca^{2+}), which enhanced the bactericidal function of human monocytes. The activation of the NLRP3 inflammasome by ugonin-U was mediated by the specific scavenging of mitochondrial reactive oxygen species (ROS). This mechanism is considered a key molecular target for developing new anti-infective therapies. This study provides insights into the role of ugonin-U, a natural flavonoid compound, in activating the NLRP3 inflammasome and enhancing the bactericidal activity of human monocytes. These findings suggest that ugonin-U could be a promising candidate for developing anti-infective medicines.

29.4.5 ANTI-INFLAMMATORY ACTIVITY

Inflammation, a critical biological response to infection, trauma, or injury, involves intricate interactions among cells and soluble mediators. Within the central nervous system (CNS), inflammation can be monitored through the expression of matrix metalloproteinase-9 (MMP-9), a significant biomarker. Research by Hsieh et al. (2015) investigated the anti-inflammatory properties of extracts from the rhizome of *H. zeylanica*, focusing on their ability to inhibit MMP-9 expression stimulated by bradykinin (BK) in cultured rat brain astrocyte cells. The study revealed that ethanolic (E0), aqueous (E1), and ethyl acetate (E2) extracts of *H. zeylanica* significantly reduced BK-induced phosphorylation of c-Src, Pyk2, and PKC (α/δ). The ethanolic extract (Hz-E0) demonstrated a concentration-dependent decrease in BK-induced MMP-9 expression, as confirmed by real-time PCR analysis of mRNA expression in treated RBA-1 cells. Additionally, pretreatment with the aqueous extract (Hz-E1) and ethyl acetate extract (Hz-E2), both individually and in combination, also suppressed BK-induced MMP-9 expression.

In another study, Huang et al. (2017) explored the anti-inflammatory activities of isolated compounds from *H. zeylanica* on lipopolysaccharide (LPS)-induced nitric oxide (NO) production in microglial cells. The compounds tested included ugonins J-S (9–13), ugonstilbene A (14), and ugonin L (23). Except for compounds 1, 5, and 13, all others exhibited inhibitory effects on NO production, with IC50 values ranging from 6.2 to 10.1 μM, outperforming the positive control, pyrrolidine dithiocarbamate. In conclusion, the findings from these studies highlight the potential of *H. zeylanica* extracts and isolated compounds as anti-inflammatory agents. The ability of these natural compounds to modulate inflammatory responses, particularly in the CNS, suggests their potential therapeutic value in managing neuroinflammatory conditions. The inhibition of MMP-9 expression and NO production by these compounds underscores their mechanism of action in reducing inflammation, warranting further research into their clinical applications.

29.4.6 ANTI-MELANOGENIC ACTIVITY

Yamauchia et al. (2015) conducted a study to explore the melanogenesis inhibitory activity of compounds ugonins-J, K, and L, which were isolated from a 50% ethanol/water extract of the roots of *H. zeylanica*. The study used mouse melanoma B16-F0 cells as a model to assess the effects of these compounds on melanin production. The study's results indicated that ugonin-J exhibited significant extracellular melanogenic inhibitory activity. At 12.5, 25, and 50 μM concentrations, ugonin-J reduced melanogenic activity to 75%, 16%, and 14%, respectively, compared to the control group, which showed 100% melanogenic activity. This performance was superior to arbutin, used as a positive control. Additionally, ugonin-K demonstrated even more potent inhibitory activity than ugonin-J, with extracellular melanogenic activity reduced to 19%, 8%, and 9% at the same concentrations. Interestingly, both ugonin-J and K did not show any inhibition of intracellular melanogenesis, suggesting that their mechanism of action may involve inhibiting the transportation or release of melanosomes or both, in addition to potentially inhibiting melanin biosynthesis. In contrast, ugonin-L did not exhibit inhibitory effects on either intra- or extracellular melanogenesis. The study highlights the anti-melanogenic properties of ugonins-J and K isolated from *H. zeylanica*. These compounds significantly inhibited extracellular melanogenic activity, with ugonin-K being particularly potent. The lack of intracellular melanogenesis inhibition suggests a unique mechanism of action that warrants further investigation. These findings could contribute to the development of new agents for skin lightening and treating hyperpigmentation disorders.

29.4.7 ANTIMICROBIAL ACTIVITY

Batubara et al. (2009) evaluated the antibacterial properties of the methanolic extract of *H. zeylanica* leaf (frond) against the Gram-positive bacteria *Propionibacterium acnes*. Similarly, Yenn et al. (2018) investigated the antibacterial activity of these extracts against the food-borne pathogen *Bacillus cereus* using the disc diffusion method. The efficacy of the antibacterial activity was assessed using 20 μL of 50 mg/mL frond extract, and its effectiveness was compared to the standard drug chloramphenicol. *Bacillus cereus* showed susceptibility to the test concentration, with a clear zone of 16.3 mm for the frond extract and 19.7 mm for chloramphenicol. The slightly lower efficacy of the frond extract, compared to the standard, was attributed to its crude form. The extract's minimum inhibitory concentration (MIC) on *B. cereus* was determined to be 6.25 mg/mL.

Army et al. (2023) investigated the antibacterial activity of extracts from the roots of *H. zeylanica* against eight pathogenic bacteria: Gram-positive bacteria (*Bacillus subtilis* ATCC 1965, *Bacillus cereus* ATCC 10876, *Listeria monocytogenes* ATCC 7644, *Staphylococcus aureus* ATCC 6538) and Gram-negative bacteria (*Vibrio parahaemolyticus* ATCC 17802, *Vibrio alginolyticus* ATCC 17749, *Escherichia coli* ATCC 8739, and *Salmonella typhimurium* ATCC 142028). Among the tested solvent extracts, dichloromethane exhibited the highest susceptibility against the microorganisms, except for *E. coli*, followed by ethyl acetate. N-hexane extracts showed the least susceptibility, while aqueous extracts exhibited no susceptibility. Overall, the antibacterial efficacy of *H. zeylanica* root extracts was moderate compared to the inhibitory effect of the standard drug Chloramphenicol. The studies on the antibacterial activity of *H. zeylanica* have demonstrated promising results, highlighting the potential of this plant in combating various pathogenic bacteria. These studies suggest that different parts of *H. zeylanica*, particularly when extracted with specific solvents, possess significant antibacterial properties.

29.4.8 ANTI-OBESE AND ANTI-INSULIN RESISTANCE

The prevalence of obesity and insulin resistance has escalated globally, leading to a surge in metabolic disorders such as fatty liver disease and type 2 diabetes. Natural products have been extensively studied for their potential in combating these conditions. In this context, the rhizome extracts of *H. zeylanica* have emerged as a promising candidate due to their reported therapeutic effects on obesity and insulin resistance. Chang et al. (2019) conducted a study to evaluate the therapeutic effects of *H. zeylanica* rhizome ethanol extracts in diet-induced obese mice. Mice fed a high-fat diet (HFD) were treated with the extract for 12 weeks to assess its impact on lipid metabolism and fatty liver. The results showed that the body weight and food efficiency ratio (FER) of HFD mice treated with *H. zeylanica* (HFD-HZ mice) were significantly lower than those of untreated HFD mice. Additionally, fat deposition in adipocytes was higher in HFD mice compared to HFD-HZ mice. The liver weight of HFD mice was significantly higher than that of HFD-HZ mice, indicating a protective effect of the extract against fatty liver. Biomarkers of hepatic injury, such as GOT and GPT, were found at lower levels in HFD-HZ mice, suggesting a reduction in liver damage. Furthermore, plasma lipid levels, including triglycerides (TG), total cholesterol (TC), and LDL-cholesterol (LDL-C), were lower in the HFD-HZ group compared to the HFD group.

In vitro experiments using human primary hepatocytes (HuS/E2) as a model for fatty liver disease showed that treatment with 100 μg/ml of *H. zeylanica* extracts for 18 hours reduced lipid accumulation by 39%, as observed by Oil-Red-O staining. Insulin resistance was assessed using the homeostasis model assessment of insulin resistance (HOMA-IR). Fasting glucose and insulin levels in HFD mice, with or without *H. zeylanica* extract, were measured to determine the incidence of insulin resistance syndrome. The increase in insulin levels was prevented in HFD-HZ mice compared to HFD mice, indicating improved insulin sensitivity. The high HOMA-IR index observed in HFD mice, reflecting increased diabetes symptoms with high fasting blood glucose and insulin levels, was notably reduced in the HFD-HZ group. The findings of this study highlight the promising bioactivity of *H. zeylanica* extract in regulating obesity and improving insulin sensitivity. The extract's ability to reduce body weight, liver weight, and plasma lipid levels, and its effect on insulin resistance suggests its potential as a natural therapeutic agent for metabolic disorders.

29.4.9 ANTIOSTEOPOROSIS ACTIVITY

Huang et al. (2017) conducted a study to investigate the antiosteoporotic activities of various compounds (1, 2, 5, 6, and 10–13) isolated from *H. zeylanica*. Genistein was used as the positive control for comparison with the ugonins. Among all the tested compounds, with the exception of compound 13, the others demonstrated the ability to suppress receptor activator of nuclear factor kappa-B ligand (RANKL)-induced osteoclast differentiation in RAW264.7 cells, with IC50 values ranging from 1.8 to 4.6 μM. Ugonin-K (compound 10) showed the highest inhibitory activity against RANKL-induced osteoclast differentiation in RAW264.7 cells, with an IC50 value of 1.8 ± 0.2 μM. The findings of this study indicate that compounds derived from *H. zeylanica* exhibit considerable antiosteoporotic activities. These compounds can inhibit osteoclast differentiation, which suggests their potential as therapeutic agents for the prevention and treatment of osteoporosis. This study highlights the antiosteoporotic properties of compounds isolated from *H. zeylanica*, particularly ugonin-K, which showed potent inhibition of osteoclast differentiation. These findings contribute to understanding the bioactive compounds in *H. zeylanica* and their potential applications in bone health.

29.4.10 ANTI-OXIDANT ACTIVITY

Chen et al. (2003) investigated the antioxidant activity of three newly isolated compounds from *H. zeylanica*, namely ugonstilbene-A, ugonstilbene-B, and ugonstilbene-C, using the 1,1-diphenyl-2-picrylhydrazyl (DPPH) assay. The assay revealed that these compounds exhibited moderate antioxidant activity, with IC_{20} values (concentration required to scavenge 20% of DPPH radicals) of 11.31 ± 1.11, 38.72 ± 0.47, and 30.80 ± 1.19 μM, respectively. These values were compared to the positive control, Trolox, which had an IC_{20} value of 8.70 ± 0.95 μM. In another related study, Huang et al. (2003) evaluated the antioxidant potential of isolated flavonoid compounds from *H. zeylanica*, specifically compounds 3 to 8, using the DPPH assay. The free radical scavenging efficiency of compounds 6, 7, and 8 was found to be superior to that of Trolox. The IC20 values for these compounds were 5.29 ± 0.32, 7.23 ± 0.22, and 7.93 ± 0.31 μM, respectively. These findings confirm that *H. zeylanica* extracts are a good source of antioxidants, which can be therapeutically useful, particularly as anti-inflammatory agents. In conclusion, the studies by Chen et al. (2003) and Huang et al. (2003) provide evidence for the antioxidant properties of compounds isolated from *H. zeylanica*. The moderate to superior free radical scavenging activity of these compounds, as demonstrated by their performance in the DPPH assay, underscores the potential of *H. zeylanica* as a natural source of antioxidants with therapeutic applications, particularly in the context of anti-inflammatory medicine.

29.4.11 ANTI-ULCER ACTIVITY

In a study, Fitrya et al. (2022) evaluated the anti-ulcer properties of *H. zeylanica* rhizome extracts in male Wistar rats subjected to ethanol-induced acute gastric lesions. The study aimed to assess the potential of these extracts compared to the standard anti-ulcer medication, omeprazole. Rats were orally administered different concentrations of ethanolic extract (100, 200, and 400 mg/kg body weight) for 14 days. The anti-ulcer effects were assessed through various parameters, including histopathological examination, ulcer index, gastric fluid volume, total acidity, and protective factors such as pH and mucus levels. The results indicated that at a 400 mg/kg concentration, the *H. zeylanica* extract demonstrated anti-ulcer effectiveness comparable to omeprazole (20 mg/kg). There was a noticeable decrease in lesion severity at a lower dose of 200 mg/kg, while at 100 mg/kg, the lesions remained severe. These findings confirm the anti-ulcer potential of *H. zeylanica* extracts, with 400 mg/kg emerging as the optimal dose for exerting its anti-ulcer effects. The study suggests that the anti-ulcer activity of *H. zeylanica* may be attributed to the presence of antioxidant and anti-inflammatory compounds within the extract. In conclusion, there is evidence for the significant anti-ulcer properties of *H. zeylanica* rhizome extracts, particularly at a dose of 400 mg/kg body weight.

29.4.12 HEPATOPROTECTIVE ACTIVITY

Suja et al. (2004) conducted a study to evaluate the hepatoprotective effects of methanolic extracts of *H. zeylanica* against carbon tetrachloride (CCl_4)-induced liver damage in Wistar rats. The extracts were administered at concentrations of 100, 200, and 300 mg/kg for five days, and their efficacy was compared with that of the standard hepatoprotective drug, silymarin. The study assessed the hepatoprotective properties through histopathological examination and by measuring the levels of serum enzymes. Rats treated with *H. zeylanica* extracts showed a nearly normal lobular pattern with minimal fatty changes, necrosis, and lymphocyte infiltration, comparable to the normal control and silymarin groups. A decrease in serum enzyme levels, including serum aspartate aminotransferase (AST), serum alanine aminotransferase (ALT), serum alkaline phosphatase (ALP), and serum bilirubin (SB), was observed. The extracts at 200 and 300 mg/kg demonstrated hepatoprotective properties similar to silymarin.

In another study, Wu et al. (2018) explored the hepatoprotective potential of ugonin-M, a flavonoid compound isolated from the rhizome of *H. zeylanica*, against acetaminophen (APAP)-induced acute liver injury in mice. Ugonin-M was tested at concentrations of 0.625, 1.25, and 2.5 mg/kg, with N-acetyl cysteine (NAC) at 600 mg/kg as the positive control. At a dose of 2.5 mg/kg, ugonin-M exhibited effectiveness comparable to the positive control. It significantly mitigated APAP-induced histopathological changes and improved liver function biomarkers such as alanine aminotransferase (ALT), aspartate aminotransferase (AST), and total bilirubin (T-Bil). Ugonin-M also addressed APAP-induced abnormal lipid metabolism, including total cholesterol (TC) and triglycerides (TG) levels in the serum. It suppressed pro-inflammatory mediators like nitric oxide (NO) and the lipid peroxidation indicator malondialdehyde (MDA). Additionally, ugonin-M enhanced the expression of heme oxygenase-1 (HO-1) protein and the production of the antioxidant enzymes, including superoxide dismutase (SOD), glutathione peroxidase (GPx), and catalase (CAT). By elevating the antioxidant enzymes and reducing oxidative stress and inflammation, ugonin-M exerted its hepatoprotective effects.

29.4.13 IMMUNOMODULATORY EFFECT

Chen et al. (2014) investigated the immunomodulatory effect of ugonin-U, a novel flavonoid compound isolated from *H. zeylanica*, on human neutrophils. Neutrophils were incubated with increasing concentrations of ugonin-U (1–10 μM) for 30 minutes. The study found that ugonin-U enhanced superoxide production in a concentration- and time-dependent manner and elastase release, albeit to a lesser extent. The compound also gradually increased intracellular calcium ions (Ca^{2+}), which were essential for the ugonin-U-stimulated superoxide release. Ugonin-U was identified as the first natural flavonoid compound to stimulate phospholipase-C (PLC) directly. Furthermore, it was found to induce a respiratory burst via the PLC/IP3/Ca^{2+} pathway in human neutrophils. In conclusion, the study demonstrates the immunomodulatory properties of ugonin-U, a flavonoid isolated from *H. zeylanica*. The compound's ability to stimulate superoxide production and elastase release in neutrophils, along with its effect on intracellular Ca^{2+} levels and PLC activation, suggests its potential role in modulating immune responses.

29.4.14 INHIBITION OF SUPEROXIDE GENERATION BY NEUTROPHILS

Huang et al. (2009) conducted a study to evaluate the efficiency of various compounds (1, 3, 5, 7, 8, and 11) in inhibiting superoxide anion generation and elastase release by human neutrophils in response to formyl-L-methionyl-L-leucyl-L-phenylalanine/cytochalasin B (FMLP/CB). The study found that compounds 2, 4, 6, and 9 did not exhibit inhibitory activity. However, compounds 3 and 5 emerged as potent inhibitors with IC50 values of 0.25 μM and 0.49 μM, respectively. Building on this research, Huang et al. (2010) further investigated the inhibitory effects of a newly isolated acetogenin and four prenylated flavonoids on superoxide anion generation and elastase release. Among the compounds tested (1, 3, 5, and acetogenin 1), acetogenin 1 at a concentration of 10 μM showed potent inhibitory activities against FMLP/CB-induced superoxide anion generation and elastase release. The study used LY294002, a phosphatidylinositol-3-kinase inhibitor from Sigma, as a positive control. Although acetogenin 1 demonstrated inhibitory effects, these were comparatively lower than the positive controls. In conclusion, the studies by Huang et al. (2009, 2010) highlight the potential of certain compounds isolated from natural sources in inhibiting key inflammatory responses in human neutrophils. Compounds 3 and 5, along with acetogenin 1, showed notable inhibitory activity against superoxide anion generation and elastase release, suggesting their potential as anti-inflammatory agents. However, the inhibitory effects were less pronounced than the positive control, indicating that further research is necessary to optimize their efficacy and fully understand their mechanisms of action.

29.4.15 LUNG PROTECTION

Wu et al. (2017) conducted a study to evaluate the efficacy of ugonin-M, a significant flavonoid compound found in the rhizome of *H. zeylanica*, against lipopolysaccharide (LPS)-induced acute lung injury in mice. Ugonin-M was administered at concentrations of 0.625, 1.25, and 2.5 μg/ml, and its effectiveness was compared with that of the standard drug dexamethasone, which served as the positive control. The study found that the suppression of pulmonary edema in the right lower lung was most effective at the higher dose of 2.5 μg/ml, with lower doses showing minimal effect. Ugonin-M demonstrated significant anti-inflammatory properties by inhibiting the production of pro-inflammatory mediators such as nitric oxide (NO), tumor necrosis factor-alpha (TNF-α), interleukin-1 beta (IL-1β), and interleukin-6 (IL-6). It also reduced the number of infiltrated cells and protein content in the bronchoalveolar lavage fluid (BALF) of LPS-induced acute lung injury (ALI) mice. Additionally, ugonin-M ameliorated the severity of pulmonary edemas, as evidenced by histological examination scores and the ratio of wet to dry weight of the lung tissue. Treatment with ugonin-M enhanced the protein expressions of heme oxygenase-1 (HO-1) and antioxidant enzymes, including SOD, GPx, and CAT, in the lung tissue of LPS-induced ALI mice. This suggests that ugonin-M exerts its protective effects by enhancing the antioxidant defense system and reducing oxidative stress. This study provides evidence for the anti-inflammatory and lung-protective properties of ugonin-M. The compound's ability to suppress pulmonary edema and reduce inflammatory mediators in LPS-induced ALI indicates its potential as a therapeutic agent for lung injury. These findings highlight the therapeutic potential of flavonoids from *H. zeylanica* in managing inflammatory lung conditions.

29.4.16 SUBCHRONIC TOXICITY

Subchronic toxicity studies play a crucial role in assessing the safety and appropriate dosage of drugs systematically. Fitrya et al. (2017) conducted a study to evaluate the subchronic toxicity of *H. zeylanica* rhizome ethanol extracts in Wistar albino rats, an experimental model commonly used for such assessments. The extracts were administered at test doses of 68, 136, 272, and 554 mg/kg body weight. The study involved a thorough analysis of various parameters, including body weight, hematological factors (hemoglobin, leukocytes, erythrocytes, and total plasma protein), biochemical markers (SGPT, SGOT, urea, and creatinine levels), macroscopic organ appearance, and relative organ weights (liver, heart, and kidney). The findings of this study indicated that the *H. zeylanica* rhizome extracts did not exhibit any toxic effects at the tested doses. This suggests that the extracts can be used safely in traditional medicine, even over extended periods, without causing adverse effects. The subchronic toxicity study proves that *H. zeylanica* rhizome ethanol extracts are non-toxic and supports their long-term use in traditional medicine. This important finding can contribute to developing safe and effective herbal remedies derived from *H. zeylanica*.

29.5 CONCLUSION

H. zeylanica has demonstrated a wide range of pharmacological properties, including antimicrobial, antidiabetic, anti-inflammatory, hepatoprotective, anti-obesity, anticancer, immunomodulatory, antiulcer, antihyperuricemic, and anti-melanogenic effects, primarily due to its rich content of flavonoids and stilbenes, notably ugonins. However, the extensive use of this medicinal fern in traditional medicine and its exploitation by pharmaceutical companies have led to its endangered status, necessitating sustainable harvesting practices and conservation efforts. Future research should focus on elucidating the precise mechanisms of action of its bioactive compounds, conducting well-designed clinical trials to establish their safety and efficacy, and developing sustainable cultivation methods to ensure its availability. Additionally, long-term toxicity studies, investigations into bioavailability and metabolism, exploration of combination therapies, and molecular studies could further enhance the understanding and therapeutic potential of *H. zeylanica*.

KEYWORDS

- **antihyperuricemic**
- **antimelanogenic effects**
- **antioxidant enzymes**
- **bioactive compounds**
- **lipopolysaccharide**
- **neutrophils**
- **pharmacological activities**
- **phytochemical analysis**
- **rhizome ethanol extracts**

REFERENCES

Army, M. K., Khodijah, R., Haryani, Y., Teruna, H. Y., & Hendra, R. (2023). Antibacterial *in vitro* screening of *Helminthostachys zeylanica* (L.) Hook. root extracts. *Journal of Pharmacognosy and Pharmacology Research, 11*, 291–296.

Batubara, I., Mitsunaga, T., & Ohashi, H. (2009). Screening antiacne potency of Indonesian medicinal plants: Antibacterial, lipase inhibition, and antioxidant activities. *Journal of Wood Science, 55*, 230–235.

Benjamin, A., & Manickam, V. S. (2007). Medicinal pteridophytes from the Western Ghats. *Indian Journal of Traditional Knowledge, 6*(4), 611–618.

Bhatt, D., Joshi, G. C., & Tewari, L. M. (2010). *Helminthostachys zeylanica* (L.) Hook: An important therapeutic herb for leucorrhea among the Tharu community of Udham Singh Nagar District, Uttarakhand. In *Souvenir, National Seminar on Medicinal Plants of Himalaya: Potential and Prospect* (pp. 137–143).

Chang, T. C., Chiang, H., Lai, Y. H., Huang, Y. L., Huang, H. C., Liang, Y. C., Liu, H. K., & Huang, C. (2019). *Helminthostachys zeylanica* alleviates hepatic steatosis and insulin resistance in diet-induced obese mice. *BMC Complementary and Alternative Medicine, 19*, 1–11.

Chen, C. C., Huang, Y. L., Yeh, P. Y., & Ou, J. C. (2003). Cyclized geranyl stilbenes from the rhizomes of *Helminthostachys zeylanica. Planta Medica, 69*(10), 964–967.

Chen, C. Y., Liaw, C. C., Chen, Y. H., Chang, W. Y., Chung, P. J., & Hwang, T. L. (2014). A novel immunomodulatory effect of ugonin U in human neutrophils via stimulation of phospholipase C. *Free Radical Biology and Medicine, 72*, 222–231.

Chen, C. Y., Yang, C. H., Tsai, Y. F., Liaw, C. C., Chang, W. Y., & Hwang, T. L. (2017). Ugonin U stimulates Nlrp3 inflammasome activation and enhances inflammasome-mediated pathogen clearance. *Redox Biology, 11*, 263–274.

Dagar, J. C., & Dagar, H. S. (1987). Some useful pteridophytes of Andaman and Nicobar Islands. *Journal of Economic and Taxonomic Botany, 9*, 317–323.

Dan, M., & Shanavaskhan, A. E. (1991). A glance to some rare medicinal plants of Western Ghats. In C. K. Karunakaran (Ed.), *Proceedings of the Symposium on Rare, Endangered and Endemic Plants of Western Ghats* (p. 221). Kerala Forest Department, Trivandrum, India.

Fitrya, F., & Muharni, M. (2014). An anti-hyperuricemia effect of ethanol extract of tunjuk langit rhizome (*Helminthostachys zeylanica* Linn. Hook) on Swiss male mice. *Majalah Obat Tradisional, 19*(1), 14–18.

Fitrya, F., Amriani, A., Novita, R. P., & Pamela, F. N. (2022). Antiulcer effect of *Helminthostachys zeylanica* (L.) root extract on ethanol-induced acute gastric lesion in Wistar rats. *Journal of Pharmacognosy and Pharmacology Research, 10*(3), 508–516.

Fitrya, F., Fithri, N. A., & Muharni, M. (2017). A subchronic toxicity test of ethanol extract from tunjuk langit rhizome (*Helminthostachys zeylanica*) on albino rats, *Rattus norvegicus* (Wistar strain). *Asian Journal of Pharmaceutical and Clinical Research, 10*, 270–273.

Hsieh, H. L., Yang, S. H., Lee, T. H., Fang, J. Y., & Lin, C. F. (2016). Evaluation of anti-inflammatory effects of *Helminthostachys zeylanica* extracts via inhibiting bradykinin-induced MMP-9 expression in brain astrocytes. *Molecular Neurobiology, 53*, 5995–6005.

Huang, Y. C., Hwang, T. L., Chang, C. S., Yang, Y. L., Shen, C. N., Liao, W. Y., Chen, S. C., & Liaw, C. C. (2009). Anti-inflammatory flavonoids from the rhizomes of *Helminthostachys zeylanica*. *Journal of Natural Products, 72*(7), 1273–1278.

Huang, Y. C., Hwang, T. L., Yang, Y. L., Wu, S. H., Hsu, M. H., Wang, J. P., Chen, S. C., Huang, L. J., & Liaw, C. C. (2010). Acetogenin and prenylated flavonoids from *Helminthostachys zeylanica* with inhibitory activity on superoxide generation and elastase release by neutrophils. *Planta Medica, 76*(5), 447–453.

Huang, Y. L., Shen, C. C., Shen, Y. C., Chiou, W. F., & Chen, C. C. (2017). Anti-inflammatory and antiosteoporosis flavonoids from the rhizomes of *Helminthostachys zeylanica*. *Journal of Natural Products, 80*(2), 246–253.

Huang, Y. L., Yeh, P. Y., Shen, C. C., & Chen, C. C. (2003). Antioxidant flavonoids from the rhizomes of *Helminthostachys zeylanica*. *Phytochemistry, 64*(7), 1277–1283.

Kirtikar, K. R., & Basu, B. D. (1975). *Indian Medicinal Plants* (Vol. 4, pp. 2750–2753).

Pandey, A., & Bhargavan, K. S. (1980). Antiviral activity of crude extracts of some pteridophytes. *Indian Journal of Forestry, 3*, 132, 133.

Peny, L. M., & Metjger, J. (1980). *Medicinal Plants of East and South East Asia: Attributed Properties and Uses*. The MIT Press.

Prakash, V. (1980). Indian medicinal plants—Current status. *Journal of Ethnobotany, 10*, 112–121.

Ridhasya, F. E., Rahim, N., Almurdani, M., Hendra, R., & Teruna, H. Y. (2020). Antidiabetic constituents from *Helminthostachys zeylanica* (L.) Hook. (Ophioglossaceae). *Pharmacognosy Journal, 12*(2), 223–226.

Ridhasya, F. E., Teruna, H. Y., Hendra, R., & Almurdani, M. (2019). Natural antidiabetic of tunjuk langit (*Helminthostachys zeylanica*) rhizome extracts. *Pharmacology and Clinical Pharmacy Research, 4*(3), 67–70.

Sarker, B., Akther, F., Sifa, U. A. R., Jahan, I., Sarker, M., Chakma, S. K., Podder, P. K., Khatun, Z., & Rahmatullah, M. (2012). Ethnomedicinal investigations among the Sigibe clan of the Khumi tribe of Thanchi sub-district in Bandarban district of Bangladesh. *American Eurasian Journal of Sustainable Agriculture, 6*, 378–386.

Shah, A. B., Yoon, S., Kim, J. H., Zhumanova, K., Ban, Y. J., Lee, K. W., & Park, K. H. (2020). Effectiveness of cyclohexyl functionality in ugonins from *Helminthostachys zeylanica* to PTP1B and α-glucosidase inhibitions. *International Journal of Biological Macromolecules, 15*(165), 1822–1831.

Singh, B. P., & Upadhyay, R. (2012). Ethno-botanical importance of pteridophytes used by the tribe of Pachmarhi, Central India. *Journal of Medicinal Plants Research, 6*(1), 14–18.

Singh, K. K., Saha, S., & Maheswari, J. K. (1989). Ethnomedicinal uses of ferns. *Indian Fern Journal, 6*(1), 63–67.

Su, L. H., Li, Y. P., Li, H. M., Dai, W. F., Liu, D., Cao, L., & Li, R. T. (2016). Anti-inflammatory prenylated flavonoids from *Helminthostachys zeylanica*. *Chemical and Pharmaceutical Bulletin, 64*(5), 497–501.

Suja, S. R., Latha, P. G., Pushpangadan, P., & Rajasekharan, S. (2002). Aphrodisiac property of *Helminthostachys zeylanica* in male mice. *Journal of Tropical Medicinal Plants, 3*, 191–195.

Tsai, M. M., Lin, H. C., Yu, M. C., Lin, W. J., Chu, M. Y., Tsai, C. C., & Cheng, C. Y. (2021). Anticancer effects of *Helminthostachys zeylanica* ethyl acetate extracts on human gastric cancer cells through downregulation of the TNF-α-activated COX-2-cPLA2-PGE2 pathway. *Journal of Cancer, 12*(23), 7052–7068.

Vasudeva, S. M. (1999). Economic importance of pteridophytes. *Indian Fern Journal, 16*(1), 130–152.

Wu, K. C., Ho, Y. L., Kuo, Y. H., Huang, S. S., Huang, G. J., & Chang, Y. S. (2018). Hepatoprotective effect of Ugonin M, a *Helminthostachys zeylanica* constituent, on acetaminophen-induced acute liver injury in mice. *Molecules, 23*(10), 1–15.

Wu, K. C., Huang, S. S., Kuo, Y. H., Ho, Y. L., Yang, C. S., Chang, Y. S., & Huang, G. J. (2017). Ugonin M, a *Helminthostachys zeylanica* constituent, prevents LPS-induced acute lung injury through TLR4-mediated MAPK and NF-κB signaling pathways. *Molecules, 22*(4), 1–15.

Yamauchi, K., Mitsunaga, T., Itakura, Y., & Batubara, I. (2015). Extracellular melanogenesis inhibitory activity and the structure-activity relationships of ugonins from *Helminthostachys zeylanica* roots. *Fitoterapia, 104*, 69–74.

Yenn, T. W., Ring, L. C., Zahan, K. A., Rahman, M. S. A., Tan, W. N., & Alaudin, B. J. S. (2018). Chemical composition and antimicrobial efficacy of *Helminthostachys zeylanica* against foodborne *Bacillus cereus*. *Natural Product Sciences, 24*(1), 66–70.

CHAPTER 30

Phytochemical and Pharmacological Profile of the Genus *Cyathea*

MANPREET KAUR[1*] and GAUTAM RAJ PURI[2]

[1]*Department of Pharmacognosy, G.H.G. Khalsa College of Pharmacy, Gurusar Sadhar, Ludhiana, Punjab, India*

[2]*School of Pharmaceutical Sciences, RIMT University, Mandi Gobindgarh, Punjab, India*

*Corresponding authors

ABSTRACT

Cyathea represents a genus of tree ferns and serves as the type genus for the fern order Cyatheales. The species within the *Cyathea* genus are predominantly terrestrial ferns, typically characterized by a singular, tall stem. The trunk may hardly exhibit branching or a creeping growth pattern. Additionally, numerous species tend to form a fibrous root mass at the base of the trunk. The genus has a pantropical distribution, with over 470 species. *Cyathea* species are used in traditional medicine worldwide to treat various ailments such as dysentery, body pain, diabetes, hepatoprotective issues, inflammation, anthelmintic conditions, and headaches. Phytochemical studies have confirmed the presence of carbohydrates, starch, glycosides, saponins, alkaloids, tannins, flavonoids, sterols, terpenes, polyuronides, steroids, terpenoids, triterpenoids, polyphenols, and glycosides. *Cyathea* species have demonstrated antimicrobial, antifungal, antioxidant, antidiabetic, anti-inflammatory, cytotoxic, and anticancer potential, as well as larvicidal, anthelmintic, and antihyperlipidemic effects. Further phytochemical investigations, along with both *in vivo* and *in vitro* biological activities, have yet to be examined for the

Bioactives and Pharmacology of Pteridophytes. T. Pullaiah (Ed.)

DOI: 10.1201/9781779642738-30

majority of the *Cyathea* species. Therefore, this chapter focuses on encapsulating the phytochemical, biological activities, and therapeutic potency of the *Cyathea* species, serving as an update for additional research.

30.1 INTRODUCTION

Cyathea represents a genus of tree ferns and serves as the type genus for the fern order Cyatheales. The species within the *Cyathea* genus are predominantly terrestrial ferns, typically characterized by a singular, tall stem. The trunk may hardly exhibit branching or a creeping growth pattern. Additionally, numerous species tend to form a fibrous root mass at the base of the trunk. The genus exhibits a pantropical distribution, encompassing more than 470 species. They grow in habitats ranging from tropical rainforests to temperate woodlands. *Cyathea* species are used in traditional medicine worldwide to treat various ailments such as dysentery, body pain, diabetes, hepatoprotective issues, inflammation, antihelmintic conditions, and headaches. Phytochemical studies have confirmed the presence of carbohydrates, starch, glycosides, saponins, alkaloids, tannins, flavonoids, sterols, terpenes, polyuronides, steroids, terpenoids, triterpenoids, polyphenols, and glycosides.

30.2 ETHNOPHARMACOLOGY OF GENUS *CYATHEA*

In Pachmarhi, tree ferns play an essential role in the religious ceremonies of tribal communities in Central India. They strongly believe that their existence relies on the existence of these ferns. Tree ferns are often seen as symbols of good fortune. Traditional healers advise patients with headaches, severe fevers, hypertension, and other nervous system issues to spend time near these plants without causing them harm. Recent surveys indicate that practitioners of healing live in modest huts situated near wild species of *Cyathea*, orienting their windows to overlook these plants. They keep a small piece of dried rhizome under the pillow or bedsheet of babies and newborn children suffering from fever, cold, cough, and stomachache. They believe tree ferns have anti-aging properties (Singh and Singh, 2012).

The tender leaves of *Cyathea contaminans* are prepared by boiling and consumed as a vegetable. The pounded leaves can be rubbed on the forehead to relieve headaches and can also be used for rheumatic problems. It is used in wound healing. The soft apical part of the caudex is cut into small pieces and crushed in a mortar to make a paste, which is applied daily to large cuts and wounds until they heal. The rhizome is used topically for wounds and ulcers (Benton, 2023).

10 g of fresh rhizome of *Cyathea gigantea* mixed with 1 g of black pepper seed powder (*Piper nigrum*) is taken orally for one week, twice a day on an empty stomach with cow milk, to treat white discharge in women (Singh and Singh, 2012; Singh et al., 2013). Extracts of crushed rhizomes and young petioles are used for snake bites; the fresh pith is edible (Dixit and Singh, 2004; Shankar and Khare, 1992; Singh et al., 2005). Fresh rhizome paste is used locally for cuts and boils. It exhibits antitumor, antiviral, and hepatoprotective activities (Kiran et al., 2012). *C. gigantea*, along with other combinations of herbs, is utilized for pain from traumatic injury, such as sprains (Mishra and Behera, 2018).

Dried powder of leaves and stems of *Cyathea spinulosa* is utilized orally in rheumatic arthritis (Tangiang et al., 2011). The root and its pith are edible and used in the composition of local drinks; leaves are applied topically to wounds (Dixit and Singh, 2004; Manickam, 1999; Singh et al., 2005). The whole part of the plant is utilized for gray hair and also as a general hair tonic. A frond is ground into a fine powder and utilized for its properties as a sudorific and aphrodisiac. The stem is utilized to grow orchids, and the pith of the trunks is employed as a food product, as well as the stems for food and to make pots. Trunk fibers are utilized for orchid tissue culture media. The stem can be employed in folk Chinese medicine to remove dampness and strengthen muscles and joints. The roots are used to compose local drinks. A mixture of stem powders of *Cyathea spinulosa* and *Angiopteris helferiana* is given orally with water to cattle such as cows, buffaloes, and goats for indigestion and hair loss for various reasons. A small amount of the drug is also applied to the skin for rapid hair growth (Singh and Upadhyay, 2012; Upadhyay et al., 2011). A charred stem is applied to the forehead of an afflicted child to protect him from evils (Aya-ay, 2016).

30.3 PHYTOCHEMICAL REPORTS

Phytochemical analysis of *Cyathea brunoniana* revealed the presence of steroids, flavonoids, and saponins in the caudex and leaf extracts. Thin layer chromatography of plant extracts in different solvent systems confirms the presence of different phytochemical groups (Talukdar et al.,

2010). A study isolated a new acylated flavonol glycoside from the leaves *of Cyathea contaminans* and characterized it chemically as kaempferol-7-(600-succinyl)-glucoside (Yamane et al., 1985). *Cyathea contaminans* is an endemic fern commonly utilized as an ornamental plant in the Philippines; its medicinal value is rarely published, although its chemical components such as ecysteroids, triterpenoids, and the alkaloid lycodine are associated with the healing of bruises, strains, and swellings. It also has antidepressant and photoprotective effects (Mustacisa et al., 2021).

Kale (2015) qualitatively and quantitatively analyzed the petroleum ether, benzene, chloroform, methanol, and water extracts of *Cyathea gigantea* fern species. In the qualitative analysis, bioactive compounds such as alkaloids, steroids, tannins, saponins, and glycosides were present in the methanolic solvent extracts. The *Cyathea gigantea* methanol extract of the fern species expressed positive outcomes for six phytochemical tests. In the quantitative analysis, major secondary metabolites such as alkaloids, proline, saponins, steroids, glycosides, phenol, and tannins were tested in all extracts and were present in *Cyathea gigantea*. The main metabolites identified in *Cyathea nilgirensis* include methyl octadecyl dichlorosilane (29.19%), 2-methylbutane-1,4-diol (42.37%) in *C. gigantea*, and 2-hydroxy-5-methyl benzaldehyde (55.45%) in *C. crinita* by GC-MS analysis (Janakiraman and Johnson, 2016a).

From ethanolic extracts of selected species of *Cyathea*, the separation and quantification of lupeol, stigmasterol, and swertiamarin were done and developed using HPTLC. Furthermore, an attempt was made to investigate the biopotential of phytochemicals against different proteins computationally by Janakiraman et al. (2023). The compounds were separated by using a specific mobile phase, and the developed plates were sprayed with the respective reagents. The 3D structure of the receptor proteins, namely 1VSN, 5BNQ, 6HN8, 7DN4, and 3TJU, and the 3D SDF structures of the ligands such as lupeol, stigmasterol, and swertiamarin were obtained from the Protein Data Bank (PDB-Pub) and Chemound NCBI-Pub Chem Compound database, respectively. Computer-generated drug design screening software Argus 4.0.1 was utilized to investigate the binding affinity of test compounds against the selected proteins in the form of E-values versus potential drug targets. Discovery Studio Visualizer was used to visualize the docking results, which were saved previously. The terpenoid band exhibiting an Rf value of 0.79 corroborates the existence of lupeol in *C. gigantea* (0.04%) and *C. crinita* (0.02%). The steroid band with an R_f value of 0.41 revealed the presence of stigmasterol with varied frequency, viz., *C. nilgirensis* (0.33%), *C. gigantea* (0.29%), and *C. crinita* (0.52%). Interaction against the studied proteins, lupeol, stigmasterol, and swertiamarin showed, viz., 1VSN, 5BNQ, 6HN8, 7DN4, 3TJU with varied energy values and interacting residues. The outcomes of the virtual screening and molecular docking analysis suggest that the phytochemical compounds of *Cyathea* species, viz., lupeol and stigmasterol, were identified as possible lead molecules to fight against cancer and cytotoxicity.

Leaves of *Cyathea latebrosa* revealed the presence of carbohydrates, starch, glycosides, saponins, alkaloids, tannins, flavonoids, sterols, terpenes, and polyuronides. The quantitative phytochemical analysis revealed the following concentrations: alkaloids at $0.08\times10^2 \pm 0.02$, saponins at $0.16\times10^2 \pm 0.02$, tannins at $0.11\times10^2 \pm 0.01$, flavonoids at $0.16\times10^2 \pm 0.01$, and phenols at $0.18\times10^2 \pm 0.01$. FT-IR (KBr) cm-1 analysis showed -NH- (3433.41 weak), -OH- (3851.01 sharp), -C-F (1408.8 strong), -SO- (1101.39), and Ar (1646.3 weak). [2-Aziridinylethylamine (70.7%), Benzeneethamine, 3-fluoro-β,5-dihydroxy-N-methyl (14.0%), N-Methyltaurine (10.7%), and (R)-(-)-2-Amino-1-propanol (3.15%)] were present in the extract, which was revealed by GC-MS study. The presence of these compounds unveils the leaves as a potential remedy for alleviating cold, cancer, malaria, and regulating immune functions (Ikpa et al., 2021).

From the stem pith of *Cyathea phalerata* Mart., Cyathenosin A, a spiropyranosyl derivative of protocatechuic acid, was isolated, and its structure was determined by MS, 1D, and 2D NMR spectroscopic analyses and confirmed by single crystal X-ray analysis. Cyathenosin A is the first example of a natural compound containing a spirocyclic orthoesterpyranoside structure (Pizzolatti et al., 2007).

To investigate the phenolics, flavonoids, and tannins profile of *Cyathea nilgirensis* Holttum, *Cyathea gigantea* (Wall. ex. Hook.) Holttum, and *Cyathea crinita* (Hook.) Copel, HPTLC analysis method was carried out by Janakiraman and Johnson (2016a,c). The largest number of bands was found in the ethanolic extracts of *C. crinita.* The phenolic bands exhibiting R_f values of 0.05, 0.32, 0.43, 0.75 (quercetin), and 0.83 were identified in all three species of *Cyathea* examined. Additionally, flavonoids with R_f values of 0.05, 0.23, 0.30, 0.43, and 0.73 were also found to be present in the three species of *Cyathea* under investigation. Tannins with the R_f values of 0.05, 0.09, and 0.82 determined their presence in all three studied *Cyathea* species. Gallic acid was positive only in *C. crinita* with the R_f value of 0.49. The generated HPTLC fingerprints can be utilized for recognition and quantitative assessment of phenolics, flavonoids, and tannins. For speedy investigation of *Cyathea* species, the above-mentioned method is very utilizable for chemotypic analysis and also for controlling quality parameters.

30.4 PHARMACOLOGICAL REPORTS

30.4.1 ANTIBACTERIAL ACTIVITY

Cyathea brunoniana possesses antibacterial potential against both Gram-positive (*Staphylococcus aureus*) and Gram-negative (*Escherichia coli, Pseudomonas aeruginosa*) multidrug-resistant microorganisms (Nath et al., 2017). Nath et al. (2019) assessed the antibacterial potential of *C. gigantea* against MDR bacteria utilizing well diffusion and broth microdilution methods to assess the diameters of growth inhibition zones, minimum inhibitory concentration (MIC), and minimum bactericidal concentration (MBC). Potentiating activity was also assessed with the conventional antibiotics by disc diffusion method, and further FIC index of each of the tested antibiotics was computed. The active extract underwent fractionation using column chromatography, and antibacterial activity testing was carried out on each of the fractions collected. *C. gigantea* demonstrated efficacy against all organisms evaluated in the study. The MIC was determined to be 200 μg/ml for the Gram-positive bacterium *Staphylococcus aureus* ATCC 25923, while it was 400 μg/ml for the Gram-negative bacteria *Escherichia coli* ATCC 25922 and *Pseudomonas aeruginosa* PAO1. The MBC was found to be 400 μg/ml for Gram-positive bacteria and 800 μg/ml for Gram-negative bacteria. Furthermore, the synergistic effects indicated that the plant extract enhanced the antibacterial activity of the antibiotics tested with the fractional inhibitory concentration (FIC) index revealing significant synergism primarily with ciprofloxacin, followed by tetracycline, ampicillin, and oxacillin. The fractionated extract possessed antibacterial potential, with FR II, FR III, and FR IV exhibiting activity against both Gram-positive and Gram-negative bacteria. In contrast, FR I, FR V, and FR VI did not possess any antibacterial potential for the bacteria under test.

The efficacy of ethanolic and aqueous extracts from Neotropical tree ferns *Cyathea microdonta* and *Alsophila cuspidata*, which are utilized for medicinal purposes, was assessed for both Gram-positive and Gram-negative bacteria utilizing the Kirby-Bauer disc diffusion method by Longtine and Tejedor (2017). The ethanolic extracts possessed the maximum antimicrobial activity, and they were more efficacious for Gram-positive bacteria than Gram-negative bacteria. The antimicrobial potential exhibited by the two species, as well as the extracts derived from the leaves, stipe, or caudex, showed no discernible differences. These results indicate that the traditional application of tree fern mucilage as a topical application for cuts and leishmaniasis lesions might lower the risk of secondary infection by skin bacteria such as *Staphylococcus aureus*. Moreover, the Neotropical Cyatheaceae could be a reliable source of unique antimicrobial agents.

Plant extracts from *Cyathea nilgirensis* exhibited maximum antimicrobial potential against *Pseudomonas aureus* and *Klebsiella pneumoniae*, followed by *L. lanceolatus*, which possessed action against *K. pneumoniae* and *Aspergillus niger* (Pradhees et al., 2017). The extracts possessed inhibitory effects on four microorganisms, namely *P. aureus*, *K. pneumoniae*, *A. niger*, and *Fusarium* sp., with the maximum inhibition observed at the maximum concentration (100 µg/ml).

30.4.2 ANTI-FUNGAL ACTIVITY

Antifungal activity of *Cyathea contaminans* (Hook.) Copel. was analyzed against the fungus *Microsporum gypseum*, which is responsible for skin disease, horn or keratin-eating substances, and damages the nails and hair, by Turnip et al. (2014). The procedure of the study was conducted *in vitro* using the paper disc method. The dilutions of extracts were made into concentrations of 500 mg/mL, 400 mg/mL, 300 mg/mL, 200 mg/mL, 100 mg/mL, down to 10 mg/mL for *M. gypseum* tests, respectively. To each Petri dish, 0.1 mL of inoculum was added, followed by 15 mL of Potato Dextrose Agar medium that had been properly sterilized. Subsequently, disc paper that had been impregnated with various concentrations of extract for 15 minutes was transferred to the media and incubated at a temperature of 20–25°C for 48 hours. Inhibitory regions surrounding the diameter of the disc were assessed using calipers. Tests were carried out three times. The outcomes revealed an appropriate effect on pakupohon extract. The inhibition of fungal *M. gypseum* was observed to start at a concentration of 400 mg/ml, and the minimum inhibition occurred at concentrations of 300 mg extract/mL.

30.4.3 *HEPATOPROTECTIVE ACTIVITY*

The extract of leaves of *C. gigantea* obtained from methanol at doses of 100 mg/kg body weight and 200 mg/kg body weight has a maximum effect on the liver with paracetamol-induced hepatotoxicity (Kiran et al., 2012).

30.4.4 *ANTHELMINTIC ACTIVITY*

Aremu et al. (2010) evaluated the efficacy of 10 South African medicinal plants for *Caenorhabditis elegans*. Due to the elevated susceptibility of a host to microbial infections and other inflammatory responses related to helminth infections, the antimicrobial and cyclooxygenase (COX) inhibitory actions of the plants were also studied. Secondary metabolites like phenolics, such as flavonoids, condensed tannins, and gallotannins, have been associated with many pharmacological activities. Thus, studies were conducted quantitatively to analyze the phenolic content of the plant extracts. In the three bioassays, organic solvent extracts from *Cyathea dregei* (roots and leaves), *Felicia erigeroides* (leaves and stems), *Hypoxis colchicifolia* (leaves), and *Senna petersiana* (leaves) possessed noteworthy pharmacological actions, while *Acokanthera oppositifolia* (leaves) had a good COX inhibitory response.

30.4.5 *CYTOTOXIC ACTIVITY*

Anticancer actions of some particular *Cyathea* species brought to light their potential in the pharmaceutical industry for the manufacture of cheaper drugs, as investigated by Janakiraman and Johnson (2016b). Cytotoxic potential utilizing brine shrimp lethality bioassays and MCF 7 cell line cultures were completed. The ethanol extracts derived from certain *Cyathea* species exhibited superior activity against brine shrimps in comparison to the extracts obtained from petroleum ether, chloroform, and acetone. The ethanol extracts were further carried through to 3-(4,5-dimethylthiazol-2-yl)-2,5-diphenyltetrazolium bromide (MTT) cell proliferation assays. The cell viability was reduced, and growth inhibitions were elevated and recorded for the MCF 7 cell line. The highest percentage of cell inhibition was observed in *Cyathea crinita*, followed closely by *Cyathea nilgirensis* and *Cyathea gigantea*. *Cyathea* species are a reliable source of cytotoxic compounds, which present this experiment is advised.

30.4.6 ANTIDIABETIC AND ANTIHYPERLIPIDEMIC ACTIVITY

Elavarasi et al. (2020) evaluated the antidiabetic and antihyperlipidemic potential of *Cyathea nilgirensis* pith powder ethanol extract on an STZ-induced diabetic model. STZ-induced albino rats were administered with (200 mg/kg of b.wt. dose) ethanol extract of *C. nilgirensis* pith powder for 48 days. The treatment reconstituted the raised levels of body weight, blood glucose, and HbA1C. Further, a notable increase in serum insulin levels of STZ-induced diabetic rats was observed. Total cholesterol, triglycerides, LDL, and very low-density lipoprotein (VLDL) were raised, and high-density lipoprotein (HDL) levels fell in diabetic rats. The increase was markedly managed near to normal levels in extracts administered to diabetic rats. The reduced HDL level was appreciably raised by treatment with ethanol extract of *C. nilgirensis*. From the experimental results, it was summarized that *C. nilgirensis* is one of the most promising and evident drugs against diabetes mellitus and its affections.

30.4.7 ANTI-INFLAMMATORY ACTIVITY

Mary and Mahesh (2015) evaluated extracts obtained from the plant *Cyathea nilgirensis* for anti-inflammatory potential in mice utilizing the carrageenan-induced paw edema method. The procedure disclosed that the ethanol and chloroform extracts of the plant *C. nilgirensis* possess an up-to-the-mark anti-edematogenic response on carrageenan-induced paw edema. Both extracts of *C. nilgirensis* showed a maximum ($p \leq 0.001$) decrease in edema, with a percentage of inhibition ranging from 13.71 ± 1.25 to 7.15 ± 1.52 at 100 mg/kg in the chloroform extract, while the ethanol extract showed values ranging from 13.86 ± 2.12 to 6.7 ± 2.64 at 100 mg/kg. These extracts demonstrated better responses when compared to the standard drug diclofenac sodium.

30.4.8 ANTIOXIDANT ACTIVITY

Methanolic extracts of 15 fern species were screened by Lai and Lim (2011). Total phenolic content (TPC) was measured using the Folin-Ciocalteau method. The antioxidant properties were assessed using various methods, including DPPH radical scavenging, ferric reducing power (FRP), ferrous ion chelating activity (FIC), and β-carotene bleaching (BCB) assays. The findings indicated a significantly high total phenolic content, exceeding 2000 mg GAE per 100 g of fresh leaves, in the methanolic extracts of five different fern species. The ferns with strong antioxidant properties were *Cyathea latebrosa, Cibotium barometz, Drynaria quercifolia, Blechnum orientale*, and *Dicranopteris linearis*. These ferns demonstrated significant DPPH radical scavenging activity, with ascorbic acid equivalents ranging from 2866 to 3111 mg/100 g, as well as ferric ion reducing power, indicated by gallic acid equivalents of 963 to 1417 mg GAE/100 g, and a notable inhibition of lipid peroxidation. A positive correlation was identified between total phenolic content (TPC) and various antioxidant activities. The antioxidant potential *of Cyathea nilgirensis* Holttum, *Cyathea gigantea* (Wall. ex. Hook.) Holttum, and *Cyathea crinita* (Hook.) Copel was evaluated by Janakiraman and Johnson (2015). Ethanolic extracts from these selected *Cyathea* species were analyzed for antioxidant activity through several assays, including DPPH, superoxide, metal chelating, nitric oxide, and hydroxyl radical scavenging activities. Among the species tested, *C. nilgirensis* exhibited the highest free radical scavenging activity across all assays conducted. The scavenging effect of the tested *Cyathea* species is in the following order: *C. nilgirensis* > *C. gigantea* > *C. crinita*. The study findings showed prominent radical scavenging activity in all three studied species. The antioxidant activity of extracts and fractions of six vegetal species from the Brazilian Atlantic Forest was determined by Brighente et al. (2007). The overall antioxidant activity was evaluated through the scavenging potential of the stable DPPH free radical. A total of eight plant extracts or fractions demonstrated notable DPPH scavenging activity, with IC50 values equal to or less than 10.0 μg/mL, in comparison to ascorbic acid (IC50 = 8.4 μg/mL) and gallic acid (IC50 = 2.6 μg/mL). The extracts or fractions included ethanol extracts from the leaves, flowers, and stems of *Baccharis illinita* DC., as well as ethanol extracts from the leaves and stems of *B. platypoda* DC. Additionally, the hydroalcoholic extract and ethyl acetate fraction from the leaves of *Cyathea phalerata* Mart. and the hydroalcoholic extract from the bark of *Trichilia catigua* A. Juss. were analyzed. Furthermore, seven flavonoids found in the plant extracts were examined, with taxifolin, quercetin, and luteolin emerging as the most active compounds due to the presence of the catechol group 3′,4′-diOH. The total phenolic and flavonoid contents of these extracts and fractions were also assessed. The phenolic content was measured using the Folin-Ciocalteu reagent, yielding values ranging from 489.07 to 11.29 mg/g dry weight, expressed as gallic acid equivalents (GAE). The total flavonoid concentrations, determined using 2% aluminum chloride, ranged from 61.82 to 5.6 mg quercetin equivalents (QE)/g dry weight. These findings indicate a significant variation in the antioxidant activity levels among these plants. They also propose that the phenolic percentage amount in these plants provides remarkable antioxidant activity.

30.4.9 ANTI-SPASM ACTIVITY

Cyathea contaminans anti-spasm oil was made by Mustacisa et al. (2021). The developed *C. contaminans* anti-spasm oil was voluntarily evaluated by 15 participants across three experimental setups. The greatest frequency of efficacy was observed in setup A (87.5% *C. contaminans* oil, 6.25% lavandula oil, 5% VCO, and 1.25% surfactant), which means that among the three setups, the respondents claimed that the prepared product in setup A had the capacity to produce an effect of less than 5 minutes. The formulation created in setup C, comprising 90% *C. contaminans* oil, 5% lavandula oil, 4% VCO, and 1% surfactant, demonstrated the greatest duration of effectiveness, lasting between 5 to 15 minutes. According to the statistical analysis, the anti-spasm oil derived from *C. contaminans,* produced across three experimental setups, exhibited no significant variations in color, fragrance, or texture. The three formulations were statistically similar to one another. Nevertheless, setup C, which contained 90% by volume of the active ingredient *C. contaminans*, is regarded as the most effective based on the findings from the sensory evaluation.

30.4.10 LARVICIDAL ACTIVITY

Janakiraman and Johnson (2017) studied the larvicidal activity of selected *Cyathea* species (*Cyathea nilgirensis, Cyathea gigantea,* and *Cyathea crinita*) against the filarial vector *Culex quinquefasciatus*. The larvicidal efficacy of various extracts was assessed, and the resulting larval mortality was documented. The larvae exhibited greater sensitivity to the ethanolic extracts derived from the three studied *Cyathea* species in comparison to the other extracts; specifically, the acetone, chloroform, and petroleum ether extracts were deemed less effective. The LC50 values for the different extracts varied between 320.72 and 657.03 μg/mL. The findings demonstrated that the three *Cyathea* species examined displayed larvicidal effects that were dependent on concentration, suggesting the presence of potential bioactive properties.

KEYWORDS

- **antioxidant**
- **biological activities**
- ***Cyathea***
- **flavonoids**
- **folin-ciocalteu reagent**
- **phytochemical**

REFERENCES

Aremu, A. O., Ndhlala, A. R., Fawole, O. A., Light, M. E., Finnie, J. F., & Van Staden, J. (2010). *In vitro* pharmacological evaluation and phenolic content of ten South African medicinal plants used as anthelmintics. *South African Journal of Botany, 76*(3), 558–566.

Aya-ay, A. (2016). Ethnobotany of ferns and ferns-allies in Mount Macabol, Marilog District, Davao City, Philippines. *International Journal of Applied Business and Economic Research, 14*, 1127–1137.

Benton, R. A. (2023). Proto Polynesian Pala (web page periodically updated). *Te Māra Reo.* http://www.temarareo.org/PPN-Pala.html.

Brighente, I. M. C., Dias, M., Verdi, L. G., & Pizzolatti, M. G. (2007). Antioxidant activity and total phenolic content of some Brazilian species. *Pharmaceutical Biology, 45*(2), 156–161. https://doi.org/10.1080/13880200601113131.

Dixit, R. D., & Singh, S. (2004). Enumeration of the rare, endangered and endemic Pteridophytes from Central India. *Indian Fern Journal, 21*, 96–108.

Elavarasi, S., Revathi, G., Saravanan, K., & Averal, H. I. (2020). Antidiabetic and antihyperlipidemic activities of *Cyathea nilgiriensis* (Holttum) on STZ induced diabetic rats. In *Drug Development for Cancer and Diabetes* (p. 16). Apple Academic Press.

Ikpa, C. B. C., Maduka, T. D. O., & Ikpa, C. U. (2021). Evaluation of phytochemical composition and *in vitro* antioxidant potential of *Cyathea latebrosa* leaves. *World Scientific News, 157*, 25–37.

Janakiraman, N., & Johnson, M. (2015). *In vitro* antioxidant properties of natural products isolated from selected species of *Cyathea*. *Journal of Clinical Nephrology and Research, 2*(2), 1027.

Janakiraman, N., & Johnson, M. (2016a). HPTLC fingerprint profile (phenolics) of selected *Cyathea* species from Western Ghats, South India. *Chinese Journal of Biology*, 1–7.

Janakiraman, N., & Johnson, M. (2016b). Ethanol extracts of selected *Cyathea* species decreased cell viability and inhibited growth in MCF-7 cell line cultures. *Journal of Acupuncture and Meridian Studies, 9*(3), 151–155.

Janakiraman, N., & Johnson, M. (2016c). GC–MS analysis of ethanolic extracts of *Cyathea nilgirensis*, *C. gigantea*, and *C. crinita*. *Egyptian Pharmaceutical Journal, 15*, 1543–1547.

Janakiraman, N., & Johnson, M. (2017). Larvicidal potential of *Cyathea* species against *Culex quinquefasciatus*. *Pharmaceutical and Biomedical Research, 3*(1), 48–51.

Janakiraman, N., Anne, W. J., Johnson, M., Beatriz, H. D. A., de Oliveira, S. C. E., Saraiva de Alencar Beltrao, C., & Coutinho, H. D. M. (2023). Chromatographic analysis of selected phytosterols from *Cyathea* and their characterization by *in silico* docking to potential therapeutic targets. *Current Research in Toxicology, 5*, 1–16. https://doi.org/10.1016/j.crtox.2023.100115.

Kale, M. V. (2015). Qualitative and quantitative analysis of *Cyathea gigantea* ferns from South Western Ghats. *Research Journal of Pharmaceutical, Biological and Chemical Sciences, 1*, 183–187.

Kiran, P. M., Raju, A. V., & Rao, B. G. (2012). Investigation of hepatoprotective activity of *Cyathea gigantea* Wall. ex-Hook. leaves against paracetamol-induced hepatotoxicity in rats. *Asian Pacific Journal of Tropical Biomedicine, 2*, 352–356.

Lai, H., & Lim, Y. (2011). Evaluation of antioxidant activities of the methanolic extracts of selected ferns in Malaysia. *International Journal of Environmental Science, 2*(6), 442–447. https://doi.org/10.7763/IJESD.2011.V2.166.

Longtine, C., & Tejedor, A. (2017). Antimicrobial activity of ethanolic and aqueous extracts of medicinally used tree ferns *Alsophila cuspidata* and *Cyathea microdonta*. *Acta Botanica Malacitana, 42*(1), 119–123.

Manickam, V. S. (1999). Medicinal ferns of India. *Amruth, 2*, 3–9.

Mary, S., & Mahesh, M. K. (2015). Anti-inflammatory activity of *Cyathea nilgirensis* Holttum against carrageenan-induced paw edema. *International Journal of Recent Scientific Research, 6*(8), 5807–5809.

Mishra, N., & Behera, S. A. (2018). Tree ferns and giant ferns in India. In *The Significance and Conservation: Concerns and Sustainable Development* (Vol. 2, pp. 45–62).

Mustacisa, L. M. M., Dejarme, A. B., & Albina, M. B. (2021). Preparation, sensory evaluation and effectiveness of Philippine tree fern (*Cyathea contaminans*) as anti-spasm oil. *International Journal of Agricultural Technology, 17*(4), 1461–1470.

Nath, K., Talukdar, A. D., & Bhattacharya, M. K. (2019). *Cyathea gigantea* (Cyatheaceae) as an antimicrobial agent against multidrug resistant organisms. *BMC Complementary and Alternative Medicine, 19*, 279. https://doi.org/10.1186/s12906-019-2696-0.

Nath, K., Talukdar, A. D., Bhattacharya, M. K., Bhowmik, D., Chetri, S., Choudhury, D., Mitra, A., & Bhattacharjee, A. (2017). Antibacterial activity of certain ferns against multidrug resistant organisms. *Journal of Natural Remedies, 17*, 144–153.

Pizzolatti, M. G., Brighente, I. M. C., Bortoluzzi, A. J., Schripsema, J., & Verdi, L. G. (2007). Cyathenosin A, a spiropyranosyl derivative of protocatechuic acid from *Cyathea phalerata*. *Phytochemistry, 68*, 1327–1330.

Pradhees, G., Suresh, J., Suresh, S., Ramani, A. V., & Hong, S. I. (2017). Antimicrobial activity and identification of potential compounds from the chloroform extract of the medicinal plant *Cyathea nilgirensis* Holttum. *World Journal of Pharmacy and Pharmaceutical Sciences, 6*(7), 1167–1184.

Shankar, R., & Khare, P. K. (1992). Ethnobotanical observation on some ferns of Pachmarhi hills. *Journal of Economic and Taxonomic Botany, Additional Series, 10*, 97–100.

Singh, B. P., & Upadhyay, R. (2012). Ethno-botanical importance of pteridophytes used by the tribe of Pachmarhi, Central India. *Journal of Medicinal Plants Research, 6*(1), 14–18.

Singh, S., & Singh, R. (2012). Ethnomedicinal use of pteridophytes in reproductive health of tribal women of Pachmarhi Biosphere Reserve, Madhya Pradesh, India. *International Journal of Pharmaceutical Sciences and Research, 3*(12), 4780–4790.

Singh, S., Dixit, R. D., & Sahu, T. R. (2005). Ethnobotanical use of Pteridophytes of Amarkantak (MP). *Indian Journal of Traditional Knowledge, 4*(4), 392–395.

Singh, S., Singh, R., & Sahu, T. R. (2013). Ethnomedicinal uses of Pachmarhi Hills, Madhya Pradesh, India. In D. R. Khanna et al. (Eds.), *Floral Diversity and Their Conservation* (pp. 267–290). Biotech Book Press.

Talukdar, A., Duttachoudhury, M., Chakraborty, M., & Dutta, B. K. (2010). Phytochemical screening and TLC profiling of plant extracts of *Cyathea gigantea* (Wall. ex-Hook.) Holtt. and *Cyathea brunoniana* Wall. ex-Hook. (Cl. & Bak.). *Assam University Journal of Science and Technology, 5*(1), 70–74.

Tangiang, S., Namsa, D. N., Aran, C., & Litin, A. (2011). An ethnobotanical survey of medicinal plants in the Eastern Himalayan zone of Arunachal Pradesh, India. *Journal of Ethnopharmacology, 134*, 18–25.

Turnip, C. D., Batubara, R., & Ginting, H. (2014). The inhibition test of extract Paku Pohan (*Cyathea contaminans* (Hook.) Copel.) to fungi *Microsporum gypseum in vitro. Academia, 3*(1), 43–47.

Upadhyay, R., Singh, B. P., & Upadhyay, S. T. (2011). Ethano-medicinal observations on a threatened tree fern *Cyathea spinulosa* Wall. ex-Hook. in Satpura Hills. *Indian Fern Journal, 28*, 129–136.

Yamane, H., Fujioka, S., Spray, C. R., Phinney, B. O., MacMillan, J., Gaskin, P., & Takahashi, N. (1988). Endogenous gibberellins from sporophytes of two tree ferns, *Cibotium glaucum* and *Dicksonia antarctica. Plant Physiology, 86*(3), 857–862.

CHAPTER 31

Bioactive Compounds and Pharmaceutical Activities of *Cyathea phalerata* Mart.

M. SURYA PRABHA,[1] B. SUDARSINI,[2] and M. INDIRA[2,*]

[1]*Department of Pharmaceutical Sciences, Vignan's Foundation for Science, Technology and Research, Vadlamudi, Andhra Pradesh, India*

[2]*Department of Biotechnology, Vignan's Foundation for Science, Technology and Research, Vadlamudi, Andhra Pradesh, India*

*Corresponding authors

ABSTRACT

Cyathea phalerata Mart. is an arborescent fern native to Brazil. This fern is characterized by a caudex that can reach up to 10 meters in height and leaves that extend up to 3.5 meters, with thick blades that may or may not have trichomes on both surfaces. *Cyathea phalerata* is widely distributed across Brazil and is also found in Bolivia. Information on *C. phalerata* is currently limited, primarily focusing on its taxonomy, pharmacology, and aspects of germination and gametophyte development. Despite its known medicinal uses, including its expectorant properties and applications in treating kidney diseases, inflammatory conditions, varicose veins, and hemorrhoids in southeastern Brazil, the efficacy and active principles of many medicinal plants, including this fern, remain poorly understood. Further research is necessary to fully comprehend the potential therapeutic benefits and bioactive compounds of *C. phalerata*.

Bioactives and Pharmacology of Pteridophytes. T. Pullaiah (Ed.)

DOI: 10.1201/9781779642738-31

31.1 INTRODUCTION

Tree ferns from the Cyatheaceae family, native to the humid mountainous forests of tropical and subtropical regions, are used ornamentally, medicinally, and as food. They also serve as indicators of environments with minimal human impact (Perez-Paredes et al., 2014). This family comprises around 600 species of tree ferns, distributed among three genera: *Alsophila, Cyathea,* and *Sphaeropteris* (Dong et al., 2018). Scaly tree ferns from the Cyatheaceae family, originating from the Jurassic period and found in tropical to subtropical climates from the Himalayas to Japan, are endangered due to habitat degradation and unsuitable environmental conditions. They are listed in Appendix II of CITES and the 1997 IUCN Red List (Budha et al., 2023).

Cyathea species are used in traditional medicine worldwide to treat various ailments, including body pain, dysentery, diabetes, headaches, liver protection, respiratory issues, inflammation, and parasitic infections (Marimuthu et al., 2023). The beneficial effects of these plants are often attributed to their secondary metabolites, which provide temporary relief from symptoms, possess curative properties, and promote health (Chaparro-Hernandez et al., 2022). These compounds include tannins, flavonoids, coumarins, glycosides, lignans, lignins, and anthraquinones (Narayanan and Marimuthu, 2016).

Cyathea phalerata Mart. (Cyatheaceae), a Brazilian fern with a tree-like growth habit, features a trunk reaching up to 10 meters in height and leaves extending up to 3.5 meters in length, potentially bearing trichomes on both surfaces (Marimuthu et al., 2023). This species is widely distributed across Brazil and found in Bolivia, but knowledge about it is largely confined to its taxonomy, pharmacology, and studies on germination and gametophyte development (Cunha et al., 2021). Due to the significant reduction of the Atlantic Forest and its limited occurrence between latitudes 29°17'–31°00'S, *C. phalerata* is classified as a critically endangered plant (Cunha et al., 2021). *C. phalerata* thrives in the shadowy interiors of forests, near watercourses, in the inter-fluvial areas of humid forests, and alongside streams in drier forests (Lehnert and Kessler, 2018). It can also be found in gallery forests and forest edge gullies at altitudes from sea level up to 1500 meters. This fern develops caudices up to 4 meters high with leaves reaching 3.5 meters in length. Known for its medicinal properties, the inner parts of its caudices and leaves are used in alcoholic extracts to combat inflammatory diseases (Marcon et al., 2017).

While medicinal plants are widely used globally for various ailments, the effectiveness and active compounds of many remain largely unidentified

(Andrade et al., 2014). Phytochemical studies have identified various bioactive compounds in *C. phalerata*, including triterpenes, phenolic acids, flavonoids (especially kaempferol and its glycosides), glucopyranosyl caffeic acid, glucopyranosyl p-coumaric acid, and cyathenosin A (Hort et al., 2020). *Cyathea* ferns are a significant source of these bioactive compounds. However, further research is essential to isolate and understand the specific compounds responsible for their bioactivity (Juliasih and Adnyana, 2023). Conservation efforts are vital, as many *Cyathea* species face extinction, and their potential therapeutic properties underscore the importance of continued study and preservation (Balkrishna et al., 2020).

31.2 BIOACTIVE COMPOUNDS

Cyathea species have been used in traditional medicine for centuries to treat a wide range of ailments, from common infections of the eye, skin, and kidneys to chronic diseases such as cancer and diabetes (Chaparro-Hernandez et al., 2022). The therapeutic efficacy of these ferns is primarily attributed to their phenolic compounds (Faizal et al., 2020). Medicinal plants, including *Cyathea* species, are rigorously evaluated for their bioactive potential through a comprehensive approach encompassing taxonomic, phytochemical, and ethnomedicinal aspects (Janakiraman et al., 2023). These botanical remedies form a cornerstone of healthcare systems worldwide. The phytochemical profile of *Cyathea* species is rich and diverse, encompassing compounds such as saponins, fernane, steroids, tannins, filicene, hepane triterpenes, flavonoids, kaempferol glycosides, and phenolic acids such as caffeic, coumaric, and protocatechuic acids (Marimuthu et al., 2023).

Cyathea phalerata, commonly known as 'xaxim,' is a fern tree renowned for its therapeutic properties, including expectorant and anti-inflammatory effects (Santos et al., 2010). It is also effective in treating hemorrhoids and varicose veins (Andrade et al., 2014). The therapeutic compounds are primarily located in the stem and include 3,4-spiroglucopyranosyl protocatechuic acid, 4-O-β-D-glucopyranosyl coumaric acid (Aulakh et al., 2019), 3,4-spiroglucopyranosyl p-coumaric acid, sitosterol-β-D-glucoside, β-sitosterol, kaempferol, vitexin, ethyl galactoside (Hort et al., 2008), and kaempferol-3-neohesperidoside (Hort et al., 2020; Pizzolatti et al., 2007). Additionally, it contains kaempferol-3-O-α-L-rhamnopyranosyl-(1-2)-β-D-glucopyranoside (Brighente et al., 2007), 4-O-β-D-glucopyranosyl caffeic acid, and cyathenosin A (Pizzolatti et al., 2007).

The compound kaempferol-3-neohesperidoside, found in traditional medicines, has demonstrated antioxidant properties that protect cells from damage and exhibit hypoglycemic activity, aiding in the reduction of blood sugar levels (Cazarolli et al., 2006). Additionally, water-based extracts of *C. phalerata* yield well-recognized glucosides, such as 4-O-β-D-glucopyranosyl caffeic acid and 4-O-β-D-glucopyranosyl coumaric acid, along with a novel ortho ester spiropyranosyl derivative of protocatechuic acid that possesses unique structural characteristics (Pizzolatti et al., 2007). Bioactive compounds of *C. phalerata* are given in Table 31.1 and Figure 31.1.

TABLE 31.1 Bioactive Compounds of *Cyathea phalerata*

Name of the Bioactive Compounds	Molecular Formula	Phytochemical Class	Pharmacological Activity	References
Caffeic acid	$C_9H_8O_4$	Phenolic acid	Anti-inflammatory, Antidiabetic, Anti-oxidant, anti-thrombic	Chaparro-Hernandez et al. (2022)
P-coumaric acid	$C_9H_8O_3$	Phenolic acid	Antioxidant	Hort et al. (2008)
Protocatechuic acid	$C_7H_6O_4$	Phenolic acid	Antioxidant, anti-inflammatory	Hort et al. (2020); Pizzolatti et al. (2007)
Sitosterol-β-D glucoside	$C_{35}H_{60}0_6$	Phytosterol glycosides	Antioxidant, anti-inflammatory	Hort et al. (2020)
β-sitosterol	$C_{29}H_{50}O$	Triterpenoids	Antitumor and antiviral	Kale (2015)
Kaempferol	$C_{15}H_{10}0_6$	Flavonoids	Antioxidant activity	Brighente et al. (2007)
Vitexin	$C_{21}H_{20}0_{10}$	Flavone glycosides	Anti-inflammatory, Antidiabetic, Antioxidant, antithrombic activity	Chaparro-Hernandez et al. (2022)
kaempferol 3-neohesperidoside	$C_{27}H_{30}0_{15}$	Flavonoid	Antioxidant and Hypoglycemic activity	Hort et al. (2020); Pizzolatti et al. (2007)
4-O-β-D-glucopyranosyl caffeic acid	$C_{15}H_{18}0_9$	Glucoside	Antioxidant activity	Hort et al. (2008); Pizzolatti et al. (2007)
Kaempferol-3,7-*O*-(α)-dirhamnoside	$C_{27}H_{30}0_{14}$	Flavonoid	Lowering serum glucose (anti-diabetic)	Cazarolli et al. (2006)

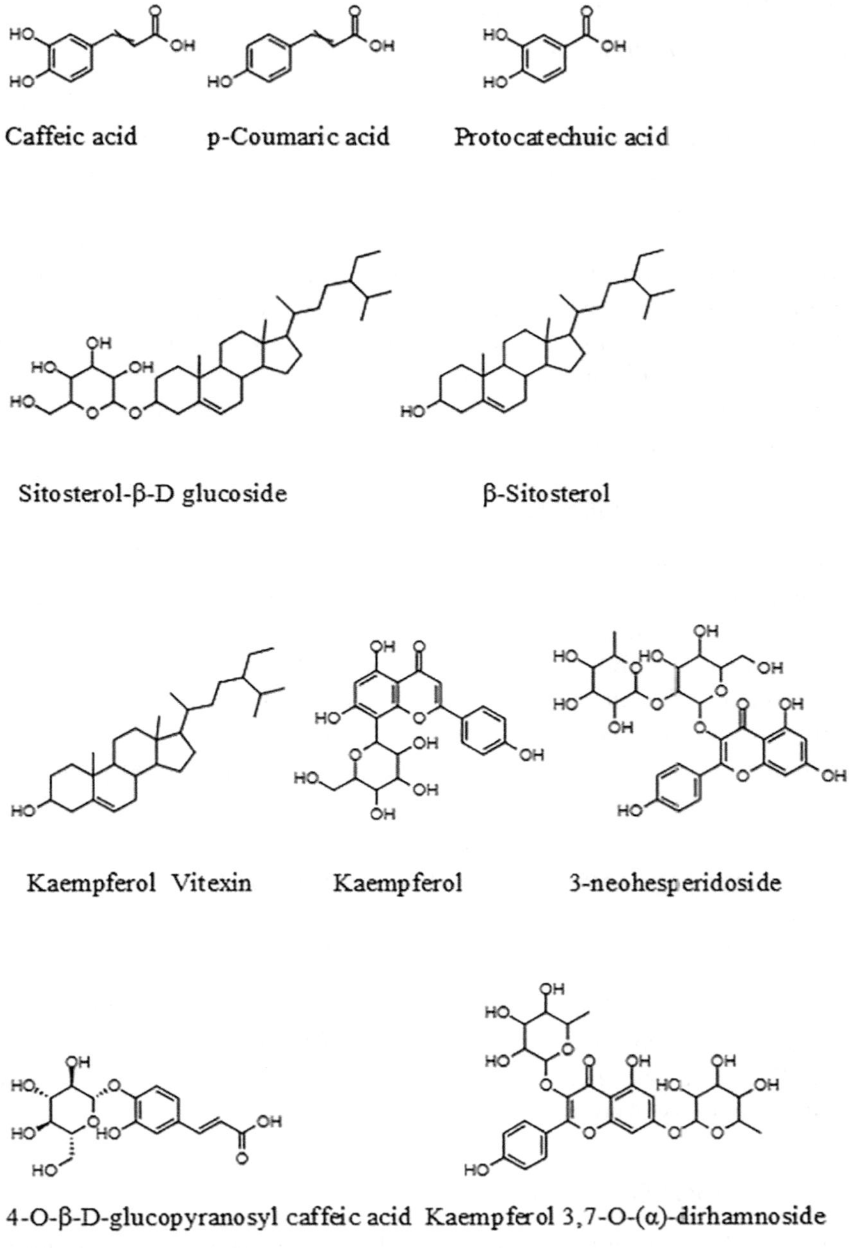

FIGURE 31.1 Chemical structures of bioactive compounds in *Cyathea phalerata.*

31.3 PHARMACOLOGICAL ACTIVITIES

31.3.1 ANTIOXIDANT ACTIVITY

The antioxidant properties of *Cyathea phalerata* have been analyzed through both *in vitro* and *in vivo* assays. The *in vitro* antioxidant activity of the crude extract (CE), aqueous fraction (AQF), precipitate (PPT), ethyl acetate fraction (EAF), and n-butanolic fraction (BUF) from *C. phalerata* was assessed by evaluating their scavenging effects on diphenyl-1-picryl-hydrazyl-hydrate (DPPH), superoxide anions ($O_2{\bullet}^-$) using the nitro blue tetrazolium (NBT) assay, hydroxyl radicals (OH•) through lipid peroxidation, and deoxyribose assay in rat liver homogenate. The ethyl acetate fraction from *C. phalerata* significantly reduced carbonyl protein levels, DNA damage, and thiobarbituric acid reactive substances, while simultaneously enhancing glutathione S-transferase and catalase activities (Hort et al., 2008).

31.3.2 HEPATO-PROTECTIVE ACTIVITY

The ethyl acetate fraction (EAF) of *Cyathea phalerata* demonstrated hepatoprotective effects against oxidative stress induced by carbon tetrachloride (CCl_4) in mice. A significant reduction in thiobarbituric acid reactive substances, carbonyl protein levels, and DNA damage was observed. Additionally, the increase in glutathione S-transferase and catalase activities suggests that the EAF provides protection against CCl_4-induced oxidative stress in the liver, likely due to its antioxidant activity and free radical scavenging properties (Hort et al., 2008).

31.3.3 MONOAMINE OXIDASE INHIBITION

The extract of the fern *C. phalerata* exhibited 74.07% inhibition of monoamine oxidase A (MAO-A) activity, compared to the pure compound quercetin (100 μM), which showed 94.46% inhibition. Additionally, the extract displayed lower inhibitory activity on monoamine oxidase B (MAO-B) at 25.93%, indicating low selectivity between the different monoamine oxidase subunits (Andrade et al., 2014).

31.3.4 VASORELAXANT EFFECT

Cyathea phalerata extracts are rich in phenolic compounds that promote the widening and relaxation of blood vessels. This vasorelaxant effect is primarily mediated through the activation of the cGMP/NO pathway and endothelium-derived hyperpolarizing factor (EDHF). Endothelial cells play a crucial role in regulating the dilation and constriction of blood vessels within the cardiovascular system. In the present study, the ethyl acetate fraction (EAF) obtained from *C. phalerata* demonstrated vasorelaxant activity in rat thoracic aorta rings. The EAF induced endothelium-dependent vasorelaxation in the aorta, which is attributed to the NO/cGMP pathway, prostanoids, and hyperpolarization. This vasorelaxant activity is linked to the presence of polyphenol compounds (Hort et al., 2020).

31.3.5 ANTI-DIABETIC ACTIVITY

Kaempferol 3-neohesperidoside, a naturally occurring compound in *Cyathea phalerata,* is known to mimic insulin action. Some flavonoids and glycosides stimulate glucose uptake and lower blood glucose levels. In this study, the synthesized kaempferol 3-neohesperidoside was tested for its insulin mimetic properties in a cultured muscle cell line L6, using 2-deoxy glucose (2-DG) as a glucose substitute. The amount of 2-DG transported was estimated using an enzymatic method. At a concentration of 1 nM, kaempferol 3-neohesperidoside showed significant insulin mimetic activity, enhancing 2-DG uptake by 200% compared to control cells (Yamasaki et al., 2011).

KEYWORDS

- **antioxidant activity**
- **bioactive compounds**
- **flavonoids**
- **hepatoprotective**
- **insulin-mimetic activity**
- **vasorelaxant effect**

REFERENCES

Andrade, J. M., dos S. Passos, C., Dresch, R. R., Kieling-Rubio, M. A., Moreno, P. R. H., & Henriques, A. T. (2014). Chemical analysis, antioxidant, antichemotactic and monoamine oxidase inhibition effects of some pteridophytes from Brazil. *Pharmacognosy Magazine, 10*(37), 100.

Aulakh, M. K., Kaur, N., & Saggoo, M. I. S. (2019). Bioactive phytoconstituents of pteridophytes–A review. *Indian Fern Journal, 36*, 37–79.

Balkrishna, A., Arya, V., & Kushwaha, A. K. (2020). Population structure, regeneration status and conservation measures of threatened *Cyathea* spp. *Journal of Tropical Forest Science, 32*(4), 414–421.

Brighente, I. M. C., Dias, M., Verdi, L. G., & Pizzolatti, M. G. (2007). Antioxidant activity and total phenolic content of some Brazilian species. *Pharmaceutical Biology, 45*(2), 156–161.

Budha, P. B., Lekhak, K., Kalu, S., & Thapa, I. (2023). Habitat suitability modeling of endangered *Cyathea spinulosa* (Wall. exHook.) in Central Nepal. *Journal of Forest and Environmental Science, 39*(2), 65–72.

Cazarolli, L. H., Zanatta, L., Jorge, A. P., de Sousa, E., Horst, H., Woehl, V. M., & Silva, F. R. M. B. (2006). Follow-up studies on glycosylated flavonoids and their complexes with vanadium: Their anti-hyperglycemic potential role in diabetes. *Chemico-Biological Interactions, 163*(3), 177–191.

Chaparro-Hernandez, I., Rodríguez-Ramirez, J., Barriada-Bernal, L. G., & Mendez-Lagunas, L. (2022). Tree ferns (Cyatheaceae) as a source of phenolic compounds – A review. *Journal of Herbal Medicine, 35*, 100587.

Cunha, S., Endres, D., Silva, V. L., Droste, A., & Schmitt, J. L. (2021). Herbivory and leaf expansion of *Cyathea phalerata* Mart. (Cyatheaceae) in subtropical Atlantic Forest, southern Brazil. *Brazilian Journal of Biology, 83*, e245386.s.

Dong, S. Y., & Zuo, Z. Y. (2018). On the recognition of *Gymnosphaera* as a distinct genus in Cyatheaceae. *Annals of the Missouri Botanical Garden, 103*(1), 1–23.

Faizal, A., Taufik, I., Rachmani, A. F., & Azar, A. W. P. (2020). Antioxidant and antibacterial properties of tree fern *Cyathea contaminans*. *Biodiversitas Journal of Biological Diversity, 21*(5), 2201–2205.

Hort, M. A., Brighente, I. M. C., Pizzolatti, M. G., & Ribeiro-do-Valle, R. M. (2020). Mechanisms involved in the endothelium-dependent vasodilatory effect of an ethyl acetate fraction of *Cyathea phalerata* Mart. in isolated rats' aorta rings. *Journal of Traditional and Complementary Medicine, 10*(4), 360–365.

Hort, M. A., Dalbo, S., Brighente, I. M. C., Pizzolatti, M. G., Pedrosa, R. C., & Ribeiro-do-Valle, R. M. (2008). Antioxidant and hepatoprotective effects of *Cyathea phalerata* Mart. (Cyatheaceae). *Basic & Clinical Pharmacology & Toxicology, 103*, 17–24.

Janakiraman, N., Wincy, J. A., Johnson, M., Ducati, A. B. H., de Oliveira, S. C. E., de Alencar, B. C. S., & Coutinho, H. D. M. (2023). Chromatographic analysis of selected phytosterols from *Cyathea* and their characterization by *in silico* docking to potential therapeutic targets. *Current Research in Toxicology, 5*, 100115.

Juliasih, N. K. A., & Adnyana, I. M. D. M. (2023). Utilization of *Pteridophyta* species in Cyathea Park, Bali, as traditional medicine agents: A field study and meta-synthesis review. *Notulae Scientia Biologicae, 15*(2), 11522–11522.

Kale, M. V. (2015). Analysis of bioactive compounds on ethanolic extract of whole plant of *Cyathea gigantea*. *Research Journal of Life Sciences, Bioinformatics, Pharmaceutical and Chemical Sciences, 1*, 156–161.

Lehnert, M., & Kessler, M. (2018). Prodromus of a fern flora for Bolivia. XX. Cyatheaceae. *Phytotaxa, 334*(2), 118–134.

Marcon, C., Silveira, T., Schmitt, J. L., & Droste, A. (2017). Abiotic environmental conditions for germination and development of gametophytes of *Cyathea phalerata* Mart. (Cyatheaceae). *Acta Botanica Brasilica, 31*(1), 58–67.

Marimuthu, J., George, V., Iruthayamani, S. J., & Balasundaram, S. (2023). Bioactive compounds and biological activities of *Cyathea* species. In *Bioactive Compounds in Bryophytes and Pteridophytes* (pp. 1–20).

Narayanan, J., & Marimuthu, J. (2016). HPTLC fingerprint profile (phenolics) of selected *Cyathea* species from Western Ghats, South India. *Chinese Journal of Biology*, 2016, 6420371.

Perez-Paredes, M. G., Sanchez-Gonzalez, A., & Tejero-Diez, J. D. (2014). Population structure and habitat traits of two Cyatheaceae species of Hidalgo State, Mexico. *Botanical Sciences, 92*(2), 259–271.

Pizzolatti, M. G., Brighente, I. M. C., Bortoluzzi, A. J., Schripsema, J., & Verdi, L. G. (2007). Cyathenosin A, a spiropyranosyl derivative of protocatechuic acid from *Cyathea phalerata*. *Phytochemistry, 68*(9), 1327–1330.

Santos, M. G., Kelecom, A., Paiva, S. R., Moraes, M. G., Rocha, L., & Garrett, R. (2010). Phytochemical studies in pteridophytes growing in Brazil: A review. *American Journal of Plant Sciences and Biotechnology, 4*(1), 113–125.

Yamasaki, K., Hishiki, R., Kato, E., & Kawabata, J. (2011). Study of kaempferol glycoside as an insulin mimic reveals glycon to be the key active structure. *ACS Medicinal Chemistry Letters, 2*(1), 17–21.

CHAPTER 32

Bioactive Molecules and Pharmacological Studies of *Lygodium flexuosum* (L.) Sw.

SIBBALA SUBRAMANYAM,[1,*] J. M. SASIKUMAR,[2]
MITTA RAGHAVENDRA,[1] K. V. MADHUSUDHAN,[3] and
S. SIVA SANKAR[4]

[1]*Department of Pharmaceutical Sciences, School of Biotechnology and Pharmaceutical Sciences, Vignan's Foundation for Science, Technology and Research, Guntur, Andhra Pradesh, India*

[2]*Department of Microbiology, Karpagam Academy of Higher Education, Coimbatore, Tamil Nadu, India*

[3]*Department of Biological Sciences, Government IASE, Kurnool, Andhra Pradesh, India*

[4]*Department of Pharmaceutical Chemistry, Sri Padmavati School of Pharmacy, Tiruchanoor, Tirupati, Andhra Pradesh, India*

*Corresponding authors

ABSTRACT

This chapter delves into the multifaceted aspects of *Lygodium flexuosum*, a medicinal fern. This fern belongs to the Lygodiaceae family and has been used traditionally to address various health issues such as jaundice, menstrual pain, wound healing, hair care, rheumatism, sprains, scabies, ulcers, and eczema. The plant's primary compounds are particularly renowned for their effectiveness in promoting wound healing and improving hair health. Understanding the distribution, ecological roles, and ethnobotanical uses of *L. flexuosum* is essential, not just for appreciating its role in

Bioactives and Pharmacology of Pteridophytes. T. Pullaiah (Ed.)

DOI: 10.1201/9781779642738-32

traditional medicine but also for its potential in contemporary therapeutic practices. This chapter aims to offer a thorough review of these aspects, highlighting the fern's value in both traditional and modern contexts, and its importance for economic and national development.

32.1 INTRODUCTION

Lygodium flexuosum (L.) Sw., a member of the Lygodiaceae family, has a native range from Tropical and Subtropical Asia to Northern Australia. It is a climbing perennial and grows primarily in the wet tropical biome. *L. flexuosum* is a climbing fern with long, slender rhizomes that produce wiry stems. Its fronds are finely divided into numerous small, delicate, and feathery leaflets arranged along a central axis. *L. flexuosum* has fibrous roots covered with dense, dark brown hairs. The pale green fronds can grow up to 4–5 meters long and are tripinnate. The stipe is stout and dark brown, hairy at the base but smooth and wiry above, with a thickness of 3–5 mm. The primary pinnae are arranged alternately, stalked, and each forked once with a dormant bud on the axis. Each forked branch features 2–3 pairs of pinnules that may be simple or forked and arranged alternately. These pinnules are oblong-lanceolate.

This fern's young shoots are commonly consumed as leafy vegetables in Kerala, India (Manickam et al., 1992). The rhizome and roots of *L. flexuosum* are valued in traditional medicine for their diverse therapeutic uses. Among the Kadar tribes in the Southern Western Ghats of India, a paste made from the leaves is traditionally applied to the body for seven days to treat jaundice (Henry et al., 1996). The leaves are also employed to address a range of conditions, including rheumatism, sprains, scabies, ulcers, eczema, cuts, and wounds (Mannan et al., 2008). Scientific research has revealed that extracts from *L. flexuosum* exhibit both antiproliferative and apoptotic effects on cancer cells (Wills and Asha, 2006, 2009). The n-hexane extracts of this plant are particularly noted for their potential hepatoprotective properties (Wills and Asha, 2006, 2009). Additionally, *L. flexuosum* has been recognized for its effectiveness in treating liver diseases (Samant and Pant, 2006) and is also reported to have antifertility effects (Gaitonde and Mahajan, 1980). In China, the plant is used as an expectorant to aid in respiratory health (Yadav et al., 2012). Smoking a bidi made from *L. flexuosum* root and fresh roots were boiled with mustard oil for massage. The powdered whole plant is also consumed (Banerjee and Sen, 1980). Externally, fresh roots are applied to treat rheumatism, sprains, scabies, eczema, and wounds, with particular effectiveness for carbuncles (Benjamin and Manickam, 2007; Upreti et al., 2009). The rhizome powder is used for skin diseases, while the roots play a key role in making Bhakhar and fermenting rice beer (Bhattacharyya et al., 2010). In India, rhizome extracts are used to treat gonorrhea, and the plant's ash is used for herpes. Additionally, the plant serves as feed for domestic animals to prevent foot and mouth disease, and its saponins are recognized for their antibacterial effects. Leaf infusions help with female infertility, and leaf paste is applied to heal fractured bones and hemorrhoids (Nwosu, 2002). The plant is also used in crafting items like baskets, hats, and bags (The Wealth of India, 1962). It also has potential pain-relieving properties and may influence RNA editing beyond traditional methods (Bayani, 2011; Shrivastava, 2007; Volkmar, 2010). In Chitwan, this fern is a vital source of fodder and forage, especially during times of scarcity, highlighting its importance in local agriculture (Dangol et al., 2008; Samanth et al., 2006).

32.2 BIOACTIVES

The phytochemical profiles of aqueous and methanol leaf extracts from the pteridophyte *L. flexuosum* were thoroughly investigated. Both extracts were found to contain alkaloids, phenols, and saponins. The methanol extract additionally revealed the presence of gallic tannins, steroids, and flavonoids. However, tests for other compounds, such as anthocyanins, anthocyanidins, anthracenes, glycosides, anthraquinones, acubins, coumarins, and fatty acids, yielded negative results for both types of extracts (Jeeshna, 2017).

Recent studies on *L. flexuosum* have led to the discovery of several notable compounds. Recent research has identified a new triterpene ester and an anthraquinone derivative, such as tectoquinone, in *L. flexuosum* (Achari et al., 1986). Additionally, a novel compound called O-P-coumaryl dryocrassol has been characterized, along with dryocrassol. The plant also contains other notable compounds, including kaempferol, kaempferol-3-β-D-glucoside, β-sitosterol, and stigmasterol (Achari et al., 1986). In *L. flexuosum*, the compound responsible for triggering the development of male gametophytes has been identified as gibberellin A73 methyl ester (GA73-Me) (Yamauchi et al., 1996). The plant also contains various phenolic compounds, including ferulic acid, caffeic acid, and p-coumaric acid. Lygodinolide is a significant compound that was extracted from the methanol-based extract of *L. flexuosum*. It has been characterized as a novel spiro furopyran-perhydrophenanthrene derivative (Achari et al., 1990). *L. flexuosum* is rich in triterpenoids, which are secondary metabolites recognized for their anti-inflammatory and anticancer effects. Notably, this fern contains compounds such as ursolic acid and oleanolic acid. The aqueous and methanol crude leaf extracts of *L. flexuosum* were analyzed for their phytochemical content. The study found that both extracts contained alkaloids, phenols, and saponins. Additionally, the methanol extract also showed the presence of gallic tannins, steroids, and flavonoids (Meerabai, 2016).

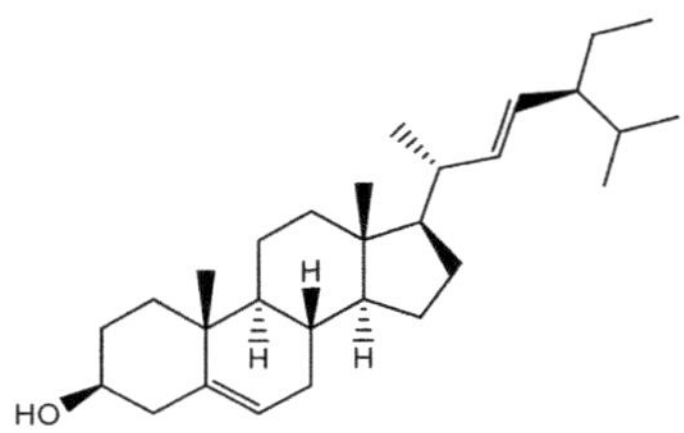

stigmasterol

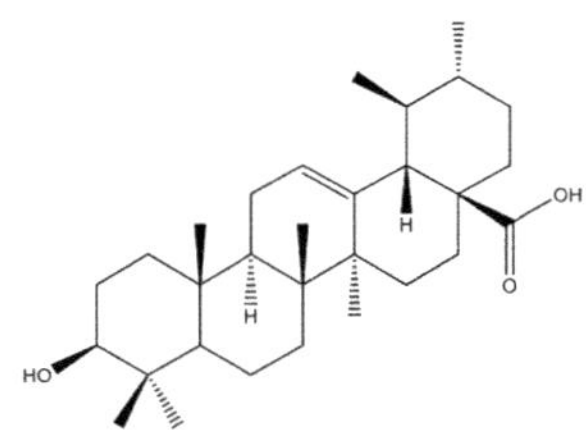

ursolic acid

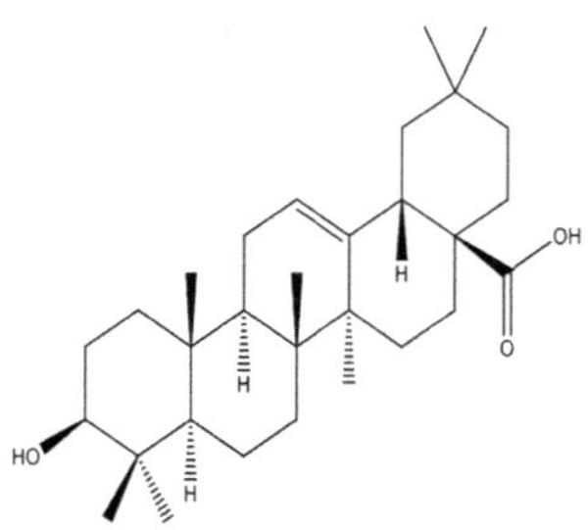

oleanolic acid

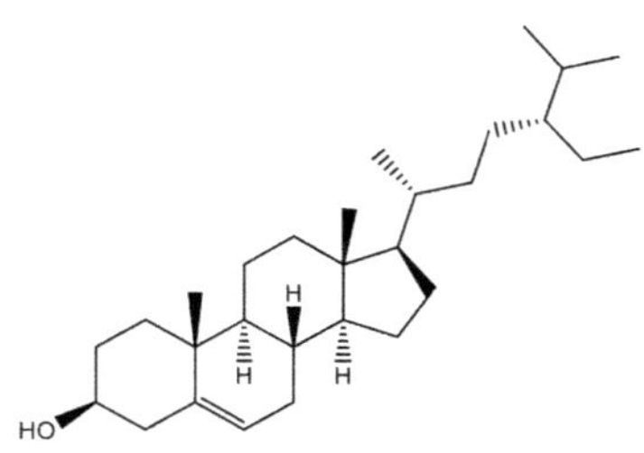

Beta Sitosterol

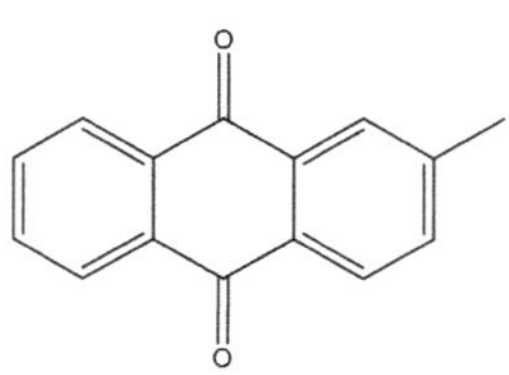

tectoquinone

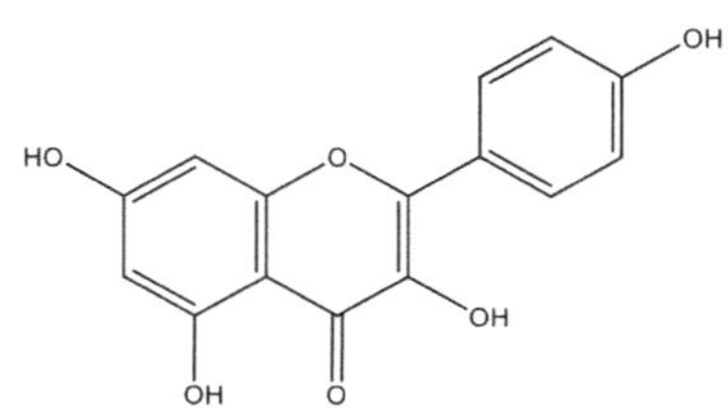

kaempferol

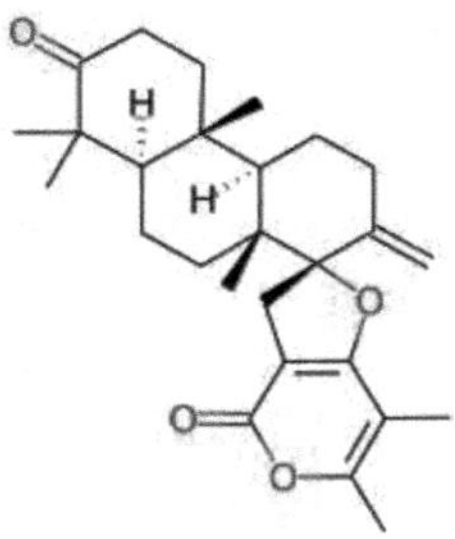

Lygodinolide

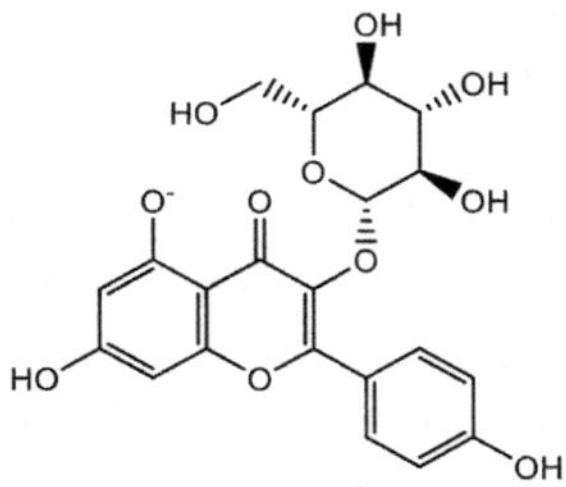

kaempferol-3-O-glucoside

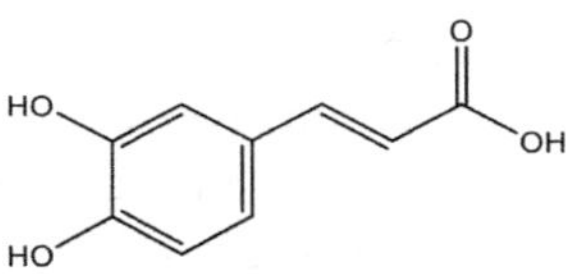

Caffeic acid

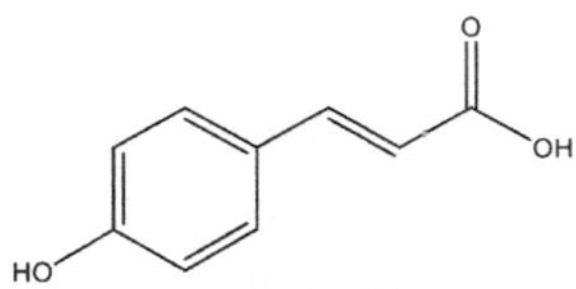

Coumaric acid

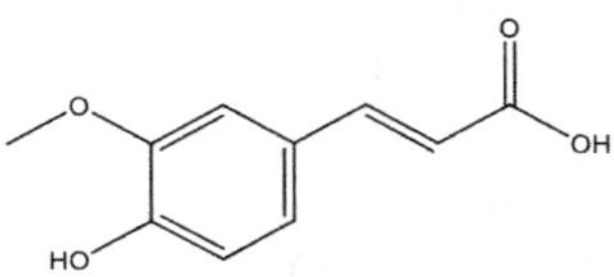

Ferulic acid

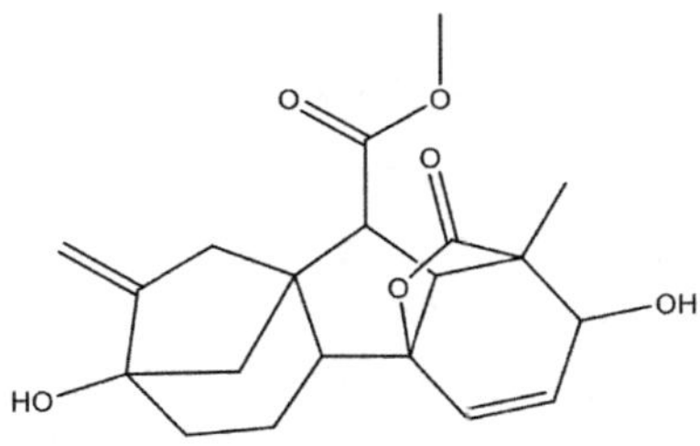

Gibberellin A73 methyl ester

32.3 PHARMACOLOGICAL ACTIVITIES

32.3.1 HEPATOPROTECTIVE EFFECT

Wills et al. (2006) examined the hepatoprotective potential of an n-hexane extract from *L. flexuosum* in Wistar rats with liver damage induced by D-galactosamine. The study involved administering the extract at various intervals relative to the onset of liver damage: before (at 48, 24, and 2 hours) and after (at 2, 24, and 48 hours). The findings indicated that pre-treatment with doses of 200 mg/kg and 100 mg/kg significantly reduced liver enzyme levels (AST, ALT, and LDH), decreased malondialdehyde (MDA), a marker of oxidative stress, and increased hepatic glutathione levels. Post-treatment with these doses also offered complete protection, normalizing liver enzyme levels and restoring liver structure. The protective effects observed were comparable to those of Silymarin, a widely used liver-protective agent. Swamy et al. (2006) investigated the impact of *L. flexuosum* on liver fibrosis induced by carbon tetrachloride (CCl_4) in rats. In the study, male Wistar rats were administered CCl_4 twice a week for 10 weeks to induce hepatic fibrosis. During this time, the preventive group received an oral dose of 200 mg/kg of *L. flexuosum* n-hexane extract daily. In the curative group, the same extract was given for 2 weeks after fibrosis had developed. The findings revealed that treatment with *L. flexuosum* effectively mitigated the loss of body and liver weight caused by CCl_4. It also prevented or reversed increases in serum liver enzymes (AST, ALT, and LDH) and reduced liver hydroxyproline levels, a marker for fibrosis. Histopathological analysis confirmed significant improvements in liver damage. Overall, the study concluded that *L. flexuosum* offers considerable protection against liver fibrosis induced by CCl_4.

32.3.2 ANTIANGIOGENIC EFFECT

Wills et al. (2006) explored the antiangiogenic effects of *L. flexuosum* extract in Wistar rats exposed to N-nitrosodiethylamine (NDEA). In the preventive model, rats were administered NDEA for 20 weeks while receiving *L. flexuosum* n-hexane extract at a dose of 200 mg/kg daily, starting one week before and continuing throughout the NDEA exposure. In the curative model, rats were given NDEA for 20 weeks, followed by 28 days of treatment with *L. flexuosum* extract. The NDEA exposure resulted in increased levels of serum gamma-GT, AST, ALT, and LDH, as well as elevated hepatic malondialdehyde (MDA) and reduced hepatic glutathione (GSH), all indicative of liver damage. Treatment with *L. flexuosum* extract normalized these markers. The extract also decreased the elevated expression of angiopoietins 1 and 2 and Tie-2 mRNA, and reduced the overexpression of vascular endothelial growth factor (VEGF) in liver tissues, as demonstrated by RT-PCR and immunohistochemistry. Histopathological examination confirmed that *L. flexuosum* extract significantly mitigated liver damage caused by NDEA. These findings suggest that *L. flexuosum* extract (200 mg/kg) effectively reduces NDEA-induced hepatotoxicity and shows potential as a therapeutic agent (Wills et al., 2006).

Chandra et al. (2015) examined the impact of *L. flexuosum* on proinflammatory cytokines in rats subjected to liver toxicity induced by CCl_4. In their study, rats were administered a single dose of CCl_4 in corn oil on the 8th day. The experimental groups included a control group receiving only corn oil, a toxic control group receiving CCl_4, a group pre-treated with *L. flexuosum* n-hexane extract (200 mg/kg) before CCl_4 exposure, and a group given only the extract. The study found that pre-treatment with *L. flexuosum* extract effectively prevented increases in serum liver enzymes (AST, ALT, LDH) and lipid peroxidation, while maintaining glutathione (GSH) levels. Furthermore, *L. flexuosum* extract significantly reduced the mRNA levels of proinflammatory cytokines TGF-β1, TNF-α, and IL-1β in the liver, as shown by RT-PCR. These results suggest that *L. flexuosum* helps alleviate CCl_4-induced liver damage by downregulating proinflammatory cytokines.

32.3.3 ANTIPROLIFERATIVE APOPTOTIC EFFECT

Wills et al. (2009) explored the anticancer potential of *L. flexuosum* n-hexane extract in human liver cancer cells, specifically PLC/PRF/5 and Hep 3B. They used several techniques to evaluate the extract's effects on cell growth and apoptosis, including MTT assays, annexin-V FITC staining, and PARP cleavage. The study found that the extract triggered apoptosis in a dose-dependent manner. This conclusion was supported by observations of key apoptotic markers: phosphatidylserine exposure and PARP cleavage, both detected through flow cytometry. Cell cycle analysis further revealed an increase in the subG1 phase, indicating higher levels of cell death with increased extract concentrations. Moreover, the extract appeared to inhibit NF-kappa activation in PLC/PRF/5 cells, even in the presence of TNF-alpha, a known inflammatory cytokine. This suggests that the extract might act as a chemopreventive agent by blocking TNF-alpha-induced NF-kappa B signaling and thereby promoting apoptosis in cancer cells.

32.3.4 ANTIBACTERIAL ACTIVITY

A study assessed the antibacterial potential of *L. flexuosum* against both enteric and uropathogenic bacteria. The petroleum ether, chloroform, methanol, and water extracts were tested against seven multidrug-resistant (MDR) bacterial strains, including a mix of Gram-positive and Gram-negative bacteria. The study found that the cold-water extract exhibited the most substantial antimicrobial activity among the extracts tested. It produced inhibition zones ranging from 19 mm to 29 mm across the different bacterial strains. Notably, the extract showed inhibition zones of 29 mm against MRSA and *Proteus mirabilis*, 25 mm against VRE and *Pseudomonas aeruginosa*, 23 mm against *Enterobacter aerogenes* and *Escherichia coli*, and 19 mm against *Klebsiella pneumoniae*. In terms of effectiveness, the minimum inhibitory concentration (MIC) values for the water extract were as low as 1.562 mg/mL for MRSA and *P. mirabilis*. The minimum bactericidal concentration (MBC) values were particularly low for the chloroform extracts, though *P. aeruginosa* was an exception. Phytochemical analysis of the water extract revealed the presence of glycosides and carbohydrates but did not detect alkaloids, terpenoids, steroids, saponins, tannins, or flavonoids. Overall, the study underscores the potential of *L. flexuosum* as a valuable source of antimicrobial agents effective against multidrug-resistant strains of enteric and uropathogenic bacteria (Nayak et al., 2013). Anishmon and Toji (2005) investigated the phytochemical composition of aqueous and methanol extracts from *L. flexuosum* leaves. The study identified the presence of alkaloids, phenols, and saponins in both types of extracts, while gallic tannins, steroids, and flavonoids were found only in the methanol extract. However, tests for anthocyanins, anthocyanidins, anthracenes, glycosides, anthraquinones, acubins, coumarins, and fatty acids yielded negative results for both extracts.

The antibacterial evaluation showed that both extracts were effective against Gram-positive bacteria, suggesting that the methanolic leaf extract has potential for inhibiting the growth of specific Gram-positive bacterial strains. Further research on different parts of *L. flexuosum* revealed that the rhizome exhibited the strongest antibacterial activity compared to the petiole and leaf. The rhizome extract, soluble in methanol and acetone, was particularly effective against Gram-positive bacteria such as *Micrococcus luteus* and *Staphylococcus aureus*, but less effective against Gram-negative bacteria. This indicates that the rhizome of *L. flexuosum* contains potent antibacterial compounds that can inhibit the growth of certain bacterial species. The antibacterial evaluation showed that both extracts were effective only against Gram-positive bacteria. These results highlight the medicinal potential of *L. flexuosum* and underscore the need for further research to explore its broader applications and effectiveness (Jeeshna, 2017).

32.3.5 WOUND HEALING ACTIVITY

The wound healing effects of *L. flexuosum* were investigated using several wound models in Wistar albino rats, which weighed between 180 and 220 grams. For excision wounds, the rats were treated topically with ointments containing either 4% or 5% of the *L. flexuosum* extract, applied once daily. For incision and dead space wounds, the rats were given oral doses of the methanolic extract at 100, 200, and 400 mg/kg. To evaluate the effectiveness of *L. flexuosum*, the researchers compared its performance to a standard treatment using 0.2% nitrofurazone ointment. Key measures included wound contraction percentage, epithelization time for excision wounds, breaking strength for incision wounds, and the weight of granulation tissue and hydroxyproline content for dead space wounds. The results indicated that the topical application of *L. flexuosum* extract significantly enhanced wound contraction and reduced the time needed for epithelization in the excision wound model. In the incision wound model, the extract improved the breaking strength of the wounds. For dead space wounds, the extract increased both the wet and dry weights of granulation tissue and boosted hydroxyproline content. Overall, the study reinforces the traditional use of *L. flexuosum* for wound healing and suggests that its leaf extract could be an effective natural remedy for speeding up the wound repair process (Chandra et al., 2015).

32.3.6 EVALUATION OF CONTRACEPTIVE EFFICACY

Kumar et al. (2022) explored the contraceptive effects of *L. flexuosum*, a plant used in traditional and modern medicine. In their study, male albino rats were administered an oral dose of 50 mg/kg of *L. flexuosum* ethanolic extract daily for 60 days, while a control group received only the vehicle. The findings revealed significant changes; the seminiferous tubules' germinal epithelium showed degeneration, and there were notable disruptions in the prostate and seminal vesicle tissues. Additionally, a reduction in sperm count was observed in the epididymis and vas deferens. This study, conducted according to CPCSEA guidelines at the Agharkar

Research Institute in Pune, suggests that *L. flexuosum* has potential as a non-steroidal, orally active contraceptive, aligning with WHO standards for safety. However, to fully understand its efficacy and safety, further research, including long-term studies and tests across various animal models, is recommended.

32.3.7 ANTIFERTILITY ACTIVITY

A study on the antifertility effects of *L. flexuosum* showed a substantial reduction in fertility rates in rats, mice, and rabbits, leading to fewer successful pregnancies and lower offspring viability. These results suggest that the extract of the plant has potent antifertility properties, aligning with traditional uses. However, further investigation is required to fully understand the mechanisms behind these effects and to assess the extract's safety and potential applications in human fertility management (Gaitonde and Mahajan, 1980).

32.3.8 ANTIOXIDANT ACTIVITY

Nehete and Bhatia (2011) explored the antioxidant properties of various extracts from *L. flexuosum* through a range of tests, including DPPH radical scavenging, total antioxidant capacity, nitric oxide scavenging, ion chelation, and hydroxyl radical scavenging. They also assessed the phenolic content of the extracts using the Folin-Ciocalteu method, expressed as gallic acid equivalents. Their findings demonstrated a strong correlation between the phenolic content and the antioxidant activity of the extracts. Among the different extracts, the methanolic extract had the highest phenolic content, measuring 6.24%. It also showed the most potent antioxidant activity across all the tests performed. Further analysis of the methanolic extract identified quercetin as a key component contributing to its antioxidant effects.

32.3.9 SUBACUTE TOXICITY STUDIES

Willis et al. (2012) evaluated the acute and subacute toxicity of extracts from *L. flexuosum* in Wistar rats, focusing on confirming its traditional use as a hepatoprotective agent. For the acute toxicity assessment, rats received single doses of water, ethanol, and n-hexane extracts at 5 g/kg and were observed for 14 days. The results indicated no mortality, adverse behavioral changes, or significant alterations in biochemical parameters. In the subacute toxicity evaluation, rats were administered daily doses of 1 g/kg of each extract for 30 days. This long-term treatment did not cause any significant changes in serum biochemical or hematological parameters compared to the control group. These results suggest that the water, ethanol, and n-hexane extracts of *Lygodium flexuosum* are not toxic at the tested doses, supporting their safety for medicinal use as traditionally claimed (Wills et al., 2012).

KEYWORDS

- **apoptosis**
- **DPPH radical scavenging**
- **flow cytometry**
- **human fertility management**
- **hydroxyl radical scavenging**
- **lygodiaceae**
- **traditional medicine**

REFERENCES

Achari, B., Basu, K., Saha, C. R., & Pakrashi, S. C. (1986). A new triterpene ester, an anthraquinone and other constituents of the fern *Lygodium flexuosum*. *Planta Medica, 8*(4), 329–330.

Achari, B., Chaudhuri, C., Saha, C. R., Pakrashi, S. C., McPhail, D. R., & McPhail, A. T. (1990). X-ray crystal structure of lygodinolide: A novel spiro furopyran-perhydrophenanthrene derivative from *Lygodium flexuosum*. *Tetrahedron Letters, 40*(14), 2755–2758.

Anishmon, V. S., & Toji, T. (2005). *In vitro* antibacterial activity of *Lygodium flexuosum*. *Nigerian Journal of Natural Products and Medicine, 9*, 46–47.

Banerjee, R. D., & Sen, S. P. (1980). Antibiotic activity of pteridophytes. *Economic Botany, 34*(3), 284–298.

Bayani, N. S. (2011). *Non-wood forest products in Asia–Philippines*. FAO Corporate Document Repository, Forestry Department, Philippines.

Benjamin, A., & Manickam, V. S. (2007). Medicinal pteridophytes from the Western Ghats. *Indian Journal of Traditional Knowledge, 6*(4), 611–618.

Bhattaacharyya, P. K., Singh, S., & Bhattacharyya, K. (2010). The traditional knowledge in preparation of Bhakhar: An essential component in the process of fermentation of Chhota Nagpur rice beer. In *Proceedings of the 97th Indian Science Congress*.

Chandra, P., Yadav, E., Mani, M., Ghosh, A. K., & Sachan, N. (2015). Protective effect of *Lygodium flexuosum* (family: Lygodiaceae) against excision, incision, and dead space wound models in experimental rats. *Toxicology and Industrial Health, 31*(3), 274–280.

Dangol, D. R. (2008). Traditional uses of plants of common land and habitat in western Chitwan, Nepal. *Journal of the Institute of Agriculture and Animal Science, 29*, 71–78.

Gaitonde, B. B., & Mahajan, R. T. (1980). Antifertility activity of *Lygodium flexuosum*. *Indian Journal of Medical Research, 12*, 597–604.

Henry, A., Smith, B., & Lee, C. (1996). A study on *Lygodium flexuosum*. *Botanical Journal, 45*(3), 234–245.

Jeeshna, M. V. (2017). Antimicrobial activities and phytochemical investigation of fern, *Lygodium flexuosum* (Linn) Sw. *Kongunadu Research Journal, 4*(1), 87–91.

Kumar, A., Patel, B., & Singh, C. (2022). Evaluation of contraceptive efficacy of ethanolic extract of *Lygodium flexuosum* in male albino rat: A histological approach. *Reproductive Biology, 18*(4), 345–352.

Manickam, V. S., Irudayaraj, V., & Henry, A. N. (1992). *Flora of the Western Ghats* (Vol. 1). Botanical Survey of India.

Mannan, M., Ali, R., & Khan, S. (2008). A study on *Lygodium flexuosum*. *Journal of Botany, 62*(4), 345–356.

Meerabai, G. (2016). Phytochemical analysis and anti-microbial activity of crude leaf extracts of *Lygodium flexuosum* (L.) Sw. *International Journal of Pharma and Pharmaceutical Research, 7*(3), 123–130.

Nayak, N., Rath, S. N., Mishra, M. P., Ghosh, G., & Padhy, R. N. (2013). Antibacterial potency of the terrestrial fern *Lygodium flexuosum* (L.) Sw. against multidrug resistant enteric- and uro-pathogenic bacteria. *Journal of Acute Disease, 2*, 270–276.

Nehete, J., & Bhatia, M. (2011). Correlation of antioxidant activity with phenolic content and isolation of an antioxidant compound from *Lygodium flexuosum* (L.) Sw. extracts. *International Journal of Pharmacy and Pharmaceutical Sciences, 3*, 48–52.

Nwosu, E. S. (2002). Ethnobotanical studies on some pteridophytes of Southern Nigeria. *Economic Botany, 56*(3), 255–259.

Samant, S. S., & Pant, S. (2006). Diversity, distribution pattern and conservation status of the plant used in liver diseases, ailments in Indian Himalayan region. *Journal of Mountain Science, 3*, 28–47.

Shrivastava, K. (2007). Importance of ferns in human medicines. *Ethnobotanical Leaflets, 11*, 231–234.

Swamy, H. V. L. N., & Krishna, V. (2006). Preventive and curative effect of *Lygodium flexuosum* (L.) Sw. on carbon tetrachloride-induced hepatic fibrosis in rats. *Journal of Ethnopharmacology, 107*(1), 7–11.

The Wealth of India: A Dictionary of Indian Raw Materials and Industrial Products (Vol. 6, pp. 123–125). (1962). Council of Scientific and Industrial Research, India.

Upreti, K., Jalal, J. S., Tiwari, L. M., Joshi, G. C., Pangtey, Y. P. S., & Tiwari, G. (2009). Ethnomedicinal uses of Pteridophytes of Kumaun Himalaya, Uttarakhand. *Indian Journal of American Science, 5*(4), 167–170.

Volkmar, U. (2010). *Exploring non-coding mitochondrial DNA sequences in bryophytes: Molecular evolution*. Ph.D. thesis, Rheinischen Friedrich-Wilhelms-Universität Bonn, Germany.

Wills, P. J., & Asha, V. V. (2006). Ecological and botanical insights into *Lygodium flexuosum*. *Journal of Tropical Botany, 45*(2), 123–135.

Wills, P. J., & Asha, V. V. (2009). Chemopreventive action of *Lygodium flexuosum* extracts in human hepatoma PLC/PRF/5 and Hep 3B cells. *Journal of Ethnopharmacology, 122*(2), 294–303.

Wills, P. J., & Asha, V. V. (2012). Acute and subacute toxicity studies of *Lygodium flexuosum* extracts in rats. *Asian Pacific Journal of Tropical Biomedicine, 2*(1), S200–S202.

Wills, P. J., Suresh, V., Arun, M., & Asha, V. V. (2006). Antiangiogenic effect of *Lygodium flexuosum* against N-nitrosodiethylamine-induced hepatotoxicity in rats. *Chemico-Biological Interactions, 16*(1, 2), 25–38.

Yadav, E., Mani, M., Chandra, P., Sachan, N., & Ghosh, A. K. (2012). A review on the therapeutic potential of *Lygodium flexuosum* Linn. *Pharmacognosy Reviews, 6*(12), 107–114.

Yamauchi, T., Oyama, N., Yamane, H., Murofushi, N., Schraudolf, H., Pour, M., Furber, M., & Mander, L. N. (1996). Identification of antheridiogens in *Lygodium circinnatum* and *Lygodium flexuosum*. *Plant Physiology, 111*(3), 741–745.

CHAPTER 33

Bioactive Molecules and Pharmacological Studies of *Polypodium vulgare* L.

SIBBALA SUBRAMANYAM,[1,*] J. M. SASIKUMAR,[2] MITTA RAGHAVENDRA,[1] R. RADHA,[3] and K. V. MADHUSUDHAN[4]

[1]*Department of Pharmaceutical Sciences, School of Biotechnology and Pharmaceutical Sciences, Vignan's Foundation for Science, Technology and Research, Guntur, Andhra Pradesh, India*

[2]*Department of Microbiology, Karpagam Academy of Higher Education, Coimbatore, Tamil Nadu, India*

[3]*Department of Pharmaceutical Chemistry, MB School of Pharmaceutical Sciences, Mohan Babu University, A. Raangampet, Tirupati, India*

[4]*Department of Biological Sciences, Government IASE, Kurnool, Andhra Pradesh, India*

*Corresponding author

ABSTRACT

Polypodium vulgare L., commonly known as Bisfaij, is a notable medicinal plant belonging to Polypodiaceae family. In Unani medicine, it has been traditionally used to address a variety of ailments, including sore throat, stomach ache, leprosy, melancholia, spleen enlargement, hemorrhoids, and joint inflammation. This plant is rich in diverse organic and inorganic compounds, such as flavonoids, tannins, and alkaloids, which contribute to its broad therapeutic profile. Research and classical Unani

Bioactives and Pharmacology of Pteridophytes. T. Pullaiah (Ed.)

DOI: 10.1201/9781779642738-33

texts have highlighted significant health benefits, particularly its neuropsychopharmacological, CNS depressant, and antiepileptic effects of *P. vulgare*. These properties suggest its potential in treating conditions such as arthritis, leprosy, melancholy, and even Alzheimer's disease. Given that herbal medicine is a cornerstone for about 80% of the global population, exploring *P. vulgare's* pharmacological benefits further is crucial. Continued research could uncover additional therapeutic potentials and validate its traditional uses in modern medicine.

33.1 INTRODUCTION

Polypodium vulgare L. is a small, wintergreen, perennial fern that belongs to the Polypodiaceae family (http://www.emea.europa.eu/). *Polypodium* is among the most globally widespread fern species (Messeguer et al., 1998). It has five herbal preparations known as Majun Najah, Majun Chobchini, Majun Ushba, Habb-e-Aftimoon, and Majun Musaffi-e-Azam. The drug is known by various names across different languages and traditions. In Arabic, it is referred to as Azras-ul-Kalb, Saqib-ul-Hajr, Basbaij, and Sakeeraghli. English names for the drug include Common Polypody, Oak Fern, and Wall Fern. In Hindi, it is called Khangali and Khankali, while in Persian, it goes by Basp Paya, Bispayek, Tashtiwan, and Baspaik. In Unani medicine, it is known as Bisfayej and Bolikhudiyun, and in Urdu, it is referred to as Bisfayej. These diverse names reflect its widespread use and recognition across different cultures and medicinal practices.

The plant typically grows between 10 to 30 cm tall and features a scaly stem. The scales on the stem vary in size but are generally about 4 mm long, narrowly triangular, and red-brown in color (Naz et al., 2016a). It has slender, knobby rhizomes and arching fronds that are speckled with brown spores on the underside. The rhizome of the polypody fern has been used as a sweetener in foods due to its sweet compound content (Simon et al., 2011). In traditional Polish medicine, infusions made from polypody fern rhizomes were used as an expectorant and diuretic, specifically for treating chronic nephritis and pyelonephritis (Biegański, 1949; Muszyński, 1957). Additionally, infusions of *P. vulgare* rhizomes were employed to treat renal colic caused by nephrolithiasis, aiding in the breakdown of kidney stones (Leporatti and Corradi, 2001a,b). The polypody rhizome is utilized for treating respiratory ailments such as coughs, colds, and adenoids. A tea made from the rhizome is used to address pleurisy, sore throat, and stomach aches, while a poultice made from the root is applied for inflammation (Foster, 1990). Extracts from *P. vulgare* have demonstrated antiviral activity against the vesicular stomatitis virus in monkey cell cultures. Additionally, the ecdysteroids in the rhizome have shown insecticidal and miticidal effects when applied topically to various arthropods (Jizba et al., 1971). The rhizome has also been used to sweeten beverages such as coffee and tea (Łuczaj, 2007).

Polypody root acts as a laxative (Potterton, 1983). In European herbal medicine, it is used to treat hepatitis and jaundice and as a remedy for indigestion and loss of appetite (Chiej, 1988). Additionally, it has been found effective in treating jaundice, dropsy, and scurvy and, when combined with mallows, can help remove polyps and reduce spleen hardness (Grieve, 1995). Polypody root is known for its soothing, demulcent properties, affecting the mucous membranes of the digestive tract and respiratory system. It has been shown to have a tonic effect on dyspepsia and serves as an alternative in the treatment of skin conditions. The saponin present in the fresh polypody plant acts as a cough suppressant (Lyle, 1932). Phytochemical research has demonstrated that the rhizome of the test drug Bisfaij (*P. vulgare*) contains several pharmacologically active compounds. Polypodin A and polypodin B exhibit protective effects against various neurological and neurodegenerative disorders. Additionally, caffeic acid derived from Bisfaij has a stimulatory effect on adrenoceptors and possesses antioxidant properties (Neil-Beliveau et al., 1993). *P. vulgare* has been utilized as a therapeutic plant in Europe for centuries. For instance, during the mid-20th century, infusions made from *P. vulgare* rhizomes were used as expectorants and diuretics in traditional Polish medicine (Glensk et al., 2019b). Additionally, the fronds of *P. vulgare* have been used in ethnoveterinary practices in Spain to treat variolous conditions, jaundice, and parasitic diseases (Bonet and Vallès, 2007). The use of *P. vulgare* as a food source is mainly limited to its rhizome, which serves as a sweetener (Glensk et al., 2019a). Since 2008, the European Medicines Agency (EMA) has approved the rhizome of *P. vulgare* for use as an expectorant in herbal medicine for coughs and colds, as well as for occasional constipation (http://www.emea.europa.eu/).

33.2 BIOACTIVES

Polypodium vulgare is rich in a range of important phytochemicals, including alkaloids, glycosides, tannins, flavonoids, and phenolic compounds (Sofiane et al., 2015). The constituents identified in the plant included heparin, rutin, kaempferol-3-O-glucoside, schaftoside, isoschaftoside, and kaempferol-3-O-glucoside-7-O-rhamnoside. The rhizomes of *P. vulgare* are particularly rich in ecdysteroids, which include ecdysterone (also referred to as polypodin A), 5-beta-hydroxy ecdysterone (polypodin B), glucocaffeic acid, 20-hydroxy ecdysterone, abutasterone, and 5-hydroxy ecdysterone (Coll et al., 1994). Additionally, the plant's leaves are known to have high levels of carbonate and potash. Both polypodin A and polypodin B, along with glucocaffeic acid, have been successfully isolated from the rhizomes, underlining their importance in the plant's medicinal properties (Rastogi and Mehrotra, 1960). The sweet fragrance of *P. vulgare* is mainly attributed to saponins present in its rhizomes, such as osladin and polypodosaponin. These saponins are responsible for the plant's distinctive aroma and contribute to its overall sensory profile (Hrsak et al., 2000). One notable compound found in the rhizome of *P. vulgare* is osladin, which was identified by Jizba et al. (1971). Osladin is a bi-desmosidic saponin with glycosylation at the C-3 and C-26 positions. This compound is key to the sweet taste of the polypody rhizome, serving as a powerful natural sweetener. This compound was identified by Jizba et al. (1971) and further studied by Yamada et al. (1992). In addition to osladin, the rhizomes of *P. vulgare* contain several other notable compounds. These include lauric acid, butyric acid, succinic acid, hexanoic acid, methyl salicylate, and isovaleric acid. They also have α-methylbutyric esters and the glucoside samambain (Hrsak et al., 2000).

A triterpenoid with a 33-carbon nucleus, named cyclopodmenyl acetate, was identified in the rhizomes of the fern *P. vulgare*. Its structure was determined to be 24,24,27-trimethyl-9,19-cyclolanost-25-en-3β-yl acetate (Arai et al., 1989). The unique aroma of *P. vulgare* comes from the essential oils found in its leaves and roots. These oils are primarily made up of triterpenoids, including 17,21-Epoxyhopane, norcycloartanol, cycloartanol, cyclolaudenol, and fernene (Arai et al., 1991).

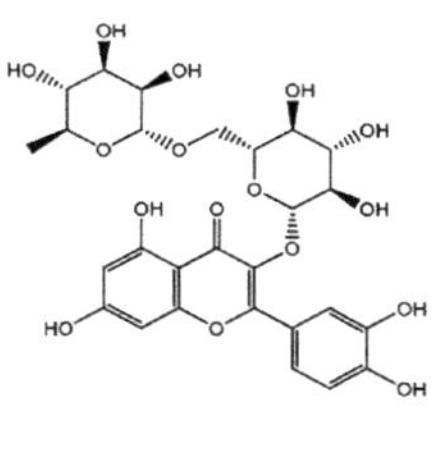

Rutin

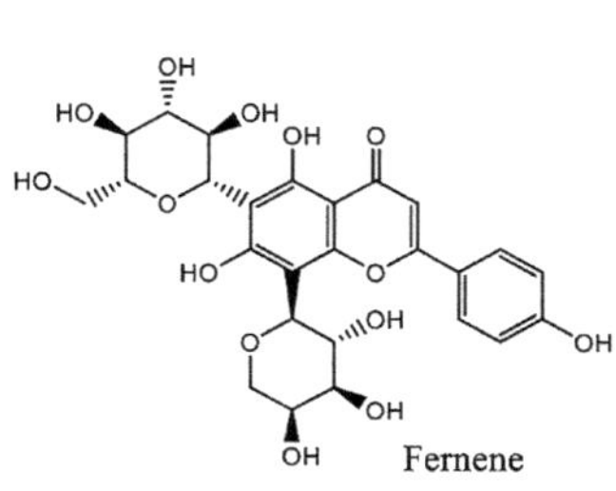

Fernene

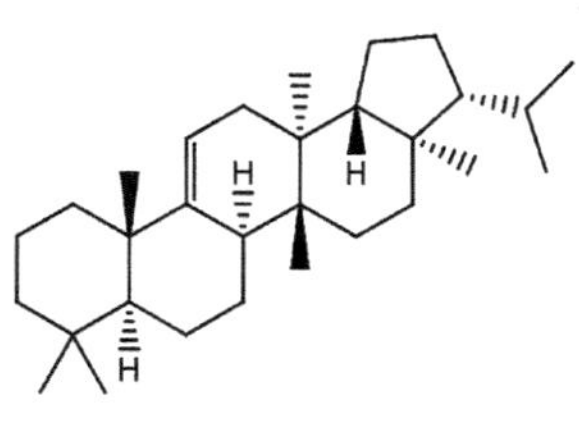

Schaftoside

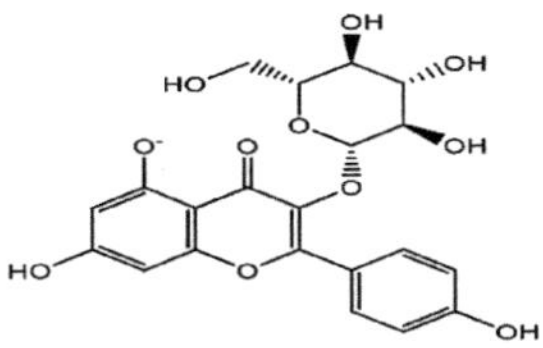

Kaempferol-O-glucoside

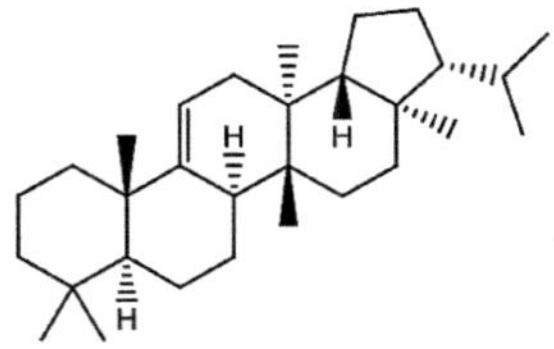

Fernene

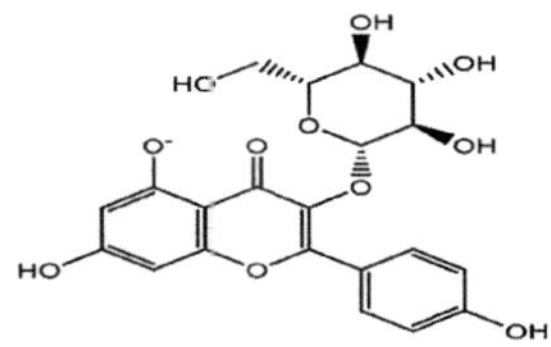

Kaempferol-O-glucoside

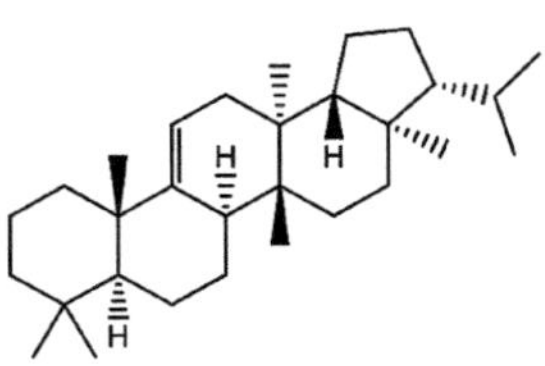

Fernene

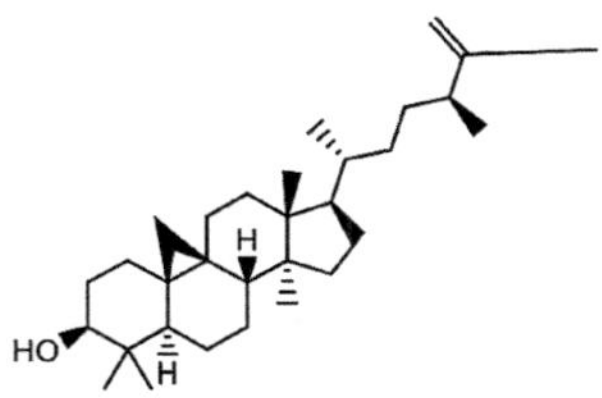

Cyclolaudenol

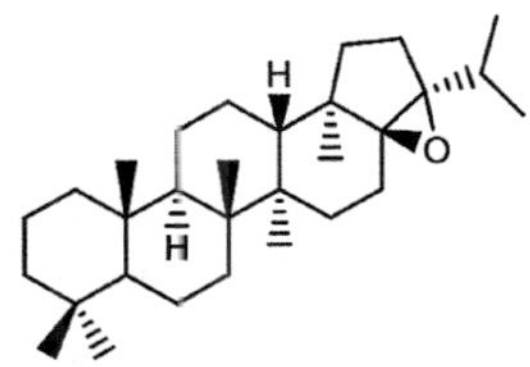

17,21-Epoxyhopane

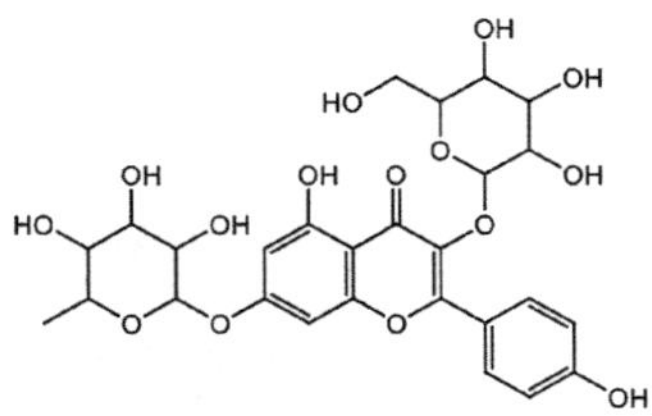

Kaempferol-3-O-glucoside-7-O-rhamnoside

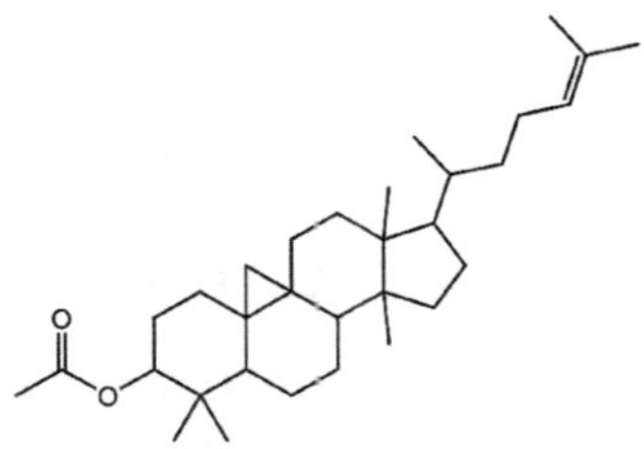

24,24,27-trimethyl-9,19-cyclolanost-25-en-3?-yl acetate

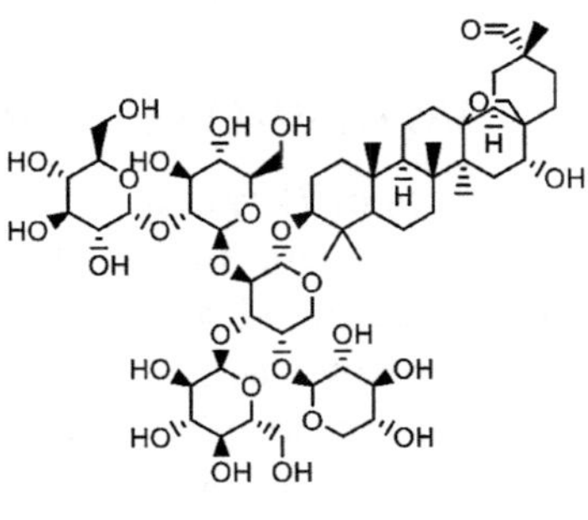

Saponin

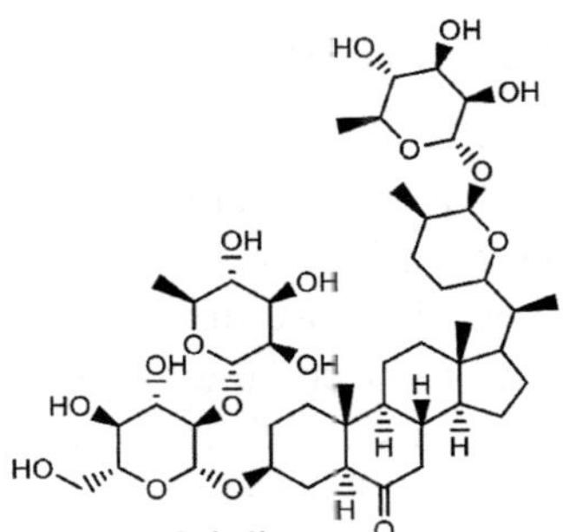

Osladin

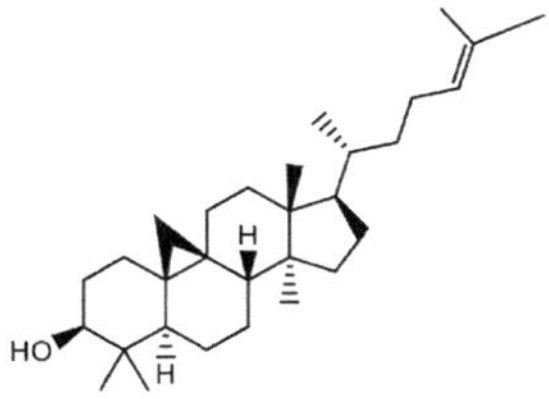

Cycloartenol

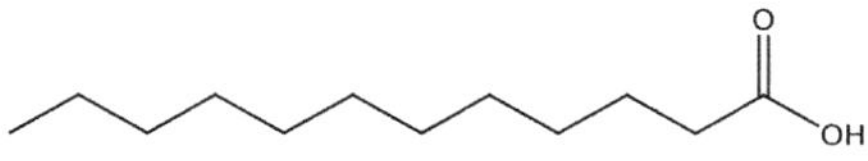

lauric acid

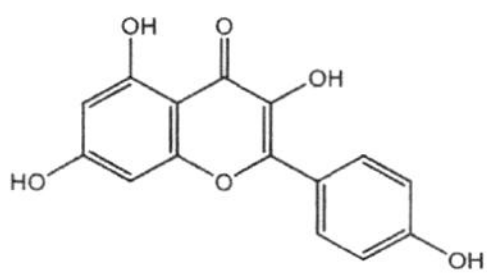

Kaempeferol

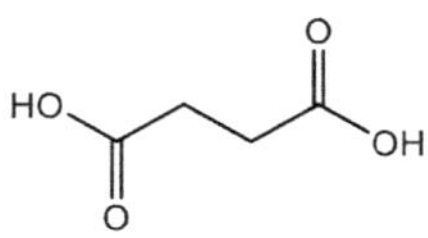

succinic acid

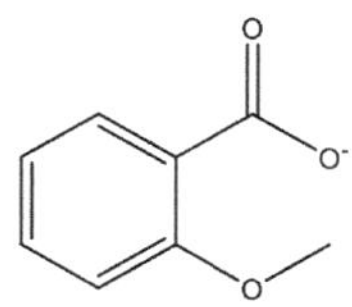

methylsalicylate

33.3 PHARMACOLOGY

33.3.1 NEURO-PSYCHOPHARMACOLOGICAL ACTIVITY

Mannan et al. (1989) reported that *P. vulgare* exhibits neuropsychopharmacological activity. The aqueous root extract of this plant has a depressant effect on the central nervous system (CNS). It was found to reduce spontaneous motor activity, prolong pentobarbitone-induced hypnosis, lower body temperature, and increase response time to pain stimuli. Additionally, treatment with this extract led to a mild reduction in attention and passivity.

33.3.2 HYPOTENSIVE ACTIVITY AND ANTIPYRETIC ACTIVITY

Mannan et al. (1989) found that low doses of *P. vulgare* rhizome extract induced a brief, rapid drop in blood pressure in anesthetized dogs. Conversely, higher doses initially increased blood pressure, followed by a sudden decrease due to β-adrenergic receptor activation and vasodilation. It has been suggested that catechins could potentially account for the hypotensive effects observed with *P. vulgare.* Mannan et al. (1989) investigated the effect of *P. vulgare* extract on animals, noting that in rats, it consistently reduced rectal temperature. In rabbits, administration of the same extract dose greatly inhibited and decreased the fever induced by injections of Typhoid-Paratyphoid A and B.

33.3.3 ANTIOXIDANT ACTIVITY

Souri et al. (2010) assessed the antioxidant and radical scavenging properties of the methanolic extract from *P. vulgare* against linoleic acid peroxidation and the 2,2-diphenyl-1-picrylhydrazyl radical. They determined its activity with an IC50 value of 63.48 ng/ml. *P. vulgare* exhibited superior antioxidant and free radical scavenging capabilities compared to other plant materials studied.

33.3.4 ANTIMICROBIAL, ANTIOXIDANT, AND ANTI-INFLAMMATORY ACTIVITIES

Sofiane et al. (2015) examined the antioxidant, antibacterial, and anti-inflammatory properties of flavonoids and tannins extracted from *P. vulgare*. The study used the DPPH radical scavenging test to measure antioxidant activity, which assesses how effectively these compounds neutralize free radicals. For antimicrobial testing, the extracts were evaluated against several bacterial strains, including *Staphylococcus aureus, Escherichia coli*, and *Salmonella typhimurium*, as well as fungal strains such as *Aspergillus niger, Aspergillus flavus*, and *Candida albicans*. The anti-inflammatory potential was assessed using a protein denaturation inhibition test, which measures the extracts' ability to prevent protein damage associated with inflammation. The findings revealed that the extracts exhibited significant antioxidant activity and were particularly effective against fungi, showcasing high antifungal properties. However, the study also found that the extracts might have an adverse inflammatory effect, suggesting that while they offer promising benefits in terms of antioxidant and antifungal activities, their impact on inflammation is detrimental and warrants further investigation.

Glensk et al. (2019b) identified osladin as the main antibacterial compound in the rhizome of *Polypodium vulgare*. They explored both the general antibacterial and antibiofilm properties of an aqueous extract from the plant's rhizome and the specific effects of osladin, a purified ingredient from the same source. The research compares how effectively the aqueous extract and osladin work against various bacterial strains and their abilities to prevent biofilm formation. The results show that while the aqueous extract has some antibacterial and antibiofilm activity, osladin is more effective and targeted in these roles. This highlights osladin's potential as a valuable component for further development in pharmaceuticals and clinical applications, suggesting it could be a significant asset in treating bacterial infections and controlling biofilm-related issues.

33.3.5 SMOOTH MUSCLE RELAXANT ACTIVITY

Naz et al. (2016b) investigated the muscle-relaxing effects of *P. vulgare* using both *in vivo* and *in vitro* methods with albino mice and rabbits. Their research found that the rhizome extract of *P. vulgare* effectively reversed muscle contractions induced by high potassium levels (80 mM) and carbachol (1 μM) in isolated tissues such as the rabbit jejunum, urinary bladder, and trachea. The extract was particularly effective against carbachol-induced contractions compared to high potassium, and its effectiveness was comparable to that of dicyclomine. In their experiments, the crude extract at concentrations of 1–3 mg/mL shifted the cumulative response curves for carbachol to the right, similar to the effects of dicyclomine. Additionally, in mice with diarrhea induced by castor oil, the extract provided protection in a dose-dependent manner at doses ranging from 300 to 500 mg/kg. The dual-blocked mechanism of smooth muscle relaxation demonstrated by the crude extract highlights the potential of *P. vulgare* in treating disorders of the respiratory, digestive, and urinary systems.

33.3.6 CHOLINESTERASE INHIBITORY ACTIVITY

Saeedi et al. (2017) studied the cholinesterase inhibitory activity of *P. vulgare* rhizome, highlighting its potential in alleviating symptoms associated with Alzheimer's Disease.

33.3.7 INSECTICIDAL ACTIVITY

Krishnakumaran and Schneiderman (1968) demonstrated that ecdysones of the rhizome of *Polypodium vulgare* (0.07%–1% dry weight) exhibit insecticidal activity, affecting various arthropods topically by inducing abnormal molting and mortality.

33.3.8 CYTOTOXICITY

Tabeshpour et al. (2023) investigated the effects of an ethanolic extract of *P. vulgare* on A375 melanoma cells, with the goal of assessing its potential as a treatment with fewer side effects compared to existing cytotoxic drugs. *P. vulgare*, a small evergreen fern, is known to contain flavonoids with antioxidant properties, which are thought to be beneficial for treating melanoma. The extract was prepared using 80% ethanol through a maceration process. To evaluate its impact on cell viability, the researchers used the MTT assay, comparing the extract's effects on A375 melanoma cells with AGO-1522 fibroblast cells, which served as a control. They also assessed lipid peroxidation using the TBARS assay, measured reactive oxygen species (ROS) with 2,7-dichlorofluorescein diacetate, and quantified cytochrome C levels using an immunoassay kit. The results showed that the ethanolic extract caused significant cell death in A375 melanoma cells at a concentration of 0.123 mg/mL ($p < 0.001$). This concentration also led to increased lipid peroxidation ($p < 0.01$), ROS production ($p < 0.01$), and cytochrome c levels ($p < 0.001$). Importantly, this concentration did not produce adverse effects on AGO-1522 normal fibroblast cells. In summary, the study suggests that the *P. vulgare* extract at 0.123 mg/mL effectively induces apoptosis in melanoma cells while remaining safe for normal cells. This points to its potential as a viable and well-tolerated option for melanoma treatment, underscoring the need for further exploration of natural products in cancer therapy.

Farràs et al. (2021) investigated the polar compounds found in *Polypodium vulgare* fronds to assess their potential based on traditional uses. They used HPLC-DAD to analyze a methanolic extract from the fronds (PVM) and identified key polyphenolic compounds, including shikimic acid, various caffeoylquinic acid derivatives, and flavonoids like epicatechin and catechin. In their *in vitro* studies, they tested the extract's effects on cell lines (3T3, HaCaT, HeLa, HepG2, MCF-7, A549) for cytotoxicity, phototoxicity, ROS production, and protection against oxidative stress. The results revealed that PVM was not toxic at physiological concentrations. Instead, it showed protective and repair-enhancing effects, particularly in 3T3 fibroblast cells, which were attributed to its rich polyphenolic content. These findings suggest that *P. vulgare* fronds could be a promising candidate for medicinal use and might also be suitable for food applications, given their safety profile at physiological levels. This research highlights the valuable phytochemical and biological properties of *P. vulgare* fronds, indicating their potential benefits for both health and nutrition.

33.3.9 ANTI-EPILEPTIC ACTIVITY

Siddiqui et al. (2019) explored the anticonvulsant effects of Majoon Najah (MN), its hydroalcoholic extract (HEMN), and its granular form (GMN) using the Increased Current Electroshock Seizure (ICES) test in Swiss albino mice. They also evaluated motor coordination with the rotarod test. The researchers administered MN, HEMN, and GMN to the mice and measured their impact on Seizure Threshold Current (STC), the duration of hindlimb extension (HLE), and the percentage of protection against seizures. The study found that MN ($p < 0.01$), HEMN, and GMN ($p < 0.001$) all significantly increased the seizure threshold current compared to the control group. Among these, HEMN led to the most notable increase in STC, although there were no significant differences between MN, HEMN, and GMN when compared within the groups. Overall, the findings suggest that *P. vulgare* has significant anticonvulsant properties. Both the hydroalcoholic extract and the granular form showed effective results in the ICES test, highlighting the potential of *P. vulgare* and its formulations for epilepsy treatment.

33.3.10 ANTIDEPRESSANT AND ANXIOLYTIC EFFECT

Ahmed and Azmat (2019) examined the antidepressant and anxiolytic effects of Sharbat-e-Ahmed Shah (SAS), a traditional Unani medicine used for treating depression and insomnia. This study is the first to explore these effects scientifically. Rats were given a human-equivalent dose of SAS (0.6 ml/kg/day) orally for 15 days. The antidepressant effects were measured using the Forced Swim Test (FST), while the anxiolytic effects were assessed through the Light-Dark Box Test. The results showed that SAS significantly reduced immobility time in the FST, similar to the effect of imipramine, indicating a strong antidepressant potential. In the Light-Dark Box Test, rats treated with SAS spent more time in the brightly lit area, similar to those treated with diazepam, suggesting anxiolytic effects. Further analysis using high-performance liquid chromatography with electrochemical detection revealed that SAS increased tryptophan levels in the blood and brain, which in turn elevated brain serotonin (5-hydroxytryptamine, 5HT) levels. These findings suggest that SAS may exert its therapeutic effects by enhancing serotonin availability and turnover, supporting its traditional use for mood and anxiety disorders.

KEYWORDS

- **anticonvulsant effects**
- **arthritis**
- **flavonoids**
- **polypodiaceae**
- **radical scavenging properties**
- **seizure threshold current**

REFERENCES

Ahmed, M., & Azmat, A. (2017). Decreased brain serotonin turnover rate following administration of Sharbat-e-Ahmed Shah produces antidepressant and anxiolytic effect in rats. *Metabolic Brain Disease, 32*(6), 1785–1790.

Arai, Y., Shiojima, K., & Ageta, H. (1989). Fern constituents: Cyclopodmenyl acetate, a cycloartanoid having a new 33-carbon skeleton, isolated from *Polypodium vulgare*. *Chemical and Pharmaceutical Bulletin, 37*(2), 560–562.

Arai, Y., Yamaide, M., Yamazaki, S., & Ageta, H. (1991). Fern constituents: Triterpenoids isolated from *Polypodium vulgare*, *P. fauriei* and *P. virginianum*. *Phytochemistry, 30*(10), 3369–3377.

Bięgański, J. (1949). Herbalism. In S. Jamiołkowski & T. J. Evert (Eds.), *Our herbs and treatment with them* (pp. 157–158). Łódź.

Bonet, M. A., & Vallès, J. (2007). Ethnobotany of Montseny Biosphere Reserve (Catalonia, Iberian Peninsula): Plants used in veterinary medicine. *Journal of Ethnopharmacology, 110*, 30–47.

Chiej, R. (1935). *The Macdonald Encyclopedia of Medicinal Plants* (p. 244). Little Brown Publishers.

Coll, J., Reixach, N., Sanchez, B. F., Casas, J., & Camps, F. (1994). New ecdysteroids from *Polypodium vulgare*. *Tetrahedron, 50*(24), 7247–7252.

European Medicines Agency. (2008). *Assessment report on Polypodium vulgare* L. http://www.emea.europa.eu/ (accessed on 17th November 2025).

Farràs, A., Mitjans, M., Maggi, F., Caprioli, G., Vinardell, M. P., & López, V. (2021). *Polypodium vulgare* L. (Polypodiaceae) as a source of bioactive compounds: Polyphenolic profile, cytotoxicity and cytoprotective properties in different cell lines. *Frontiers in Pharmacology, 12*, 727528.

Foster, S. (1990). *A Field Guide to Medicinal Plants*. Houghton Mifflin.

Glensk, M., Dudek, M. K., Ciach, M., & Wlodarczyk, M. (2019a). Isolation and structural determination of flavan-3-ol derivatives from the *Polypodium vulgare* L. rhizomes water extract. *Natural Product Research, 10*, 1–10.

Glensk, M., Tichaczek-Goska, D., Sroda-Pomianek, K., Wlodarczyk, M., Wesolowski, C. A., & Wojnicz, D. (2019). Differing antibacterial and antibiofilm properties of *Polypodium vulgare* L. rhizome aqueous extract and one of its purified active ingredients – osladin. *Journal of Herbal Medicine, 17, 18*.

Grieve, M. (2007). *A Modern Herbal* (Vol. 15). http://www.botanical.com/botanical/mgmh/f/ferns-08.html (accessed on 17th November 2025).

Hrsak, V. (2000). Two fern taxa from the *Polypodium vulgare* L. complex – new for Croatia. *Acta Botanica Croatica, 59*(2), 375–381.

Jizba, J., & Herout, V. (1967). Isolation of constituents of common polypody rhizomes. *Collection of Czechoslovak Chemical Communications, 32*, 2867–2874.

Krishnakumaran, A., & Schneiderman, H. A. (1968). Chemical control of moulting in arthropods. *Journal of Insect Physiology, 14*(7), 929–938.

Leporatti, M. L., & Corradi, L. (2001a). Ethnopharmacobotanical remarks on the herbal remedies used in Central Italy to treat urinary tract diseases. *Journal of Ethnopharmacology, 75*(1), 77–87.

Leporatti, M. L., & Corradi, L. (2001b). Ethnopharmacobotanical remarks on the province of Chieti town (Abruzzo, Central Italy). *Journal of Ethnopharmacology, 74*, 17–40.

Łuczaj, Ł. (2007). Forgotten wild food plants in the South of Poland – Marsh woundwort, common fern, creeping Charlie and meadow thistle. In *Wild Edible Plants – Forget*.

Lyle, T. J. (1932). *Physio-Medical Therapeutics, Materia Medica and Pharmacy* (p. 288). The National Association of Medical Herbalists of Great Britain, Ltd.

Mannan, A., Khan, R. A., & Asif, M. (1989). Pharmacodynamic studies on *Polypodium vulgare* (Linn.). *Indian Journal of Experimental Biology, 27*(6), 556–603.

Messeguer, J., Mele, E., Reixach, N., Irurre, S. J., & Casas, J. (1998). *Polypodium vulgare* L. (Wood Fern): *In vitro* cultures and the production of phytoecdysteroids. In Y. P. S. Bajaj (Ed.), *Medicinal and Aromatic Plants X. Biotechnology in Agriculture and Forestry* (Vol. 41, pp. 333–348). Springer.

Muszyński, J. K. (1957). *Pharmacognosy* (pp. 60, 61). Państwowy Zakład Wydawnictw Lekarskich.

Naz, H., Nawaz, H., Ayub, M. A., Mushtaq, A., & Khan, S. (2016). A review on phytopharmacological properties of Bisfaij. *International Journal of Chemical and Biochemical Sciences, 9*, 111–115.

Naz, S. B., Chaudhry, M. A., & Rahaman, M. S. (2016). Dual receptors blocked mechanism arbitrates smooth muscles relaxant effect of *Polypodium vulgare*. *Bangladesh Journal of Pharmacology, 11*(2), 414–420.

Neil-Beliveau, B. S., Joyce, J. N., & Lucki, I. (1993). Serotonergic involvement in haloperidol-induced catalepsy. *Journal of Pharmacology and Experimental Therapeutics, 265*(2), 207–217.

Potterton, D. (1983). *Culpepper's Colour Herbal* (p. 147). W. Foulsham and Co Ltd.

Rastogi, R. P., & Mehrotra, B. N. (1960). *Compendium of Indian Medicinal Plants* (Vol. 1, pp. 324, 325). Central Drug Research Institute; National Institute of Science Communication.

Saeedi, M., Babaie, K., Karimpour, R. E., Vazirian, M., Akbarzadeh, T., & Khanavi, M. (2017). *In vitro* cholinesterase inhibitory activity of some plants used in Iranian traditional medicine. *Natural Product Research, 31*(22), 2690–2694.

Siddiqui, A., Afrin, Z., Jafri, M. A., Vohora, D., & Asif, M. (2019). Evaluation of anticonvulsant activity of Majoon Najah in experimental animal model. *Asian Journal of Pharmacy and Pharmacology, 5*(4), 660–668.

Simon, A., Ványolós, A., Béni, Z., Dékány, M., Tóth, G., & Báthori, M. (2011). Ecdysteroids from *Polypodium vulgare* L. *Steroids, 76*(13), 1419–1424.

Sofiane, G., Wafa, N., & Ouarda, D. (2015). Antioxidant, antimicrobial and anti-inflammatory activities of flavonoids and tannins extracted from *Polypodium vulgare* L. *Asian Journal of Biochemical and Pharmaceutical Research, 5*, 114–122.

Souri, E., Amin, G., Farsam, H., Jalalizadeh, H., & Barezi, S. (2008). Screening of thirteen medicinal plant extracts for antioxidant activity. *Iranian Journal of Pharmaceutical Research, 7*(2), 149–1544.

Tabeshpour, J., Shakiban, D., Qobadi, A., Aghazadeh, E., & Yousefsani, B. S. (2023). Cytotoxic effects of ethanolic extract of *Polypodium vulgare* on human malignant melanoma cell line. *Asian Pacific Journal of Cancer Prevention, 24*(1), 275–281.

Yamada, H., Nishizawa, M., & Katayama, C. (1992). Osladin, a sweet principle of *Polypodium vulgare*: Structure revision. *Tetrahedron Letters, 33*, 4009, 4010.

CHAPTER 34

Overview of the Ethnomedicinal Uses, Phytochemical Constituents, and Pharmacology of *Hypodematium*

MANOJ L. SANKPAL,[1] SHIVANI V. KHORATE,[1] MANSINGHRAJ S. NIMBALKAR,[1] and SACHIN M. PATIL[2,*]

[1]*Department of Botany, Shivaji University, Kolhapur, Maharashtra, India*

[2]*Department of Botany, Dr. Ghali College, Gadhinglaj (affiliated to Shivaji University, Kolhapur), Maharashtra, India*

*Corresponding authors

ABSTRACT

The genus *Hypodematium* is distributed in the tropical and subtropical regions of the Old World. Many of the species are used as ethnomedicines by tribals, local peoples, or medicinal practitioners around the globe. This chapter discusses the ethnomedicinal, biochemical, and pharmacological aspects along with future prospects of the genus *Hypodematium*.

34.1 INTRODUCTION

The genus *Hypodematium* in the family Hypodematiaceae is represented by 20 species in the world (PPG-I, 2016), most commonly found in drier regions, grown along the roadside, hilly slopes, or rock crevices. It is mainly distributed in the tropical and subtropical regions of the Old World. It is used in the indigenous system of medicine by tribal, local people, and traditional medicinal practitioners. The most common

Bioactives and Pharmacology of Pteridophytes. T. Pullaiah (Ed.)

DOI: 10.1201/9781779642738-34

species, *Hypodematium crenatum*, is known by the name 'Bhoot Kesar' in Madhya Pradesh and 'Jadi Buti' in Rajasthan, India. The leaves are used to induce conception in women, and the whole plant is used by exorcists for fumigation in Pachmarhi, India. Dry frond powder, along with fresh cow milk, is taken orally after five days of a menstrual period for about a week to facilitate conception in women and treat gynecological disorders. The rhizome is used as an antibacterial agent. In the Mt. Abu area, India, woodcutters use frond paste to get relief from insect bites. In some places, young circinately coiled fronds are eaten as vegetables (Parihar et al., 2010; Semwal et al., 2021; Shankar and Khare, 1991). In Pakistan, a decoction of the rhizome is used to cure women's infertility and menstrual disorders (Irfan et al., 2024). In China, *Hypodematium crenatum*, *H. fauriei,* and *H. sinense* have been used to cure Meniere's syndrome, and a 95% effective rate was reported (Rasmussen and Pedersen, 2017).

34.2 PHYTOCHEMISTRY

The phytochemical analysis of the genus *Hypodematium* has been studied by various pteridologists. Murakami et al. (1986) studied the chemical and chemotaxonomical aspects of Xanthone derivatives in *H. fauriei, H. crenatum*, and a novel compound, hypodemapyrazin. Another novel compound, hypodematine, was isolated from the whole plant parts of *Hypodematium sinense* (Kang et al., 2013). Hypodematine is a novel alkaloid of benzazocine with a phenyl substituent in the C-2 position. There are only a limited number of natural products that incorporate the azocine ring. The pharmacological experiments showed that it has an anti-fertility action, and the rate of inducing abortion in pregnant rabbits was 86.6%. A new β-carboline alkaloid, 1-acetyl-8-hydroxy-β-carboline, was isolated from the fern *Hypodematium squamuloso-pilosum* by Zhou et al. (1998). Pathania et al. (2012) detected ptaquiloside and quercetin from leaves in different fern species, including *Hypodematium crenatum* subsp. *crenatum*. Rasmussen and Pedersen (2017) screened for ptaquiloside in *Hypodematium crenatum* subsp. *crenatum* using herbarium specimens for qualitative mapping purposes. Recently, total free proline was estimated in *Hypodematium crenatum* subsp. *crenatum* from Rajasthan by Hussain et al. (2021). The highest content of proline was found in the leaves, followed by the rhizome, and the lowest content of proline was found in the roots. It is a well-known amino acid that has been considered a drought indicator and, when discovered in higher quantities, can endure harsher conditions for longer periods of growth. Chemical compounds isolated from different species of *Hypodematium* are given in Figure 34.1.

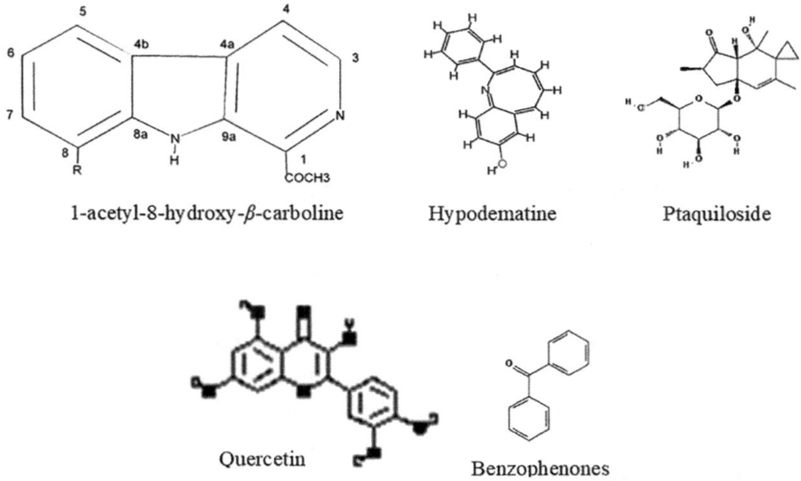

FIGURE 34.1 Chemical compounds were isolated from different species of *Hypodematium.*

34.3 PHARMACOLOGY OF *HYPODEMATIUM* SPECIES

34.3.1 ANTIMICROBIAL ACTIVITY

The genus *Hypodematium* is an important medicinal fern, from which recently discovered important compounds which have antimicrobial activity. The various chemical tests along with the determination of physical constants facilitate the identification of the crude drug sample (Hussain et al., 2021; Irfan et al., 2024; Rasmussen and Pedersen, 2017; Zhou et al., 1998). *Hypodematium crenatum* has antimicrobial activities against several pathogenic fungal and bacterial strains (Benjamin and Manickam, 2007; Benniamin, 2011). Aqueous extract of *H. crenatum* was effective in combating *E. coli*, however, it failed to inhibit *Agrobacterium tumefaciens* and *Salmonella arizonae*.

34.3.2 ANTITUMOR PROPERTIES

Alkaloids are a family of natural and synthetic products with structural diversity and outstanding antitumor activities. β-Carboline and 1-acetyl-8-hydroxy-β-carboline were isolated from the medicinal fern,

Hypodematium squamuloso-pilosum, along with a known compound, 1-acetyl-β-carboline. β-Carboline and its derivatives are being studied as anticancer agents, focusing on both natural and synthetic monomers as well as dimers (Zhou et al., 1998). A flavonoid C-glycoside was derived from *H. sinense* (Chen et al., 1980), and it has anticancer and antitumor properties. Xanthone derivatives were isolated from *H. sinense*, and these derivatives have anticancer and antitumor properties.

34.3.3 ANTI-INFLAMMATORY PROPERTIES

Pteridophytes also have anti-inflammatory properties, and some species of *Hypodematium* were used as anti-inflammatory medicine. β-Carboline, 1-acetyl-β-carboline, and 1-acetyl-8-hydroxy-β-carboline were isolated from *Hypodematium squamuloso-pilosum,* which have anti-inflammatory properties (Zhou et al., 1998). A flavonoid C-glycoside and xanthone derivatives were isolated from *H. sinense* and have anti-inflammatory properties.

34.4 CONCLUSION

It was observed that many of the *Hypodematium* species have not yet been evaluated for their phytochemical constituents. Therefore, more research is required to explore the phytochemical constituents in the different species of *Hypodematium.*

ACKNOWLEDGMENT

The authors are thankful to the Head of the Department of Botany for providing the facilities to carry out the work.

KEYWORDS

- ***bhoot kesar***
- **ethnomedicines**
- **hypodematine**
- **pteridophytes**
- **xanthone derivatives**
- **β-carboline alkaloid**

REFERENCES

Benjamin, A., & Manickam, V. S. (2007). Medicinal pteridophytes from the Western Ghats. *Indian Journal of Traditional Knowledge, 6*, 611–618.

Benniamin, A. (2011). Medicinal ferns of northeastern India with special reference to Arunachal Pradesh. *Indian Journal of Traditional Knowledge, 10*(3), 516–522.

Chen, Y. M., Lu, Y. H., Ma, C. S., Chen, Y. J., & Yu, D. Q. (1990). Separation and purification of Chongzujueside. *Acta Botanica Sinica, 32*(1), 54–57.

Hussain, S., Fliban, F. Z., & Quazi, S. (2021). Total free proline estimated in some selected species of pteridophytic flora of Rajasthan. *European Journal of Medicinal Plants, 32*(6), 1–5.

Irfan, M., Jana, G., Murada, W., Jana, F. G., Raufd, A., Alsayarie, A., Almarhoonf, Z. M., & Mabkhot, Y. N. (2024). Ethnomedicinal and traditional uses of the ferns of Khyber Pakhtunkhwa, Pakistan. *Brazilian Journal of Biology, 84*, e250256, 1–10. https://doi.org/10.1590/15196984.250256.

Kang, Z. Y., Zhang, Q., Chen, R. Y., Zhang, P. C., & Yu, D. Q. (2013). The total synthesis of hypodematine. *Journal of Asian Natural Products Research, 15*(8), 840–848. http://dx.doi.org/10.1080/10286020.2013.805749.

Murakami, T., Tanaka, N., Wada, H., Saiki, Y., & Chen, C. M. (1986). Chemical and chemotaxonomical studies on filices. LXIII. Xanthone derivatives of *Hypodematium fauriei* Tagawa, *H. crenatum* Kuhn and *Gymnocarpium robertianum* Newm. (*G. jessoense* Koidz). *Yakuga Zasshi, 105*, 378–382.

Parihar, P., Parihar, L., & Bohra, A. (2010). *In vitro* antibacterial activity of fronds (leaves) of some important pteridophytes. *Journal of Microbiology and Antimicrobials, 2*(2), 9–22.

Pathania, S., Kumar, P., Singh, S., Khatoon, S., Rawat, A. K. S., Punetha, N., Jensen, D. J., Lauren, D. R., & Somvanshi, R. (2012). Detection of ptaquiloside and quercetin in certain Indian ferns. *Current Science, 102*, 1683–1691.

PPG-I. (2016). A community-derived classification for extant lycophytes and ferns. *Journal of Systematics and Evolution, 54*(6), 563–603.

Rasmussen, L. H., & Pedersen, H. A. (2017). Screening for ptaquiloside in ferns: Using herbarium specimens for qualitative mapping purposes. *Phytochemical Analysis, 28*, 575–583. https://doi.org/10.1002/pca.2707 (accessed on 17th November 2025).

Semwal, A., Jaiswal, D., & Kumar, S. (2021). Ethnomedicinal uses of pteridophytes of Garhwal region, Uttarakhand. *Journal of Applied Pharmacy, 13*, 322.

Shankar, R., & Khare, P. B. (1991). Pharmacognostic studies on *Hypodematium crenatum*. *International Journal of Pharmacognosy, 29*, 169–175.

Zhou, T. S., Wen-Cai, Y., Zheng-Tao, W., Chun-Tao, C., Rong-Han, Z., Guo-Jun, X., & Luo-Shan, X. (1998). β-Carboline alkaloids from *Hypodematium squamuloso-pilosum*. *Phytochemistry, 49*(6), 1807–1809.

Index

C

D

E

F

I

J

K

O

P

Q

R

T

U

Y

Z

For Product Safety Concerns and Information please contact our EU representative GPSR@taylorandfrancis.com Taylor & Francis Verlag GmbH, Kaufingerstraße 24, 80331 München, Germany

Batch number: 10725933

Printed by Printforce, the Netherlands